Bacterial Physiology

A Molecular Approach

Walid El-Sharoud

Editor

Bacterial Physiology

A Molecular Approach

 Springer

Dr. Walid M. El-Sharoud
Faculty of Agriculture
Mansoura University
Mansoura
Egypt
wmel_sharoud@mans.edu.eg

Cover Illustration: Image shows exponentially growing *Bacillus subtilis* cells expressing GFP-MreB (upper panel). Second panel shows DNA stain, third panel overlay of GFP-MreB (green) and membrane stain (red, indicating the septa between cells, which grow in chains), lower panel overlay of GFP-MreB (green) and DNA stain (blue). Image courtesy of Hervé Joël Defeu Soufo and Peter L. Graumann, University of Freiburg, Germany.

ISBN: 978-3-540-74920-2   e-ISBN: 978-3-540-74921-9

Library of Congress Control Number: 2007934673

© 2008 Springer-Verlag Berlin Heidelberg

Cover design: WMX Design GmbH, Heidelberg

Printed on acid-free paper

9 8 7 6 5 4 3 2 1

springer.com

*Dedicated to my father, Mahmoud El-Sharoud,
who inspired my love for writing!!
W. M. El-Sharoud*

Preface

How well do we understand bacteria? How true is the hypothesis of Jacques Monod that if we understand *Escherichia coli*, we will be able to understand elephant? It is striking that the more we learn regarding bacteria the less we can be convinced of this hypothesis. For a long time, scientists thought that bacteria were just simple single-celled organisms that could be used as model systems for higher, complex, and multi-celled eukaryotic organisms. In a rather simplified view, bacteria were considered as self-contained packets of enzymes that had little interaction with each other or the surrounding environments. Conversely, it is now appreciated that bacteria can show community features of communication between individual cells and interaction with surrounding environments. As with higher organisms, a community of bacteria can display individual phenotypic variations; this is the case even within a clonal population of identical genetic composition (see Chap. 12). Such individual diversity seems to promote communication between cells that aids adaptation in the environment (see Chap. 9). The tremendous diversity between different bacterial species is also striking; suffice it to say that the genetic distance between two bacterial groups may be of a similar order to the average genetic distance between plants and animals.

Although the significance of the early contributions of pioneering microbiologists cannot be overemphasised, it is unquestionable that the study of bacteria has been revolutionized by the application of molecular methodologies. These methods allowed direct manipulation and monitoring of bacterial cells' components rather than hypothesizing on them. This was not an option for an old bacteriologist, given the limitation of traditional methodology in microbiology, and it has been the massive advancement in genetics, biochemistry, biophysics, bioinformatics, etc. that made molecular techniques available. The use of molecular methods including the "omics" (genomics, transcriptomics, proteomics) and fluorescence-based techniques not only allowed new discoveries, but has also changed our way of researching and thinking of bacteria. Modern bacteriologists do now adopt "a molecular approach," in which cell structure, function, and behaviour are interpreted on molecular basis. This approach has broadened and deepened the level of study of bacterial cells, but also raised concerns regarding a number of the past's theories.

Within this context comes the present book, which presents the impact of applying the molecular approach to the study of bacterial physiology. The first chapter discusses recent discoveries of subcellular organisation that have been made possible by the use of molecular techniques. The author exploits such sophisticated structural findings in reforming our ideas regarding the basic physiological processes of transcription, translation, cell division, etc. The second chapter reports on the presence of cytoskeletal elements in bacteria, a structural property that was thought to be restricted to eukaryotic cells. This is an interesting and significant discovery, given the involvement of bacterial cytoskeleton in shaping cells, cell division, chromosome segregation, and cell motility. The third chapter is also on newly discovered structural phenomena related to the cytoplasmic membrane. The authors provide comprehensive review of the presence of mechanosensitive channels that form large pores in the cytoplasmic membrane that switch between open and closed states, aiding cell survival during environmental stress. In relation to this, the authors also describe recent findings showing the dynamic structural nature of bacterial cell wall. The fourth chapter considers an interesting aspect of one of the basic physiological processes: respiration. The author discusses the phenomenon of respiratory flexibility in bacteria, where cells use a range of different electron donors and acceptors in response to different environmental pressures. Chapter 5 describes protein secretion systems, a novel research area with particularly potential medical applications. Chapter 6 discusses the regulation of gene expression by DNA supercoiling. This is an unorthodox view of gene regulation, usually thought to be mediated by primary DNA sequences, activators, repressors, etc. Here, the author shows that DNA topology is significantly important for regulating gene expression. Chapters 7 through 9 provide reports on different systems used by bacteria to sense changes in the surrounding environment. These chapters emphasize the ability of bacterial cells to interact with each other and with surrounding environments, a trait that enables adaptation and survival under different environmental conditions. Chapters 10 and 11 further describe other cellular mechanisms to cope with environmental stress. Chapter 10 reports on ribosome modulation factor, whose binding to bacterial ribosomes has been shown to aid cell survival during stress, whereas Chapter 11 demonstrates diverse aspects of the so-called "stress master regulator," RpoS, which is a sigma factor protein mediating the transcription of stress-responsive genes. Apart from structural and functional issues discussed in the previous chapters, the last chapter explains the striking phenomena of phenotypic switching and bistability in bacteria. The authors provide an account of the molecular basis of these phenomena, in which individual cells with identical genotypes may display different phenotypes under identical conditions within the same clonal population.

As mentioned above, bacteria display a high degree of structural and functional diversity. Since much of our knowledge of bacteria has been gained through the study of relatively few species, such as *E. coli* and *Bacillus subtilis*, this raises the question of how applicable our current understanding is to the physiology of the rest of bacterial species. It is interesting to see in this book several examples of knowledge generated with bacterial species other than the previous model ones. This will certainly help provide better and thorough understanding of the physiology of bacterial cells. As we will also see in most chapters, there is a concluding section showing potential applications of the aspects discussed in each chapter. It could be realized from these sections that significant applications in biotechnology and drug discovery can be made effective using the wealth of basic knowledge in bacterial physiology.

This book has been developed to suit readers of diverse backgrounds. While the text serves as a reference for researchers pursuing work in areas highlighted by the book chapters, it is also intended to be useful to undergraduates and postgraduates majoring in microbiology and to microbiologists who wish to be familiar with advances in other areas of microbiology. I am very grateful to the colleagues who contributed chapters to this book, dedicating time and sincere effort for such a project. I would also like to thank all of them for their patience with me during the review process. My greatest appreciation to the following professors who kindly contributed to reviewing the book chapters: Peter Graumann, Frank Mayer, Paul Williams, Wolfgang Schumann, Regin Hengge, Eberhard Klauck, Tracy Palmer, and Mattew Hicks. I would like also to thank Dr. Christina Eckey, Ms. Ursula Gramm, Ms. Alice Blanck, and the rest of the editorial team at Springer for their support and help during the development of this book.

Indeed, I am most appreciative to Dr. Gordon Niven for significantly contributing to my development as a scientist. My deepest thanks also to Dr. Bernard Mackey, Prof. Robin Rowbury, Prof. Martin Adams, and Prof. David White for their continued support and advice. Finally, all the kind words cannot express my warmest gratitude to my mother and father, with whom I feel the joy of kindness and tenderness. My love and sincere appreciation to my lovely wife, Sherine Mostafa, who turned my life into enjoyable times. At last, and never at least, I am writing this preface while awaiting the birth of my new daughter Merna, to whom I also dedicate this book.

Aga, July 2007

Walid M. El-Sharoud

Contents

Contributors

Ian R. Booth
School of Medical Sciences, Institute of Medical Sciences, University of Aberdeen, Aberdeen AB25 2ZD, UK

Bronwyn G. Butcher
Department of Microbiology, Cornell University, Ithaca, NY 14853-8101, USA

Todd Cameron
Biological Sciences Department, California State Polytechnic University, USA

Tao Dong
Department of Biology, McMaster University, Canada

Charles J Dorman
Department of Microbiology, Moyne Institute of Preventive Medicine, Trinity College Dublin, Dublin 2, Ireland, cjdorman@tcd.ie

Michelle D. Edwards
School of Medical Sciences, Institute of Medical Sciences, University of Aberdeen, Aberdeen AB25 2ZD, UK, m.d.edwards@abdn.ac.uk

Walid M. El-Sharoud
Food Safety and Microbial Physiology (FSMP) Laboratory, Mansoura University, Egypt

John D. Helmann
Department of Microbiology, Cornell University, Ithaca, NY 14853-8101, USA, jdh9@cornell.edu

Charlie Joyce
Department of Biology, McMaster University, Canada

Malavika Kamath
Department of Molecular Medicine and Pathology, University of Auckland, Auckland, New Zealand

Oscar P. Kuipers
Department of Genetics, University of Groningen, Groningen Biomolecular
Sciences and Biotechnology Institute, Kerklaan 30, 9751NN, Haren,
The Netherlands

Peter J. Lewis
School of Environmental and Life Sciences, University of Newcastle, Callaghan,
NSW 2300, Australia, Peter.Lewis@newcastle.edu.au

Thorsten Mascher
Department of General Microbiology, Georg-August-University, 37077 Göttingen,
Germany

Katharine A. Michie
Medical Research Council Laboratory of Molecular Biology, Hills Rd, Cambridge
CB2 0QH, UK, kmichie@mrc-lmb.cam.ac.uk

Samantha Miller
School of Medical Sciences, Institute of Medical Sciences, University of
Aberdeen, Aberdeen AB25 2ZD, UK

Gordon W. Niven
Dstl, Porton Down, Salisbury, Wiltshire SP4 0JQ, UK, gwniven@dstl.gov.uk

Akiko Rasmussen
School of Medical Sciences, Institute of Medical Sciences, University of
Aberdeen, Aberdeen AB25 2ZD, UK

Tim Rasmussen
School of Medical Sciences, Institute of Medical Sciences, University of
Aberdeen, Aberdeen AB25 2ZD, UK

David J. Richardson
School of Biological Sciences, University of East Anglia, Norwich NR4 7TJ, UK,
d.richardson@uea.ac.uk

Robin J. Rowbury
Biology Department, University College London, Gower Street London WC1E
6BT, UK and 3, Dartmeet Court, Poundbury, Dorchester, Dorset DT1 3SH, UK,
rrowbury@tiscali.co.uk

Maria C. Rowe
Department of Molecular Medicine and Pathology, University of Auckland,
Auckland, New Zealand

Herb E. Schellhorn
Department of Biology, McMaster University, Canada, schell@mcmaster.ca

Wiep Klaas Smits
Department of Genetics, University of Groningen, Groningen Biomolecular
Sciences and Biotechnology Institute, Kerklaan 30, Haren, The Netherlands,
and Department of Biology, Building 68-530, Massachusetts Institute of
Technology, Cambridge, MA 02139, USA, smitswk@gmail.com

Christos Stathopoulos
Biological Sciences Department, California State Polytechnic University, USA,
stathopoulos@csupomona.edu

Simon Swift
Department of Molecular Medicine and Pathology, University of Auckland,
Auckland, New Zealand, s.swift@auckland.ac.nz

Casey Tsang
Biological Sciences Department, California State Polytechnic University, USA

Jan-Willem Veening
Department of Genetics, University of Groningen, Groningen Biomolecular
Sciences and Biotechnology Institute, Kerklaan 30, 9751NN, Haren, The
Netherlands and Department of Biology, Building 68-530, Massachusetts Institute
of Technology, Cambridge, MA 02139, USA

Yihfen T. Yen
Biological Sciences Department, California State Polytechnic University, USA

1
Subcellular Organisation in Bacteria

Peter J. Lewis

Abstract After the first formal description of "animacules" by Antonie van Leeuwenhoek in 1683, we became aware of a whole new hidden microscopic world, of which one component is the bacteria. Technological advances in subsequent years allowed us to examine these organisms in ever greater detail, to the point that protein synthesis could be clearly observed using the electron microscope (Miller et al. 1970). Despite all this activity, apart from the observation of large subcellular particles such as glycogen granules, the bacterial cell was considered to be a bag of undifferentiated cytoplasm. Cells were so small, all the necessary biochemical reactions needed for survival would be able to occur through simple diffusion. Since the mid-1990s, it has become increasingly clear that small though they are, there is a considerable level of subcellular organisation within bacterial cells, and this is important in many processes, including cell division and chromosome segregation. We are just beginning to scratch the surface and are beginning to use an increasing number of sophisticated techniques to probe bacterial cell structure. This chapter focuses

Peter J. Lewis
School of Environmental and Life Sciences, University of Newcastle,
Callaghan, NSW 2300, Australia
Peter.Lewis@newcastle.edu.au

W. El-Sharoud (ed.) *Bacterial Physiology: A Molecular Approach.*
© Springer-Verlag Berlin Heidelberg 2008

1

on the subcellular organisation of chromosomes and their segregation, cell division, transcription, translation, membranes, and the integrated systems involved in asymmetric division. Most of what we know regarding bacterial subcellular organisation is focused on the model Gram-negative *Escherichia coli* and Gram-positive *Bacillus subtilis* cells, although the intrinsically asymmetric *Caulobacter crescentus* has also proved a fertile subject for study.

1.1 Chromosome Organisation

The largest single structure within the cytoplasm of a cell will be the chromosome, which has to be copied and faithfully segregated before the cell can divide. Most of the well-characterised bacteria have single circular chromosomes, although there are also plenty of examples of organisms with multiple chromosomes, plasmids, linear chromosomes, etc. Nevertheless, the problem for all of these organisms remains the same: how to organise their genetic material and how to ensure it is faithfully segregated. Electron micrographs indicated that the chromosome occupies the middle of the cell and is often visible as a lighter staining region of cytoplasm in negatively stained images. Sometimes, fibrous regions could be identified, but no readily identifiable regular organisation could be attributed to any of the structures. Systematic attempts were made to carefully reconstruct chromosomes from serial sections of cryo-substituted cells, but the resulting structure failed to illuminate our understanding of organisation with respect to genetic loci or of how segregation could be efficiently performed. What did appear, however, was a chromosome, in which many loops emanated into the cytoplasm from a central core (Hobot et al. 1985). Maybe these loops represented regions of DNA containing actively transcribed genes, because their organisation would enable efficient coupling of transcripts with ribosomes?

Although electron microscopy is still undoubtedly the approach that is able to provide unparalleled resolution, traditional approaches have all suffered from the fact that samples are heavily fixed and dehydrated so that they can be viewed. In recent years, vitrification and cryo sectioning of unfixed cells combined with tomography and cryo electron microscopy have helped to circumvent these problems. Nevertheless, despite the fact that there is no nuclear membrane in bacteria, electron micrographs do show that there is a phase separation between the chromosome and the rest of the cytoplasm, and so there is an effective segregation of many metabolic activities within the cell.

The recent advances in our understanding of chromosome structure, replication, and segregation all come from approaches involving fluorescence microscopy, and, particularly, the use of fluorescent protein fusions, which permit the examination of *dynamic* behaviour in live cells. Two main approaches have been used. In the first, a specific protein is labelled by creating a genetic fusion of its gene with that of a fluorescent protein. The resulting fluorescently tagged protein can then be directly visualised in live cells and its dynamics can be monitored. In the second, a genetic locus is tagged so that the location and movement of a specific region of the chromosome can be monitored in live cells. To tag a specific chromosomal site, both the *lac* and *tet* repressor systems have been used (e.g., Lau et al. 2003). A fluorescent protein fusion is made to the repressor protein, which is then able to bind to a large tandem array (up to 256 repeats of the *lac* operator) and, thus, mark up a specific chromosomal locus. Use of spectral variants of fluorescent proteins permits colocalisation studies either of two different loci using strains containing both the *lac* and *tet* repressor systems (Lau et al. 2003), or using strains containing one of the repressor systems along with an additional tagged protein involved in replication/segregation (e.g., Berkmen and Grossman 2006).

Before reviewing chromosome replication and segregation, it is important to bear in mind some key aspects of bacterial physiology. As with all organisms, the size of a bacterial cell is very small compared with the length of its genome. For example, the contour length of the 4.3-Mb *Escherichia coli* genome is approximately 1.8 mm, whereas an *E. coli* cell is often not much more than 1- to 2-μm in length. To compact and fold DNA so that the chromosome can fit, cells use a combination of topoisomerase activity combined with DNA binding proteins that help compact the chromosome (Woldringh and Nanninga 2006) (see Chap. 6). In addition to the direct activity of these proteins, the very high cytoplasmic concentration of proteins, RNA, and other molecules (possibly as high as 340 mg/ml) helps to keep the chromosome highly compacted (Woldringh and Nanninga 2006) (see Chap. 6). In addition, there is no separation of cell growth, cell division, and chromosome replication cycles in bacteria, and, therefore, it is important to consider the effect of the cell cycle when formulating models to describe various events. To simplify matters, most studies on chromosome segregation have used growth conditions, in which the vast majority of the cells are undergoing a single round of DNA replication, but many of the bacteria we are familiar with are able to undergo multifork replication, in which more than one round of replication could be underway and, therefore, any models for segregation have to also be able to account for this event. In a circular chromosome, replication is initiated at a single site, known as the origin or *oriC*. Replication forks then move bidirectionally around the chromosome until the forks meet at a point almost exactly opposite *oriC* called the terminus or *terC* (Duggin et al. 2005; Neylon et al. 2005). Once at *terC*, replication forks fuse and chromosomes are decatenated so that they can be effectively segregated into daughter cells on cell division (Lewis 2001; Lemon et al. 2001).

Two key observations in the late 1990s, using the model Gram-positive organism *Bacillus subtilis*, led to a resurgence in interest in chromosome replication and segregation in bacteria. The first was that chromosome segregation in bacteria, just as in eukaryotes, was dependent on a mitotic-like event, and the second was that, contrary to belief, chromosomes moved through a fixed assembly of replication proteins (a "replication factory") rather than vice versa (Glaser et al. 1997; Lemon and Grossman 1998; Webb et al. 1998). Members of the Losick lab in the United

States labelled origin and terminus regions using the *lac* repressor system and monitored their localisation in both vegetative and sporulating cells. Origin regions were found to be oriented toward the cell poles during single-round vegetative growth and were juxtaposed with the cell poles during the initial stages of sporulation when a highly asymmetric division event occurs in which one chromosome is moved into a small polar prespore compartment (Webb et al. 1998). Conversely, terminus regions were found to localise toward the middle of the cell and were not subject to such large translocations within the cell as origin regions. In a separate study, members of the Errington lab in the United Kingdom fluorescently labelled a protein called Spo0J, which possesses similarity to the Par proteins involved in plasmid partitioning. Although mutations in *spo0J* resulted in a sporulation phenotype, it was also observed that they also resulted in an approximately 100-fold increase in the frequency of anucleate vegetative cells (Ireton et al. 1994). Although this phenotype was still mild (change from 0.02 to 1.4% anucleate cells), these results did suggest a potential role for Spo0J in chromosome segregation. On examination of fluorescently labelled cells, the Spo0J fusion was observed to assemble into foci that colocalised with the *oriC* region (Glaser et al. 1997; Lewis et al. 1997; Lin et al. 1997), and examination of the dynamics of Spo0J foci suggested that an active partition system was involved in the movement of *oriC* regions that was independent of cell growth (Fig. 1.1) (Glaser et al. 1997). Furthermore, it was observed that Spo0J foci moved to the cell poles during the initial stages of sporulation, suggesting that the mitotic behaviour of this protein was important in the segregation of prespore chromosomes. Thus, these two pieces of work suggested that chromosome segregation was an active mitotic-like process

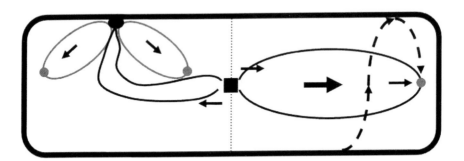

Fig. 1.1 Bacterial chromosome segregation. DNA replication is performed in replication factories (*black circle*) that seem to be located adjacent to the cell membrane. Origin regions (*dark grey circles*) then rapidly segregate to ¼ and ¾ positions along the cell length during a single round of replication. Origin movement is facilitated by active segregation processes including Spo0J–Soj partitioning complexes and MreB filaments. Bulk chromosome segregation involves SMC and RNAP activity, whereas topoisomerase activity, DNA binding proteins, and molecular crowding help maintain the chromosome in a compacted form (*large black arrow*). Terminus regions (*light grey circles*) are resolved by the activity of *dif*/XerC/XerD/FtsK (*black square*), which relies on the invaginating division septum (*dotted line*)

in bacteria, and studies that are more recent have shown that this system operates to segregate DNA regions in the same sequential order as they occur in the chromosome (Viollier et al. 2004).

The second discovery, that chromosome replication occurs by DNA movement through a fixed replication factory, has also been important in helping us understand the mechanisms of chromosome segregation along with the coordination of DNA replication and cell division cycles. During single-round replication, fluorescently tagged replication proteins (DNA polymerase III and the τ subunit) were observed to form a single focus in the mid-cell region. If the replication forks were moving around the chromosome, then two foci should have been observed for a significant portion of the cell cycle, which did not occur (Lemon and Grossman 1998). When cells were grown in a medium supporting multifork replication, a focus was still observed at mid-cell, along with two additional foci at the ¼ and ¾ positions, which represented new factories performing the second round of DNA replication (Fig. 1.1). The location of these replication foci at ¼ and mid-cell positions also suggested that the sites of DNA replication could help mark the sites for assembly of the cell division apparatus because, on completion of a round of DNA replication and disassembly of a replication factory, the cell division machinery was free to use this site.

Subsequent studies have provided more information on this process in other organisms as well as refining our understanding of the process in *B. subtilis*. In both *E. coli* and *Caulobacter crescentus*, DNA replication seems to occur toward one pole of the cell and then rapidly migrates to the mid-cell region (Gordon et al. 1997; Jensen et al. 2001; Li et al. 2002). Such an arrangement provides further evidence of a mitotic apparatus because one newly replicated origin region will need to remain close to the cell pole where it originated, whereas the other has to rapidly migrate to the opposite cell pole. In *E. coli*, a single DNA polymerase III heterodimer is responsible for both leading and lagging strand synthesis, but in *B. subtilis* and many other organisms, one enzyme is responsible for leading strand (PolC) and another for lagging strand (DnaE) synthesis. Studies have shown that both leading and lagging strand synthesis occur in replication factories (Dervyn et al. 2001). An additional study examining the precise positioning of replication factories showed that although their average position was at mid-cell, they were able to migrate around this position and were not tightly fixed there (Migocki et al. 2004). This was important because it showed that replication factory positioning was much less precise than that of the cell division apparatus and although replication could "mark" the mid-cell region, it could not account for the precise positioning of the division septum. It is now also clear that, although replication occurs at defined regions within the cell, the enzymes are not tightly linked into a super-assembly comprising both leading and lagging strand synthesis of both replication forks, because replication foci are often observed to split in two before reforming a single focus (Migocki et al. 2004; Berkmen and Grossman 2006). As such, replication factories probably represent a lose aggregation of the proteins involved in DNA replication into a replication zone rather than a fixed, defined, macromolecular assembly.

What is the mitotic apparatus? Most likely, there are several complementary and overlapping mechanisms that all contribute to effective chromosome segregation. In addition to the discovery of dynamic and mitotic-like origin movement along with localised DNA replication, SMC proteins have also been identified and characterised in bacteria during the last 10 years or so. SMC (structural maintenance of chromosomes) proteins are well known in eukaryotes and are involved in many aspects of chromosome biology, including cohesion before mitosis (Ghosh et al. 2006). They are large proteins that form dimers with a characteristic structure. There is a globular head region containing an ATP binding site, a long coiled–coil domain, and a globular hinge region. Each half of the SMC dimer folds back on itself so that each subunit forms one half of a V-shaped molecule, with the two hinge regions interacting to form the apex of the V (Ghosh et al. 2006). E. coli do not contain a clear SMC homologue, but do contain the protein MukB, which is structurally and functionally similar to SMCs. Many other bacteria, such as B. subtilis, do contain clear SMC homologues (Volkov et al. 2003). It has been postulated that SMC proteins could aid chromosome segregation through capturing newly duplicated DNA emerging from a replication factory and condensing it toward opposite poles of the cell (the extrusion-capture model; Lemon and Grossman 2001). Indeed, mutations in SMC cause decondensation of chromosomes along with aberrant segregation, and SMC proteins have been found to concentrate with origin regions, suggesting that they are certainly capable of helping condense newly replicated DNA consistent with the extrusion–capture model (Britton et al. 1998). The fact that smc null mutations are still viable (albeit, very sick) suggests that SMC is important, but not the only factor involved in chromosome segregation.

Mitosis, as we understand it in eukaryotes, involves reorganisation of clearly defined cytoskeletal elements, but bacteria were not known to possess such structures until recently (see Chap. 2). A class of proteins with an actin-like fold (mainly called MreB) have been shown to form highly dynamic cytoskeletal structures (Jones et al. 2001; Carbalido-Lopez 2006). The dynamic turnover of these filaments suggested that they were ideal candidates for the effectors of bacterial mitosis. Although these proteins were originally implicated in having a role in regulating cell shape, and are essential under normal growth conditions, they have also been implicated in mitotic events (Carbalido-Lopez 2006). In addition, MreB from E. coli has been shown to bind to RNA polymerase (RNAP), and may be able to aid segregation through association with this abundant DNA binding protein (Fig.1.1) (Kruse et al. 2006). Despite the fact that MreB proteins fulfil many of the criteria required for the mitotic apparatus, they are not present in many of the sequenced bacterial genomes, and, therefore, cannot represent a universal mechanism. Rather, any chromosome segregation properties of these proteins will have arisen during the more recent evolution of bacteria.

There is one enzyme that is universally conserved in bacteria and has been implicated in facilitating chromosome segregation: RNAP (Dworkin and Losick 2002). When large amounts of genome sequences were becoming available, it was noticed that many genes were oriented in the same direction for transcription as for the movement of DNA polymerase during chromosome replication (Brewer 1988). This polarity was assumed to help minimise head-on collisions between transcription and replication complexes. However, the polarity may also facilitate chromosome segregation. RNAP is one of the most powerful molecular motors characterised thus far (Wang et al. 1998), and transcription elongation complexes comprising RNAP and associated transcription factors (see Section 1.3) may be too large to freely diffuse through the cytoplasm. In an elegant set of experiments, Elowitz and coworkers showed that molecules larger than approximately 400 kDa were not able to diffuse through the cytoplasm and had very little free mobility (Elowitz et al. 1999). Thus, if RNAP is effectively fixed in space within the cytoplasm (or tethered as it is in eukaryotic systems) (Cook 1999), it could act as a very powerful molecular motor driving segregation of DNA as it transcribes (Fig. 1.1). Dworkin and Losick took advantage of the fact that in *B. subtilis*, induction of the stringent response causes the arrest of DNA replication forks at sites approximatley 130 kb from *oriC* called *Ster* sites (Levine et al. 1995). On induction of the stringent response, replication forks that had not progressed beyond the *Ster* sites would be arrested, and a fluorescently tagged *oriC* proximal marker (*dacA*) would appear as a single fluorescent spot in a cell. On release of the stringent response, replication was able to restart and the newly duplicated *dacA* foci could be seen to rapidly segregate. However, on addition of a specific inhibitor of transcription elongation (streptolydigin), no segregation of foci was observed after relief of the stringent response. This effect could be specifically attributed to transcription because, when a mutant strain resistant to streptolydigin was used, *dacA* segregation was observed after relief of the stringent response (Dworkin and Losick 2002). Although it may not be the only system involved in chromosome segregation, the universal conservation of RNAP makes it a very strong contender for being an important component of bulk DNA segregation in bacteria.

Chromosome segregation is a remarkably faithful event; daughter cells virtually always contain fully and correctly segregated nucleoids. Thus far, several systems have been discussed, all of which make important contributions to the movement of specific regions or bulk movement of the chromosome, but what happens at the end of this process? When replication forks meet, they must fuse, chromosome catenae must be resolved, and the terminus regions of the newly completed chromosomes must be segregated before closure of the division septum. Nucleoids that have been bisected by a division septum are only reliably observed in strains carrying particular mutations. Localisation studies show that the terminus regions of chromosomes tend to reside around the mid-cell region, close to where a division septum will form (Teleman et al. 1998; Jensen et al. 2001; Lau et al. 2003). Chromosomes carry specific DNA sequences called *dif* loci that help give the terminus regions polarity (Higgins 2007). As the division septum closes, the highly conserved ATPase FtsK acts as a directional DNA translocase along with the *dif*/XerC/XerD system to move DNA to the appropriate side so that, on septum closure, no DNA is trapped and sister nucleoids are fully segregated (Fig. 1.1) (Aussel et al. 2002).

Although most of the bacteria that have been characterised in detail contain single chromosomes, there are some exceptions, including *Vibrio cholera*, which contains two nonidentical chromosomes (Trucksis et al. 1998). It seems that segregation of chromosome 1 is similar to that in *C. crescentus*, where a polar origin is observed to duplicate, with one origin remaining close to the original pole and the other one segregating to the opposite pole. In contrast, the origin region of chromosome 2, which segregates much later in the cell cycle, is localised around the middle of the cell, and, on segregation, duplicate origins migrate to ¼ and ¾ positions so that they are at mid-cell on cell division (Fogel and Waldor 2005). In contrast to eukaryotic systems, in which there is a single segregation apparatus for all chromosomes, it seems that in bacteria containing multiple chromosome (or at least the *V. cholera* system), each chromosome is segregated by a different mechanism (Fogel and Waldor 2005; Fiebig et al. 2006). The use of different mechanisms of segregation may be because of the evolution of chromosome 2, which seems to be of plasmid origin and may have retained a plasmid-specific segregation apparatus (Fogel and Waldor 2005).

1.2   Cell Division

Once chromosome segregation is underway, the process of cell division can begin. Although it is widely thought that chromosome replication needs to be complete before assembly of the cell division apparatus can occur, this is certainly not the case in *B. subtilis*, and probably many other organisms, because it has been shown that once approximately 70% of a chromosome replication cycle has occurred, the cell is able to undergo division (McGinness and Wake 1981; Wu et al. 1995a). In model systems such as *E. coli* and *B. subtilis*, cell division occurs very precisely at mid-cell, exactly equidistant between the two cell poles (Harry et al. 2006). The situation is a bit different in organisms such as *C. crescentus*, in which division is slightly asymmetric because of the differentiation between the swarmer and stalk cells, but it is, nevertheless, precisely asymmetric (Shapiro and Losick 2000)!

How is the division septum positioned so precisely? Many models have been proposed, and we certainly know some of the mechanisms involved, but it is unlikely that we know the full story. We do know that several cytoskeletal elements are involved in cell division and division site selection, including the master division protein and tubulin homologue FtsZ, and the site selector and a member of a subfamily of actin called MinD (see Chap. 2). However, although FtsZ is almost universally conserved, MinD and its associated proteins are not, and, just as with chromosome segregation, there are probably several mechanisms involved in these processes. In the early 1990s, Woldringh and colleagues formulated the nucleoid occlusion model following observations that the division septum always formed between segregating nucleoids and not over regions of cytoplasm containing high concentrations of DNA (Woldringh et al. 1995). Thus, it seemed that some property of the nucleoids inhibited the formation of active division rings. However, the proteins involved in nucleoid occlusion are also not universally conserved and are not related in the two organisms, in which they have been described thus far (Wu and Errington 2004; Bernhardt and de Boer 2005), suggesting that there are multiple possible mechanisms that permit cell division to occur with appropriate accuracy in different organisms. Nevertheless, there is substantial subcellular organisation involved in ensuring the correct placement of the division septum, and I will focus on the best characterised systems from *E. coli* and *B. subtilis* that show how even when a system contains some of the same proteins, they can be induced to perform their function in more than one way.

Cell division occurs when FtsZ polymerises to form a ring at mid-cell (Harry et al. 2006). As the ring assembles, or shortly after assembly, a suite of additional proteins are then recruited to this site (see Chap. 2) (Harry et al. 2006). Cell division occurs when the FtsZ ring contracts followed by simultaneous synthesis of septal peptidoglycan by some of the newly recruited division proteins. Although FtsZ has been assumed to be a soluble cytoplasmic protein that only polymerises at the site of cell division, there is some evidence that it may actually be present as a rapidly turned-over filament that condenses into a ring structure at sites of cell division (Ben-Yehuda and Losick 2002; Thanedar and Margolin 2004; Peters et al. 2007). If this is the case, and FtsZ is predisposed to polymerisation within the cytoplasm, there needs to be a mechanism that ensures that FtsZ rings do not condense into a ring structure at an inappropriate site. One such system involves the Min proteins. The Min system involves the MinC, MinD, and MinE proteins in *E. coli* and MinC, MinD, and DivIVA in *B. subtilis*. The purpose of this system is to inhibit FtsZ polymerisation at sites other than the mid-cell region. MinC inhibits FtsZ ring assembly, and, therefore, the system has evolved to minimise MinC activity at mid-cell and maximise it at cell poles.

Surprisingly, although both *E. coli* and *B. subtilis* division selection systems use the same MinCD complex, the topological specifying component is different and the mechanism of division site selection varies between these two organisms. In both organisms, the MinCD complex acts to prevent FtsZ ring assembly and causes depolymerisation of FtsZ filaments (de Boer et al. 1992; Marston and Errington 1999). The MinCD complex is closely juxtaposed with the inner side of the cytoplasmic membrane. In *E. coli*, the MinCD complex oscillates from one pole to the other every 10 to 30 seconds, creating a net gradient of Z-ring inhibitor with a minimum concentration at mid-cell and maximal concentration at the poles, predisposing assembly of the division apparatus at mid-cell (Raskin and de Boer 1999b). This pole-to-pole oscillation is dependent on the activity of the MinE topological regulator (Hu and Lutkenhaus 1999; Raskin and de Boer 1999a). MinE forms a ring at mid-cell and then progresses toward one pole. On reaching the pole, the ring reassembles at mid-cell and moves down to the opposite pole. The periodicity of this ring migration is the same as that of MinCD oscillation, and dual fluorescent protein labelling experiments have shown that as MinE moves from mid-cell toward a pole, MinD signal becomes restricted toward the same pole, indicating that MinE causes displacement of the MinCD inhibitor (Raskin and de Boer 1999a).

Conversely, no oscillatory movement of any of the Min proteins has been observed in *B. subtilis*. Rather, DivIVA seems to recognise cell poles nucleating MinCD complex assembly in those regions (Edwards and Errington 1997). Although DivIVA seems to be restricted to the poles, the MinCD complex forms a gradient of diminishing concentration from the poles toward mid-cell, leaving the mid-cell region free (or largely free) of MinCD and, therefore, able to support FtsZ division ring assembly (Marston et al. 1998). Remarkably, although DivIVA is conserved in Gram-positive organisms, it is still able to localise at cell poles in organisms that do not contain DivIVA homologues, including *E. coli* and even the eukaryotic fission yeast, *Schizosaccharomyces pombe*, suggesting that its localisation is caused by some specific feature of cell poles (Edwards et al. 2000).

Just as FtsZ has been found to be present as a filamentous structure within the cytoplasm, it also seems that MinD is also able to form helical cytoplasmic filaments (Shih et al. 2003). This discovery implies that division site selection could be caused by the sequential depolymerisation and repolymerisation of MinD from one end of a filament to another. It is possible that the use of filamentous structures for this type of process is highly efficient and economical for a cell.

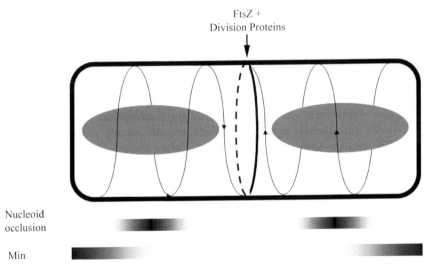

Fig. 1.2 Mid-cell assembly of cell division rings. The tubulin homologue, FtsZ, naturally assembles into dynamic filaments (*arrowed lines*). If filaments are able to nucleate into a ring, additional cell division proteins will be recruited, and the cell will undergo division. The precise location of the division septum must be carefully controlled to ensure equipartitioning. In the best-characterised organisms, *E. coli* and *B. subtilis*, two systems operate to ensure that division only occurs at mid-cell. The nucleoid occlusion system involves specific DNA binding proteins acting to inhibit formation of FtsZ rings in regions of high DNA concentration (*grey ovals*), preventing division events that could bisect the chromosome. The Min system involves an FtsZ-ring inhibitory complex forming a decreasing gradient away from the cell pole that prevents division occurring close to the cell poles

A large amount of Min protein would be required to coat the inside of the cytoplasmic membrane, preventing FtsZ ring assembly, whereas a helical filament that runs along the long axis of a cell will be able to interact with a ring that forms at any point along the cytoplasmic membrane, inhibiting inappropriate ring formation, but would require fewer molecules to perform the task (Fig. 1.2).

In addition to the Min system, there is also an additional mechanism for ensuring that cell division occurs at mid-cell while at the same time preventing accidental bisection of segregating nucleoids by the division septum, called nucleoid occlusion. In the late 1980s, the concept of nucleoid occlusion was formed based on the observation that division septa did not form over the centrally located nucleoid in cells carrying mutations in genes involved in DNA replication or segregation (Mulder and Woldringh 1989; Woldringh et al. 1991). However, nucleoid occlusion systems are not sufficient to regulate division septum placement because they are unable to prevent polar division events, which are a target for the Min system. Therefore, if there is a system that forms from the poles to the middle and another that is active over DNA that is located centrally, how do cells ever form a division septum? The nucleoid occlusion effect seems (at least partially) to be dependent on local DNA concentrations because septa will form between segregating nucleoids where DNA is still present, but at a much lower concentration than within the bulk nucleoids. Recently, it has been found that nucleoid occlusion is not a property inherent in the DNA, but is caused by specific proteins. In *B. subtilis*, this protein is called Noc (_n_ucleoid _oc_clusion) whereas in *E. coli*, it is called SlmA (Wu and Errington 2004; Bernhardt and de Boer 2005). Noc and SlmA are not similar proteins, indicating that nucleoid occlusion has evolved at least twice, and not all organisms possess identifiable Noc or SlmA homologues. The low G+C Gram-positive organisms do seem to possess Noc, whereas the Proteobacteria contain SlmA homologues. There may be other unidentified occlusion proteins, and there may be alternative mechanisms preventing bisection of nucleoids by division septa.

Nucleoid occlusion systems seem to work in the same way as Min systems in that they prevent FtsZ ring formation. Careful examination of micrographs of strains containing fluorescently tagged SlmA indicate that it does not bind uniformly over the DNA, and this becomes much more apparent as nucleoid segregation occurs. Noc and SlmA both seem to be more concentrated on the pole-most ends of the nucleoids, forming a gradient of reducing concentration toward the mid-cell region where chromosome segregation occurs (Wu and Errington 2004; Bernhardt and de Boer 2005). As the concentration of DNA and occlusion proteins lowers at mid-cell, nucleoid occlusion is relieved and FtsZ division rings can form and initiate cell division (Fig. 1.2).

In most organisms, cell division results in the formation of equally (or approximately equally) sized daughter cells. However, there are some exceptions in which highly asymmetric events occur, the paradigm being sporulation in the low G+C Gram-positive organisms. During sporulation, cells undergo a highly asymmetric cell division event that results in the formation of a small prespore and large mother cell compartment. Each compartment then undergoes a highly coordinated programme of differential gene expression that results in the formation and release of a highly stable and resistant spore (Piggot and Hilbert 2004). The change from a pattern of symmetric to highly asymmetric division requires large-scale reorganisation within the cytoplasm. The first morphological event that can be readily detected is the formation of a highly elongated nucleoid structure called an axial filament (Bylund et al. 1993). RacA helps in the formation of the axial filament and attaches the chromosomal origin regions to the cell poles via DivIVA and is dependent on the chromosome partition proteins Soj and Spo0J (Ben-Yehuda and Losick 2002; Wu and Errington 2003). Once the origin-proximal regions of the chromosome are attached to the cell poles, FtsZ helices spiral toward both cell poles and form rings at each end of the cell, just below the poles (Ben-Yehuda and Losick 2002). One of the rings becomes division competent, and the prespore septum is formed around the axial filament, trapping approximately 30% of a chromosome in the prespore compartment (Wu and Errington 1994). As the prespore septum closes around the axial filament, a remarkable event occurs in which the cell division/chromosome segregation protein SpoIIIE (FtsK) assembles around the trapped chromosome and translocates the remaining 70% of the chromosome into the prespore (Wu and Errington 1994; Wu et al. 1995b). The discovery of the activity of SpoIIIE/FtsK was an immensely significant finding, and opened the door to research regarding the role of this highly conserved protein in vegetative chromosome segregation.

1.3 Transcription

Transcription also seems to be highly organised within the cell. RNAP is a highly abundant and conserved protein (Ebright 2000). In bacteria, all transcription is carried out by a single multisubunit RNAP, whereas in eukaryotes there are three polymerases with specific activities. PolI is responsible for ribosomal RNA (rRNA), PolII for messenger RNA (mRNA), and PolIII for transfer RNA (tRNA) and 5S RNA transcription. In bacteria, RNAP comprises two α subunits that form a scaffold for the assembly of the catalytic β and β' subunits. The ω subunit is thought to aid in the assembly of the enzyme. This core $\alpha_2\beta\beta'\omega$ enzyme is catalytically active, but is unable to initiate transcription at the correct sites. The crystal structure of RNA has been solved and shows the enzyme to possess a "crab-claw" structure formed by the β and β' subunits, which carry the catalytic activity of the enzyme (Zhang et al. 1999). The channel formed by the crab-claw structure is just wide enough to accommodate double-stranded DNA, and the catalytic site of the enzyme is at the apex of the crab-claw structure (Fig. 1.3). In the schematic shown in Fig. 1.3b, DNA is bent approximately 90° as it moves through RNAP, with transcription moving left to right across the page. Downstream DNA is to the right of the enzyme, and upstream sequences approximately above, delineating the upstream and downstream sides of the enzyme that will be referred to frequently in this section. Nucleotide triphosphates (NTPs) required for RNA synthesis are

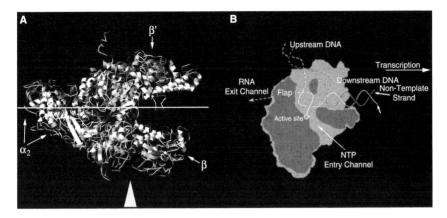

Fig. 1.3 Structure of RNAP. The structure of RNAP (*B. subtilis* homology model; MacDougall et al. 2005) is shown in (**a**), with the various subunits labelled. DNA fits in the "crab claw" formed by the β and β′ subunits. Key features and the arrangement of DNA and RNA with respect to the enzyme are shown in (**b**). The view of the enzyme is cut away approximately along the line shown in (**a**) and up through the β subunit, as indicated by the *arrowhead*. The direction of transcription is indicated along with upstream and downstream sides of the enzyme. NTP substrates enter through the NTP entry channel and RNA exits between the β-flap and β′ subunit through the RNA exit channel

presumed to enter the enzyme and move to the active site via a channel located on the downstream side of the enzyme called the nucleotide entry channel or secondary channel. RNA chain elongation occurs at approximately a 90° angle to the template DNA, and the transcript exits the enzyme down another channel formed between the upstream side of the enzyme and a structure called the β-flap (see Fig. 1.3). Modulation of the aperture of the β-flap is used to control processes such as pausing (Nudler and Gottesman 2002).

An additional factor, σ, is required to ensure the correct binding to promoter regions and initiation of transcription. The resulting complex containing a σ factor is called the holoenzyme (HE). σ factors comprise a large and diverse family of DNA binding proteins with a series of common motifs (Paget and Helmann 2003). *E. coli* possesses seven different σ factors, *B. subtilis* 17, and *Streptomyces coelicolor* an astonishing 63! In all bacteria, there is a single highly conserved σ factor that is responsible for the bulk of transcription during "normal" vegetative growth, and it is this factor that recognises the classical −35 and −10 promoter regions. The remaining factors are responsible for the transcription of subsets of genes and are often only activated in response to specific signals such as heat shock, salt stress, starvation, etc. These factors each recognise a different promoter sequence, directing RNAP to transcribe genes in specific response to a change in environmental stimuli. However, the job of all σ factors is the same in that they direct the sequence-specific interaction of RNAP with a promoter region, bend the DNA into the crab-claw channel formed by the β and β′ catalytic subunits, cause local unwinding of the DNA just downstream of the "−10" sequence, leading to the initiation of RNA synthesis (Murakami and Darst 2003).

Once RNAP clears the promoter region, σ is no longer required and can dissociate from the enzyme, where it can be recycled to help initiate a new round of transcription. This process involves the dissociation of σ that has many contact points with RNAP and is a multistep process that can occur very slowly (Murakami and Darst 2003) and, therefore, some σ may be still bound to RNAP that has cleared promoters and is considered to be in the elongation phase. This σ is consequently able to cause elongating RNAP to become stalled at sequences that resemble −10 sequences (Ring et al. 1996). In addition, there is some evidence that in some cases σ can remain associated with RNAP throughout the transcription cycle (or certainly well into the elongation phase), and this may be important in aiding transcription of genes in which the binding of σ to form HE is a rate-limiting step (Bar-Nahum and Nudler 2001; Mukhopadhyay et al. 2001).

RNAP is a highly abundant protein within the cell and represents one of the major DNA binding proteins within the bacterial nucleoid. It is estimated that there are approximately 7 to 10,000 molecules of RNAP per cell in both *E. coli* and *B. subtilis* cells during exponential growth (Bremer and Dennis 1996; Davies et al. 2005). RNAP is involved in a complex cycle of events. It is a large enzyme (~400 kDa) whose synthesis and assembly takes time; the enzyme can bind DNA and scan in two dimensions before locating and binding a promoter region. Once bound to the promoter in a closed complex, the DNA must be moved into the active site and unwound to form a transcription-competent open complex. During the initiation stages of transcription, multiple abortive initiation events occur in which short transcripts, approximately 6 nt, are produced without the enzyme leaving the promoter. Finally, a long enough transcript is produced, permitting the enzyme to clear the promoter and enter the elongation phase. Transcription must be terminated either by a mechanism dependent on the RNA helicase ρ, or through formation of a ρ-independent terminator structure in the transcript. Finally, and probably after multiple rounds of transcription cycles, the enzyme needs to be turned over by the cytoplasmic proteasomes. Because of the large number of events involved in transcription, there are various estimates regarding how much RNAP is actually involved in the process of transcription elongation at any one time, with some estimates as low as approximately 20% (Bremer and Dennis 1996).

How is transcription organised within the cell? RNAP has been localised in both *E. coli* and *B. subtilis* using immunofluorescence and fluorescent protein tagging (Azam et al. 2000; Lewis et al. 2000; Cabrera and Jin 2003), and, as expected, was found to localise exclusively within the nucleoid. Although the immunofluorescence studies revealed information regarding the localisation of RNAP, because cells were fixed before staining, no information on the dynamics of the localisation was obtained (Azam et al. 2000). We now have good information regarding the dynamics of localisation with both *B. subtilis* and *E. coli* because of the construction of fluorescent protein fusions to the β′ subunit in both organisms (Lewis et al. 2000; Cabrera and Jin 2003). What has become clear from these studies is that the localisation of RNAP is highly sensitive to growth conditions and the enzyme is redistributed through the nucleoid in a growth rate-dependent fashion (Lewis et al. 2000; Cabrera and Jin 2003). Initial studies were performed in *B. subtilis*, and the subsequent studies in *E. coli* have produced data consistent with the former. At low growth rates, very little structure can be seen other than that RNAP seems to be exclusively localised to the nucleoid (Lewis et al. 2000), and this is consistent with data obtained from the levels of free cytoplasmic RNAP in studies using minicell

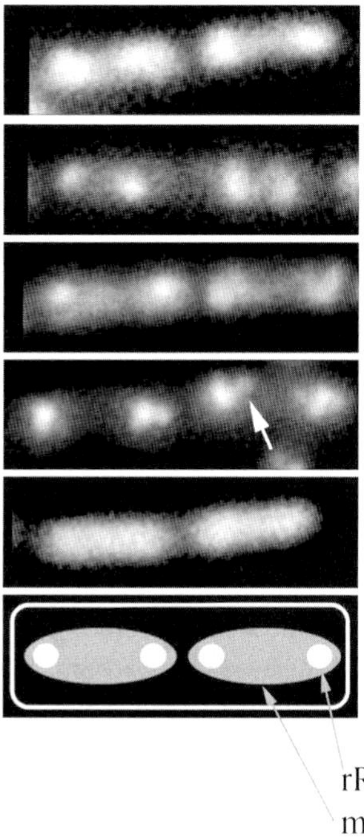

		Molecs/Cell	Molecs/RNAP rRNA	Molecs/RNAP mRNA
	RNAP	7500	1.0	1.0
	NusA	8900	1.6	1.0
	NusG	6050	0.8	0.8
	NusB	3400	1.0	0.3
	GreA	13800	-	>2.0

	RNAP	NusA	NusG	NusB/E	GreA
rRNA	1	2	1	1	-
mRNA	1	1	1	0	>2

Fig. 1.4 Subcellular organisation of transcription. Micrographs of GFP-tagged RNAP and transcription elongation factors in *B. subtilis* are shown on the left-hand side of the figure. A cartoon representation of rapidly growing cell containing two nucleoids and four origin regions is shown below the micrographs for comparison. RNAP localises within the nucleoid (*grey ovals, cartoon*) and concentrates into regions called transcription foci (TF; *light grey circles, cartoon*). Transcription elongation factors associated with RNAP localise according to their function. General factors such as NusA and NusG have patterns of distribution very similar to that of RNAP. NusB is primarily localised to TFs, but small additional foci (*white arrow*, NusB micrograph) are thought to represent rRNA elongation complexes active at one of the origin-distal rRNA operons (see text for details). GreA does not localise to TF. Details of the levels of RNAP and transcription factors are given on the right-hand side and include the total number of molecules (*molecs*) per cell as well as the relative ratio of transcription factor to RNAP involved in rRNA and mRNA elongation complexes. A summary of the localisation and quantification data is shown at the bottom, demonstrating the distribution of different transcription complexes within the cell

mutants infected with phage (Shepherd et al. 2001). However, as the growth rate increases, although RNAP signal remains distributed throughout the nucleoid, it also becomes concentrated into subnucleoid regions that have been termed transcription foci (TF) (Fig. 1.4) (Lewis et al. 2000). Initial studies with *E. coli* also indicated the presence of TF, although they seemed to be much smaller and punctuate than those observed in *B. subtilis* (Cabrera and Jin 2003). This difference may have been caused by the approach used to visualise samples because *B. subtilis* cells were viewed live on agarose pads, whereas the *E. coli* cells were fixed and attached to plain slides. Subsequent data has revealed that TF in *E. coli* are remarkably similar in distribution and appearance to those first reported in *B. subtilis* (Jin and Cabrera 2006).

Comparison of the frequency of TF with that of Spo0J-green fluorescent protein (GFP) foci showed they were very similar, suggesting that TF formed on DNA in close proximity to *oriC* (Lewis et al. 2000). In addition, time-lapse microscopy showed that TF duplicate and segregate in a remarkably similar manner to origin regions/Spo0J foci. Given the growth-rate dependence of TF formation and the correspondence of their localisation with *oriC* regions, it seemed likely that they represented the sites of rRNA synthesis within the cell, because 7 of the 10 rRNA operons in *B. subtilis* lie within approximately 200 kb of *oriC*. rRNA promoters are very heavily transcribed during rapid growth because of the cell's demand for large numbers of ribosomes. Indeed, using very elegant electron microscopy studies, it has been shown that during rapid growth there is one RNAP every 65 to 90 nt on rRNA operons, representing almost fully saturated DNA (French and Miller 1989), suggesting that TF may well represent the loading of RNAP onto rRNA operons. To determine whether this was the case, advantage was taken of the stringent response, which results in a massive down-regulation of rRNA transcription caused by the RelA-dependent production of the alarmone ppGpp in response to uncharged tRNAs binding to the ribosome (Cashel et al. 1996). It is possible to induce the stringent response using amino acid analogues such as arginine hydroxamate. After induction of the stringent response with arginine hydroxamate in *B. subtilis*, a rapid and large reduction in RNA synthesis was detected using tritiated uridine, and this was also correlated with a rapid disappearance of TF (Lewis et al. 2000). Data from *E. coli* is consistent with this finding, because induction of the stringent response also causes TF to disappear, and this effect can be relieved in a *relA* mutant unable to respond to the stringent response (Cabrera and Jin 2003).

The formation of TF raised an interesting question: If RNAP is concentrated within subregions of the cell to produce rRNA, do structures akin to eukaryotic nucleoli exist in bacteria? In eukaryotes, rRNA genes are translocated to subsites within the nucleus, called nucleoli, containing Pol I, where they are transcribed. Thus, rRNA synthesis in eukaryotes involves the translocation of DNA regions to specific RNAP factories, rather than RNAP diffusing though the nucleoplasm to transcribe rRNA genes. Because of the localised clustering of the seven rRNA operons, *rrnA, G, H, I, J, O,* and *W* in a small origin-proximal region of the chromosome, it is possible to envisage assembly of a nucleolus-type structure in *B. subtilis*. However, the remaining three rRNA operons, *rrnB, D,* and *E* are distributed over nearly half of the chromosome (representing ~2,000 kb of DNA) and, therefore, their incorporation into a bacterial nucleolus would require significant chromosomal rearrangements that would also have to permit effective DNA segregation during ongoing rounds of replication (see Section 1.1). Although not as tightly clustered, a similar scenario in *E. coli* could also be envisaged, with *rrnA, B, C,* and *E* residing within a 270-kb region, but with *rrnD, G,* and *H* being distributed over a much larger portion of the genome.

The possibility that bacterial nucleoli could form was addressed by Davies and Lewis (2003), in which the *lacO* system described above was used to tag rRNA operons, permitting their colocalisation with *oriC* regions indirectly marked using a Spo0J-CFP fusion. In this work, careful attention was paid to the cell cycle as the copy number of individually tagged rRNA operons, and their ratio compared with Spo0J foci would depend on whether or not they had been duplicated, and how far from *oriC* they were located. Because excellent cell cycle models were available (Sharpe et al. 1998), cells were grown under conditions in which only single rounds of DNA replication took place, and populations of cells were individually examined and classified according to cell length (which is directly proportional to the progress through cell cycle; Sharpe et al. 1998). The level of colocalisation of fluorescent foci was determined in cells in which the copy number of *oriC*-tagged *rrn* was equal. In strains in which an *rrn* situated adjacent to *oriC* was labelled, there was a high level of overlap of signals, but in strains containing tagged *rrns* distant from *oriC*, the level of signal overlap was no greater than in a control strain containing a tagged mRNA gene (*yvfS*) (Davies and Lewis 2003). Therefore, there is no evidence that nucleoli-like structures exist in bacteria.

So why can we see TF and what are they if they are not nucleoli? As indicated (see page 16), at high growth rates, the loading of RNAP onto rRNA operons is extremely high, with one RNAP loaded approximately every 65 to 90 nt. In contrast, the loading on mRNA is considerably less, with some evidence suggesting that it may correspond to one RNAP every 10 to 20,000 nt or so (French and Miller 1989). Therefore, the simple effect of very heavily loading an rRNA operon with RNAP will cause the appearance of a concentration of fluorescent signal in tagged cells. Because of the low resolution of the light microscope, a cluster of genes within a region of approximately 300 kb would seem to have a similar subcellular location, and, therefore, loading of RNAP onto the seven origin-proximal rRNA operons in *B. subtilis* would result in the appearance of a single bright focus.

Although transcription is efficient in vitro using only purified RNAP, nucleotide triphosphates, and DNA, this is not sufficient in vivo, and cells contain a suite of factors involved in regulating the rate of transcription elongation and aiding RNAP's negotiation of situations such as DNA lesions, protein roadblocks, etc. The best-characterised transcription elongation factors are the Nus factors that were originally discovered because of their involvement in the delayed early round of transcription in the phage λ lytic cycle. To initiate delayed early transcription, production of the immediate early gene product N is required. N is an RNA binding protein that forms part of a nucleoprotein complex comprising a series of host-encoded factors called Nus factors (*N* *u*tilisation *s*ubstance). Once associated with RNAP, an anti-termination complex is formed that is resistant to ρ-dependent terminators and is able to read through those at the end of the N and cro genes and, therefore, go on to transcribe the delayed early genes.

Nus factors are also highly conserved among the bacteria and play important roles in the regulation of transcription in all cells. It seems that the N-anti-termination system used by phage λ has been modified from the host cells' rRNA transcription system. Because of their length, and the high levels of secondary structure of their products, rRNA transcription complexes have a different structure and activity to those involved in mRNA synthesis. It is very important that rRNA is transcribed efficiently because of the cellular requirement for ribosomes, particularly at high growth rates, and it is very important that no premature termination occurs on transcribing these genes because a functional ribosome requires a complete set of 16S, 23S, and 5S rRNAs. Not only are rRNA operons transcribed at high RNAP densities with high fidelity, they are also transcribed at approximately twice the rate of an mRNA gene (~90 versus 50 nt/s; Vogel and Jensen 1995). Although rRNA operons are transcribed at high rates with very little or no premature termination, mRNA genes are transcribed relatively slowly and are highly prone to pausing and premature termination (Richardson and Greenblatt 1996). The slower transcription rates and tendency to pause may be important in helping to ensure efficient coupling with the translation apparatus (Landick et al. 1996).

It is unlikely that we have identified all transcription elongation factors, but I will outline the roles of the factors currently known before discussing how their known biochemical functions relate to their subcellular distribution. NusA was the first Nus factor identified (Friedman and Baron 1974), and was identified through its required incorporation in the λ anti-termination system. NusA is a highly conserved protein that is present in most bacteria. It is probably an essential protein in $B. subtilis$ because, to date, it has not been possible to inactivate the gene (Ingham et al. 1999), and unless combined with mutations in rho, NusA is also essential in $E. coli$ (Zhang and Friedman 1994). The structure of NusA is known and shows that it is an elongated molecule with several domains related to its function. The N-terminal domain is involved in binding RNAP (Mah et al. 1999), and this is followed by a flexible linker region that connects to an S1 RNA binding domain and a KH1 KH2 RNA binding domain. Although for many years NusA was assumed to be an RNA binding protein, it had always proved difficult to show this unequivocally. However, NusA from $Mycobacterium tuberculosis$ has recently been shown to bind to a specific RNA sequence called $boxC$ (see NusA binding to boxC page 20), and the structure of this complex has also been determined showing the role in this process of the S1 and KH1 KH2 domains (Arnvig et al. 2004; Beuth et al. 2005). In $E. coli$ and other closely related bacteria, there is an additional C-terminal domain that is able to interact with the α subunit of RNAP and is also involved in N-dependent anti-termination (Mah et al. 1999).

NusA is thought to bind to the same region of RNAP as σ factors (the N-terminal region of the β′ subunit; Traviglia et al. 1999), suggesting that there could be a cycle involving σ dissociation on completion of transcription initiation, followed by NusA association for progress through elongation, and finally NusA dissociation and σ reassociation on termination of transcription before re-initiation (Gill et al. 1991). As mentioned (see page 14), the reality is likely to be more complicated than this because we know that σ does not have to dissociate from RNAP for NusA to bind and exert its activity (Bar-Nahum and Nudler 2001; Mukhopadhyay et al. 2001; Mooney and Landick 2003). Nevertheless, models have been presented in which the N-terminal region of NusA binds to RNAP, permitting the RNA binding domains to interact with a transcript as it exits the enzyme through the β-flap and, thus, exert a regulatory effect on transcription (Borukhov et al. 2005). NusA is important in both mRNA and rRNA synthesis, but seems to have opposing effects on RNAP, depending on which class of gene is being transcribed. During mRNA synthesis, NusA has been shown to be important in promoting pausing, and this is thought to be important in ensuring efficient coupling with translation (Landick et al. 1996) as well as aiding the formation of RNA secondary structures in attenuation processes (Yakhnin and Babitzke 2002). Conversely, NusA is also an essential component of anti-termination complexes, being required for full anti-termination activity, indicating that, in these complexes, the role of NusA is to prevent pausing activity (Richardson and Greenblatt 1996; Nudler and Gottesman 2002). It is possible that mRNA and rRNA complexes contain different amounts of NusA that could be important in these alternate activities, because some data suggests that phage λ anti-termination and *B. subtilis* rRNA complexes contain two molecules of NusA, whereas mRNA complexes only contain a single molecule (Horwitz et al. 1987; Davies et al. 2005). Models have been proposed that account for the contradictory activity of NusA in which either two molecules are required for anti-termination (Nudler and Gottesman 2002) or a single molecule is sufficient for anti-termination (Borukhov et al. 2005).

NusG is another well-characterised transcription elongation factor. NusG seems to exert an opposing effect to NusA on RNAP, in that it is important in reducing the level of pausing, thus, increasing the overall rate of transcription (Richardson and Greenblatt 1996; Burns et al. 1998). There is genetic evidence suggesting a close interaction between NusG and the transcription terminator ρ, and, therefore, it may help prevent premature termination events through this interaction. NusG is essential in *E. coli*, but not in *B. subtilis*, and this may be related to the fact that there is very little ρ-dependent termination in the latter organism (Ingham et al. 1999).

NusB and NusE form a heterodimer that binds both specific RNA sequences and RNAP, and is important in the formation of anti-termination complexes (Grieve et al. 2005). NusE was identified through genetic screens as being involved in λ anti-termination before it was discovered that this protein was also ribosomal protein S10. Thus, there seems to be a direct link between anti-termination complexes (rRNA transcription) and cellular levels of ribosomal proteins. NusB has also been implicated in translation regulation (Squires and Zaporojets 2000) and, therefore, it is possible that these proteins provide an important regulatory link between transcription and protein synthesis.

The RNA sequence is also important in formation of anti-termination complexes. rRNA operons have a very specific promoter structure comprising two promoters,

P_1 and P_2, which is followed immediately by the *boxBAC* region. P_1 seems to be responsible for the bulk of transcription and is sensitive to changes in growth rate, whereas P_2 is responsible for a fairly constant basal level of transcription (Krasny and Gourse 2004; Paul et al. 2004). The *boxBAC* region contains three elements involved in formation of the anti-termination complex. Although *boxB* from *E. coli* is bound by protein N in the formation of the phage λ anti-termination complex, it is not known whether any protein binds *boxB* during rRNA transcription. *boxB* forms a hairpin structure, and this is conserved among bacteria, although the actual sequence is not conserved (Berg et al. 1989). Therefore, it seems that the structure but not the sequence is important with respect to the role of *boxB* in rRNA anti-termination. *boxA* is highly conserved and binds the NusB/E heterodimer (Nodwell and Greenblatt 1993). The formation of the *boxA*–NusB–NusE complex seems to be an important nucleation event in the establishment of an anti-termination complex, stabilising the interaction of other factors with RNAP, and also important in the dissociation of the complex on completion of transcription of an rRNA operon, where incorporation of the NusE (S10) component of the anti-termination complex is incorporated into the ribosome assembling on the rRNA transcript and the NusB is recycled for a fresh round of rRNA synthesis (Greive et al. 2005).

Gre proteins are also involved in transcription elongation, and form part of a growing family of structurally related proteins with, it is thought, a common binding site on RNAP. In *E. coli*, the transcription initiation factor DksA has been shown to be important in controlling the initiation of rRNA synthesis (Paul et al. 2004), whereas GreA and GreB are principally involved in restarting stalled elongation complexes, although they may also play a limited role in some initiation events (Susa et al. 2006; Rutherford et al. 2007). During transcription elongation, if RNAP is induced to pause, it can sometimes backtrack over the piece of DNA that has just been transcribed. When this occurs, the 3′-OH end of the transcript can slide out of the active site and out of the enzyme along the nucleotide entry channel (Borukhov et al. 2005). Subsequently, one of the two Gre proteins binds to RNAP and induces a change in activity of the active site so that the polymerase is temporarily converted to a nuclease and the backtracked transcript is cleaved. This will put a new 3′-OH group back into the active site, allowing re-initiation of transcription (Toulme et al. 2000). GreA and GreB perform the same role, but act on backtracked substrates of slightly different lengths (Toulme et al. 2000). In *B. subtilis* and other Gram-positive organisms, there is no known DksA homologue and only a single gene encoding a Gre protein, which is called *greA*. It seems that this single GreA protein is able to accommodate the activities of both GreA and GreB that are found in Gram-negative organisms. A further protein Gfh1 from *Thermus aquaticus* has been structurally and biochemically characterised. This protein is also related to the Gre family of proteins, but inhibits rather than stimulates transcript cleavage (Lamour et al. 2006). The structure of GreB bound to RNAP has also been solved and this helps illuminate how the protein performs its biochemical function (Opalka et al. 2003). The protein is shaped rather like a revolver, and the barrel of the revolver is able to fit into the nucleotide entry channel, where highly conserved residues at the tip of the barrel are involved in inducing the conversion from polymerase to nuclease in RNAP.

Surprisingly, mutations in *gre* genes are not lethal, although when combined with mutations in a gene encoding an additional transcription factor called *mfd*, cells cannot survive (Trautinger et al. 2005). Mfd is a large protein that is involved in transcription-coupled repair to damaged DNA, and structural analysis has again indicated how this protein could perform its function (Deaconescu et al. 2006). RNAP will stall after encountering a DNA lesion or a protein bound tightly to the DNA (called a roadblock). Mfd binds on the upstream side of RNAP and is able to physically lever RNAP either over the lesion or off the DNA, permitting access to the lesion for repair enzymes.

The dynamic distribution of transcription factors has also been investigated, along with a systematic approach to quantifying their levels so that we can better understand the factors regulating RNAP activity in vivo (Davies et al. 2005; Doherty et al. 2006). To establish any dynamic reorganisation of factors, their localisation was determined under at least two different growth conditions; a slow growth medium in which TF are not normally visible for RNAP, and a defined rich medium, in which nearly all cells contain visible RNAP TF. Transcription factors were also quantified using standardised procedures when grown in the defined rich medium so that their levels could be directly compared with those of RNAP (Davies et al. 2005; Doherty et al. 2006). Where possible, it was also established that the fusion to the fluorescent protein tag had no detrimental effect on viability or growth rate.

After initial examination of cells grown in defined rich medium, the NusA distribution seemed to be very similar to that of RNAP, with localisation throughout the nucleoid, and concentration into foci analogous to TF (Fig. 1.4). As with RNAP, NusA foci rapidly disappeared on induction of the stringent response, indicating that NusA also concentrates at site of rRNA synthesis. However, there was a striking difference after examination of cells grown in minimal medium. Whereas there was a very low frequency of TF observed for RNAP (~7%), NusA foci were visible in nearly 70% of cells (Davies et al. 2005). This result confirmed the importance of NusA in rRNA transcription. The levels of NusA were also quantified and compared with that of RNAP. Surprisingly, NusA is highly abundant and is present at a very similar level to RNAP (~9,000 molecules of NusA to ~7,500 molecules of RNAP). Subsequent determination of the level of fluorescence in TF showed a greater level of NusA compared with RNAP, resulting in a ratio of approximately 1.6 molecules of NusA to 1 RNAP in TF, whereas both proteins were present at a 1:1 ratio in regions of the nucleoid outside TF. Quantitative analysis of isolated transcription elongation complexes confirmed rRNA anti-termination complexes contained two molecules of NusA per molecule of RNAP, supporting the data of Horwitz et al. (1987) regarding the amount of NusA required for the formation of a λ anti-termination complex.

NusG localisation seems very similar to that of RNAP (Fig. 1.4). In defined rich medium, the proportion of cells with NusG TF is very high (92%), but in minimal medium, very few cells with NusG TF could be detected (17%) (Doherty et al. 2006). Determination of the cellular levels of NusG showed there was slightly less than RNAP (~6,000 molecules), and integration of this data with the proportion of fluorescence in TF was consistent with NusG interacting with RNAP in a 1:1 ratio in both mRNA and rRNA complexes.

NusB localisation seemed very different to that of RNAP, NusA, or NusG, with the bulk of fluorescence restricted to foci. Induction of the stringent response caused these foci to rapidly disperse, indicating that they also corresponded to the sites of rRNA synthesis within the cell. In addition, it was noticed that these foci appeared more "punctuate" than those normally observed with RNAP, NusA, and NusG (Fig. 1.4). This punctuate pattern was attributed to the distribution of rRNA operons within the chromosome, with the brightest focus corresponding to NusB localised to sites of rRNA synthesis on the origin-proximal operons *rrnA, I, H, G, J, O,* and *W,* whereas the smaller foci corresponded to the origin-distal operons *rrnB, D,* and *E.* Determination of the level of NusB within the cell indicated there were slightly more than 3,000 molecules of NusB, which corresponded to a NusB: RNAP ratio of 1:1 in TF and 0.3:1 in regions outside TF. These results confirmed the primary role for NusB as being the regulation of rRNA synthesis, and suggested that there was very little NusB available for any direct involvement in the regulation of translation (Doherty et al. 2006).

GreA localisation was also different from that of all the other proteins thus far mentioned. In cells expressing a GreA–GFP fusion, fluorescence was restricted to the nucleoid, but there was no evidence of TF formation under any growth conditions (Fig. 1.4) (Doherty et al. 2006). This result suggested that GreA has little if any involvement in rRNA synthesis, and is mainly restricted to regulation of mRNA synthesis. This would be consistent with the known role of GreA in elongation, where it is involved in the restarting of stalled and backtracked elongation complexes. Because of the formation of rRNA anti-termination complexes that are highly resistant to pausing, there would be little anticipated need for a restart molecule such as GreA. Conversely, in mRNA elongation complexes that are prone to pausing and premature termination, there would be a high requirement for GreA activity. Nevertheless, Gre proteins have been implicated in the initiation of transcription, including from rRNA promoters (Susa et al. 2006; Rutherford et al. 2007), and it is possible that GreA might still be involved in this process, although the bulk of the protein is clearly not recruited to regions of high rRNA synthesis (Fig. 1.4). After quantifying the levels of GreA within the cell, it was found that levels were surprisingly high, with approximately 14,000 molecules of GreA per cell. This corresponds to a GreA:RNAP ratio of approximately 2:1 (Doherty et al. 2006). If we make the assumption that GreA plays little role in rRNA synthesis, the ratio of GreA:RNAP in mRNA elongation complexes would be even higher.

This result was unexpected. A structure of GreB in complex with RNAP from *E. coli* has been published (Opalka et al. 2003) in which GreB bound within the nucleotide entry channel. The positioning of the single molecule of GreB was able to account for all the known biochemical properties of the protein, and, therefore, there would not seem to be a need for additional binding sites. So, is GreA in *B. subtilis* all bound to RNAP? Gre proteins are known to bind to RNAP, but not to RNA or DNA and, therefore, the colocalisation of GreA with RNAP within the nucleoid suggests that it is bound to it. Evidence supporting this statement was provided after treating cells containing a GreA–GFP fusion with high concentrations of chloramphenicol. This causes nucleoids to collapse into small, dense structures in the middle of the cell (Woldringh et al. 1995). On nucleoid collapse, all of the GreA–GFP fluorescence remained localised within the nucleoid, indicating that it was bound to RNAP (Doherty et al. 2006). In addition, the particular strain used in this study contained a copy of wild-type *greA* under the control of a xylose-inducible promoter. On addition of xylose to the medium, and production of wild-type GreA, the GreA–GFP fusion was partially dislodged from RNAP and fluorescence could be seen throughout the whole cell, not just in the nucleoid. This effect was also observed on chloramphenicol-induced nucleoid collapse, suggesting that, in the absence of xylose, all of the GreA–GFP fusion was bound specifically to RNAP.

Is there more than one binding site for Gre proteins on RNAP? At this stage, we still do not know, but there is some circumstantial evidence that suggests there could be. First, in a study by Traviglia et al. (1999) using protein footprinting techniques, a binding site for Gre proteins was reported in the region around the nucleotide entry channel, where these proteins have been shown to bind (Opalka et al. 2003), and also on a fragment of the β subunit. This binding site maps to a region of on the bottom of the β subunit, and is not close to a region of the enzyme on which GreA or B could exert its known activity (see Fig. 1.3). Studies that are more recent indicate that the only region to which GreA from *B. subtilis* binds is around the nucleotide entry channel (G. Doherty and P. Lewis, unpublished data). This data supports other work using *E. coli* proteins in which it was suggested that although one molecule of Gre protein is able to bind in the nucleotide entry channel, a second molecule may also bind in the same region to stabilise the binding of the first molecule (Koulich et al. 1998). Because there are two Gre proteins in Gram-negative organisms, GreA and GreB, it seems reasonable to assume that there may well be binding sites for both proteins on RNAP that allow insertion of the relevant protein into the nucleotide entry channel on transcription pausing. Determination of the cellular levels of GreA and GreB in *E. coli* are also consistent with an overall Gre protein:RNAP ratio of approximately 2:1. However, unless there is an actual requirement for additional Gre proteins to stabilise interactions with GreA bound in the nucleotide entry channel, it is not clear why there should be so much GreA in the cell and why it is all bound to RNAP.

As stated in the preceeding paragraph, the Gre proteins form part of a family of which two additional members, DksA and Gfh1, are currently known. In addition to GreA and GreB, DksA is also present in *E. coli* and has been modelled binding within the nucleotide entry channel, where it interacts with ppGpp close to the active site of RNAP (Perederina et al. 2004). DksA seems to be predominantly involved in the initiation of transcription, but this still implies that GreA, GreB, and DksA could all bind RNAP simultaneously if their localisation turns out to be similar to that of GreA in *B. subtilis*.

This leads us to an interesting point concerning the levels of transcription factors compared with RNAP levels within the cell. Not all RNAP is involved in transcription elongation. Some is being synthesised, some is being degraded, some is scanning for promoters, some is involved in formation of closed complexes, some in open complexes, some in abortive initiation, and some in termination, in addition to the enzyme involved in elongation. It has been estimated that there may be as little as approximately 20% of RNAP actively involved in transcription elongation within the cell. This means that only approximately 1,500 molecules of RNAP per cell would be involved in a process requiring a transcription elongation factor, and yet we find them present at levels similar to or in excess of the total cellular levels of RNAP. Probably more than 20% of RNAP is involved in elongation, but even if the level is 50% (and as high as 65% (Ishihama 2000)), there would still seem to be a large excess of transcription factors in the cell. A study by Mooney and Landick (2003) examining the effects of covalently tethering σ^{70} to RNAP showed that the enzyme behaved in a very similar manner to wild-type untethered enzymes. This suggested that transcription factors are present at levels within the cell that ensure their ready accessibility to the enzyme to enable rapid response to what would otherwise be rate-limiting events. Therefore, it is possible that transcription factors are present within the cell at much higher local concentrations than would seem to be necessary from biochemical experiments, and, at these concentrations, probably exceed their dissociation constants, suggesting that they are bound to, or very closely associated with, RNAP throughout the transcription cycle, enabling rapid and effective responses to changing conditions.

1.4   Translation

The traditional view of the arrangement of transcription and translation in bacteria is one in which the two processes are very tightly linked. Compelling evidence for this view was provided in the early 1970s by Miller and coworkers, who were able to isolate coupled transcription and translation complexes from *E. coli* and observe them directly by electron microscopy (Miller et al. 1970) (Fig. 1.5a). Their micrographs showed ribosomes assembled and synthesising protein on mRNA molecules that were still being transcribed by RNAP on the DNA. Clearly coupled transcription and translation does occur, and there are regulatory processes within the cell that depend on it. The classic example is that of the regulation of the *trp* operon by an attenuation mechanism in *E. coli* whereby ribosome stalling or progress through

Fig. 1.5 (continued) ribosome distribution than in the absence of DAPI, suggesting that IF3 is part of a relatively large complex that is unable to penetrate the spaces between DNA strands after nucleoid collapse. **c** The different possible pathways for assembling translation complexes. Although presented as separate pathways, there is no reason why all of these pathways could not overlap, and this is considered the most likely scenario. On the left, translation complexes have assembled directly onto a transcript as it is being generated by RNAP. In the middle, CSPs and Csh bind to the transcript to prepare single-stranded substrate suitable for translation. These proteins may be associated directly with ribosomes because their localisation is restricted to the same regions of the cytoplasm as ribosomes, despite their small size. On the right, initiation complexes bind the transcript as it is formed (including deep within the nucleoid) and are able to resolve any secondary structures adjacent to the translation start codon to enable the formation of a translation competent 70 S ribosome in the extra-nucleoid cytoplasm. See text for further details

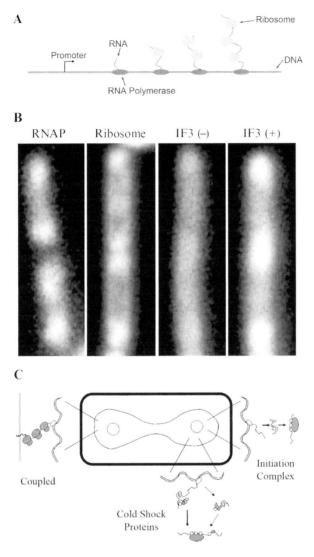

Fig. 1.5 Translation organisation. **a** Representation of the images observed by Miller and coworkers that provided compelling evidence for the coupling of transcription and translation. In the micrographs, ribosomes could be seen bound to incomplete transcripts that were still attached to RNAP in the process of transcribing a gene. **b** Micrographs of GFP-labelled RNAP, ribosomes, and IF3 in *B. subtilis* illustrating how transcription and translation are largely segregated within the cell. RNAP is concentrated within the middle of the cell, whereas ribosomes are concentrated toward cell poles, inter-nucleoid spaces, and adjacent to the cytoplasmic membrane. The localisation of an IF3–GFP fusion in the absence of the DNA stain, DAPI, is shown in the panel marked *IF3 (−)*. Although some concentration of signal can be observed at the cell poles and inter-nucleoid spaces, fluorescence is observed throughout the cell. The localisation of the IF3–GFP fusion in the presence of DAPI is shown in the panel marked *IF3 (+)*. After nucleoid collapse, the proportion of IF3 localised to the cell poles and inter-nucleoid spaces bears a much closer resemblance to
(continued)

a leader peptide sequence controls the formation of a ρ-independent terminator, ensuring that transcription of the operon only occurs if both the tryptophan (via the *trp*-repressor protein) and trp–tRNA (via ribosome progress on the leader peptide) levels are at suitably low levels (Yanofsky 2004). Examples of attenuation systems continue to appear in the literature, and, in recent years, the incredible diversity of attenuation systems present has become clear (Henkin and Yanofsky 2002). Regulation of tryptophan synthesis occurs also by attenuation in *B. subtilis* but involves a completely different system, whereby formation of the ρ-independent terminator is dependent on the binding of the TRAP 11-mer complexed with tryptophan to the transcript (Anston et al. 1999). Formation of the TRAP 11-mer complex is regulated by the activity of the anti-TRAP protein, whose synthesis is also controlled by a separate attenuation system that does require a ribosome (Yanofsky 2004). In addition, the T-box system is dependent on the ratio of charged:uncharged tRNAs and their ability to interact with the leader transcript (Henkin and Yanofsky 2002), and there are several systems that involve the direct binding of a small molecules to the transcript (Nudler and Mironov 2004).

If ribosomes are going to interact with a transcript as it is formed by RNAP, they must be able to have rapid and easy access to transcripts. Because of their large size, it is unlikely that ribosomes are free to diffuse through the cytoplasm (Elowitz et al. 1999) or through the nucleoid. Most electron micrographs of negatively stained cells show large granular particles that represent ribosomes in the cytoplasm, but very few are visible in the less densely stained regions of the cell occupied by the nucleoid. This would suggest that, for closely coupled transcription and translation to occur, transcription needs to occur on regions of the nucleoid/cytoplasm boundary. Three-dimensional analysis of the distribution of RNAP within the nucleoid of both *B. subtilis* and *E. coli* suggests that it is distributed throughout the nucleoid, with no evidence for any enrichment at the nucleoid cytoplasm boundary (Fig. 1.5b) (Lewis et al. 2000; Jin and Cabrera 2006).

Ribosome distribution has been examined in detail in *B. subtilis* and ribosomes were found to localise mainly to the cytoplasm surrounding the nucleoid and were also particularly abundant at cell poles (Fig. 1.5b) (Lewis et al. 2000; Mascarenhas et al. 2001). Colocalisation of RNAP and ribosomes also indicated there was little overlap of the signals, suggesting that the level of co-transcription and translation is likely to be relatively low (Lewis et al. 2000). Treatment of cells with antibiotics or prevention of cell division indicated that the polar localisation of ribosomes was caused by nonspecific effects, and that large ribosomal and polysomal structures simply occupy the regions of the cytoplasm where there is most room away from the regions occupied by the nucleoid (Lewis et al. 2000). More recently, the subcellular distribution of ribosomes in *Spiroplasma melliferum* has been examined by cryo-electron tomography (Ortiz et al. 2006). In this technique, cells are rapidly frozen to prevent the formation of ice crystals (vitrified), to preserve biological structures. As long as a cell is not too thick (<less than>1 µm), it is possible to obtain a three-dimensional view of the cytoplasm by taking multiple images of the cell at different tilt angles. Such approaches are extremely specialised, requiring cryo-electron microscopes fitted with tilting stages, careful limitation of the electron dosage of the sample to prevent radiation damage, and specialised image-processing techniques to analyse the very low signal:noise images and reconstruct three-dimensional volumes. In the study by Oritz et al. (2006), it was possible to identify individual ribosomes in the reconstructed images through matching micrograph densities to ribosome structures (template matching). In these cells, it was estimated that ribosomes occupied approximately 5% of the total cell volume, approximately 40% of the ribosomes were probably in polysomes, and approximately 15% were involved in synthesis of integral membrane/extracellular protein synthesis. Although the nucleoid was not directly visualised in these experiments, the authors did comment that there was no region of the cell where the ribosome concentration seemed to be low and, therefore, it is not clear whether the separation of ribosome-rich cytoplasm and the nucleoid observed in many electron micrographs and directly in *B. subtilis* is a universal arrangement in bacteria. Colocalisation of ribosomes with nucleoid-specific structures in *S. melliferum* by tomography should resolve this issue.

Translation factors are required for the assembly of a translation competent ribosome–mRNA–fMet–tRNA complex as well as the processive elongation of peptides, and examination of their distribution with reference to that of ribosomes should help us understand how translation is organised within the cell in more detail. To date, we have very little information regarding the localisation of translation factors because it has proved difficult to obtain functional fluorescent protein fusions to translation factors. However, as has been shown with transcription factors that remain closely juxtaposed to RNAP, we would expect that the distribution of translation factors would closely mimic that of ribosomes. To date, we only have preliminary data regarding the localisation of initiation factor (IF)-3. The role of IF-3 is to bind across the P/E site on the 30 S subunit, preventing the physical association of the 30 S and 50 S subunits by blocking sites of interaction (Dallas and Noller 2001). After formation of a mature 30 S initiation complex in which IF-2 brings an fMet–tRNA to the P-site, IF-1 and IF-3 can dissociate, permitting formation of a 70 S initiation complex (Ramakrishnan 2002).

When considering the localisation of translation factors with ribosomes, it is also important to consider their cellular levels. Approximately 80% of ribosomes are thought to be present as 70 S translating structures, and are present at approximately 50,000 copies per cell during rapid growth (Forchhammer and Lindahl 1971; Bremer and Dennis 1996; Asai et al. 1999). IF-3 is present at approximately 10% of the level of ribosomes, and close to equimolar with the estimated level of free 30 S subunits within the cytoplasm. The distribution of IF-3 in the cell was similar to that of ribosomes, but not identical. Although there was a clear concentration of signal toward the cell poles, there were also large amounts of signal throughout the cytoplasm (Fig. 1.5b, IF3−). Thus, it seems that IF-3 is free to diffuse throughout the cytoplasm. This is not surprising because IF-3 is a small molecule (19.6 kDa), and even with a GFP tag should still be free to diffuse within the cytoplasm (Elowitz et al. 1999). However, after addition of the DNA-specific dye, DAPI, to cells to colocalise IF-3 and nucleoid signals, a dramatic change in localisation was observed, whereby much of the IF-3 signal became excluded from the nucleoid and the pattern of distribution bore a much closer resemblance to that of ribosomes (Fig. 1.5b, IF3+). The addition of DAPI to cells does cause compaction of the nucleoid, and these results have been interpreted as showing IF-3 exclusion from the nucleoid after its collapse. Because such behaviour is not observed for other fluorescent protein fusions, many of which are much larger than the IF-3–GFP fusion, this was a very surprising observation. It is possible that IF-3, as part of a 30 S initiation complex, is able to penetrate the nucleoid, where it can pick up transcripts and, after association of the 50 S subunit, the 70 S translating ribosome is excluded and restricted to the nucleoid-free cytoplasm. It is possible that such complexes may be important in restricting the level of mRNA secondary structure within the nucleoid because it has been shown that initiation complexes are able to resolve secondary structures around the Shine–Dalgarno sequence that inhibit the formation of translation initiation complexes (Studer and Joseph 2006). It will be interesting to obtain a more comprehensive set of data with additional initiation and elongation factors to further dissect the subcellular organisation of translation.

Cold-shock proteins (CSPs) are also important translation factors. These represent a class of small, highly conserved single-stranded DNA/RNA binding proteins (El-Sharoud and Graumann 2007). CSPs have been implicated in a multitude of metabolic functions, including DNA folding and transcription anti-termination, in addition to aiding efficient translation (El-Sharoud and Graumann 2007). Although some CSPs are induced after cold shock, many are also present and required for normal vegetative growth, indicating that they play an important role in general cellular metabolism. Localisation of CSPs showed that their distribution was very similar to that of ribosomes (Weber et al. 2001a), and this localisation strengthens the argument that the major function of these proteins is as RNA chaperones within the cell. The colocalisation of CSPs and ribosomes was shown to be dependent on the transcriptional activity of the cell, leading to the hypothesis that CSPs may act as chaperones helping to transport mRNA transcripts to the ribosome. Importantly, CSP binding to RNA prevents the formation of secondary structures that inhibit translation and, therefore, these proteins are likely to be important in ensuring the prevention of such structures, ensuring efficient translation. It has also been shown that IF-1 from *E. coli*, which has a similar structure to CSPs, can complement the loss of CSPs in *B. subtilis* (Weber et al. 2001b). IF-1 probably binds to mRNA in the A site of 30 S initiation complexes (Ramakrishnan 2002), and may also be important in the prevention of secondary structure formation in 30 S initiation complexes (Studer and Joseph 2006).

The formation of secondary structures in RNA is more prevalent at low temperatures, which may account for the cold-induced expression of some of the CSPs, but no matter what the temperature, it is important that the cell has systems to deal with secondary structures when they do form. Although CSPs bound to RNA prevent the formation of secondary structures, they cannot resolve these structures if they have already formed. Once an RNA secondary structure has formed, the activity of cold-shock helicases (Csh) is required to regenerate single-stranded substrates for the CSPs. Csh have also been shown to colocalise with CSPs and ribosomes, and protein-interaction analyses indicate that Csh and CSPs are able to interact in vivo

(Hunger et al. 2006). Thus, there seems to be a clear mechanism within the cell that permits the resolution of RNA secondary structures that inhibit translation. If such a well conserved and abundant system exists in bacteria, it seems reasonable to suggest that it exists to aid the translation of transcripts that are not tightly coupled to the translation apparatus. The localisation of the very small CSPs (~7.5 kDa) to the region of the cytoplasm occupied by ribosomes suggests that they may be peripherally bound to ribosomes, because otherwise they would be expected to be able to diffuse freely through the cytoplasm. A loose association with ribosomes would permit CSPs and Csh to bind mRNA substrates and process them to ensure that they were in a suitable single-stranded form for translation. The localisation pattern of these proteins does not suggest that they are able to migrate into the nucleoid to bind and transport transcripts from sites of synthesis to sites of translation and, therefore, there may be several overlapping pathways to ensure efficient assembly of translation complexes within the cell (Fig. 1.5c). The first pathway would be dependent on 30 S initiation complexes that are able to penetrate the nucleoid, binding mRNA transcripts and resolving any secondary structure close to the ribosome binding site. On binding a 50 S subunit, the complex would be excluded to the nucleoid-free cytoplasm. In such complexes, significant secondary structure could still form in the downstream transcript and, therefore, Csh and CSP activity could be important in resolving secondary structure and maintaining single-stranded RNA for translation (Fig. 1.5c). Alternatively, some transcripts may form high levels of secondary structure within the nucleoid and diffuse to the ribosome-rich areas where Csh and CSPs closely associated with ribosomes are able to process the transcript into a linear form suitable for translation (Fig. 1.5c). The compaction of the transcript through formation of secondary structure may be important in ensuring that the transcript does not become entangled around the DNA and may aid its diffusion to ribosome-rich regions of the cell.

Finally, a proportion of transcription will occur near the nucleoid periphery and ribosomes will be able to readily bind transcripts, thus, coupling transcription and translation. Such complexes may also require the activity of Csh and CSPs, but transcription elongation factors such as NusA have also been implicated in modulating the rate of transcription to ensure that it occurs at a similar rate to translation, helping to maintain close coupling of the events (Landick et al. 1996) (Fig. 1.5c).

1.5   Membranes and Chemotaxis

A substantial proportion of the proteins within a cell is peripherally associated with, or is embedded in, the cytoplasmic membrane. Bacterial membranes, like all biological membranes, are a mosaic of phospholipid and protein domains. Many integral membrane proteins are also closely associated with specific phospholipids, creating small proteolipid domains. The phospholipid content of membranes differs greatly between species, classes, orders, and phyla, but there are several commonly conserved components, including phosphatidylethanolamine (PE), phosphatidylg-lycerol (PG), and cardiolipin (Kadner 1996; de Medoza et al. 2002). The various lipids do not mix freely, but form separate microdomains, often called lipid rafts. PE domains tend to be enriched for proteins, and cardiolipin-rich areas are often associated with cell poles and division septa (Fishov and Woldringh 1999; Vanounou et al. 2003). Clearly, membranes are not homogenous structures.

Membrane protein localisation has been performed in a variety of organisms, but several studies in *B. subtilis* illustrate many of the major observations to date. Johnson et al. (2004) reported the localisation of the ATP synthase and succinate dehydrogenase complexes and their dynamics. Both protein complexes were found to be distributed heterogeneously throughout the membrane, but three-dimensional reconstruction of ATP synthase localisation indicated that this did not follow any particular pattern, such as MreB/Mbl-directed spirals (Johnson et al. 2004). Additional experiments also indicated that the protein-rich domains were free to diffuse within the two dimensions of the cytoplasmic membrane. Dual-labelling experiments also showed that although there was some coincidence of ATP syn-thase and succinate dehydrogenase signals, there were also regions of mutual exclusion and partial overlap, indicating that there was no correlation between the localisation of the two complexes within the membrane.

A separate study by Meile et al. (2006) to investigate the localisation of proteins that had been shown to interact with the DNA replication machinery in a compre-hensive yeast two-hybrid screen indicated there are three main patterns of protein localisation to the membrane. The first comprised a heterogeneous pattern, the same as that reported previously by Johnson et al. (2004), and included the ATP synthase complex as well as YkcC (similar to dolichol phosphate mannose syn-thase). The second pattern was a homogenous localisation of proteins around the membrane that could either include or exclude the division septum, and was repre-sented by the protein YhaP (similar to unknown proteins). The third pattern com-prised localisation only to cell poles and/or the division septum and included the methyl-accepting chemotaxis protein TlpA, and the component of the twin-arginine translocation pathway, TatCY. There were several other minor variations on these basic themes. However, the results clearly show that a multitude of processes con-trol the distribution of proteins within the membrane. Despite its lack of glamour, perhaps the most unusual distribution pattern is that of homogenously distributed proteins. Because the membrane environment itself is heterogeneous on a macro-scopic scale (Matsumoto et al. 2006) and because proteins are known to preferably segregate to PE-rich domains, it is surprising that some proteins do not seem to be influenced by the local lipid composition.

The localisation of proteins involved in peptidoglycan synthesis has also been examined (Scheffers et al. 2004). Once again, a variety of localisation patterns was observed and some of these could be directly attributed to the function of the pro-tein. Penicillin-binding proteins (PBP) 3 and 4a localise along the long axis of the cell in foci, indicating a role in cell growth. PBP5 and possibly PBP4, although exhibiting slightly different localisation patterns, were also attributed roles in cell growth (Scheffers et al. 2004). The localisation of some of these proteins to foci along the long axis of the cell prompted the analysis of the three-dimensional distribution of these proteins to determine whether they are localised in a spiral pattern consistent with an interaction with one of the actin-like cytoskeletal proteins, but no evidence was found to suggest that this was the case. Interestingly, similar studies in *C. crescentus* have suggested that there is a possible correlation with PBP2 and MreB (Figge et al. 2004), and it has been shown in *B. subtilis* using fluorescent vancomycin derivatives that lateral wall growth is in spirals, consistent with the distribution of the cytoskeletal cell shape-determining proteins (Daniel and Errington 2003). Other PBPs, such as PBP1 and PBP2b, were found to be localised to the septum, consistent with their role in septal biosynthesis, which requires a 90° change in direction from lateral to septal growth in rod-shaped organisms. Thus, by and large, the distribution of PBPs within the membrane tends to reflect the role of the protein in cell wall synthesis.

Proteins involved in chemotaxis tend to be localised to one cell pole, as was shown in the pioneering study by Maddock and Shapiro (1993). The restriction of the chemoreceptors to one pole of the cell is a simple and effective way of enabling a rapid and directed response to chemical stimuli. Through classic signal transduction pathways, the chemoreceptor localised at one pole is able to direct the activity of the flagellar motor at the other pole. However, the organisms in which polar orientation is most comprehensively studied is *C. crescentus*, which is inherently asymmetric because of its life cycle. A stalk cell immobilised to a substrate grows and divides to produce a motile swarmer cell, which moves away to colonise a separate area as it settles and develops into a stalk cell. In this system, it is crucial that there are effective mechanisms to help the cell distinguish between "stalk end" and "swarmer end."

Some of the regulatory circuits that control this process are now known and involve a complex interplay of temporally and spatially integrated signals that link DNA replication, cell division, and flagellar synthesis (Fig. 1.6) (Viollier and Shapiro 2004). In swarmer cells that are flagellated, motile, and involved in colonisation, processes such as DNA replication and cell division are repressed because cell division is not desirable until the swarmer cell has found a suitable new niche to inhabit. Once the swarmer cell settles, it sheds its flagellum and develops a stalk that adheres it firmly to the surface. As the stalk cell develops, cell growth and division is initiated, resulting in the formation of a new swarmer cell, which moves away at cell division (Fig. 1.6). Thus, in a swarmer cell, it is necessary to repress genes involved in cell growth and duplication, and once the cell differentiates into a stalked cell, repression must be lifted. How is this achieved?

In swarmer cells, the phosphorylated version of the response regulator CtrA (CtrA-P) is abundant and regulates transcription from genes such as *ftsZ* (cell division) and *pilA* (pilus formation), and blocks the origin of chromosome replication. This protein is highly abundant, and, once the cell settles and initiates differentiation into a stalked cell, must be rapidly degraded by the ClpXP proteasome (Viollier and Shapiro 2004). CtrA-P represses expression of another master regulator protein, GcrA, and as CtrA-P is degraded, GcrA is produced, which leads to activation of transcription of genes such as *dnaA*, *dnaB*, and *dnaQ* involved in DNA replication,

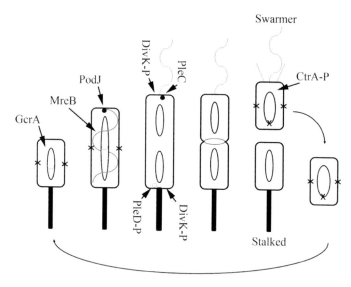

Fig. 1.6 Control of asymmetry in *C. crescentus*. The *C. crescentus* cell cycle involves two cell types, a stalked cell and a swarmer cell. Proteins that control the generation of asymmetry at various stages of the cell cycle and their localisation are indicated. Chromosomes are represented by the *oval shapes*. *Xs* indicate inhibition of cell division (on cell wall) and chromosome replication (on chromosomes). See text for additional details

and *plec* and *podJ*, which are involved in polar morphology (McAdams and Shapiro 2003).

PodJ is a polar localisation factor that is found at the swarmer cell pole, and it seems that this localisation may be dependent on MreB cytoskeletal treadmilling toward the swarmer cell pole (Fig. 1.6) (Viollier and Shapiro 2004). Thus, PodJ associates directly or indirectly with MreB filaments, which transport it to the swarmer cell pole. In turn, localisation of the membrane histidine kinase PleC is dependent on previous localisation of PodJ. PleC, and another membrane histidine kinase, DivJ, interact with the response regulators, DivK and PleD (Fig. 1.6). Both DivK and PleD are found throughout the cytoplasm, but after phosphorylation by DivJ/PleC, become localised to cell poles. DivK-P localises to both stalk and incipient swarmer cell poles, whereas PleD-P is restricted to the stalk cell pole (Fig. 1.6). PleD controls morphological events at the cell pole, and, after activation, produces the signalling molecule cyclic-di-GMP (Viollier and Shapiro 2004).

One of the key functions of the signal transduction circuits in this process is to generate asymmetry (McAdams and Shapiro 2003). As indicated (see Fig 1.6), the master regulators of this process are CtrA-P and GcrA, and their cyclical regulation of gene expression. The evidence from accurate cell cycle analysis of CtrA and GcrA levels suggests that, after cell division, the old stalk cell contains GcrA, whereas the new swarmer cell contains CtrA (Holtzendorff et al. 04). Thus, not only must the subcellular levels of these transcription regulators be carefully controlled, but it also seems that they are segregated within the predivisional cytoplasm before completion of the division septum (Fig. 1.6) (Viollier and Shapiro 2004).

Throughout this chapter, the localisation of specific proteins to regions of the cell has been discussed. Many of the protein localisation patterns are dependent on larger subcellular structures, such as cytoskeletal filaments, chromosomes, and the cytoplasmic membrane. It also seems that large protein complexes are too large to freely diffuse through the cytoplasm (Elowitz et al. 1999) and, therefore, some form of segregation event is required for their even distribution at cell division. It is possible that segregation of CtrA and GcrA in *C. crescentus* is caused by their assembly into subcellular complexes that are too large to readily diffuse, enabling their segregation into swarmer and stalk cell, respectively. Of course, this begets the question of how would such complexes be restricted to swarmer or stalk cell cytoplasm in the first place? Ribosomes and polysomes are too large to diffuse, but are readily segregated at cell division (Lewis et al. 2000). In this situation, ribosomes are highly abundant, and it was assumed that ribosome mobility is related to bulk movements that occur as the cytoplasm flows around segregating nucleoids. Clearly there are mechanisms for moving and segregating chromosomes, and structures such as magnetosomes (see Chap. 2), but what about proteins?

An astonishing finding has recently been reported regarding the localisation of the transducer-like protein TlpT chemoreceptor clusters in the bacterium *Rhodobacter sphaeroides* (Thompson et al. 2006). TlpT assembles into large cytoplasmic complexes so that, in a nondividing cell, a single large cluster is visible in the middle of the cell when tagged with yellow fluorescent protein (YFP). On progression through the cell cycle, the single focus duplicates and segregates to ¼ and ¾ positions so that, after cell division, each newborn cell contains a single mid-cell TlpT cluster (Thompson et al. 2006). Segregation of TlpT foci was found to be dependent on the product of a gene located downstream in the operon encoding TlpT, called *ppfA* (for *p*rotein *p*artitioning *f*actor). PpfA is related to Type I DNA partitioning factors, which include the ParA and MinD families of ATPases. In the absence of PpfA, TlpT foci failed to duplicate and segregate, indicating that this required the presence of an active partitioning system. PpfA homologues are found in many sequenced bacterial genomes, indicating that this mechanism of protein complex segregation may well be commonplace and that distinct and active mechanisms exist for the segregation of both chromosomes and large protein clusters within the cell.

1.6 Microcompartments

Although bacteria are generally considered to be organelle free, there are a few examples of large subcellular structures called microcompartments, in which a specific subset of biochemical reactions take place (Bobik 2006). The best-known examples are the microcompartments called carboxysomes from cyanobacteria such as *Synechocystis*, which contain most, if not all, of the enzyme complex RuBisCO that converts bicarbonate to CO_2. Carboxysomes are huge subcellular structures, reminiscent of icosahedral viral particles in both size and shape, that envelope the RuBisCO complex within a highly organised protein coat. The protein coat packs together in a series of interlocking hexamers that contain pores through which bicarbonate and CO_2 are thought to pass, enabling compartmentalisation of the enzymatic activity of RuBisCO (Kerfeld et al. 2005).

Although carboxysomes are the best-known examples of microcompartments, other examples are emerging. *Salmonella enterica* has also been shown to contain microcompartments involved in the metabolism of 1,2-propanediol that are also polyhedral proteinaceous compartments (Bobik 2006). These *pdu* organelles may be restricted to the γ proteobacteriaceae, although the presence of specific polyhedra to two distantly related groups of microorganisms (cyanobacteria and enterobacteriaceae) suggests that microcompartments may well be widespread throughout the bacteria. Indeed, there is tantalising new evidence emerging to support this notion. The *B. subtilis* stressosome involved in regulation of the σ^B-dependent stress response is a huge multi-protein complex (Chen et al. 2003) that also seems to be icosahedral in shape (Marles-Wright and Lewis 2007).

The presence of such huge complexes within cells leads me to speculate that, as with the *Rhodobacter* TlpT chemosensory complexes (see page 33), specific organelle segregation mechanisms will be required to ensure the efficient partitioning of these huge structures in cell division.

1.7 Concluding Remarks

We have come a long way in a short time in our efforts to understand the subcellular organisation of bacterial cells. Despite their small size, it is now clear that there is an pressing need for the organisation and coordination of structures and processes within bacteria, as much as there is in larger eukaryotic cells. It must also be borne in mind that our traditional view of bacteria as small and eukaryotic cells as large is not strictly true (Schulz and Jorgensen 2001). Some bacteria are truly enormous, with *Thiomargariata namibiensis* being a similar size to a fruit fly head! However, such giants are the result of massive deposition of sulphur into vacuoles, and the actual volume of cytoplasm in the cells may be relatively modest. However, bacteria such as *Epulopiscium fishelsonii* truly are giant bacteria that grow up to 0.8-mm long and are visible to the naked eye (Angert et al. 1993). These bacteria are distantly related to sporulating Gram-positive organisms, and grow and reproduce by a mechanism that seems to have evolved from a sporulation-like mechanism that was characterised in the smaller, but still very large, *Megabacterium polyspora* (Angert and Losick 1998). To coordinate their cell cycles, considerable subcellular organisation and, presumably, cytoskeletal elements will be necessary.

Nevertheless, even in bacteria that are in the 1- to 5-μm range, there is considerable subcellular organisation. Chromosomes are precisely packaged, duplicated, and segregated to ensure the efficiency of the cell cycle; cytoskeletal elements regulate cell division, morphology, division septum positioning (in association with nucleoid occlusion proteins), and the segregation of large nondiffusible protein clusters; transcription complexes are segregated within the nucleoid, creating localised regions rich in rRNA synthesis; ribosomes and their factors synthesise proteins in the nucleoid-free regions of the cytoplasm; and membrane proteins segregate to distinct subregions that may be dynamic around the cell periphery, or fixed at the cell poles. Most of these advances during recent years have been made by exploiting the ability to genetically tag proteins using fluorescent protein genes, and this will remain an important tool because of the ability to study protein dynamics in live cells. However, the development of higher resolution approaches, such as cryo-tomography will ultimately permit the characterisation of the bacterial cytoplasm to the level of resolution of individual proteins and complexes. It is likely that we have many more surprises in store before we are satisfied with our atlas of the bacterial cell.

Highly Recommended Readings

Ben-Yehuda S, Losick R (2002) Asymmetric cell division in *B. subtilis* involves a spiral-like intermediate of the cytokinetic protein FtsZ. Cell. 109:257–266

Fogel MA, Waldor MK (2005) Distinct segregation dynamics of the two *Vibrio cholerae* chromosomes. Mol. Microbiol. 55:125–136

Jones LJ, Carballido-Lopez R, Errington J (2001) Control of cell shape in bacteria: helical, actin-like filaments in *Bacillus subtilis*. Cell. 104:913–922

Lemon KP, Grossman AD (1998) Localization of bacterial DNA polymerase: evidence for a factory model of replication. Science. 282:1516–1519

Lewis PJ, Thaker SD, Errington J (2000) Compartmentalization of transcription and translation in *Bacillus subtilis*. EMBO J. 19:710–718

McAdams HH, Shapiro L (2003) A bacterial cell-cycle regulatory network operating in time and space. Science. 301:1874–1877

Miller OL Jr, Hamkalo BA, Thomas CA (1970) Visualisation of bacterial genes in action. Science. 169:392–395

Ortiz JO, Forster F, Kurner J, Linaroudis AA, Baumeister W (2006) Mapping 70S ribosomes in intact cells by cryoelectron tomography and pattern recognition. J. Struct. Biol. 156:334–341

Raskin DM, de Boer PA (1999) Rapid pole-to-pole oscillation of a protein required for directing division to the middle of *Escherichia coli*. Proc. Natl. Acad. Sci. USA. 96:4971–4976

Thompson SR, Wadhams GH, Armitage JP (2006) The positioning of cytoplasmic protein clusters in bacteria. Proc. Natl. Acad. Sci. USA. 103:8209–8214

References

Angert ER, Clements KD, Pace NR (1993) The largest bacterium. Nature. 362:239–241

Angert ER, Losick RM (1998) Propagation by sporulation in the guinea pig symbiont *Metabacterium polyspora*. Proc. Natl. Acad. Sci. USA. 95:10218–11023

Antson AA, Dodson EJ, Dodson G, Greaves RB, Chen X, Gollnick P (1999) Structure of the trp RNA-binding attenuation protein, TRAP, bound to RNA. Nature. 401:235–242

Arnvig KB, Pennell S, Gopal B, Colston MJ (2004) A high-affinity interaction between NusA and the *rrn nut* site in *Mycobacterium tuberculosis*. Proc. Natl. Acad. Sci. USA. 101:8325–8330

Asai T, Condon C, Voulgaris J, Zaporojets D, Shen B, Al-Omar M, Squires C, Squires CL (1999) Construction and initial characterization of Escherichia coli strains with few or no intact chromosomal rRNA operons. J. Bacteriol. 181:3803–3809

Aussel L, Barre F-X, Aryoy M, Stasiak A, Stasiak AZ, Sherratt DJ (2002) FtsK is a DNA motor protein that activates chromosome dimer resolution by switching the catalytic state of the XerC and XerD recombinases. Cell. 108:195–205

Azam TA, Hiraga S, Ishihama A (2000) Two types of localization of the DNA-binding proteins within the *Escherichia coli* nucleoid. Genes Cells. 5:613–626

Bar-Nahum G, Nudler E (2001) Isolation and characterization of σ^{70}-retaining transcription elongation complexes from *Escherichia coli*. Cell. 106:443–451

Ben-Yehuda S, Losick R (2002) Asymmetric cell division in *B. subtilis* involves a spiral-like intermediate of the cytokinetic protein FtsZ. Cell. 109:257–266

Berkmen MB, Grossman AD (2006) Spatial and temporal organization of the *Bacillus subtilis* replication cycle. Mol. Microbiol. 62:57–71

Bernhardt T G, de Boer PAJ (2005). SlmA, a nucleoid-associated, FtsZ binding protein required for blocking septal ring assembly over chromosomes in *E. coli*. Mol. Cell. 18:555–564

Berg KL, Squires C, Squires CL (1989) Ribosomal RNA operon antitermination: function of leader and spacer region *boxB-boxA* sequences and their conservation in diverse micro-organisms. J. Mol. Biol. 209:345–358

Beuth B, Pennell S, Arnvig KB, Martin SR, Taylor IA (2005) Structure of a *Mycobacterium tuberculosis* NusA-RNA complex. EMBO J. 24:3576–3587

Bobik, TA (2006) Polyhedral organelles compartmenting bacterial metabolic processes. Appl. Microbiol. Biotechnol. 70: 517–525

Borukhov S, Lee J, Laptenko O (2005) Bacterial transcription elongation factors: new insights into molecular mechanism of action. Mol. Microbiol. 55:1315–1324

Bremer H, Dennis PD (1996) Modulation of chemical composition and other parameters of the cell by growth rate. In *Escherichia coli* and *Salmonella typhimurium*: Cellular and Molecular Biology, 2nd ed. Neidhardt FC in chief. (Washington, D.C. ASM Press), pp. 1553–1569

Brewer BJ (1988) When polymerases collide: Replication and the transcriptional organization of the *E. coli* chromosome. Cell. 53:679–686

Burns CM, Richardson LV, Richardson JP (1998) Combinatorial effects of NusA and NusG on transcription elongation and Rho-dependent termination in *Escherichia coli*. J. Mol. Biol. 278:307–316

Britton RA, Lin DC, Grossman AD (1998) Characterization of a prokaryotic SMC protein involved in chromosome partitioning. Genes Dev. 12:1254–1259

Bylund JE, Haines MA, Piggot PJ, Higgins ML (1993) Axial filament formation in *Bacillus subtilis*: induction of nucleoids of increasing length after addition of chloramphenicol to exponential -phase cultures approaching stationary phase. J. Bacteriol. 175:1886–1890

Cabrera JE, Jin DJ (2003) The distribution of RNA polymerase in *Escherichia coli* is dynamic and sensitive to environmental cues. Mol. Microbiol. 50:1493–1505

Carballido-Lopez R (2006) The bacterial actin-like cytoskeleton. Microbiol. Mol. Biol. 70:888–909

Cashel MD, Gentry DR, Hernandez VJ, Vinella D (1996) The stringent response. In *Escherichia coli* and *Salmonella typhimurium*: Cellular and Molecular Biology, 2nd ed. Neidhardt FC in chief. (Washington, D.C. ASM Press), pp. 1458–1496

Chen CC, Lewis RJ, Harris R, Yudkin MD, Delumeau O (2003) A supramolecular complex in the environmental stress signalling pathway of *Bacillus subtilis*. Mol. Microbiol. 49:1657–1669

Cook PR (1999) The organization of replication and transcription. Science. 284:1790–1795

Dallas A, Noller HF (2001) Interaction of translation initiation factor 3 with the 30S ribosomal subunit. Mol. Cell. 8:855–864

Daniel RA, Errington J (2003) Control of cell morphogenesis in bacteria: two distinct ways to make a rod-shaped cell. Cell. 113:767–776

Davies KM, Lewis PJ (2003) Localisation of rRNA Synthesis in *Bacillus subtilis*: Characterization of loci involved in transcription focus formation. J. Bacteriol. 185:2346–2353

Davies KM, Dedman AJ, van Hork S, Lewis PJ (2005) The NusA:RNA polymerase ratio is increased at sites of rRNA synthesis in *Bacillus subtilis*. Mol. Microbiol. 57:366–379

de Boer PA, Crossley RE, Rothfield LI (1992) Roles of MinC and MinD in the site-specific septation block mediated by the MinCDE system of *Escherichia coli*. J. Bacteriol. 174:63–70

Deaconescu AM, Chambers AL, Smith AJ, Nickels BE, Hochschild A, Savery NJ, Darst SA (2006) Structural basis for bacterial transcription-coupled DNA repair. Cell. 124:507–520

de Mendoza D, Schujman GE, Aguilar PS (2002) Biosynthesis and function of membrane lipids. In *Bacillus subtilis* and its Closest Relatives: from Genes to Cells. Edited by Sonenshein AL, Hoch JA, Losick R. (Washington, DC: American Society for Microbiology Press) pp. 43–56.

Dervyn E, Suski C, Daniel R, Bruand C, Chapuis J, Errington J, Janniere L, Ehrlich SD (2001) Two essential DNA polymerases at the bacterial replication fork. Science. 294:1716–1719

Doherty GP, Meredith DH, Lewis PJ (2006) Subcellular partitioning of transcription factors in *Bacillus subtilis*. J. Bacteriol. 188:4101–4110

Duggin IG, Matthews JM, Dixon NE, Wake RG, Mackay JP (2005) A complex mechanism determines polarity of DNA replication fork arrest by the replication terminator complex of *Bacillus subtilis*. J. Biol. Chem. 280:13105–13113

Dworkin J, Losick R (2002) Does RNA polymerase help drive chromosome segregation in bacteria? Proc. Natl. Acad. Sci. USA. 99:14089–14094

Ebright RH (2000) RNA polymerase: structural similarities between bacterial RNA polymerase and eukaryotic RNA polymerase II. J. Mol. Biol. 304:687–698

Edwards DH, Errington J (1997) The *Bacillus subtilis* DivIVA protein targets to the division septum and controls the site specificity of cell division. Mol. Microbiol. 24:905–915

Edwards DH, Thomaides HB, Errington J (2000) Promiscuous targeting of *Bacillus subtilis* cell division protein DivIVA to division sites in *Escherichia coli* and fission yeast. EMBO J. 19:2719–2727

Elowitz MB, Surette MG, Wolf PE, Stock JB, Leibler S (1999) Protein mobility in the cytoplasm of *Escherichia coli*. J. Bacteriol. 181:197–203

El-Sharoud WM, Graumann PL (2007) Cold shock proteins aid coupling of transcription and translation in bacteria. Sci. Prog. 90:15–27

Fiebig A, Keren K, Theriot JA (2006) Fine-scale time-lapse analysis of the biphasic, dynamic behaviour of the two *Vibrio cholerae* chromosomes. Mol. Microbiol. 60:1164–1178

Figge RM, Divakaruni AV, Gober JW (2004) MreB, the cell shape-determining bacterial actin homologue, coordinates cell wall morphogenesis in *Caulobacter crescentus*. Mol. Microbiol. 51:1321–1332

Fishov I, Woldringh C (1999) Visualization of membrane domains in *Escherichia coli*. Mol. Microbiol. 32:1166–1172

Fogel MA, Waldor MK (2005) Distinct segregation dynamics of the two *Vibrio cholerae* chromosomes. Mol. Microbiol. 55:125–136

Forchhammer J, Lindahl L (1971) Growth rate of polypeptide chains as a function of the cell growth rate in a mutant of *Escherichia coli* 15. J. Mol. Biol. 55:563–568

French SL, Miller OL Jr (1989) Transcription mapping of the *Escherichia coli* chromosome by electron microscopy. J. Bacteriol. 171:4207–4216

Friedman DI, Baron LS (1974) Genetic characterization of a bacterial locus involved in the activity of the N function of phage lambda. Virology. 58:141–148

Ghosh SK, Hajra S, Paek A, Jayaram M (2006) Mechanisms for chromosome and plasmid segregation. Annu. Rev. Biochem. 75:211–241

Gill SC, Weitzel SE, Von Hippel PH (1991) *Escherichia coli* sigma 70 and NusA proteins. I. Binding interactions with core RNA polymerase in solution and within the transcription complex. J. Mol. Biol. 220:307–324

Glaser P, Sharpe ME, Raether B, Perego M, Ohlsen K, Errington J (1997) Dynamic mitotic-like behaviour of a bacterial protein required for accurate chromosome partitioning. Genes Dev. 11:1160–1168

Gordon GS, Sitnikov D, Webb CD, Teleman A, Straight A, Losick R, W. Murray AW, Wright A (1997) Chromosome and low copy plasmid segregation in *E. coli*: visual evidence for distinct mechanisms, Cell 90:1113–1121

Greive SJ, Lins AF, von Hippel PH (2005) Assembly of an RNA-protein complex. Binding of NusB and NusE (S10) proteins to *boxA* RNA nucleates the formation of the antitermination complex involved in controlling rRNA transcription in *Escherichia coli*. J. Biol Chem. 280:36397–36408

Harry E, Monahan L, Thompson L (2006) Bacterial cell division: the mechanism and its precision. Int. Rev. Cytol. 253:27–94

Henkin TM, Yanofsky C (2002) Regulation by transcription attenuation in bacteria: how RNA provides instructions for transcription termination/antitermination decisions. Bioessays. 24:700–707

Higgins, NP (2007) Mutational bias suggests that replication termination occurs near the *dif* site, not at Ter sites: what's the Dif? Mol. Microbiol. 64:1–4

Hobot JA, Viliger W, Escaig J, Maeder M, Ryter A, Kellenberger E (1985) Shape and fine structure of nucleoids observed on sections of ultrarapidly frozen and cryosubstituted bacteria. J. Bacteriol. 162:960–971

Horwitz RJ, Li J, Greenblatt J (1987) An elongation control particle containing the N gene transcription anti-termination protein of bacteriophage lambda. Cell. 51:631–641

Hu Z, Lutkenhaus J (1999) Topological regulation of cell division in *Escherichia coli* involves rapid pole to pole oscillation of the division inhibitor MinC under the control of MinD and MinE. Mol. Microbiol. 34:82–90

Hunger K, Beckering CL, Wiegeshoff F, Graumann PL, Marahiel, MA (2006) Cold-induced putative DEAD box RNA helicases CshA and CshB are essential for cold adaptation and interact with cold shock protein B in *Bacillus subtilis*. J. Bacteriol. 188:240–248

Ingham CJ, Dennis J, Furneaux PA (1999) Autogenous regulation of transcription termination factor Rho and the requirements for Nus factors in *Bacillus subtilis*. Mol. Microbiol. 31:651–663

Ireton K, Gunther NW IV, Grossman AD (1994) spo0J is required for normal chromosome segregation as well as the initiation of sporulation in *Bacillus subtilis*. J. Bacteriol. 176:5320–5329

Ishihama A (2000) Functional modulation of *Escherichia coli* RNA polymerase. Annu. Rev. Microbiol. 54:499–518

Jensen RB, Wang SC, Shapiro L (2001) A moving DNA replication factory in *Caulobacter crescentus*. EMBO J. 20:4952–4963

Jin DJ, Cabrera JE (2006) Coupling the distribution of RNA polymerase to global gene regulation and the dynamic structure of the bacterial nucleoid in *Escherichia coli*. J. Struct. Biol. 156:284–289

Johnson AS, van Horck S, Lewis PJ (2004) Dynamic localization of membrane proteins in *Bacillus subtilis*. Microbiology. 150:2815–2824

Jones LJ, Carballido-Lopez R, Errington J (2001) Control of cell shape in bacteria: helical, actin-like filaments in *Bacillus subtilis*. Cell. 104:913–922

Kadner RJ (1996). Cytoplasmic membrane. In *Escherichia coli* and *Salmonella*: Cellular and Molecular Biology 2nd ed. Edited by Neidhardt FC et al. (Washington DC: American Society for Microbiology) pp 58–87

Kerfeld CA, Sawaya MR, Tanaka S, Nguyen CV, Phillips M, Beeby M, Yeates TO (2005) Protein structures forming the shell of primitive bacterial organelles. Science 309:936–938

Koulich D, Nikiforov V, Burokhov S (1998) Distinct functions of N and C-terminal domains of GreA, an *Escherichia coli* transcript cleavage factor. J. Mol. Biol. 276:379–389

Krasny L, Gourse RL (2004) An alternative strategy for bacterial ribosome synthesis: *Bacillus subtilis* rRNA transcription regulation. EMBO J. 23:4473–4483

Kruse T, Blagoev B, Lobner-Olesen A, Wachi M, Sasaki K, Iwai N, Mann M, Gerdes K (2006) Actin homolog MreB and RNA polymerase interact and are both required for chromosome segregation in *Escherichia coli*. Genes Dev. 20:113–124

Lamour V, Hogan BP, Erie DA, Darst SA (2006) Crystal structure of *Thermus aquaticus* Gfh1, a Gre-factor paralog that inhibits rather than stimulates transcript cleavage. J. Mol. Biol. 356:179–188

Landick R, Turnbough C Jr, Yanofsky C (1996) Transcription attenuation. In *Escherichia coli* and *Salmonella*: Cellular and Molecular Biology. Neidhardt, FC, Curtiss R III, Ingraham JL, Lin ECC, Low KB, Magasanik B, et al. (eds) (Washington, DC: American Society for Microbiology Press), pp. 822–848

Lau IF, Filipe SR, Søballe B, Okstad ØA, Barre F-X, Sherratt DJ (2003) Spatial and temporal organization of replicating *Escherichia coli* chromosomes. Mol. Microbiol. 49:731–743

Lemon KP, Grossman AD (1998) Localization of bacterial DNA polymerase: evidence for a factory model of replication. Science. 282:1516–1519

Lemon KP, Grossman AD (2001) The extrusion capture model for chromosome partitioning in bacteria. Genes Dev. 15:2031–2041

Lemon KP, Kurster I, Grossman AD (2001) Effects of replication termination mutants on chromosome partitioning in *Bacillus subtilis*. Proc. Natl. Acad. Sci. USA. 98:212–217

Levine A, Autret S, Seror SJ (1995) A checkpoint involving RTP, the replication terminator protein, arrests replication downstream of the origin during the Stringent Response in *Bacillus subtilis*. Mol. Microbiol. 15:287–295

Lewis PJ (2001) Bacterial chromosome segregation. Microbiology. 147:519–526

Lewis PJ, Errington J (1997) Direct evidence for active segregation of *oriC* regions of the *Bacillus subtilis* chromosome and co-localization with the SpoOJ partitioning protein. Mol. Microbiol. 25:945–954

Lewis PJ, Thaker SD, Errington J (2000) Compartmentalization of transcription and translation in *Bacillus subtilis*. EMBO J. 19:710–718

Li Y, Sergueev K, Austin S (2002) The segregation of the *Escherichia coli* origin and terminus of replication. Mol. Microbiol. 46:985–996

Lin DC, Levin PA, Grossman AD (1997) Bipolar localization of a chromosome partition protein in *Bacillus subtilis*. Proc. Natl. Acad. Sci. USA. 94:4721–4726

McAdams HH, Shapiro L (2003) A bacterial cell-cycle regulatory network operating in time and space. Science. 301:1874–1877

McGinness T, Wake RG (1981) A fixed amount of chromosome replication needed for premature division septation in *Bacillus subtilis*. J. Mol. Biol. 146:173–177

Maddock JR, Shapiro L (1993) Polar location of the chemoreceptor complex in the *Escherichia coli* cell. Science. 259:1717–1723

Mah TF, Li J, Davidson AR, Greenblatt J (1999) Functional importance of regions in *Escherichia coli* elongation factor NusA that interact with RNA polymerase, the bacteriophage lambda N protein and RNA. Mol. Microbiol. 34:523–537

Marles-Wright J, Lewis RJ (2007) The structure of the *Bacillus subtilis* stressosome revealed by cryo-electron microscopy and single particle analysis. 4th conference on functional genomics of gram-positive organisms, Tirrenia, Italy. T74

Marston AL, Thomaides HB, Edwards DH, Sharpe ME, Errington, J (1998) Polar localization of the MinD protein of Bacillus subtilis and its role in selection of the midcell division site. Genes Dev. 12:3419–3430

Marston AL, Errington J (1999) Selection of the midcell division site in *Bacillus subtilis* through MinD-dependent polar localization and activation of MinC. Mol. Microbiol. 33:84–96

Mascarenhas J, Weber MHW, Graumann PL (2001) Specific polar localization of ribosomes in *Bacillus subtilis* depends on active transcription. EMBO R. 2:685–689

Matsumoto K, Kusaka J, Nishibori A, Hara H (2006) Lipid domains in bacterial membranes. Mol. Microbiol. 61:1110–1117

Meile JC, Wu LJ, Ehrlich SD, Errington J, Noirot P (2006) Systematic localisation of proteins fused to the green fluorescent protein in *Bacillus subtilis*: identification of new proteins at the DNA replication factory. Proteomics. 6:2135–2146

Migocki MD, Wake RG, Lewis PJ, Harry EJ (2004) The midcell replication factory in *Bacillus Subtilis* is highly mobile: implications for positioning the division site. Mol. Microbiol. 54:452–463

Miller OL Jr, Hamkalo BA, Thomas CA (1970) Visualisation of bacterial genes in action. Science. 169:392–395

Mooney RA, Landick R (2003) Tethering sigma70 to RNA polymerase reveals high in vivo activity of sigma factors and sigma70-dependent pausing at promoter-distal locations. Genes Dev. 17:2839–2851

Mukhopadhyay J, Kapanidis AN, Mekler V, Kortkhonjia E, Ebright YW, Ebright RH (2001) Translocation of σ^{70} with RNA polymerase during transcription: fluorescence resonance energy transfer assay for movement relative to DNA. Cell. 106:453–463

Mulder E, Woldringh, CL (1989) Actively replicating nucleoids influence positioning of division sites in Escherichia coli filaments forming cells lacking DNA. J. Bacteriol. 171:4303–4314

Murakami KS, Darst SA (2003) Bacterial RNA polymerases: the whole story. Curr. Opin. Struct. Biol. 13:31–39

Neylon C, Kralicek AV, Hill TM, Dixon NE (2005) Replication termination in Escherichia coli: structure and antihelicase activity of the Tus-Ter complex. Microbiol. Mol. Biol. Rev. 69:501–526

Nodwell JR, Greenblatt J (1993) Recognition of boxA antiterminator RNA by the E. coli antitermination factors NusB and ribosomal protein S10. Cell. 29:261–268

Nudler E, Gottesman ME (2002) Transcription termination and anti-termination in E. coli. Genes Cells. 7:755–768

Nudler E, Mironov AS (2004) The riboswitch control of bacterial metabolism. Trends Biochem. Sci. 29:11–17

Opalka N, Chlenov M, Chacon P, Rice PW, Wriggers W, Darst SA (2003) Structure and function of the transcription elongation factor GreB bound to bacterial RNA polymerase. Cell. 114:335–345

Ortiz JO, Forster F, Kurner J, Linaroudis AA, Baumeister W (2006) Mapping 70 S ribosomes in intact cells by cryoelectron tomography and pattern recognition. J. Struct. Biol. 156:334–341

Paget MS, Helmann JD (2003) The sigma70 family of sigma factors. Genome Biol. 4:203

Paul BJ, Ross W, Gaal T, Gourse RL (2004) rRNA transcription in Escherichia coli. Annu. Rev. Genet. 38:749–770

Peters PC, Migocki MD, Thoni C, Harry EJ (2005) A new assembly pathway for the cytokinetic Z ring from a dynamic helical structure in vegetatively growing cells of Bacillus subtilis. Mol Microbiol. 64:487–499

Perederina A, Svetlov V, Vassylyeva MN, Tahirov TH, Yokoyama S, Artsimovitch I, Vassylyev DG (2004) Regulation through the secondary channel—structural framework for ppGpp-DksA synergism during transcription. Cell. 118:297–309

Piggot PJ, Hilbert DW (2004) Sporulation of Bacillus subtilis. Curr. Opin. Microbiol. 7:579–586

Ramakrishnan V (2002) Ribosome structure and the mechanism of translation. Cell. 108:557–572

Raskin DM, de Boer PA (1999a) MinDE-dependent pole-to-pole oscillation of division inhibitor MinC in Escherichia coli. J. Bacteriol. 181:6419–6424

Raskin DM, de Boer PA (1999b) Rapid pole-to-pole oscillation of a protein required for directing division to the middle of Escherichia coli. Proc. Natl. Acad. Sci. USA. 96:4971–4976

Richardson J P, Greenblatt J (1996) Control of RNA chain elongation and termination. In Escherichia coli and Salmonella: Cellular and Molecular Biology. Neiderhardt FC, Curtiss R III, Ingraham JL, Lin ECC, Low KB, Magasanik B, et al. (eds) (Washington, DC: American Society for Microbiology Press), pp. 822–848

Ring B, Yarnell W, Roberts J (1996) Function of E. coli RNA polymerase σ factor σ^{70} in promoter-proximal pausing. Cell. 86:485–493

Rutherford ST, Lemke JJ, Vrentas CE, Gaal T, Ross W, Gourse RL (2007) Effects of DksA, GreA, and GreB on transcription initiation: insights into the mechanisms of factors that bind in the secondary channel of RNA polymerase. J. Mol. Biol. 366:1243–1257

Shapiro L, Losick R (2000) Dynamic spatial regulation in the bacterial cell. Cell. 100:89–98

Sharpe ME, Hauser PM, Sharpe RG, Errington J (1998) Bacillus subtilis cell cycle as studied by fluorescence microscopy: constancy of cell length at initiation of DNA replication and evidence for active nucleoid partitioning. J. Bacteriol. 180:547–555

Shepherd N, Dennis P, Bremer H (2001) Cytoplasmic RNA polymerase in *Escherichia coli*. J. Bacteriol. 183:2527–2534

Scheffers D-J, Jones LJF, Errington J (2004) Several distinct localisation patterns for penicillin-binding proteins in *Bacillus subtilis*. Mol. Microbiol. 51:749–764

Schulz HN, Jorgensen BB (2001) Big bacteria. Annu. Rev. Microbiol. 55:105–137

Shih Y-L, Le T, Rothfield L (2003) Division site selection in *Escherichia coli* involves dynamic redistribution of Min protein within coiled structures that extend between the two cell poles. Proc. Natl. Acad. Sci. USA. 100:7865–7870

Squires CL, Zaporojets D (2000) Proteins shared by the transcription and translation machines. Annu. Rev. Microbiol. 54:775–798

Studer SM, Joseph S (2006) Unfolding of mRNA secondary structure by the bacterial translation initiation complex. Mol. Cell. 22:105–115

Susa M, Kubori T, Shimamoto N (2006) A pathway branching in transcription initiation in *Escherichia coli*. Mol. Microbiol. 59:1807–1817

Teleman AA, Graumann PL, Lin DC, Grossman AD, Losick R (1998) Chromosome arrangement within a bacterium. Curr. Biol. 8:1102–1109

Thanedar S, Margolin W (2004) FtsZ exhibits rapid movement and oscillation waves in helix-like patterns in *Escherichia coli*. Curr. Biol. 14:1167–1173

Thompson SR, Wadhams GH, Armitage JP (2006) The positioning of cytoplasmic protein clusters in bacteria. Proc. Natl. Acad. Sci. USA. 103:8209–8214

Toulme F, Mosrin-Huaman C, Sparkowski J, Das A, Leng M, Rahmouni AR (2000) GreA and GreB proteins revive backtracked RNA polymerase in vivo by promoting transcript trimming. EMBO J.19:6853–6859

Trautinger BW, Jaktaji RP, Rusakova E, Lloyd RG (2005) RNA Polymerase modulators and DNA repair activities resolve conflicts between DNA replication and transcription. Mol. Cell. 19:247–258

Traviglia SL, Datwyler SA, Yan D, Ishihama A, Meares CF (1999) Targeted protein footprinting: where different transcription factors bind to RNA polymerase. Biochemistry. 38:15774–15778

Trucksis M, Michalski J, Deng YK, Kaper JB (1998) The *Vibrio cholerae* genome contains two unique circular chromosomes. Proc. Natl. Acad. Sci. USA. 95:14464–14469

Vanounou S, Parola AH, Fishov I (2003) Phosphatidylethanolamine and phosphatidylglycerol are segregated into different domains in bacterial membrane. A study with pyrene-labelled phospholipids. Mol. Microbiol. 49:1067–1079

Viollier PH, Shapiro L (2004) Spatial complexity of mechanisms controlling a bacterial cell cycle. Curr. Opin. Microbiol. 7:572–578

Viollier PH, Thanbichler M, McGrath PT, West L, Meewan M, McAdams HH, Shapiro L. Rapid and sequential movement of individual chromosomal loci to specific subcellular locations during bacterial DNA replication. Proc. Natl. Acad. Sci. USA. 101:9257–9262

Vogel U, Jensen KF (1995) Effects of the antiterminator *boxA* on transcription elongation kinetics and ppGpp inhibition of transcription elongation in *Escherichia coli*. J. Biol. Chem. 270:18335–18340

Volkov A, Mascarenhas J, Andrei-Selmer C, Ulrich HD, Graumann PL (2003) A prokaryotic condensin/cohesin-like complex can actively compact chromosomes from a single position on the nucleoid and binds to DNA as a ring-like structure. Mol. Cell Biol. 23:5638–5650

Wang MD, Schnitzer MJ, Yin H, Landick R, Gelles J, Block SM (1998) Force and velocity measured for single molecules of RNA polymerase. Science. 282:902–907

Webb C, Teleman A, Gordon S, Straight A, Belmont A, Lin D, Grossman A, Wright A, Losick R (1997) Bipolar localization of the replication origin regions of chromosomes in vegetative and sporulating cells of *B. subtilis*. Cell. 88:667–674

Webb CD, Graumann PL, Kahana JA, Teleman AA, Silver PA, Losick R (1998) Use of time-lapse microscopy to visualize rapid movement of the replication origin region of the chromosome during the cell cycle in *Bacillus subtilis*. Mol. Microbiol. 28:883–892

Weber MH, Volkov AV, Fricke I, Marahiel, MA, Graumann PL (2001a) Localization of cold shock proteins to cytosolic spaces surrounding nucleoids in *Bacillus subtilis* depends on active transcription. J. Bacteriol. 183:6435–6443

Weber MH, Beckering CL, Marahiel MA (2001b) Complementation of cold shock proteins by translation initiation factor IF1 in vivo. J. Bacteriol. 183:7381–7386

Woldringh CL, Mulder E, Huls PG, Vischer N (1991) Toporegulation of bacterial division according to the nucleoid occlusion model. Res. Microbiol. 142:309–320

Woldringh CL, Jensen PR, Westerfho HV (1995) Structure and partitioning of bacterial DNA: determined by a balance of compaction and expansion forces? FEMS Microbiol. Lett. 131:235–242

Woldringh CL, Nanninga N (2006) Structural and physical aspects of bacterial chromosome segregation. J. Struct. Biol. 156:273–283

Wu L-J, Errington J (1994) *Bacillus subtilis* SpoIIIE protein required for DNA segregation during asymmetric cell division. Science 264:572–575

Wu LJ, Franks AH, Wake RG (1995a) Replication through the terminus region of the *Bacillus subtilis* chromosome is not essential for the formation of a division septum that partitions the DNA. J. Bacteriol. 177:5711–5715

Wu L-J, Lewis PJ, Allmansberger R, Hauser PM, Errington J (1995b) A conjugation-like mechanism for prespore chromosome partitioning during sporulation in *Bacillus subtilis*. Genes Dev. 9:1316–1326

Wu LJ, Errington J (2003) RacA and the Soj-Spo0J system combine to effect polar chromosome segregation in sporulating *Bacillus subtilis*. Mol. Microbiol. 49:1463–1475

Wu LJ, Errington J (2004) Coordination of cell division and chromosome segregation by a nucleoid occlusion protein in *Bacillus subtilis*. Cell. 117:915–925

Yakhnin AV, Babitzke P (2002) NusA-stimulated RNA polymerase pausing and termination participates in the *Bacillus subtilis trp* operon attenuation mechanism in vitro. Proc. Natl. Acad. Sci. USA. 99:11067–11072

Yanofsky C (2004) The different roles of tryptophan transfer RNA in regulating trp operon expression in *E. coli* versus *B. subtilis*. Trends Genet. 20:367–374

Zhang G, Campbell EA, Minakhin L, Richter C, Severinov K, Darst SA (1999) Crystal structure of *Thermus aquaticus* core RNA polymerase at 3.3 Å resolution. Cell. 98:811–824

Zheng CH, Friedman DI (1994) Reduced Rho dependent termination permits NusA independent growth of *E. coli*. Proc. Natl. Acad. Sci. USA. 91:7543–7547

2
Molecular Components of the Bacterial Cytoskeleton

Katharine A. Michie

Abstract It is only relatively recently that a prokaryotic cytoskeleton akin to that in eukaryotes has been identified, revealing a much higher order of cellular complexity than was previously thought. The proteins that form these bacterial cytoskeletal elements not only carry out similar roles to their eukaryotic counterparts, but they also have related protein folds, suggesting an ancient evolutionary relationship and

Katharine A. Michie
Medical Research Council Laboratory of Molecular Biology, Hills Rd,
Cambridge CB2 0QH, UK
kmichie@mrc-lmb.cam.ac.uk

W. El-Sharoud (ed.) *Bacterial Physiology: A Molecular Approach.*
© Springer-Verlag Berlin Heidelberg 2008

the conservation of fundamental mechanisms. This chapter will introduce to the reader what is known at the molecular level regarding the proteins that comprise eubacterial and, in some cases, archaeal cytoskeletal elements.

2.1 Introduction

2.1.1 What Is a Cytoskeleton?

In eukaryotes, the definition of the cytoskeleton has come to encompass several types of filamentous structures within the cell, some of which are dynamic structures, whereas others are more stable. Each of these filament types is largely composed of a single protein component that can assemble into polymers in vivo and also in vitro (a schematic showing the way these types of proteins assemble into filaments is shown in Fig. 2.1). The systems of these filaments contribute significantly to cellular organisation and are responsible for determining and maintaining cell shape (as well as contributing to mechanical strength); for the movement of molecules, vesicles, and organelles; and for cell division.

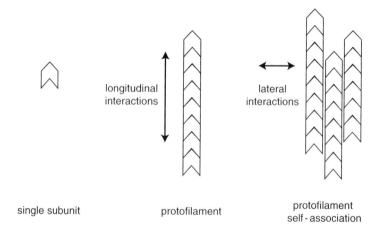

Fig. 2.1 The assembly of protein monomers into polymeric structures. Schematic description of polymerisation of cytoskeletal proteins. A single protein subunit is depicted on the *left*, and protofilament assembly is represented in the *middle*, showing the axis of longitudinal filament extension. On the *right*, one type of protofilament self-association (sheet formation) and the axis of lateral interactions is shown. Note that protofilament self-association may occur in many different ways, including the antiparallel alignment of filaments. Lateral interactions may also arise in all directions perpendicular to the axis of filament extension to form bundles, tubes, and asters. Accessory proteins may also mediate lateral interactions between protofilaments

In eukaryotes, a cytoskeletal filament is constructed from a protein belonging to one of the three cytoskeletal protein superfamilies: actin, tubulin (which form dynamic actin filaments and microtubules, respectively), and intermediate filaments (IFs), including the keratins, lamins, and other specialised proteins that form more static filaments.

2.1.2   The Cytoskeleton in Bacteria

Bacteria have historically proven problematic to the cell biologist studying cellular organisation, mainly because of the generally small size of their cells. Their tiny dimensions stretch the resolution of optical microscopes to the limit, and, for decades, bacteria were thought not to possess cytoskeletal elements. Only a few internal structures had been observed in bacterial cells, and these were apparently organism specific and obscure.

However, ideas regarding prokaryotic cellular organisation began to change after important results were reported in 1991. Using immunoelectron microscopy, Bi and Lutkenhaus discovered that a well-conserved protein (called FtsZ) that was linked to cell division localised with a unique pattern at the mid-cell site before any observable septum invagination in *Escherichia coli*. Furthermore, during cell

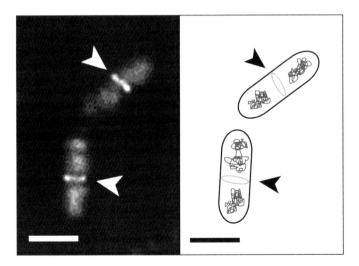

Fig. 2.2 FtsZ assembles into a structure at the middle of the cell in most eubacteria and archaea. *Left*, subcellular localisation of FtsZ and in DNA *Bacillus subtilis* cells as visualised by an overlay of immunofluorescently labelled FtsZ and DAPI-stained DNA. FtsZ localisation is indicated by the *arrows*. The *right* schematic depicts the cell membrane (*oval outline*), the position of the Z ring (shown by the *gray ring* structure denoted by the *arrows*), and DNA (represented by the *twisted lines*) close to the poles of the cell. Scale bar represents 1 μm

division, FtsZ remained at the leading edge of the enclosing septum, with a pattern consistent with a constricting ring structure (Bi and Lutkenhaus 1991). This was one of the first observations of a highly organised intracellular structure that assembles within bacterial cells. Since then, FtsZ rings have been observed in many bacteria (Fig. 2.2) and are thought to form the basic cytoskeletal structure underpinning the division apparatus.

The progress of our understanding of bacterial cell organisation has been accelerated by advances in optical microscopy methods, including the use of green fluorescent protein (GFP) fusion technologies, time-lapse imaging, and deconvolution analysis. More recently, the development of cryo-electron tomography techniques promises to reveal even greater detail (Lucic et al. 2005; Briegel et al. 2006).

2.1.3   Bacteria May Have Many Families
of Cytoskeletal Proteins

We now know that bacteria have considerable intracellular organisation, with several cytoskeletal elements, including the cell division apparatus, also called the divisome or septasome. In fact, all three of the known eukaryotic cytoskeletal proteins (actin, tubulin, and IFs) have counterparts in eubacteria that form filamentous structures with cytoskeletal roles. In this chapter, the prokaryotic cytoskeletal proteins are discussed, with a focus on the biophysical and biochemical qualities of these proteins, which include a tubulin homologue called FtsZ; two specialist tubulin homologues (BtubA and BtubB) found in *Prosthecobacter* species; a range of bacterial actin-like proteins, including MreB, which helps maintain cell shape in many rod-shaped bacteria; some actin-like specialist elements, such as ParM and MamK filaments; and a single IF-like protein represented by crescentin, which is involved in determining cell shape in *Caulobacter crescentus*. Recently, a potential new class of cytoskeletal proteins called the WACA proteins, for _W_alker _A_ _c_ytoskeletal _A_TPase, has been described in bacteria, with yet-unidentified counterparts in eukaryotes. The protein members of this family include ParA, MinD, Soj, and MipZ.

2.2   The Tubulin Superfamily

In this section, the bacterial tubulin-like proteins are described. To introduce the reader to these types of proteins, the first section covers the molecular characteristics and some interesting biochemical properties of eukaryotic tubulin, followed by a summary of what is known regarding FtsZ. A brief description of the proteins linked to cell division follows. Finally, two specialist tubulin-like proteins (BtubA and BtubB) are briefly discussed.

2.2.1 Eukaryotic Tubulin

Eukaryotic cells express several tubulin proteins, including α-, β-, γ-, δ-, and ε-tubulin. The best-characterised tubulins are α- and β-tubulin, which are the main components of microtubules. Microtubules are essential cytoskeletal elements that assemble in all eukaryotes, and are required for many intracellular transport events and for cell division. For example, microtubules form the mitotic spindle that provides the framework for separation of daughter chromatids toward opposite cell poles during mitosis.

Microtubules are generally comprised of 13 filaments. Each filament is made up of longitudinally end-to-end-associated heterodimers of $\alpha\beta$-tubulin that laterally associate into a tube-like structure (see Fig. 2.1 for a schematic representation of protein polymer formation). The core structures of α- and β-tubulin are composed of two β-sheets surrounded by α-helices (Nogales et al. 1998b), making up two functional domains. The N-terminal of the two domains has a Rossmann fold similar to that of many ATPases, and it contains a GTP binding site. The C-terminal domain is structurally homologous to the family of chorismate mutase-like proteins, and carries some of the catalytic residues for GTP hydrolysis (Nogales et al. 1998a).

The tubulin proteins are GTPases, with the active site formed at the interface between subunits, using essential amino acid residues from both subunits. Thus, GTPase activity only occurs when two or more subunits associate, where the N-terminal domain of one subunit provides the nucleotide-binding site and the C-terminal domain of another subunit provides the "T7 loop" that has the residues responsible for nucleotide hydrolysis.

Microtubules are assembled from $\alpha\beta$-tubulin heterodimers joined end-to-end so that α and β subunits of tubulin alternate, with GTP binding pockets between each (See Fig. 2.3 for the arrangement of α and β subunits of tubulin within a filament). This arrangement results in a distinct polarity, with β-tubulin always present at one end (designated the plus end) and α-tubulin at the other end (called the minus end).

Microtubules exhibit dynamic instability, whereby the filaments may grow or shrink rapidly. This characteristic arises from three biochemical features. First, subunit exchange within the filament does not occur and the filaments can only assemble and disassemble from the ends. Second, the nucleotide-binding pocket between monomers of tubulin is occluded and nucleotide exchange is prohibited within the filament. Third, it has been proposed that GTP hydrolysis induces a destabilising conformational change within the filament, causing a bent or curved morphology. The GDP "bent" form of the filament is unstable and, if unrestrained, the filament disassembles rapidly. However, at the end of the microtubule, a GTP cap can stabilise the filament, restraining the filaments in the "straight" conformation, enabling the filaments to grow. If the GTP-cap is hydrolysed to GDP, the filaments are free to spontaneously disassemble. Thus, the state of this cap has a dramatic effect on whether microtubules grow or shrink (Desai and Mitchison 1997).

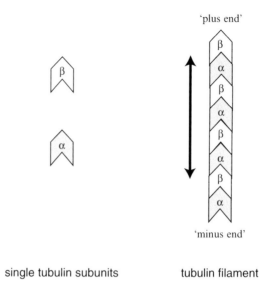

single tubulin subunits   tubulin filament

Fig. 2.3 Assembly of α and β subunits of tubulin within a single protofilament. Note that all of the subunits align in the same orientation along the length of the protein, but alternate

2.2.2 FtsZ

In bacteria, the first known step in cell division is the localisation of FtsZ into a ring structure (the Z ring) at the nascent division site (Bi and Lutkenhaus 1991). FtsZ is one of a number of proteins (see Sect. 2.2.3 below) that are involved in cell division. Many of these cell division proteins were identified by the isolation of conditional mutants that have filamentous cellular phenotypes when grown at a nonpermissive temperature. Thus, many of the cell division proteins have been termed *fts*, for *f*ilamentous *t*emperature *s*ensitive proteins (Rothfield et al. 1999). Z ring formation is followed by the localisation of a series of other cytoplasmic and membrane-bound proteins that also play a part in cell division. It is the final structure of all these proteins built up around the Z ring that is the apparatus that carries out cell division that is referred to as the "divisome" or "septasome."

Much of the research focusing on the molecular processes in bacterial cell division and the roles of the divisome has been centred on understanding the function of FtsZ. This is for several reasons. FtsZ is the first protein known to localise to the future division site and it is essential for the division process. In addition, FtsZ is extremely well conserved among most bacteria and archaea, suggesting that it performs a fundamental biological role. FtsZ is thought to be absent only from the Crenarchaeota (Kawarabayasi et al. 1999; Vaughan et al. 2004), *Ureaplasma urealyticum* (Glass et al. 2000), a *Pirellula* species (Glockner et al. 2003), and the *Chlamydiaceae* family (Stephens et al. 1998; Brown and Rockey 2000; Read et al. 2000). Finally, consistent with the endosymbiont model for the origins of chloroplasts and mitochondria, FtsZ is also found in these organelles in some eukaryotes (Osteryoung and Vierling 1995; Beech et al. 2000), where it retains a conserved function in organelle division (Osteryoung and McAndrew 2001; Vitha et al. 2001).

In the early 1990s, a conserved tubulin motif was identified within the FtsZ sequence (Mukherjee et al. 1993; Bermudes et al. 1994), leading to the idea of an ancient evolutionary relationship between FtsZ and tubulin. This was surprising because the proteins only share an average of 10 to 18% sequence identity in amino acid sequence alignments (de Pereda et al. 1996). However, in 1998, the debate was settled when the three-dimensional protein structure of FtsZ was reported and FtsZ was shown to have the same two-domain fold as tubulin, with a conserved GTPase binding pocket in the N-terminal domain, and the residues responsible for GTP hydrolysis residing on a flexible loop (called the T7 loop) in the C-terminal domain (Löwe and Amos 1998; Nogales et al. 1998a).

Consistent with its relationship to tubulin, FtsZ assembles in vivo into a highly ordered structure, although its precise molecular architecture is undefined. It is only very recently that structures suggested to be FtsZ filaments have been observed by cryo-electron tomography, although more information is required to confirm this interpretation (Briegel et al. 2006). The Z-ring structure is extremely dynamic and, in vivo, it is remodelled or rebuilt continually with FtsZ subunits exchanging between the ring and the cytoplasm in a timeframe of seconds, as observed by _flu_oresence _r_ecovery _a_fter _p_hotobleaching (FRAP) experiments (Stricker et al. 2002; Anderson et al. 2004). There are also multiple reports of moving ring and helical structures assembled from FtsZ in vivo, as observed by time-lapse live-cell microscopic imaging of FtsZ–GFP fusions in both wild-type and mutant bacterial strains (Ben-Yehuda and Losick 2002; Thanedar and Margolin 2004; Grantcharova et al. 2005; Michie et al. 2006). The dynamic behaviour of the Z ring suggests that it is a flexible and adaptable structure.

In vitro, FtsZ self-assembles into protofilaments (protofilaments are linear chains of associating molecules). Several types of protofilaments have been observed, including rings and straight and curved forms. FtsZ protofilaments also assemble into higher-order structures, including sheets, tubules, asters, and bundles. All of these reported assemblies form under a wide range of conditions and in almost all known nucleotide-bound states, including the GTP-, GDP-, and the nonhydrolysable GMPCPP-bound forms (GMPCPP: guanylyl-(α,β)-methylene diphosphate) (Bramhill and Thompson 1994; Mukherjee and Lutkenhaus 1994; Erickson et al. 1996; Yu and Margolin 1997; Mukherjee and Lutkenhaus 1998; Löwe and Amos 1999; Löwe and Amos 2000; Lu et al. 2000; Oliva et al. 2003). The role of GTP hydrolysis in FtsZ function is still debated, and in vivo FtsZ mutants with reduced GTPase activity are able to support cell division (Phoenix and Drapeau 1988; Lu et al. 2001; Stricker et al. 2002), suggesting that either the intrinsic GTPase activity of FtsZ is modulated or it is much higher than required to support division.

The sheer number of different polymer morphologies and the wide range of conditions that support FtsZ self-assembly argue in favour of a model whereby

FtsZ assembles in vivo unassisted by specific proteins. However, we do not know which in vitro polymer form of FtsZ most closely resembles the native form, and the situation is further complicated by the large number of proteins that are known to interact with FtsZ either directly or indirectly during assembly of the divisome, which could alter the biochemical properties of FtsZ. Fortunately, structural determination at the atomic level of protofilaments formed by FtsZ in vitro has provided insight into the likely arrangement of individual FtsZ subunits within the polymer (Oliva et al. 2004). Monomers of FtsZ within protofilaments assemble in a head-to-tail arrangement similar to those observed for αβ-tubulin within microtubules. In a protofilament, each FtsZ monomer maintains longitudinal contacts with an almost identically arranged FtsZ molecule (with a 10° tilt) above and below (Oliva et al. 2004), such that all the molecules are aligned in the same orientation. Thus, the main interactions that form the basis of single protofilament formation are longitudinal interactions between FtsZ molecules. It seems likely that this will be the case in vivo. Sandwiched between two associating FtsZ molecules is a nucleotide-binding pocket, contributed by one molecule of FtsZ, and a catalytic loop for GTP hydrolysis that is contributed by an adjacent FtsZ molecule in the protofilament. The catalytic loop contains two conserved aspartate residues and a glutamine residue that coordinates a magnesium ion. This arrangement precipitates a special characteristic whereby nucleotide hydrolysis cannot occur unless FtsZ self-associates, and, thus, GTPase activity reflects FtsZ–FtsZ interaction (Nogales et al. 1998b; Scheffers and Driessen 2002; Oliva et al. 2004).

The nucleotide-binding pocket of FtsZ shows important differences to the nucleotide-binding pocket of tubulin. The dynamic instability of tubulin assembly is critically linked to the fact that the nucleotide-binding pockets of tubulin are occluded in the microtubule-assembled form, and nucleotide exchange is prohibited. In contrast, the nucleotide-binding pocket of FtsZ is likely to be solvent accessible, which would more readily allow nucleotide exchange (Oliva et al. 2004). This characteristic could critically affect the dynamic behaviour of FtsZ filaments. Analysis by atomic force microscopy has shown that FtsZ filaments continuously rearrange in vitro (Mingorance et al. 2005). End-to-end joining of FtsZ filaments and depolymerisation of FtsZ from within the middle of filaments have been observed in vitro, further suggesting that nucleotide exchange occurs at internal sites at least in a single protofilament (Mingorance et al. 2005).

Whether the Z ring is one protofilament thick or comprised of a bundle of filaments laterally associated has not been experimentally determined, but it is generally thought that the Z ring is comprised of several to many filaments, and the concentration of FtsZ in the cell is sufficiently high (3000–15000 FtsZ molecules per cell) to wind around the cellular circumference many times (Lu et al. 1998; Feucht et al. 2001; Rueda et al. 2003). Using data collected in the FRAP experiments described previously, it was estimated that approximately 30% of the cellular FtsZ seems to be present in the Z-ring structure at any one time (Stricker et al. 2002), which is consistent with the multifilament model for Z rings.

If the Z ring were comprised of multistranded FtsZ filaments, then lateral interactions between adjacent protofilaments would probably be important for the assembly of the Z ring and for maintaining its stability. Indeed, in the case of tubulin, lateral interactions between protofilaments are essential for forming microtubules and are also important for tubulin's interactions with accessory proteins. It is likely that FtsZ lateral interactions are also important for interacting with other proteins and also with the cell membrane. Although several proteins are known to interact directly with FtsZ, including FtsA, ZapA, and MinC, little is known regarding the molecular nature of these interactions, and further investigation into the binding surfaces of FtsZ and its partner proteins should be very illuminating.

Finally, it is worth noting that several bacterial and archaeal genomes encode multiple *ftsZ* genes. This may provide some clues regarding the evolutionary history of the tubulin superfamily. In addition, *ftsZ* genes were recently discovered in plasmids of various *Bacillus* species (Scholle et al. 2003; Tang et al. 2006; Tinsley and Khan 2006). This is surprising because all of these organisms carry a chromosomal copy of FtsZ and are viable without these plasmids. This raises the question of what role the plasmid-encoded FtsZ plays. Preliminary research indicates that, at least for one such plasmid gene, it is required for the survival of the plasmid in its host, and, thus, thought to be involved in the replication and/or segregation of the plasmid (Tinsley and Khan 2006). Such a function would be more reminiscent of the function of tubulin than FtsZ.

2.2.3 Proteins of the Divisome

The proteins involved in cell division can generally be divided into two classes: those probably performing a direct role in the function of the divisome, i.e., building the division septum, and those responsible for regulating the divisome, i.e., determining when and where the Z ring assembles (see Chap. 1). To give the reader an appreciation for the complicated nature of the structure that assembles with complete dependence on the FtsZ cytoskeletal structure, the following section very briefly summarises divisome assembly and regulation (for a detailed description, see Margolin 2005; Møller-Jensen and Löwe 2005; Harry et al. 2006; Shih and Rothfield 2006; Lutkenhaus 2007).

After FtsZ localisation, a large number of proteins assemble at the mid-cell site. These proteins include the cytoplasmic proteins FtsA, ZipA, ZapA, SepF, EzrA, and the transmembrane proteins FtsK, FtsQ/DivIB, FtsB/DivIC, FtsL, FtsW, FtsI/ PBP 2B/ PBP3, and FtsN. In *E. coli*, protein assembly at mid-cell occurs largely in hierarchical steps and is dependent on FtsZ assembly. The cytoplasmic proteins assemble first, followed by membrane-bound proteins. However, the localisation orders of homologous proteins vary between species, and some of the membrane-bound proteins that assemble late in the localisation hierarchy are interdependent on each other for assembly and are thought to form a complex outside of the divisome (Daniel et al. 1998; Daniel and Errington 2000; Robson et al. 2002; Buddelmeijer and Beckwith 2004). Some of the later-assembling proteins, such as FtsI/PBP2B, are known to be involved in cell wall (peptidoglycan) biosynthesis

(Yanouri et al. 1993), whereas others, such as FtsK in *E. coli*, are involved in coordinating chromosome segregation with cell division by facilitating the complete separation of chromosomes into the newly forming cells created by the division septum (Liu et al. 1998).

The divisome needs cues that allow it to assemble at the correct time and place in the cell. A number of proteins are known to interact with FtsZ in a manner presumed to regulate its localisation. With the exception of SulA, the molecular details of how these proteins exert control over FtsZ are unknown. SulA is a cell division inhibitor that is expressed as a part of the *E. coli* SOS response to DNA damage. SulA functions by binding to the FtsZ polymerisation interface and titrating away monomeric FtsZ, thus, inhibiting Z-ring assembly (Cordell et al. 2003).

The second best-characterised proteins known to affect Z-ring assembly are the Min proteins (for a detailed review, see Lutkenhaus 2007). The Min proteins form a part of a system that, in rod-shaped bacteria, is responsible for the inhibition of inappropriate assembly of the divisome near the poles of the cell. MinC is the component of the system that interacts directly with FtsZ to inhibit Z-ring assembly, although the nature of the MinC–FtsZ interaction has yet to be defined (Bi and Lutkenhaus 1990; de Boer et al. 1990; Hu et al. 1999). The subcellular localisation of MinC at the cell poles is established by the other Min proteins, thereby, only inhibiting cell division at the cell poles (de Boer et al. 1989; Fu et al. 2001; Hale et al. 2001; Shih et al. 2002).

Other proteins, including EzrA, SlmA, ZapA, Noc, and MipZ, have stabilising or destabilising roles, probably by interacting directly with FtsZ, yet the mechanisms by which they do this are largely unknown. Both ZapA (Gueiros-Filho and Losick 2002) and MipZ (Thanbichler and Shapiro 2006b) are thought to stabilise the Z ring. In contrast, EzrA (Levin et al. 1999), SlmA (Bernhardt and de Boer 2005), and Noc (Wu and Errington 2004) are thought to destabilise the Z ring. Whether these proteins work at the level of FtsZ polymerisation, at the level of protofilament bundling, or by some other indirect mechanism remains to be investigated.

2.2.4 *BtubA and BtubB*

FtsZ is not the only tubulin-like protein in prokaryotes. The bacterial genus *Prosthecobacter* expresses two unusual tubulin homologues called BtubA and BtubB (Jenkins et al. 2002). These proteins do not exist in other bacterial species, suggesting that they probably assemble into a specialist cytoskeletal element. Both proteins show a closer relationship to eukaryotic tubulin than to FtsZ, although the crystal structures of BtubA and BtubB revealed that each protein has mixed characteristics of α- and β-tubulin and cannot be assigned to either α- or β-tubulin (Schlieper et al. 2005). The function of BtubA and BtubB is unknown, however, they self-assemble in vitro, assuming a filamentous form similar to both αβ-tubulin and FtsZ. Their low divergence from eukaryotic tubulin suggests that they might be products of horizontal gene transfer events (Schlieper et al. 2005).

2.3 The Actin-Like Superfamily

The actin family of ATPases is very diverse in sequence and function. The members of this family share a conserved ATPase fold and a conserved set of sequence motifs involved in nucleotide binding. Not all members of this family are classified as cytoskeletal proteins, because they do not all share the ability to polymerise, nor do they all have roles in defining the shape of the cell. Within prokaryotes, there exist a number of actin-like proteins, most of which were first identified in a sequence homology search based on the catalytic core shared by actin, hexokinase, and the hsp70 proteins (Bork et al. 1992). The actin-like proteins in prokaryotes include MreB, MreB-like proteins (Mbl and MreBH), ParM (StbA), FtsA, ActA, DnaK (hsp70), and the hexokinases. Of this group, DnaK (hsp70) and the hexokinases are not cytoskeletal proteins, and these proteins will, thus, not be discussed further. The case for FtsA is currently ambiguous and is discussed briefly in Section 2.3.5. More recently, another protein called MamK was identified with similarity to actin and MreB (Komeili et al. 2006). MamK is apparently a specialised protein only found in magnetotactic organisms.

MreB, ParM, and MamK are thought to have important roles in cell shape determination, plasmid segregation, and the assembly of specialist cytoskeletal elements. These proteins are described in detail in the sections after a brief description of actin.

2.3.1 Eukaryotic Actin

Actin is a highly abundant protein, found in almost all eukaryotic cells. It forms a dynamic network that various motor proteins (which transport molecules, vesicles, and organelles) use to track around the cell. In some cell types, actin is largely responsible for determining the shape of the cell and enabling cell locomotion. For example, actin filaments are directly involved in the formation of pseudopodia, central to the mechanism enabling the locomotion of amoeba. Actin, together with myosin, forms a core part of the machinery that enables contraction in muscle cells.

The two domains, named I and II, that comprise actin can be divided into two subdomains: A and B. The larger two of these, designated IA and IIA, comprise a five-strand β-sheet enclosed by three α-helices. IB and IIB, the two smaller subdomains, show variation in both size and structure across the actin family and impart some of the distinct properties of each protein. Between the two domains lies a highly conserved ATP-binding pocket containing essential aspartate residues that, together with either Mg^{2+} or Ca^{2+}, bind and hydrolyse ATP, which is central to disassembly of F-actin filaments.

Actin shows cooperative assembly kinetics, with a slow nucleation step requiring nucleation factors in vivo to stimulate and control actin filament assembly. The actin subunits themselves assemble in a head-to-tail arrangement, forming a dynamic helical polymer known as F-actin (similar to the linear filaments depicted in Fig. 2.1). After polymerisation, each actin subunit undergoes structural changes (Holmes et al. 1990), facilitated in part by rotation of domains I and II with respect to each other. Because of the head-to-tail arrangement of the monomers within the filament, actin has a distinct asymmetry, and the ends of the filaments have different biochemical qualities. Actin displays treadmilling behaviour that occurs because the asymmetrical ends of the actin filament have different affinities for polymerisation. Actin subunits preferentially assemble at one end (called the barbed end), and, after ATP hydrolysis and phosphate release, subunits dissociate from the nonpreferred end, called the "pointed end" (Korn et al. 1987). This leads to a net movement of subunits through the filament. Filaments maintain constant length only when assembly and disassembly rates are equivalent. Thus, the filament length (growth and shrinkage) and the rate that actin subunits pass along the filament while treadmilling are controlled by the rates of monomer addition and dissociation. Many accessory proteins that affect the assembly, disassembly, and rearrangement of actin filaments in vivo have been identified in eukaryotes (Schmidt and Hall 1998).

2.3.2   MreB

For a long time, a cluster of genes (known as *mre* for murein cluster e) was known to be important for determining the cell shape in many of the more complex-shaped bacteria, including the rod-shaped *Bacillus subtilis* and *E. coli*, as well as the differentiating *C. crescentus*. These genes are also present in some mollicutes and archaea, but are largely absent from coccoid bacteria, and are also absent from some rod-shaped bacteria (Daniel and Errington 2003). The *mre* cluster contains the *mreB*, *mreC*, and *mreD* genes. Some organisms have several *mreB*-related genes, such as *B. subtilis*, which also has *mrebl* (*mreB*-like) and *mrebh* (*mreB* homologue). *mreB* seems to be essential because mutation or depletion of MreB and MreB-like proteins results in severe defects in normal cell morphology, normally resulting in cell death (Varley and Stewart 1992; Figge et al. 2004; Kruse et al. 2005). A common feature of *mreB* mutants and strains depleted of MreB is abnormalities in cell size, particularly cell width (Jones et al. 2001). *mbl* mutant strains of *B. subtilis* display distorted twisted and bent morphologies (Abhayawardhane and Stewart 1995). These phenotypes are strongly suggestive of the disruption of key cytoskeletal structures.

Supportive of a cytoskeletal role, in vivo, MreB and the other MreB-like proteins form a helical filament close to the cytoplasmic face of the cell membrane (Jones et al. 2001). It is now thought from a number of experiments that the MreB family of proteins play essential roles in localising cell wall synthesis machinery along the length of the lateral cell wall. Initial experiments performed in *B. subtilis* using a fluorescent derivative of vancomycin that binds to newly synthesised peptidoglycan revealed that new peptidoglycan is inserted in a helical pattern along the length of the cell. This pattern of vancomycin localisation was dependent on the MreB-like protein, Mbl (Daniel and Errington 2003), although these results have been questioned (Tiyanont et al. 2006). Experiments testing the localisation dependencies of a component of the peptidoglycan synthesis machinery (penicillin-binding proteins [PBP]-2) in *C. crescentus* demonstrated that localisation was dependent on MreB (Figge et al. 2004) and that the cell wall hydrolase enzyme, LytE, in *B. subtilis* has a localisation pattern dependent on MreBH (Carballido-Lopez et al. 2006).

Data suggesting that MreB has roles in the segregation of DNA (Kruse et al. 2003; Soufo and Graumann 2003) has caused great controversy and currently remains a point of contention (Hu et al.; Gitai et al. 2004; Formstone and Errington 2005; Gitai et al. 2005; Kruse et al. 2006; Defeu Soufo and Graumann 2006).

The three-dimensional atomic structure of MreB, solved by X-ray crystallography, revealed that MreB shares the same fold with actin (van den Ent et al. 2001). Fortunately, MreB crystallised in a polymerised form, providing insight into the likely arrangement of MreB self-association within protofilaments. Both MreB and actin show very similar protein–protein contacts with similar subunit repeats being 51 Å and 55 Å in MreB and F-actin, respectively. They also show similar orientations, with both proteins forming two-stranded filaments (van den Ent et al. 2001). One notable difference is in the amount of rotation occurring within these double filaments; MreB shows very little twist (or axial rotation) in comparison with F-actin, where the filaments twist around one another.

In vivo, MreB and MreB-like filaments are highly dynamic. Time-lapse imaging of fluorescently labelled MreB in *E. coli* and *C. crescentus* and of MreB, Mbl, and MreBH in *B. subtilis* have revealed a variety of helical localisation patterns, with different filament pitches adopted within a population of cells. This most likely relates to the observation that the localisation of these proteins changes dynamically throughout the cell cycle. For example, in synchronised *C. crescentus* cells, MreB has been observed to localise as a helical structure along the length of the cell that collapses into a single band at mid-cell at a time coinciding with Z-ring formation (this was shown to be dependent on Z-ring assembly). During the progression of cell division, the MreB "band" then expands out again until it stretches along the length of both daughter cells (Gitai et al. 2004). FRAP experiments with Mbl in *B. subtilis* have shown that the Mbl filaments are continuously remodelled, with a half-life of approximately 8 min. Surprisingly, this technique also revealed that, unlike F-actin, there is no apparent polarity in Mbl filaments with Mbl turnover occuring along the length of the filaments (Carballido-Lopez and Errington 2003). The number of Mbl molecules in the cell is approximately 10-fold higher than what would be required to assemble a single helical protofilament, assuming the Mbl filaments were similar to MreB filaments (Jones et al. 2001), and it has been suggested that Mbl might assemble into short protofilaments that are able to bundle together without a uniform polarity into a multistranded "cable" (Carballido-Lopez and Errington 2003). The dynamic nature of such a cable might then arise from the continuous replacement with the cytoplasmic supply of Mbl subunits dissociating from the ends of each short protofilament within the bundle (Carballido-Lopez and

Errington 2003). An important piece of the puzzle finally came together when it was revealed that single MreB molecules within the MreB cable exhibit treadmilling behaviour in vivo in *C. crescentus*, whereas the overall cable shows no polarity (Kim et al. 2006). Thus, it seems that the MreB family of proteins assemble into short protofilaments, each of which is capable of treadmilling while assembled into a bundled cable that shows no overall polarity.

In vitro studies of MreB filaments have identified straight and curved protofilaments as well as small ring-like structures (with circumferences too small to span the diameter of the cell), and bundles of filaments. As in the case for the filament morphologies observed for FtsZ, the biological relevance of such structures has not been determined and some of them are probably artefactual. It is interesting to note that bundled filaments of MreB show significant increases in rigidity, which might be very important if the MreB cables in vivo carry out a mechanical role. It is also possible that auxiliary proteins regulate the assembly of MreB in vivo, either by mediating MreB–MreB interactions, or MreB–membrane interactions.

Although MreB filament formation is nucleotide dependent (as is the case for F-actin), with ATP and GTP inducing filament formation (van den Ent et al. 2001), the assembly dynamics for MreB are much faster than those observed for F-actin (Esue et al. 2005). MreB has a critical concentration for assembly (~3 nM), 100-fold lower than actin, and, unlike actin, MreB does not have a pronounced nucleation step (Esue et al. 2005).

In eukaryotes, actin filaments serve as "tracks" for use by motor proteins. In addition, a large number of proteins are known to interact with actin in a regulatory context. It was anticipated that similar factors would be identified for MreB. However, despite much searching, no candidate motor proteins that interact with MreB have yet been identified in prokaryotes.

2.3.3 ParM

ParM (also called StbA) is a specialist cytoskeletal element, because it is only required for the correct partition of the R1 low-copy number plasmids in *E. coli*, and, as such, is not essential for normal cell function.

R1 plasmids are actively partitioned by the *par* system, comprised of three components that are sufficient to move plasmids to opposite ends of the cell rapidly after replication (Jensen and Gerdes 1999). The *par* system is comprised of ParM and two other components (ParR and *parC*). In vivo, ParM assembles into dynamic filaments that orient along the length of the cell (Møller-Jensen et al. 2002). Dual-labelling immunofluorescence microscopy experiments demonstrated that the R1 plasmids localise to opposite ends of the ParM filament structures (Møller-Jensen et al. 2003). The association of the ParM filament with R1 plasmids requires ParR. The ParR protein cooperatively binds to the cis-acting centromere-like *parC* DNA sequence encoded on the R1 plasmid (Jensen and Gerdes 1997). Thus, ParM was suggested (Møller-Jensen et al. 2002), and recently shown in vitro (Garner et al.

2007), to function like a rudimentary mitotic spindle, moving the newly replicated plasmids toward opposite poles of the cell.

ParM was known to have similarities to the actin family from early sequence analysis (Bork et al. 1992). When the structure was determined in 2002, it was obvious that the core ATPase domains of actin were conserved, but differences arise in the domains IA, IB, and IIB in the form of strand insertions, absent helices, and extended loops (van den Ent et al. 2002). Although ParM filaments are structurally similar to those of actin, being two-stranded, and winding helically around each other, differences arise in the helical repeats of the filaments, with a full turn occurring every 300 Å in ParM filaments, whereas the actin repeat occurs every 360 Å (van den Ent et al. 2002). Further comparison of ParM with actin reveals that the largest regions of difference correlate with the interaction faces of actin for protofilament contacts, suggesting that ParM might exhibit differences in assembly.

Indeed, ParM has three biochemical properties distinguishing it significantly from actin. First, ParM shows rapid self-assembly, with a nucleation rate 300 times faster than actin, and does not require special nucleation factors to assemble in vivo like actin does (Garner et al. 2004). Second, whereas actin assembles unidirectionally, ParM filaments assemble bidirectionally in a symmetrical fashion, but disassemble unidirectionally (Garner et al. 2004). The filaments themselves are able to self-assemble with a dependence on Mg^{2+} and either ATP, ADP, or either of the nonhydrolysable ATP analogues ATP-γ-S or AMPPNP, although the ADP form is extremely unstable (Møller-Jensen et al. 2002; Garner et al. 2004). Third, ParM filaments exhibit dynamic instability and are able to switch between periods of rapid growth and catastrophic disassembly, whereas actin exhibits a more steady-state tread-milling behaviour (Garner et al. 2004).

Although the molecular arrangement of the ParM filament in vivo is not precisely known, ParM is highly expressed (~15000–18000 molecules per cell), suggesting that either the functional in vivo ParM filament is likely to be comprised of more than one subunit in thickness (Møller-Jensen et al. 2002), or that the high concentration of protein is required to promote filament nucleation and polymerisation. In vitro, the three components of the Par system are able to self-assemble into a machine capable of providing mechanical force suitable for plasmid partition (Garner et al. 2007). ParM filaments were observed to self-assemble into filaments, nucleating at ParR/*parC* complexes. The ParM filaments exhibited dynamic instability, growing and shrinking rapidly but only from free-ends not bound to ParR/*parC* complexes, suggesting that the ParR/*parC* complex functions to stabilise ParM filaments. When ParM filaments were bivalently bound to ParR/*parC* complexes at each end, a stable "spindle" was formed, protected from catastrophic disassembly. This "spindle" was able to push apart ParR/*parC* complexes by the addition of new ParM subunits into the ParM filament. Surprisingly, the addition of ParM monomers occurred at the ends of the filaments and not in the middle of the filaments, leaving an unresolved question regarding the mechanism by which ParM extends the ends of the filaments while the ParR/*parC* complex is attached.

Although ParM is restricted to some low-copy plasmids in *E. coli*, another actin-like protein, called AlfA, has recently been shown to be required for the segregation of a low-copy number plasmid in *B. subtilis*. AlfA assembles in vivo into long filaments with no apparent polarity and its ATPase activity is linked to plasmid partitioning (Becker et al. 2006). Sequence analysis revealed that AlfA lies between MreB and ParM on a phylogenetic tree, suggesting that variations of polymerising actin-like proteins might be a common theme for low-copy plasmid segregation across bacteria.

2.3.4   MamK

Magnetotactic bacteria are a diverse group of aquatic bacteria that are able to orient themselves within (geo)magnetic fields. It is thought that this ability helps in the search for favourable habitats. To perform this, these bacteria have special organelles called magnetosomes that are aligned within the cell by a customised cytoskeletal element to form a structure that has a function akin to a compass needle. The magnetosomes are formed by the invagination of the inner cell membrane (Komeili et al. 2006) around particles of magnetite (Fe_3O_4) or greigite (Fe_3S_4) (Bazylinski and Frankel 2004), and the mechanisms that orchestrate their formation are still being elucidated. Of relevance to this chapter, however, are the proteins that arrange the magnetosomes into a linear structure. Gene mutation studies have identified two proteins, MamK and MamJ, that are essential for the correct assembly of the magnetosomes into a linear structure (Komeili et al. 2006). Our current understanding of these proteins suggests that MamK is the filament-forming cytoskeletal protein, whereas MamJ is probably an adapter protein that tethers the magnetosomes to the MamK filament (Scheffel et al. 2006). MamK shares sequence homology with both MreB and ParM, and all the known MamK proteins cluster into a distinct family of bacterial actin-related proteins that probably have phylogenetic and functional properties differing significantly from both ParM and MreB.

Using cryo-electron tomography, Komeili et al. (2006) observed filaments with a thickness of 6 nm running parallel to and closely associated with the magnetosomes. In a strain in which the *mamK* gene was deleted the magnetosomes lost their linear organisation and no comparable filaments were observed. Supplementing this strain with GFP–MamK restored magnetosome localisation and caused the reappearance of filaments in some cells, suggesting that MamK may be the cytoskeletal protein responsible for aligning the magnetosomes. Immunogold labelling experiments by Pradel et al. (2006) revealed that MamK is indeed a part of the filament, but, most importantly, they demonstrated that GFP–MamK from *Magnetospirillum magneticum* self-assembles into filaments spontaneously in *E. coli* in the absence of any other gene from *M. magneticum*. This finding suggests that MamK could exist as an independent self-assembling cytoskeletal element (Pradel et al. 2006). Coexpression in *E. coli* of fluorescently labelled MreB from *E. coli* and MamK from *M. magneticum* revealed that both proteins have independent localisation patterns and coexist as independent structures. Little is known regarding the biochemistry of MamK, and experiments directed at investigating MamK assembly have not yet clearly resolved how MamK filaments might nucleate, or whether ATP binding and hydrolysis are required for assembly (Pradel et al. 2006).

2.3.5   Other Actin-Like Proteins: Ta0583 and FtsA

Archaeal actin-like genes are highly divergent and scattered sporadically across the different orders, suggesting that they do not form fundamental functions in these organisms, but have more specialised roles, and have probably been acquired by lateral gene transfer. One such protein of unknown function, Ta0583, is encoded by *Thermoplasma acidophilum*. Sequence analysis of Ta0583 places it approximately equidistant from all members of the actin family, although it has a slightly higher similarity to ParM than either MreB or actin. As predicted from its primary sequence, the crystal structure of Ta0583 revealed an actin-like fold with conservation of all subdomains (Roeben et al. 2006). Although biochemical analysis revealed that Ta0583 is an active ATPase, polymerisation assays failed to observe Ta0583 self-assembly. Interestingly, addition of 5% glycerol to a solution of recombinant Ta0583 resulted in the formation of crystalline sheets of protein with a longitudinal repeat of 51 Å (Roeben et al. 2006), which is the same as the repeat reported for MreB (van den Ent et al. 2001). However, the similarities in repeat distance are only circumstantial evidence of a cytoskeletal function and the protein sheets might not be biologically relevant, because there is no in vivo data supporting filament formation. Also, the levels of Ta0583 in vivo were found to be low—less than 0.04% of total cellular protein (Roeben et al. 2006), unlike the high expression levels of other actin-like proteins that form filaments in vivo.

The actin-like protein FtsA, which interacts directly with FtsZ, linking the Z ring to the membrane, might also be debated to be a cytoskeletal protein. Although attempts to induce FtsA self-assembly in vitro have largely failed, FtsA from *Streptococcus pneumoniae* self-assembles in the absence of nucleotide into very stable corkscrew-like filaments (Lara et al. 2005). In a strain of *E. coli* carrying a mutation in the *ftsA* gene that causes a deletion of the membrane targeting sequence, long cytoplasmic filaments have been observed (Gayda et al. 1992). These filaments were suggested to have formed from aberrant FtsA. Further fluorescent labelling of similar *ftsA* mutants revealed localisation in the shape of a rod along the length of the cell, consistent with the cytoplasmic filaments being comprised, at least in part, of mutant FtsA (Pichoff and Lutkenhaus 2005). The composition and biological relevance of these filaments has yet to be shown, and FtsA is generally considered as an accessory protein with important regulatory roles in cell division. It is possible, however, that in vivo FtsA undergoes some form of polymerisation that may be surface assisted (similar to the WACA proteins discussed in section 5) by the membrane or by a protein such as FtsZ.

2.4   Proteins from the IF Family

2.4.1   IFs in Eukaryotes

Intermediate filaments, so named because they form filaments with a diameter
between that of F-actin and microtubules, are abundant, very stable filaments that
have roles in providing mechanical support in a wide range of eukaryotic cell types.
They have no roles in cell motility, they are unable to undergo treadmilling, and
there are no known motor proteins that use IFs for tracking. Instead, IFs tend to take
on structural roles that are more permanent, such as mediating cell–cell and cell–
matrix contacts, but they do exhibit some dynamic features (Helfand et al. 2004).

   IFs are comprised of proteins that are extremely α-helical. The proteins in this
generic family have conserved structural features, being comprised of a central
coiled–coil motif with varied N and C termini. IF proteins are highly divergent in
sequence and vary considerably in molecular weight. In addition, no characteristic
conserved sequence motifs that have facilitated the identification of tubulin and
actin homologues have been identified in IF proteins. Instead, IF proteins are comprised of large regions of coiled coil, which is common in many other proteins.

   Why might bacteria possess IF proteins? Both eubacteria and archaea are capable
of assuming a wide range of shapes (for a review on bacterial cell shape, see Young
2006). Precisely how they do this is unknown, and filaments formed by IF proteins
might provide the cell with structural restraints suitable for establishing cell morphology, as happens in eukaryotic cells. To date, there has only been one bacterial
protein ascribed to the IF family, crescentin.

2.4.2   Crescentin

A clue to how cell shape is controlled has been obtained from *C. crescentus* during a
screen for insertion mutants that affect cell morphology. *C. crescentus* is a vibrioid-
shaped bacterium that was found to form rod-shaped cells on the disruption of
the *creS* gene (Ausmees et al. 2003). The gene product of *creS* is crescentin,
which is thought to be at least a functional homologue of IFs in eukaryotes (Ausmees
et al. 2003).

   Crescentin has an amino acid sequence comprised of heptad repeats (a heptad
repeat is a structural motif of seven amino-acids with hydrophobic residues occuring every one and four amino acids and polar residues at all other positions)—
consistent with coiled–coil structures and the IF family of proteins. However, as
discussed, the coiled–coil repeat is a poor criterion to judge homology, and the IF
proteins do not have any known enzymatic or nucleotide-binding capability clearly
marking their role.

   However, there are several reasons why crescentin is thought to be related to IFs.
Firstly, crescentin is able to assemble in vitro into filaments under conditions very

similar to those required for IF assembly. These filaments are stable and do not require nucleotide or cofactors. Furthermore, the crescentin filaments have physical dimensions very similar to IF filaments (Ausmees et al. 2003).

More support comes from in vivo data, in which crescentin is observed to form a filament with a long helical turn that localises to the concave side of the cell. Interestingly, *C. crescentus* cells, when left for a long time in stationary culture, exhibit a helical twist morphology with a similar pitch to the twist observed in the crescentin filaments formed in vitro. Unfortunately, the molecular three-dimensional structures of IF proteins have proven remarkably hard to solve and currently there is no atomic–structural data known for crescentin or for IF proteins, making it difficult to resolve whether these proteins share similar protein folds.

2.5 A Fourth Cytoskeletal Family: The Walker A Cytoskeletal ATPases

Bacteria express proteins belonging to a family of deviant Walker A ATPases that can be described as a fourth family of cytoskeletal elements that have no known counterpart of eukaryotic cytoskeletal proteins. These proteins (the _W_alker _A_ _c_ytoskeletal _A_TPases [WACA] proteins) are widely distributed, and most bacteria encode one or more members of the family, which includes ParA, MinD, Soj, SopA ParF, IncC, and probably MipZ. The WACA family is characterised by a high conservation of primary sequence (including a deviant Walker A motif), a conserved three-dimensional structure, and ATP-dependent dimer formation. A conserved characteristic of this family is their mutual ATPase stimulation brought about by a companion "activation" protein. WACA proteins exhibit increased ATPAse activity when their activation proteins are present. For example, MinD is modulated by MinE, ParB activates both ParA (Radnedge et al. 1998) and MipZ (Thanbichler and Shapiro 2006a), and the short N-terminal tail of SpoOJ activates Soj (Radnedge et al. 1996; Leonard et al. 2005). Although these proteins share overall sequence and structural homology, their biological roles differ, being involved in DNA segregation, plasmid partitioning, and the positioning and timing of Z ring assembly. It has been proposed that they may be molecular switches (Leonard et al. 2005), yet the molecular mechanisms by which they function remain largely unknown.

Of the WACA members that have been examined, all have the ability to polymerise in vitro, suggesting that these proteins might form cytoskeletal elements. However, the nature of the polymers formed by these proteins is unusual, with the proteins apparently binding to the entire surface of their substrate by some form of surface-assisted polymerisation. For example, MinD in vitro binds phospholipids vesicles with high density (Hu et al. 2002), and Soj completely coats DNA (Leonard et al. 2004; Leonard et al. 2005). Further evidence for the cytoskeletal nature of these proteins comes from in vivo localisation patterns, which are varied across the group but include helices, gradients, and discrete patches (Hu and Lutkenhaus 1999; Marston and Errington 1999; Raskin and de Boer 1999; Shih et al. 2003).

Interestingly, these localisation patterns show dynamic and time-dependent behaviour, in which the pattern changes over time. As examples, Soj localisation alternates erratically as discrete patches between nucleoids and around the same nucleoid (Marston and Errington 1999). MinD oscillates between the cell poles, moving from one end of the cell to the other in *E. coli* (Hu and Lutkenhaus 1999; Raskin and de Boer 1999), but, in *B. subtilis*, MinD forms a fixed gradient, with the highest concentration spreading from the poles (Marston et al. 1998). The periodicities of altered localisation patterns also vary significantly. The time between Soj movements ranges from minutes to up to 1 hour (Marston and Errington 1999), whereas MinD oscillations occur rapidly and rhythmically, taking approximately a minute to move over the entire length of the cell (Hu and Lutkenhaus 1999; Raskin and de Boer 1999).

In general, the precise roles of this class of proteins and the molecular mechanisms behind these roles are poorly understood. Many questions remain unanswered, such as how these proteins are able to dynamically change their localisation patterns. Nor is it clear in many cases why their localisation patterns change.

2.6   Other Cytoskeletal Elements

Over the years, many obscure filamentous structures have been observed in bacteria and archaea (Hixon and Searcy 1993; Izard et al. 1999). For example, the highly expressed elongation factor GTPase EF-Tu protein (Beck et al. 1978; Schilstra, 1984) has been reported to form a "cytoskeletal web" within the cell (Mayer 2006). This idea is controversial, and further characterisation is required to determine the role, if any, of EF-Tu as a cytoskeletal protein (Vollmer 2006; Nanninga 1998; Löwe et al. 2004).

Although some of the obscure filaments previously observed may represent specialist cytoskeletal systems unique to a specific organism (such as MamK), and others that have been reported may be observations of structures we are already familiar with, i.e., FtsZ, MreB and related helices, and crescentin, it is also possible that there are further cytoskeletal elements that are ubiquitous and have yet to be identified.

Recent information suggests that a dynamin homologue (called _b_acterial _d_ynamin-_l_ike _p_rotein [BDLP]) is present in many bacteria (Low and Löwe 2006). Although dynamin is not considered a cytoskeletal element in eukaryotes, dynamin (and BDLP) self-assembles into regular structures in the presence of lipid (it tubulates membranes) in vitro. The role of BDLP in bacteria is unknown, however, dynamin-like proteins present in mitochondria and chloroplasts of many eukaryotes have essential roles in organelle division. Dynamin division rings similar to Z rings assemble at the division plane during organelle division. Because chloroplasts and mitochondria have endosymbiotic bacterial origins and the division of these organelles bears some resemblance to prokaryotic cell division, it is tempting to speculate that BDLP might assemble into a cytoskeletal element to assist with division in bacteria.

2.7 Conclusions—The Future

The discovery of cytoskeletal elements in bacteria has changed the way biologists think about bacteria. The old paradigm of the "bag of enzymes" bacterial cell is being left to rest, while the new era of microbiology is revealing highly ordered and complex structures that regulate essential biological processes in prokaryotes.

Many of the cytoskeletal proteins have overlapping roles with those of eukaryotes based on ancient phylogenetic relationships. However, it is fascinating that, across the domains of life, some of these cytoskeletal proteins seem to have switched biological functions during evolution. For example, the tubulin homologue FtsZ has a central role in bacterial cell division, whereas actin contributes a similar role in eukaryotes. Likewise, tubulin provides the scaffold for mitosis in eukaryotes, whereas actin homologues mediate plasmid segregation mechanisms in prokaryotes.

The conserved properties of cytoskeletal proteins reveal the basic characteristics for minimal self-assembling mechanical systems. Understanding how the minimal system works is important if we plan to make practical use of such systems. For instance, the development of complex engineered systems that require structural order might be successful if we understand the fundamental properties of the self-ordering and self-assembling cytoskeletal systems. From a medical perspective, the specific design of antibacterial drugs targeting cytoskeletal proteins will rely on a detailed knowledge of the important similarities and differences between the cytoskeletal proteins of the bacterial pathogen and eukaryotic host.

An increasing amount of research is focused on identifying agents that disrupt the normal function of the bacterial cytoskeleton. Such agents would be useful not only as tools to facilitate our understanding of cytoskeletal systems, but also as potential therapeutic antimicrobial agents, and, with the increased incidence of antibiotic-resistant bacteria, the need to develop novel antibacterial agents is becoming more pressing.

Existing antimicrobial agents target a relatively narrow range of cellular functions—largely being those involved in cell wall, protein, and DNA synthesis. The bacterial cytoskeletal elements provide more targets that might be exploited for antimicrobial development because the cytoskeletal components are essential for cell viability and the proteins within these systems are often highly conserved across bacteria while remaining significantly different to eukaryotic systems, allowing for the development of selective and specific agents.

Because the discovery of the prokaryotic cytoskeleton has been relatively recent, the development of agents inhibiting cytoskeletal function is still in its infancy, however there are already promising leads. Currently, FtsZ has been the most intensively targeted protein, with a range of methods used to identify lead compounds (Paradis-Bleau et al. 2004; Margalit et al. 2004; White et al. 2002;

Stokes et al. 2005). To a lesser extent, MreB, ZipA, and FtsA have also been targeted (Gitai et al. 2005; Iwai et al. 2002; Kenny et al. 2003; Paradis-Bleau et al. 2005; Sutherland et al. 2003) and it is likely that many more of the proteins involved in the prokaryotic cytoskeleton will eventually provide useful avenues for developing antimicrobial agents.

Currently, we still know very little regarding the in vivo nature of the macromolecular arrangement of the cytoskeletal filaments. The molecular mechanisms that regulate these cytoskeletal structures are also not yet discernible and, thus, much basic research into the bacterial cytoskeleton is required. With advances in electron tomography of filaments within bacterial cells, and further cell biology experiments, most of the cytoskeletal proteins will eventually be characterised at the molecular and macromolecular level.

Acknowledgments I thank Iain Duggin, Jan Löwe, and Chris Mercogliano for valuable comments on the manuscript. I was supported during the writing of this chapter by a Marie Curie International Fellowship within the 6th European Community Framework Programme.

Highly Recommended Readings

Anderson DE, Gueiros-Filho FJ, Erickson HP (2004) Assembly dynamics of FtsZ rings in *Bacillus subtilis* and *Escherichia coli* and effects of FtsZ-regulating proteins. J Bacteriol 186:5775–5781

Ausmees N, Kuhn JR, Jacobs-Wagner C (2003) The bacterial cytoskeleton: an intermediate filament-like function in cell shape. Cell 115:705–713

Garner EC, Campbell CS, Weibel DB, Mullins RD (2007) Reconstitution of DNA segregation driven by assembly of a prokaryotic actin homolog. Science 315:1270–1274

Jones LJ, Carballido-Lopez R, Errington J (2001) Control of cell shape in bacteria: helical, actin-like filaments in *Bacillus subtilis*. Cell 104:913–922

Schlieper D, Oliva MA, Andreu JM, Löwe J (2005) Structure of bacterial tubulin BtubA/B: Evidence for horizontal gene transfer. Proc Natl Acad Sci USA 102: 9170–9175

Shih YL, Le T, Rothfield L (2003) Division site selection in *Escherichia coli* involves dynamic redistribution of Min proteins within coiled structures that extend between the two cell poles. Proc Natl Acad Sci USA 100:7865–7870

References

Abhayawardhane Y, Stewart GC (1995) *Bacillus subtilis* possesses a second determinant with extensive sequence similarity to the *Escherichia coli mreB* morphogene. J Bacteriol 177:765–773

Anderson DE, Gueiros-Filho FJ, Erickson HP (2004) Assembly dynamics of FtsZ rings in *Bacillus subtilis* and *Escherichia coli* and effects of FtsZ-regulating proteins. J Bacteriol 186:5775–5781

Ausmees N, Kuhn JR, Jacobs-Wagner C (2003) The bacterial cytoskeleton: an intermediate filament-like function in cell shape. Cell 115:705–713

Bazylinski DA, Frankel RB (2004) Magnetosome formation in prokaryotes. Nat Rev Microbiol 2:217–230

Becker E, Herrera NC, Gunderson FQ, Derman AI, Dance AL, Larsen RA, Pogliano J (2006) DNA segregation by the bacterial actin AlfA during *Bacillus subtilis* growth and development. EMBO J 25:5919–5931

Beck BD, Arscott PG, Jacobson A (1978) Novel properties of bacterial Elongation Factor Tu. Proc Natl Acad Sci USA 75: 1250–1254

Beech P, Nheu T, Schultz T, Herbert S, Lithgow T, Gilson PR, McFadden GI (2000) Mitochondrial FtsZ in a chromophyte alga. Science 287:1276–1279

Ben-Yehuda S, Losick R (2002) Asymmetric cell division in *B. subtilis* involves a spiral-like intermediate of the cytokinetic protein FtsZ. Cell 109:257–266

Bermudes D, Hinkle G, Margulis L (1994) Do prokaryotes contain microtubules? Microbiol Rev 58:387–400

Bernhardt TG, de Boer PA (2005) SlmA, a nucleoid-associated, FtsZ binding protein required for blocking septal ring assembly over chromosomes in *E. coli*. Mol Cell 18:555–564

Bi E, Lutkenhaus J (1990) Interaction between the *min* locus and *ftsZ*. J Bacteriol 172:5610–5616

Bi E, Lutkenhaus J (1991) FtsZ ring structure associated with division in *Escherichia coli*. Nature 354:161–164

Bork P, Sander C, Valencia A (1992) An ATPase domain common to prokaryotic cell cycle proteins, sugar kinases, actin, and hsp70 heat shock proteins. Proc Natl Acad Sci USA 89:7290–7294

Bramhill D, Thompson CM (1994) GTP-dependent polymerization of *Escherichia coli* FtsZ protein to form tubules. Proc Natl Acad Sci USA 91:5813–5817

Briegel A, Dias DP, Li Z, Jensen RB, Frangakis AS, Jensen GJ (2006) Multiple large filament bundles observed in *Caulobacter crescentus* by electron cryotomography. Mol Microbiol 62:5–14

Brown WJ, Rockey DD (2000) Identification of an antigen localized to an apparent septum within dividing *Chlamydiae*. Infect Immun 68:708–715

Buddelmeijer N, Beckwith J (2004) A complex of the *Escherichia coli* cell division proteins FtsL, FtsB and FtsQ forms independently of its localization to the septal region. Mol Microbiol 52:1315–1327

Carballido-Lopez R, Errington J (2003) The bacterial cytoskeleton: in vivo dynamics of the actin-like protein Mbl of *Bacillus subtilis*. Dev Cell 4:19–28

Cordell SC, Robinson EJ, Löwe J (2003) Crystal structure of the SOS cell division inhibitor SulA and in complex with FtsZ. Proc Natl Acad Sci USA 100:7889–7894

Daniel R, Errington J (2000) Intrinsic instability of the essential cell division protein FtsL of *Bacillus subtilis* and a role for DivIB protein in FtsL turnover. Mol Microbiol 36:278–289

Daniel RA, Errington J (2003) Control of cell morphogenesis in bacteria: two distinct ways to make a rod-shaped cell. Cell 113:767–776

Daniel RA, Harry EJ, Katis VL, Wake RG, Errington J (1998) Characterization of the essential cell division gene *ftsL* (*ylID*) of *Bacillus subtilis* and its role in the assembly of the division apparatus. Mol Microbiol 29:593–604

de Boer PAJ, Crossley RE, Rothfield LI (1989) A division inhibitor and a topological specificity factor coded for by the minicell locus determine proper placement of the division septum in *E. coli*. Cell 56:641–649

de Boer PAJ, Crossley RE, Rothfield LI (1990) Central role for the *Escherichia coli minC* gene product in two different cell division-inhibition systems. Proc Natl Acad Sci USA 87:1129–1133

de Pereda JM, Leynadier D, Evangelio JA, Chacon P, Andreu JM (1996) Tubulin secondary structure analysis, limited proteolysis sites, and homology to FtsZ. Biochem 35:14203–14215

Defeu Soufo HJ, Graumann PL (2006) Dynamic localization and interaction with other *Bacillus subtilis* actin-like proteins are important for the function of MreB. Mol Microbiol 62:1340–1356

Desai A, Mitchison TJ (1997) Microtubule polymerization dynamics. Annu Rev Cell Dev Biol 13:83–117

Erickson HP, Taylor DW, Taylor KA, Bramhill D (1996) Bacterial cell division protein FtsZ assembles into protofilament sheets and minirings, structural homologs of tubulin polymers. Proc Natl Acad Sci USA 93:519–523

Esue O, Cordero M, Wirtz D, Tseng Y (2005) The assembly of MreB, a prokaryotic homolog of actin. J Biol Chem 280:2628–2635

Feucht A, Lucet I, Yudkin MD, Errington J (2001) Cytological and biochemical characterization of the FtsA cell division protein of *Bacillus subtilis*. Mol Microbiol 40:115–125

Figge RM, Divakaruni AV, Gober JW (2004) Mreb, the cell shape-determining bacterial actin homologue, co-ordinates cell wall morphogenesis in *Caulobacter crescentus*. Mol Microbiol 51:1321–1332

Formstone A, Errington J (2005) A magnesium-dependent *mreB* null mutant: implications for the role of *mreB* in *Bacillus subtilis*. Mol Microbiol 55:1646–1657

Fu X, Shih Y-L, Zhang Y, Rothfield LI (2001) The MinE ring required for proper placement of the division site is a mobile structure that changes its cellular location during the *Escherichia coli* division cycle. Proc Natl Acad Sci USA 98:980–985

Garner EC, Campbell CS, Mullins RD (2004) Dynamic instability in a DNA-segregating prokaryotic actin homolog. Science 306:1021–1025

Garner EC, Campbell CS, Weibel DB, Mullins RD (2007) Reconstitution of DNA segregation driven by assembly of a prokaryotic actin homolog. Science 315:1270–1274

Gayda RC, Henk MC, Leong D (1992) C-shaped cells caused by expression of an *ftsA* mutation in *Escherichia coli*. J Bacteriol 174:5362–5370

Gitai Z, Dye N, Shapiro L (2004) An actin-like gene can determine cell polarity in bacteria. Proc Natl Acad Sci USA 101:8643–8648

Gitai Z, Dye NA, Reisenauer A, Wachi M, Shapiro L (2005) MreB actin-mediated segregation of a specific region of a bacterial chromosome. Cell 120:329–341

Glass JI, Lefkowitz EJ, Glass JS, Heiner CR, Chen EY, Cassell GH (2000) The complete sequence of the mucosal pathogen *Ureaplasma urealyticum*. Nature 407:757–762

Glockner FO, Kube M, Bauer M, Teeling H, Lombardot T, Ludwig W, Gade D, Beck A, Borzym K, Heitmann K, Rabus R, Schlesner H, Amann R, Reinhardt R (2003) Complete genome sequence of the marine planctomycete *Pirellula* sp. strain 1. Proc Natl Acad Sci USA 100:8298–8303

Grantcharova N, Lustig U, Flardh K (2005) Dynamics of FtsZ assembly during sporulation in *Streptomyces coelicolor* A3(2). J Bacteriol 187:3227–3237

Gueiros-Filho FJ, Losick R (2002) A widely conserved bacterial cell division protein that promotes assembly of the tubulin-like protein FtsZ. Genes Dev 16:2544–2556

Hale CA, Meinhardt H, de Boer PA (2001) Dynamic localization cycle of the cell division regulator MinE in *Escherichia coli*. EMBO J 20:1563–1572

Harry E, Monahan L, Thompson L (2006) Bacterial cell division: the mechanism and its precison. Int Rev Cytol 253:27–94

Helfand BT, Chang L, Goldman RD (2004) Intermediate filaments are dynamic and motile elements of cellular architecture. J Cell Sci 117:133–141

Hixon WG, Searcy DG (1993) Cytoskeleton in the archaebacterium *Thermoplasma acidophilum*? Viscosity increase in soluble extracts. Biosystems 29:151–160

Holmes KC, Popp D, Gebhard W, Kabsch W (1990) Atomic model of the actin filament. Nature 347:44–49

Hu B, Yang G, Zhao W, Zhang Y, Zhao J MreB is important for cell shape but not for chromosome segregation of the filamentous cyanobacterium *Anabaena* sp. PCC 7120. Mol Microbiol 63:1640–1652

Hu Z, Gogol E, Lutkenhaus J (2002) Dynamic assembly of MinD on phospholipid vesicles regulated by ATP and MinE. Proc Natl Acad Sci USA 99:6761–6766

Hu Z, Lutkenhaus J (1999) Topological regulation of cell division in *Escherichia coli* involves rapid pole to pole oscillation of the division inhibitor MinC under the control of MinD and MinE. Mol Microbiol 34:82–90

Hu Z, Mukherjee A, Pichoff S, Lutkenhaus J (1999) The MinC component of the division site selection system in *Escherichia coli* interacts with FtsZ to prevent polymerization. Proc Natl Acad Sci USA 96:14819–14824

Iwai N, Nagai K, Wachi M (2002) Novel S-benzylisothiourea compound that induces spherical cells in Escherichia coli probably by acting on a rod-shape-determining protein(s) other than penicillin-binding protein 2. Biosci Biotechnol Biochem 66:2658–2661

Izard J, Samsonoff WA, Kinoshita MB, Limberger RJ (1999) Genetic and structural analyses of cytoplasmic filaments of wild-type *Treponema phagedenis* and a flagellar filament-deficient mutant. J Bacteriol 181:6739–6746

Jenkins C, Samudrala R, Anderson I, Hedlund BP, Petroni G, Michailova N, Pinel N, Overbeek R, Rosati G, Staley JT (2002) Genes for the cytoskeletal protein tubulin in the bacterial genus *Prosthecobacter*. Proc Natl Acad Sci USA 99:17049–17054

Jensen RB, Gerdes K (1997) Partitioning of plasmid R1. The ParM protein exhibits ATPase activity and interacts with the centromere-like ParR-parC complex. J Mol Biol 269:505–513

Jensen RB, Gerdes K (1999) Mechanism of DNA segregation in prokaryotes: ParM partitioning protein of plasmid R1 co-localizes with its replicon during the cell cycle. EMBO J 18:4076–4084

Jones LJ, Carballido-Lopez R, Errington J (2001) Control of cell shape in bacteria: helical, actin-like filaments in *Bacillus subtilis*. Cell 104:913–922

Kawarabayasi Y, Hino Y, Horikawa H, Yamazaki S, Haikawa Y, Jin-no K, Takahashi M, Sekine M, Baba S, Anka A, Kosugi H, Hosoyama A, Fukui S, Nagai Y, Nishijima K, Nakazawa H, Takamiya M, Masuda S, Funahashi T, Tanaka T, Kudoh Y, Yamazaki J, Kushida N, Oguchi A, Kikuchi H (1999) Complete genome sequence of an aerobic hyper-thermophilic crenarchaeon, *Aeropyrum pernix* K1. DNA Res 6:83–101

Kenny CH, Ding W, Kelleher K, Benard S, Dushin EG, Sutherland AG, Mosyak L, Kriz R, Ellestad G (2003) Development of a fluorescence polarization assay to screen for inhibitors of the FtsZ/ZipA interaction. Anal Biochem 323:224–233

Kim SY, Gitai Z, Kinkhabwala A, Shapiro L, Moerner WE (2006) Single molecules of the bacterial actin MreB undergo directed treadmilling motion in *Caulobacter crescentus*. Proc Natl Acad Sci USA 103:10929–10934

Komeili A, Li Z, Newman DK, Jensen GJ (2006) Magnetosomes are cell membrane invaginations organized by the actin-like protein MamK. Science 311:242–245

Korn ED, Carlier MF, Pantaloni D (1987) Actin polymerization and ATP hydrolysis. Science 238:638–644

Kruse T, Blagoev B, Lobner-Olesen A, Wachi M, Sasaki K, Iwai N, Mann M, Gerdes K (2006) Actin homolog MreB and RNA polymerase interact and are both required for chromosome segregation in *Escherichia coli*. Genes Dev 20:113–124

Kruse T, Bork-Jensen J, Gerdes K (2005) The morphogenetic MreBCD proteins of *Escherichia coli* form an essential membrane-bound complex. Mol Microbiol 55:78–89

Lara B, Rico AI, Petruzzelli S, Santona A, Dumas J, Biton J, Vicente M, Mingorance J, Massidda O (2005) Cell division in cocci: localization and properties of the *Streptococcus pneumoniae* FtsA protein. Mol Microbiol 55:699–711

Leonard TA, Butler PJ, Löwe J (2005) Bacterial chromosome segregation: structure and DNA binding of the Soj dimer—a conserved biological switch. EMBO J 24:270–282

Leonard TA, Butler PJG, Löwe J (2004) Structural analysis of the chromosome segregation protein Spo0J from *Thermus thermophilus*. Mol Microbiol 53:419–432

Levin PA, Kurtser IG, Grossman AD (1999) Identification and characterization of a negative regulator of FtsZ ring formation in *Bacillus subtilis*. Proc Natl Acad Sci USA 96:9642–9647

Liu G, Draper GC, Donachie WD (1998) FtsK is a bifunctional protein involved in cell division and chromosome localization in *Escherichia coli*. Mol Microbiol 29:893–903

Low HH, Löwe J (2006) A bacterial dynamin-like protein. Nature 444:766–769

Löwe J, Amos LA (1998) Crystal structure of the bacterial cell-division protein FtsZ. Nature 391:203–206

Löwe J, Amos LA (1999) Tubulin-like protofilaments in Ca^{2+}-induced FtsZ sheets. EMBO J 18:2364–2371

Löwe J, Amos LA (2000) Helical tubes of FtsZ from *Methanococcus jannaschii*. Biol Chem 381:993–999

Löwe J, van den Ent F, Amos LA (2004) Molecules of the bacterial cytoskeleton. Annu Rev Biophys Biomol Struct 33:177–198

Lu C, Reedy M, Erickson HP (2000) Straight and curved conformations of FtsZ are regulated by GTP hydrolysis. J Bacteriol 182:164–170

Lu C, Stricker J, Erickson HP (1998) FtsZ from *Escherichia coli*, *Azotobacter vinelandii*, and *Thermotoga maritima*—quantitation, GTP hydrolysis, and assembly. Cell Motil Cytoskel 40:71–86

Lu C, Stricker J, Erickson HP (2001) Site-specific mutations of FtsZ – effects on GTPase and *in vitro* assembly. BMC Microbiol 1:7 doi:10.1186/1471-2180-1-7

Lucic V, Förster F, Baumeister W (2005) Structural studies by electron tomography: from cells to molecules. Annu Rev Biochem 74:833–865

Lutkenhaus J (2007) Assembly dynamics of the bacterial MinCDE system and spatial regulation of the Z ring. Annu Rev Biochem 76:539–562

Margalit DN, Romberg L, Mets RB, Hebert AM, Mitchison TJ, Kirschner MW, RayChaudhuri D (2004) Targeting cell division: Small-molecule inhibitors of FtsZ GTPase perturb cytokinetic ring assembly and induce bacterial lethality. Proc Natl Acad Sci USA 101:11821–11826

Margolin W (2005) FtsZ and the division of prokaryotic cells and organelles. Nat Rev Mol Cell Biol 6:862–871

Marston A, Errington J (1999) Dynamic movement of the ParA-like soj protein of *B. subtilis* and its dual role in nucleoid organization and developmental regulation. Mol Cell 5:673–682

Marston AL, Thomaides HB, Edwards DH, Sharpe ME, Errington J (1998) Polar localization of the MinD protein of *Bacillus subtilis* and its role in selection of the mid-cell division site. Genes Dev 12:3419–3430

Mayer F (2006) Cytoskeletal elements in bacteria *Mycoplasma pneumoniae*, *Thermoanaero bacterium* sp., and *Escherichia coli* as revealed by electron microscopy. J Mol Microbiol Biotechnol 11:228–243

Michie KA, Monahan LG, Beech PL, Harry EJ (2006) Trapping of a spiral-like intermediate of the bacterial cytokinetic protein FtsZ. J Bacteriol 188:1680–1690

Mingorance J, Tadros M, Vicente M, Gonzalez JM, Rivas G, Velez M (2005) Visualization of single *Escherichia coli* Ftsz filament dynamics with atomic force microscopy. J Biol Chem 280:20909–20914

Møller-Jensen J, Borch J, Dam M, Jensen RB, Roepstorff P, Gerdes K (2003) Bacterial mitosis: ParM of plasmid R1 moves plasmid DNA by an actin-like insertional polymerization mechanism. Mol Cell 12:1477–1487

Møller-Jensen J, Jensen RB, Löwe J, Gerdes K (2002) Prokaryotic DNA segregation by an actin-like filament. EMBO J 21:3119–3127

Møller-Jensen J, Löwe J (2005) Increasing complexity of the bacterial cytoskeleton. Curr Opin Cell Biol 17:75–81

Mukherjee A, Dai K, Lutkenhaus J (1993) *Escherichia coli* cell division protein FtsZ is a guanine nucleotide binding protein. Proc Natl Acad Sci USA 90:1053–1057

Mukherjee A, Lutkenhaus J (1994) Guanine nucleotide-dependent assembly of FtsZ into filaments. J Bacteriol 176:2754–2758

Mukherjee A, Lutkenhaus J (1998) Dynamic assembly of FtsZ regulated by GTP hydrolysis. EMBO J 17:462–469

Nanninga N (1998) Morphogenesis of *Escherichia coli*. Microbiol Mol Biol Rev 62:110–129

Nogales E, Downing KH, Amos LA, Löwe J (1998a) Tubulin and FtsZ form a distinct family of GTPases. Nat Struc Biol 5:451–458

Nogales E, Wolf SG, Downing KH (1998b) Structure of the ab tubulin dimer by electron crystallography. Nature 391:199–203

Oliva MA, Cordell SC, Löwe J (2004) Structural insights into FtsZ protofilament formation. Nat Struc Mol Biol 11:1243–1250

Oliva MA, Huecas S, Palacios JM, Martin-Benito J, Valpuesta JM, Andreu JM (2003) Assembly of archaeal cell division protein FtsZ and a GTPase-inactive mutant into double-stranded filaments. J Biol Chem 278:33562–33570

Osteryoung KW, McAndrew RS (2001) The plastid division machine. Annu Rev Plant Physiol Plant Mol Biol 52:315–333

Osteryoung KW, Vierling E (1995) Conserved cell and organelle division. Nature 376:473–474

Paradis-Bleau C, Sanschagrin F, Levesque RC (2004) Identification of *Pseudomonas aeruginosa* FtsZ peptide inhibitors as a tool for development of novel antimicrobials. J Antimicrob Chemother 54:278–280

Paradis-Bleau C, Sanschagrin F, Levesque RC (2005) Peptide inhibitors of the essential cell division protein FtsA. Prot Eng Design Select 18:85–91

Phoenix P, Drapeau GR (1988) Cell division control in *Escherichia coli* K-12: some properties of the *ftsZ84* mutation and suppression of this mutation by the product of a newly identified gene. J Bacteriol 170:4338–4342

Pichoff S, Lutkenhaus J (2005) Tethering the Z ring to the membrane through a conserved membrane targeting sequence in FtsA. Mol Microbiol 55:1722–1734

Pradel N, Santini C-L, Bernadac A, Fukumori Y, Wu L-F (2006) Biogenesis of actin-like bacterial cytoskeletal filaments destined for positioning prokaryotic magnetic organelles. Proc Natl Acad Sci USA 103:17485–17489

Radnedge L, Davis MA, Austin SJ (1996) P1 and P7 plasmid partition: ParB protein bound to its partition site makes a separate discriminator contact with the DNA that determines species specificity. EMBO J 15:1155–1162

Radnedge L, Youngren B, Davis M, Austin S (1998) Probing the structure of complex macromolecular interactions by homolog specificity scanning: the P1 and P7 plasmid partition systems. EMBO J 17:6076–6085

Raskin DM, de Boer PA (1999) Rapid pole-to-pole oscillation of a protein required for directing division to the middle of *Escherichia coli*. Proc Natl Acad Sci USA 96:4971–4976

Read TD, Brunham RC, Shen C, Gill SR, Heidelberg JF, White O, Hickey EK, Peterson J, Utterback T, Berry K, Bass S, Linher K, Weidman J, Khouri H, Craven B, Bowman C, Dodson R, Gwinn M, Nelson W, DeBoy R, Kolonay J, McClarty G, Salzberg SL, Eisen J, Fraser CM (2000) Genome sequences of *Chlamydia trachomatis* MoPn and *Chlamydia pneumoniae* AR39 Nucl Acids Res 28:1397–1406

Robson SA, Michie KA, Mackay JP, Harry EJ, King GF (2002) The *Bacillus subtilis* cell division proteins FtsL and DivIC are intrinsically unstable and do not interact with one another in the absence of other septasomal components. Mol Microbiol 44:663–674

Roeben A, Kofler C, Nagy I, Nickell S, Hartl FU, Bracher A (2006) Crystal structure of an archaeal actin homolog. J Mol Biol 358:145–156

Rothfield L, Justice S, Garcia-Lara J (1999) Bacterial cell division. Annu Rev Genet 33:423–448

Rueda S, Vicente M, Mingorance J (2003) Concentration and assembly of the division ring proteins FtsZ, FtsA, and ZipA during the *Escherichia coli* cell cycle. J Bacteriol 185:3344

Scheffel A, Gruska M, Faivre D, Linaroudis A, Plitzko JrM, Schüler D (2006) An acidic protein aligns magnetosomes along a filamentous structure in magnetotactic bacteria. Nature 440:110–114

Scheffers DJ, Driessen AJ (2002) Immediate GTP hydrolysis upon FtsZ polymerization. Mol Microbiol 43:1517–1521

Schilstra MJ, Slot JW, van der Meide PH, Posthuma G, Cremers AF, Bosch L (1984) Immunocytochemical localization of the elongation factor Tu in *E. coli* cells. FEBS Lett 165(2):175–179

Schlieper D, Oliva MA, Andreu JM, Löwe J (2005) Structure of bacterial tubulin BtubA/B: Evidence for horizontal gene transfer. Proc Natl Acad Sci USA 102: 9170–9175

Schmidt A, Hall MN (1998) Signaling to the actin cytoskeleton. Ann Rev Cell Dev Biol 14:305–338

Scholle MD, White CA, Kunnimalaiyaan M, Vary PS (2003) Sequencing and characterization of pBM400 from *Bacillus megaterium* QM B1551. Appl Environ Microbiol 69:6888–6898

Shih Y-L, Fu X, King GF, Le T, Rothfield L (2002) Division site placement in *E. coli*: mutations that prevent formation of the MinE ring lead to loss of the normal midcell arrest of growth of polar MinD membrane domains. EMBO J 21:3347–3357

Shih Y-L, Rothfield L (2006) The bacterial cytoskeleton. Microbiol Mol Biol Rev 70:729–754

Shih YL, Le T, Rothfield L (2003) Division site selection in *Escherichia coli* involves dynamic redistribution of Min proteins within coiled structures that extend between the two cell poles. Proc Natl Acad Sci USA 100:7865–7870

Soufo HJ, Graumann PL (2003) Actin-like proteins MreB and Mbl from *Bacillus subtilis* are required for bipolar positioning of replication origins. Curr Biol 28;1916–1920

Stephens RS, Kalman S, Lammel C, Fan J, Marathe R, Aravind L, Mitchell W, Olinger L, Tatusov RL, Zhao Q, Koonin EV, Davis RW (1998) Genome sequence of an obligate intracellular pathogen of humans: *Chlamydia trachomatis*. Science 282:754–759

Stokes NR, Sievers J, Barker S, Bennett JM, Brown DR, Collins I, Errington VM, Foulger D, Hall M, Halsey R, Johnson H, Rose V, Thomaides HB, Haydon DJ, Czaplewski LG, Errington J (2005) Novel inhibitors of bacterial cytokinesis identified by a cell-based antibiotic screening assay. J Biol Chem 280:39709–39715S

Stricker J, Maddox P, Salmon ED, Erickson HP (2002) Rapid assembly dynamics of the *Escherichia coli* FtsZ-ring demonstrated by fluorescence recovery after photobleaching. Proc Natl Acad Sci USA 99:3171–3175

Sutherland AG, Alvarez J, Ding W, Foreman KW, Kenny CH, Labthavikul P, Mosyak L, Petersen PJ, Rush TS 3rd, Ruzin A, Tsao D, Wheless KL (2003) Structure-based design of carboxybi-phenylindole inhibitors of the ZipA-FtsZ interaction. Org Biomol Chem 1:4138–4140

Tang M, Bideshi DK, Park H-W, Federici BA (2006) Minireplicon from pBtoxis of *Bacillus thuringiensis* subsp. *israelensis*. Appl Environ Microbiol 72:6948–6954

Thanbichler M, Shapiro L (2006a) Chromosome organization and segregation in bacteria. J Struct Biol 156:292–303

Thanbichler M, Shapiro L (2006b) MipZ, a spatial regulator coordinating chromosome segrega-tion with cell division in *Caulobacter*. Cell 126:147–162

Thanedar S, Margolin W (2004) FtsZ exhibits rapid movement and oscillation waves in helix-like patterns in *Escherichia coli*. Curr Biol 14:1167–1173

Tinsley E, Khan SA (2006) A novel FtsZ-like protein is involved in replication of the anthrax toxin-encoding pXO1 plasmid in *Bacillus anthracis*. J Bacteriol 188:2829–2835

Tiyanont K, Doan T, Lazarus MB, Fang X, Rudner DZ, Walker S (2006) Imaging peptidoglycan biosynthesis in *Bacillus subtilis* with fluorescent antibiotics. Proc Nat Acad Sci USA 103:11033–11038

van den Ent F, Amos LA, Löwe J (2001) Prokaryotic origin of the actin cytoskeleton. Nature 413:39–44

van den Ent F, Møller-Jensen J, Amos LA, Gerdes K, Löwe J (2002) F-actin-like filaments formed by plasmid segregation protein ParM. EMBO J 21:6935–6943

Varley AW, Stewart GC (1992) The *divIVB* region of the *Bacillus subtilis* chromosome encodes homologs of *Escherichia coli* septum placement (MinCD) and cell shape (MreBCD) determi-nants. J Bacteriol 174:6729–6742

Vaughan S, Wickstead B, Gull K, Addinall SG (2004) Molecular evolution of FtsZ protein sequences encoded within the genomes of archaea, bacteria, and eukaryota. J Mol Evol 58:19–29

Vitha S, McAndrew RS, Osteryoung KW (2001) FtsZ ring formation at the chloroplast division site in plants. J Cell Biol 153:111–120

Vollmer W (2006) The prokaryotic cytoskeleton: a putative target for inhibitors and antibiotics? Appl Microbiol Biotechnol 73:37–47

White EL, Suling WJ, Ross LJ, Seitz LE, Reynolds RC (2002) 2-Alkoxycarbonylaminopyridines: inhibitors of *Mycobacterium tuberculosis* FtsZ. J Antimicrob Chemother 50:111–114

Wu LJ, Errington J (2004) Coordination of cell division and chromosome segregation by a nucleoid occlusion protein in *Bacillus subtilis*. Cell 117:915–925

Yanouri A, Daniel RA, Errington J, Buchanan CE (1993) Cloning and sequencing of the cell division gene *pbpB*, which encodes penicillin-binding protein 2B in *Bacillus subtilis*. J Bacteriol 175:7604–7616

Young KD (2006) The selective value of bacterial shape. Microbiol Mol Biol Rev 70:660–703

Yu XC, Margolin W (1997) Ca^{2+}-mediated GTP-dependent dynamic assembly of bacterial cell division protein FtsZ into asters and polymer networks in vitro. EMBO J 16:5455–5463

3
Mechanosensitive Channels: Their Mechanisms and Roles in Preserving Bacterial Ultrastructure During Adaptation to Environmental Changes

Ian R. Booth, Samantha Miller, Akiko Rasmussen, Tim Rasmussen, and Michelle D. Edwards(⊡)

Abstract The integrity of bacterial cells has long been identified with the possession of the peptidoglycan cell wall. However, the presence of "natural" disruptive forces has been recognized for almost 60 years. Mitchell determined that bacteria possess an outwardly directed turgor pressure of greater than 4 atmospheres. Other early experiments indicated that the substantial pools of amino acids that are retained by bacteria could be released very rapidly by hypoosmotic shock. More recently, two new elements have been added to the equation, one of which is universal and the other may have more limited distribution. The first is the discovery of mechanosensitive channels that open rapidly, producing large holes in the cytoplasmic membrane, in response to increases in membrane tension. The second is that the cell wall is a dynamic structure, in which changes are required during adaptation to stress. The interplay between the two phenomena is examined in this chapter, with much emphasis being placed on the structure and function of mechanosensitive channels.

Michelle D. Edwards
School of Medical Sciences, Institute of Medical Sciences, University of Aberdeen,
Aberdeen AB25 2ZD, UK
m.d.edwards@abdn.ac.uk

W. El-Sharoud (ed.) *Bacterial Physiology: A Molecular Approach.*
© Springer-Verlag Berlin Heidelberg 2008

3.1 Introduction

Bacterial cells derive their shape from the peptidoglycan (PTG) that forms a major component of the cell wall. This has been known for longer than 50 years and is amply supported by electron microscope images of isolated cell walls. The profile of a bacillary organism is still retained when the cytoplasm and membranes have been digested away using enzymes and detergents (Holtje 1998). Holtje has described PTG as "a spectacular three-dimensional structure, a hollow body that completely surrounds the bacterial cell." In essence, the shape of the bacterial cell is defined by the bonding patterns laid down in the PTG, which is simultaneously the shape-defining "molecule" and the stress-bearing polymer. Bacterial cells maintain an outwardly directed turgor pressure from the cytoplasm that is greater than 10 times the force that would be needed to rupture a lipid bilayer, such as the cytoplasmic membrane. For *Escherichia coli*, the turgor pressure is approximately 4 atm, but only approximately 0.2 atm is required for rupture of the lipid bilayer (Strop et al. 2003). Yet, this force of the turgor pressure is exerted against the cytoplasmic membrane of the cell, and the survival of cell integrity is dependent primarily on the PTG (Tomasz 1979; Holtje 1998). One of the best manifestations of this fact is that the most successful group of antibiotics, the penicillins and cephalosporins, target the cross-linking of the bacterial PTG (Denome et al. 1999). The unique contribution of PTG to bacterial cell walls is what makes these antibiotics so specifically antibacterial.

Despite all of the foregoing, the molecular structure of PTG is still controversial, the structural elements that result in cell integrity are still poorly defined, and the actual "in-cell" dynamics that allow the cell to survive large changes in the environment are still to be elucidated (Holtje 1998). Much of the enzymology of the bacterial PTG synthesis has been elucidated, but there are considerable questions remaining. In particular, the mechanisms that control the deployment of enzymes to specific locations and the systems that control their activities are poorly understood. A significant new opportunity for beginning to understand cell wall synthesis and dynamics arose with the discovery of mechanosensitive (MS) channels and their role in cell physiology (Martinac et al. 1987; Levina et al. 1999). Application to bacterial protoplasts of patch clamping, an electrophysiological technique developed for understanding the properties of ion channels in mammalian cells, led to the discovery of several classes of MS channels in *E. coli* (Martinac et al. 1987). Subsequent extensions of this technique demonstrated similar channels in Gram-positive bacteria and in the archaea (Pivetti et al. 2003). During the last few years, the properties and significance of MS channels have become clearer. The critical discovery was that mutant cells lacking the channels often lyse when exposed to an extreme decrease in external osmolarity (hypoosmotic shock) (Levina et al. 1999). In such experiments (equivalent in the natural world to a bacterium in a "dry" spot suddenly encountering rainwater), the turgor pressure increases threefold to fourfold greater than the normal levels in a few seconds. This means that the pressure is now approximately 30 times that needed to lyse the cytoplasmic membrane. The cell wall can provide some resistance, but alone is not sufficient to withstand such forces. MS channels gate in response to the increase in tension in the membrane consequent upon the increase in cell turgor, and transiently create large pores in the membrane that release hydrated solutes from the cytoplasm (Levina et al. 1999; Sukharev et al. 1999; Sukharev 2002). This lowers the osmotic gradient and diminishes the increase in cell turgor. Mutants that lack MS channels lyse when subjected to large increases in turgor during hypoosmotic shock. MS channel activation is, thus, a requirement for some bacterial cells for the maintenance of their structural integrity.

The best-characterized examples of MS channels are MscL and MscS from *E. coli* (Booth et al. 2007c; Steinbacher et al. 2007). The availability of crystal structures, electrophysiological techniques, and substantial biochemical and genetic protocols has led to models for structural rearrangements that occur during MS channel gating. The functions and mechanisms elucidated to date for these channels are discussed below.

3.2   The Bacterial Cell Wall

Bacterial cell wall structures fall into one of two distinct forms: Gram positive and Gram negative. The major difference between these two forms is the possession by Gram-negative cells of a thin layer of PTG surmounted by an outer membrane containing a high density of lipopolysaccharide (LPS) molecules. The outer membrane is held against the PTG by lipoproteins that are covalently attached to peptides within the PTG (Fig. 3.1a). Many functions for the LPS layer have been documented or proposed, of which, two major functions are (1) limiting the access of hydrophobic molecules and enzymes to the cytoplasmic membrane and (2) providing significant structural strength to the wall (Nikaido 1996). At the base of the LPS chains are phosphate groups that are linked by Ca^{2+} and Mg^{2+} ions. It has been shown that mutations affecting LPS structure render cells somewhat osmotically fragile during normal growth, but also permeable to a range of hydrophobic molecules. Similarly, chelating the Ca^{2+} and Mg^{2+} ions with EDTA has the effect of weakening the wall and increasing its permeability to hydrophobic molecules.

The basic structure of PTG is well known (Holtje 1998). The backbone consists of aminosugars; alternating molecules of *N*-acetylglucosamine (NAG) and *N*-acetylmuramic acid (NAM) linked together by β1–4 linkages. Each NAM has a peptide attached via the lactyl group. The cross-linking of the peptides gives the PTG the characteristic of mesh containing pores of different sizes (see below). The cross-links between peptides are a major contributor to the strength of the PTG. They are also the major source of species-specific diversity in PTG because, although the sugars are largely invariant in their structure, the amino acid composition of the peptides can vary significantly. One of the most significant features is the presence of a dibasic amino acid, *meso*-diaminopimelic acid (m-A_2pm), which is essential for the peptide bond to be formed. Cross-bridges are formed between

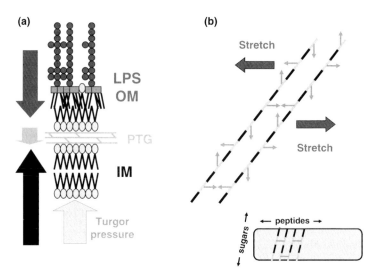

Fig. 3.1 Schematic depicting the cell envelope structure of Gram-negative bacterial cells. **a** A cross-sectional view of the layers surrounding the cell of a Gram-negative bacterium and their related pressure orientations. The inner membrane (*IM*) experiences the force of the turgor pressure from the cytoplasm and, thus, presses outward. This is counterbalanced by the strength of the PTG matrix and the outer membrane (*OM*)/LPS layer. See text for more details. **b** PTG chains. The PTG backbone consists of a repeating pattern of two aminosugars, NAG (*black bars*) and NAM (*light gray bars*). Cross-links are formed between peptides that are attached to each NAM unit. Each NAM molecule is rotated 90° to the previous NAM and, therefore, the peptides point in the plane of the PTG chains but also upward and downward (as indicated by the *small mid-gray arrows*). The planar peptides may be cross-linked between adjacent chains, but not all will form bridges and, thus, although strong, an irregular mesh is created. This structure can withstand some stretch in the longitudinal direction of the bacterial cell (as indicated by *large dark gray arrows*) but not circumferentially. *Inset* shows the orientation of PTG chains around a bacterium

m-A$_2$pm of one peptide and the penultimate D-ala residue of a peptide attached to NAM in an adjacent PTG chain. This reaction takes place in the periplasm, whereas synthesis of the NAG–NAM peptides takes place in the cytoplasm, followed by translocation across the membrane into the periplasm. Each successive sugar residue is orientated 90° to the previous residue, such that the peptides project above and below (axial) and to each side of the sugar chain (planar) (Fig. 3.1b). Planar peptides from adjacent PTG strands are cross-linked to form a net-like structure, whereas the m-A$_2$pm residues of axial peptides above the plane of the PTG serve to anchor lipoproteins that are inserted into the outer membrane.

The PTG is under pressure exerted from the cytoplasm and, therefore, exists in a stretched conformation that has the potential to create a grid of pores (Isaac and Ware 1974). Detailed biochemical analysis of the PTG from *E. coli* has revealed that such a regular structure is unlikely to exist except under conditions of extensive cross-link formation, which most often takes place in stationary phase (Holtje

1998). Growing cells have a much more dynamic cell wall and exhibit a much lower level of cross-linking. From these studies, we know that rather than a single continuous chain of alternating NAG and NAM residues, the PTG is created from many short sugar strands, the majority of which have only 5 to 10 NAG–NAM pairs, but with some larger strands of approximately 60 to 100 sugar pairs. The average for *E. coli* is approximately 29 disaccharide units. Considering a cell of approximately 1,000-nm diameter (circumference 3142 nm) and given that a NAG–NAM pair is approximately 1 nm in length, it is clear that to span the whole circumference of the cell at a specific point requires approximately 3,000 disaccharide units. If the average length is approximately 29 NAG–NAM, then approximately 100 chain fragments are needed per circumference. We know that the majority of strands are composed of 5 to 10 disaccharide units and, therefore, the actual number probably increases threefold to sixfold. Because the cylindrical portion of the cell may be up to 3,000 nm in length and the peptide bridging two chains is approximately 4 nm in length, then approximately 750 rings of NAG–NAM polymers are required to make a single cell. Taken together, the cylindrical section of the cell requires between 75,000 and 450,000 independent PTG units! Cross-linking is clearly of critical importance in the structural integrity of the wall. Analysis has revealed that, of the potential peptide cross-links, the actual number varies with the phase of growth and possibly other factors (Vollmer and Holtje 2001; Vollmer and Holtje 2004). If there are only approximately 30% cross-links, then the structure is extremely loose and will contain a significant diversity of "pore sizes." The degree of cross-linking increases as cells enter stationary phase, in which cells become smaller and are more rounded (Glauner et al. 1988; Santos et al. 1999). However, there may be significant inter-cell heterogeneity in a growing population.

PTG is a dynamic molecule. Approximately half of the wall is recycled every generation because of hydrolysis and recycling of the molecules that comprise the NAG–NAM peptides (Goodell 1985). A sophisticated network of enzymes and transporters ensures the recovery of the material released by cell wall lytic enzymes. Recycling is an active component of the process of growing the cell wall (Templin et al. 1999). Moreover, there is a complex maturation process that includes changes in the nature of the cross-links and a progressive lowering of the average glycan chain length. These observations serve to reinforce the impression that the cell wall is not concrete, but, rather, a dynamic network that is continuously being broken, expanded, reshaped, and ligated to create a strong structural entity that if ruptured, leads to bursting of the cytoplasmic membrane.

3.3 Osmoregulation in Bacterial Cells

Virtually all bacterial and archaeal cells have evolved a common strategy for dealing with variations in external osmolarity (Booth et al. 1988; Poolman et al. 2004). The specific details vary from one organism to another, particularly in the nature of the solutes accumulated in response to high osmolarity (Imhoff 1986). All bacterial cells maintain an outwardly directed turgor pressure that arises from water entry into the cell down the osmotic gradient (Mitchell and Moyle 1956). The osmotic gradient arises from the cell's accumulation of certain solutes to very high concentrations relative to the environment. To some extent, this reflects the need to accumulate metabolic intermediates in the cytoplasm to concentrations compatible with enzyme function at high rates. Estimates have suggested that *E. coli* accumulates osmotically active anions to approximately 200 mM even when growing at a moderately low osmolarity (approximately 220 mOsm) (Roe et al. 1998). Anions cannot be accumulated in the cytoplasm without replacement of their accompanying proton by another cation, usually K^+. The uncompensated accumulation of 200 mM protons would cause protein denaturation and DNA damage because of the severe acidification of the cytoplasm (Booth 1985). However, metabolic anions cannot alone account for the high cytoplasmic ion concentrations.

Cell growth requires the expansion of the volume of the cell. Biosynthesis of new proteins requires some expansion of the cell and this is achieved by coordinated synthesis of phospholipid to increase the surface area of the membrane. Driving this expansion is the entry of water down the osmotic gradient. Bacteria have evolved highly regulated K^+ transporters to achieve the controlled accumulation of this cation (Epstein 1986). In general, *E. coli* cells are relatively impermeant to K^+ ions. This is indicated by the observation that mutants lacking the major K^+ transport systems (Kdp, Trk, and Kup) require 40 mM concentrations of K^+ in the environment to sustain growth (Epstein and Davies 1970; Buurman et al. 2004). In contrast, strains that possess all three transport systems can grow rapidly at even 10 μM K^+. Cells also exhibit an amazing capacity to retain K^+; when cells are transferred to K^+-free media, the cytoplasmic pool is maintained at approximately 300 mM for many hours, indicating that the cell membrane is relatively impermeable to this cation (Bakker et al. 1987). In fact, the cell membrane can sustain gradients of K^+ close to 10^6-fold.

Water is much more freely permeable across the membrane than are ions (see discussion in Steinbacher et al. 2007). The response to an increase in external osmolarity is an initial loss of water from the cytoplasm caused by the change in the osmotic gradient (Booth et al. 1988). *E. coli* cells respond by taking up K^+, either by activating their constitutive Trk K^+ transporter or by inducing their highly regulated K^+-scavenging K^+-specific ATPase system, Kdp (Epstein 1986). It should be noted that the Kdp system is a high-affinity transport system that is only induced/derepressed when the external K^+ is low (Rhoads and Epstein 1978). The accumulation of K^+ is well-controlled such that the increase in cytoplasmic K^+ concentration parallels the change in the external osmolarity. The accumulation of K^+ is accompanied by some net proton extrusion, raising the cytoplasmic pH, but the principal balancing of the cation is achieved by the synthesis (or transport) of glutamate (Mclaggan et al. 1994). At very high external osmolarities, the accumulation of K^+ and glutamate in the cytoplasm can reach close to molar levels and this causes impaired functioning of enzymes. Consequently, cells have evolved a second strategy that enables rapid growth at high osmolarity without sacrificing the cell turgor that is needed for cell expansion.

Compatible solutes were first defined as those compounds that restored enzyme activity when added to enzymes incubated in high salt (Imhoff 1986). These solutes have the capacity to change the hydration state of proteins. In *E. coli*, a two-stage adaptation to high external osmolarity is observed. First, cells accumulate K$^+$ glutamate but then progressively they exchange these ions for compatible solutes. The most effective compatible solutes for *E. coli*, and many other bacteria, are betaine, proline (and its close relatives), and ectoines. However, in addition, *E. coli* and some other bacteria have the ability to synthesize the disaccharide trehalose and to use it as a compatible solute (Strom and Kaasen 1993). Compatible solute accumulation is highly regulated both at the level of the transport systems and through regulation of the expression of the transport systems. Betaine, proline, and ectoine accumulate in the cell in response to osmotic stress, and their cytoplasmic concentrations reflect the external osmolarity, increasing proportionately to the elevation of external osmolarity (Koo and Booth 1994). The accumulation of compatible solutes is achieved at relatively constant cell turgor, because these solutes displace K$^+$ glutamate from the cell. Thus, growth stimulation by compatible solutes is achieved by a combination of the beneficial effects of these solutes on protein folding and the lowering of the pools of the inhibitory ions.

3.4   Mechanosensitive Channels

First indications that cells might possess mechanisms to release solutes nondiscriminately arose from the observed rapid loss of amino acid pools on hypoosmotic shock (Britten and McClure 1962). As described above, bacteria that have been grown, or incubated, at high osmolarity accumulate high concentrations of solutes in the cytoplasm as a mechanism of generating turgor. The cell membrane is selectively permeable—a low passive permeability to solutes and a high permeability to water. Consequently, when cells adapted to high osmolarity encounter a lower osmolarity, water moves rapidly across the membrane into the cell, faster than any initial movement of solute, simply because of the relative permeabilities. A cell passing into a medium that is 300 mOsm (approximately 0.15 M NaCl) lower in osmolarity experiences an immediate increase in cell turgor of approximately 7 atm (Kung 2005). The immediate inrush of water should be followed by expansion of the cell, but this is resisted by the membrane, which has limited expansive capacity. What little stretching takes place is immediately transferred to the cell wall, increasing the expansion of the wall (approximately 20–30%) (Holtje 1998). Activation of MS channels provides the much sought-after relief for the cell (Fig. 3.2).

MS channels create transient large pores (8–35 Å effective diameter) in the cytoplasmic membrane (Sukharev et al. 2001; Bass 2002). The large diameter and the generally hydrophilic nature of the lining of the open channels essentially preclude selectivity between different solutes except on the basis of size. Thus, any cytoplasmic molecule less than 300 to 400 Da will readily pass through most MS channels that have been studied, although it is possible that the smaller channels

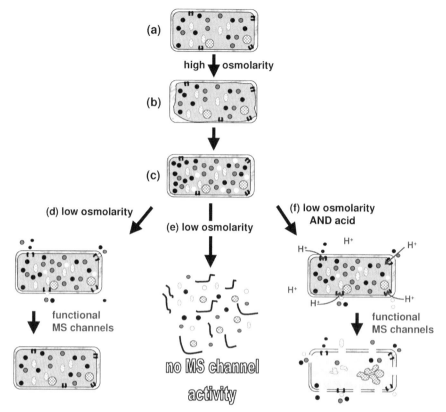

Fig. 3.2 The effects of hypoosmotic stress and acid in the absence and presence of MS channels. Bacterial cells growing in low osmolarity medium will accumulate 300- to 400-mM ions in the cytoplasm, e.g., potassium (*gray dots*) and glutamate (*black dots*) (**a**). Transfer from medium of low osmolarity (**a**) to one of high osmolarity initially causes efflux of water and cell shrinkage (**b**). To prevent loss of the turgor pressure necessary for growth and maintaining cell shape, cells accumulate more potassium (*gray dots*) and glutamate (*black dots*) as well as compatible solutes (*white dots*) (**c**). After transfer of cells from high to low osmolarity, MS channels in the cytoplasmic membrane are activated by the rapid increase in membrane tension caused by the entry of water. These large-diameter nonspecific channels mediate the immediate release of ions and compatible solutes, saving cells from lysis (**d**). If cells lack MS channels or the channels fail to gate, cell integrity will be lost and lysis may occur (**e**). However, if the hypoosmotic medium is acidic, when functional MS channels gate to release ions and solutes, protons will enter down their concentration gradient, acidifying the cytoplasm, denaturing proteins and DNA, and compromising cell viability (**f**)

may impose some greater restriction on the solutes that are released. This means that amino acids, organic acids that are intermediates in metabolism, cofactors, disaccharides, and nucleotides can be rapidly lost from the cell in a hydrated state (Berrier et al. 1992; Schleyer et al. 1993). Removal of structured water that is hydrogen bonded to organic molecules and ions is a slow process and, consequently, to achieve rapid release of solutes, the passage of hydrated solutes through the channel pore is essential. In contrast, there are also ion-specific MS channels in many higher organisms (Kung 2005). In these channels, atomic groups within the pore replace the water around the ions (Mackinnon 2000) and it is possible that these channels are more associated with cell signaling than with relief of osmotic stress.

Rapid release of solutes, on a millisecond time scale, eliminates the osmotic gradient that is driving water influx and, thereby, relieves the stress on the cell wall. Of course, opening these channels immediately creates different stresses, namely loss of nutrients, cofactors, and intermediates; loss of cytoplasmic homeostasis (cation and anion balance and pH); and alteration of membrane potential (Levina et al. 1999). The former is countered by enzymes in the periplasm that breakdown compounds into their component parts, which are then substrates for transport systems, thus, ensuring their recapture by the cell. Homeostasis is restored by pumping K^+ back into the cell, expelling ions that are not required, and reestablishing control over cytoplasmic pH. The cell can cope with pH transients in the neutral to mildly acidic nature, but will lose viability if the environmental pH is too acidic (Fig. 3.2) (Levina et al. 1999). Thus, opening MS channels is a highly risky strategy undertaken only to cope with an extreme form of stress that, if not relieved, compromises the structural integrity of the cell.

3.4.1 Structures and Gating of Mechanosensitive Channels

MS channels have been characterized by electrophysiology in various bacterial species (Martinac et al. 1987; Berrier et al. 1992; Szabo et al. 1992; Szabo et al. 1993; Kloda and Martinac 2001). Frequently, these channels exhibit large conductance values greater than 0.1 nanosiemens. From calculations based on these conductance measurements, we know that the potential pores created by open MS channels can be several angstroms in diameter. To place this in context, the porins in the outer membrane of *E. coli*, such as OmpF and OmpC, have a fixed pore that is approximately 6 Å diameter (Nikaido and Vaara 1985). The geometry of porin pores is often significantly asymmetric, such that they can operate both with a degree of selectivity and at reduced ion conductance relative to that expected from the diameter observed in crystal structures. However, small molecules, such as antibiotics and sugars, can block the pore transiently during transfer across the membrane (Nestorovich et al. 2002). The MS channels must be significantly larger than this to avoid any restriction arising in the passage of solutes through the pore.

The most highly characterized MS channels are MscL and MscS from *E. coli* (Steinbacher et al. 2007). The understanding of their properties arises from extensive electrophysiological analysis, matched by molecular genetics, biochemistry, and biophysics (Sukharev et al. 1994b; Blount et al. 1996; Blount et al. 1997; Ou et al. 1998; Levina et al. 1999; Sukharev et al. 1999; Moe et al. 2000; Li et al. 2002;

McLaggan et al. 2002; Perozo et al. 2002a; Perozo et al. 2002b; Sukharev 2002; Powl et al. 2005). Crystal structures for MscL and MscS were obtained from *Mycobacterium tuberculosis* and *E. coli*, respectively (Chang et al. 1998; Bass et al. 2002). The two proteins differ in size, complexity, and mechanism, despite both having the function of creating large transient pores. Initially, there was some doubt regarding the precise location of the MS channel proteins, but it is now clear that they are components of the cytoplasmic membrane. However, one class of MS channel protein, the MscK family, has the potential to interact with either periplasmic proteins or those in the outer membrane through its extended amino terminal peripheral domain, which has a periplasmic location (Levina et al. 1999; Li et al. 2002; McLaggan et al. 2002). As far as it is known, no ancillary proteins are required for MS channel activity and, thus, the transmission of the tension signal that gates these channels is transmitted wholly through the lipid bilayer (Sukharev et al. 1994a; Sukharev et al. 1999; Sukharev 2002).

3.4.1.1 MscL—Mechanosensitive Channel of Large Conductance

MscL channels are widely distributed among bacteria. MscL is a homopentamer of approximately 17- to 20-kDa subunits, each of which generally contains between 120 and 156 amino acid residues (Chang et al. 1998; Pivetti et al. 2003). Some higher organisms have been found to possess proteins that contain sequences highly similar to the transmembrane (TM) helices of MscL, but no detailed analysis of these proteins has been reported. At this time, archaea have not been reported to possess MscL homologs in their genomes. The open MscL channel exhibits the largest conductance of known channels, approximately 3 nanosiemens. In *E. coli* membrane patches, the opening of this channel takes place at pressures greater than 150 mmHg (1 atm = 760 mmHg), and this pressure range lies close to the lytic pressure of the membrane patch. The opening takes place within $3 \mu s$ of the pressure change, and each channel opening is of short duration, approximately 10 ms (Sukharev et al. 1999; Shapovalov and Lester 2004). The channel transitions between closed and open states as long as the pressure remains above the threshold for gating. As far as it is known, bacterial cells possess just a few of these channels in their cytoplasmic membrane, possibly as few as three to five channels per cell. Again, the abundance of these channels has not been analyzed systematically, although it is known that the expression of the genes is controlled both by osmolarity and by growth phase, allowing some adaptive range to the abundance of channels (Stokes et al. 2003).

Each MscL subunit consists of two TM domains, designated TM1 and TM2, separated by a variable length periplasmic loop and flanked by amino-terminal and carboxy-terminal α-helices (Chang et al. 1998). The proteins are relatively conservative with respect to length, with the most significant variations taking place in the periplasmic loop. Recent revisions of the MscL structure have suggested that this sequence forms a β-sheet hairpin that has the potential to line the rim of the open channel (Steinbacher et al. 2007). Each TM α-helix is approximately 30

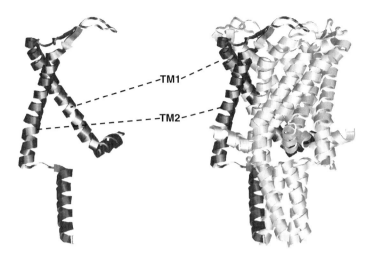

Fig. 3.3 The three-dimensional structure of MscL. The MscL protein from *Mycobacterium tuberculosis* was crystallized to a resolution of 3.5 Å in 1998 (Chang et al.). MscL is a homopentamer and the figure shows the ribbon representation of a single subunit alongside the full pentameric structure, with one subunit defined in *black*. TM1, which line the channel pore, and TM2, which interact with the TM1s and the membrane lipids, are indicated. The figures were created using RasTop (http://sourceforge.net/projects/rastop/)

residues in length, with approximately 80% embedded in the membrane bilayer. Five TM1 helices form an "inverted teepee" with the splayed ends toward the periplasm and the cytoplasmic ends meeting to form the pore seal at the cytoplasmic side of the membrane (Fig. 3.3). In the recently revised crystal structure of the *Mycobacterium* channel, these helices are tilted at an angle of approximately 28° to the pore axis, which is perpendicular to the membrane plane (Chang et al. 1998). The TM2 helices adopt a peripheral location in the closed channel, display a pronounced kink at the cytoplasmic side of the bilayer, and interact predominantly with amino termini of adjacent TM1 helices (which project outward from the pore) and with the lipid bilayer (Fig. 3.3) (Steinbacher et al. 2007). The transition from the closed to the open state involves rotation of the TM1 helices (and, to a lesser extent, the TM2 helices) and tilting of the helices so that they cross the membrane at a more acute angle than in the closed state. The carboxy-terminal α-helices form a bundle below the plane of the membrane on the cytoplasmic side and are largely dispensable to the structure and function of the channel because deletion mutants lacking this region remain functional (Blount et al. 1996). The newly revised structure of MscL indicates that each amino-terminal α-helix lies in the plane of the membrane and makes contact with the TM2 helix of the next but one subunit— thus, potentially forming a resistance to the transition from closed to open. This proposal is supported by previous findings that indicated that only short deletions of the amino-terminal α-helix were tolerated (Blount et al. 1996).

Crystal structures provide clear and testable models for the structure and the gating of channels, but are less informative regarding any alternative conformations that are adopted during the gating transition. The MscL crystal structure was proposed to be the closed state, although it has also been suggested that it represents a structure that is in transition to the open state and is imperfectly closed (Bartlett et al. 2004). Less clear is the structure of the open state. What is known has been informed by electron paramagnetic resonance (EPR) studies using the *E. coli* protein, into which single cysteine (Cys) residues have been introduced (Perozo et al. 2001; Perozo et al. 2002a; Perozo et al. 2002b; Perozo and Rees 2003; Bartlett et al. 2004). The native protein has no Cys residues, and studies have suggested that the protein is tolerant of Cys residues at almost all positions throughout its sequence (this contrasts with MscS, in which the stability of Cys-substituted proteins is dependent on the position of the inserted residue). After purification of the Cys-containing protein, it is reacted with a spin-label that attaches covalently to the Cys residue, and the modified protein is then used for EPR analysis (Perozo et al. 2001). EPR makes use of the phenomenon that the signal emitted by an unpaired electron is dependent on the environment. The EPR signal has a characteristic shape that is perturbed by the environment in which the spin label is located and its location can be further probed by examining its accessibility by reagents that are either lipophilic or hydrophilic. Purified protein can be reconstituted into liposomes, which are closed vesicles created by dissolving specific lipids in buffer and allowing them to form bilayers (Sukharev et al. 1993). The gating of the MS channels can be caused either by application of a pressure differential across a patch of the liposome in a patch-clamp electrode, or by adding a lipophilic reagent to entire liposomes in an EPR tube. Both of these treatments create a change in tension in the bilayer, which is responsible for gating the channel. Lysophosphatidyl choline (LPC) is the preferred agent used to gate MS channels in liposomes (Perozo et al. 2002a). LPC is a derivative of a phospholipid that has a single fatty acid attached to the phosphocholine. It is unable to form bilayers, but can insert into a liposome. The rate at which the LPC can translocate from the outer exposed leaflet to the inner leaflet is slow and, therefore, insertion of LPC generates a membrane area asymmetry between the outer and inner leaflet that distorts the membrane, increasing tension that gates the channel (see also Martinac et al. 1990). Using this technique, Perozo and colleagues were able to study the conformation of the MscL protein in the closed, intermediate, and open states. Analysis of the EPR signals of Cys residues at different positions led to a predicted structure for the open state (Perozo et al. 2002a).

Genetic studies have been important to identify residues that are critical to the closed state. Blount and colleagues defined selection regimes that allowed them to characterize mutant channels as gain-of-function (GOF), i.e., having lost the ability to maintain the closed conformation when the cell is under nonstressful conditions (Blount et al. 1996; Blount et al. 1997; Ou et al. 1998). As described above, MS channels are in the cytoplasmic membrane and must be maintained closed for most of the growth and division cycle of the organism. The large diameter of the pores combined with their lack of discrimination in the solutes that they pass means that an open channel seriously perturbs cellular homeostasis (Levina et al. 1999). The cytoplasmic pools of K^+ and amino acids will tend to decline, other ions that are normally excluded may enter the cells down their concentration gradient, cytoplasmic pH will change toward that of the environment, and cofactors for enzyme reactions may be lost to the environment (Schleyer et al. 1993). For any MS channel, there is a specific relationship between bilayer tension and the open probability (Sukharev 2000). In the cell, the net pressure across the membrane (resultant of turgor and resistance by the cell wall) is translated into tension in the bilayer that is maintained at a value just below that required to activate the MS channel with the lowest threshold, so that the dominant state of the MS channels is the closed form. Blount and colleagues reasoned that a mutant channel that had lost the ability to maintain the closed state under the normal growth conditions of the cell would inhibit cell growth (Blount et al. 1996; Blount et al. 1997). Indeed, when they screened a large population of MscL GOF mutants, they observed that a number of mutant proteins were inhibitory to cell growth. Subsequent electrophysiological analysis showed that these mutant channels generally exhibited a shifted pressure threshold to lower values.

The discovery of the structural gene for the MscS channel by Booth and colleagues defined a physiological assay for channel function (Levina et al. 1999). In *E. coli* cells, MscL and MscS are redundant—to observe a phenotype associated with loss of function in the channel protein, one must create a mutant strain that lacks both MscL and MscS. Such cells fail to survive hypoosmotic shock and are observed to lyse in response to severe shock (Fig. 3.2). Cells expressing both channels, or either MscL or MscS, survive and do not exhibit significant lysis. This has provided the basis for a number of assays based on the phenotype of the double mutant that lacks MscL and MscS, for example, demonstrations of channel activation in vivo (Batiza et al. 2002). Channels that have lost function because of a mutation are less effective than the parent channel in protecting against hypoosmotic shock. Using this method, a number of MscL mutants were characterized and were shown not to gate when subjected to high imposed TM pressure induced by hypoosmotic stress (Yoshimura et al. 2004). In patch-clamp experiments, it is generally difficult to show that such mutant channels gate at higher pressure because the MscL channel gates at pressures that are close to the lytic limit of the membrane bilayer. In contrast, channels bearing a mutation that causes a GOF phenotype in growing cells usually gate at lower pressure than the wild-type channels and can be more easily assayed in patch-clamp experiments (Blount et al. 1997).

Physiological assays are useful for showing channel function in the cellular context, because all other assays are dependent on either isolated membrane patches from cells treated with an antibiotic or use purified proteins reconstituted into artificial lipid bilayers. The cell-based assays have generally confirmed the properties observed in the more refined subcellular preparations. However, it must be appreciated that the outcome of the assay is dependent on the level of expression of the channel protein (Booth et al. 2007a). In *E. coli* cells, the *mscL* and *mscS* genes are subjected to transcriptional regulation by osmotic pressure and growth phase (Stokes et al. 2003). At least one of the mechanisms regulating expression is the osmotic stress-dependent shift from the σ^{70}-driven RNA polymerase to the σ^S-driven enzyme. The *rpoS* gene encodes sigma factor S (σ^S) that associates with the core RNA polymerase to recognize and transcribe genes involved in stress adaptation during the stationary phase and other adverse conditions. Expression of σ^S protein is highly regulated at both transcriptional and translational levels by various conditions, particularly osmolarity, growth phase, pH, and temperature (Hengge-Aronis 2002a; Hengge-Aronis 2002b) (see Chap. 11 in this book for further information). When bound to RNA polymerase, σ^S transcribes a large number of gene products that are associated with stress resistance. MscL and MscS are two of these gene products and they undergo twofold to threefold elevation during early stationary phase and during growth at high osmolarity. This change in channel abundance is on the same scale as that observed for the expression of the cloned *mscS* gene to convert a MscL⁻MscS⁻ mutant cell from being sensitive to hypoosmotic shock to being resistant (Stokes et al. 2003). The precise phenotype of a cell expressing a channel that has acquired one or more mutations is entirely dependent on the stability of the channel protein in the membrane. A channel protein that is very unstable will not accumulate in the membrane and, consequently, the worst effects of the gating change will be ameliorated by the presence of only a few channels in the membrane. Similarly, a mild loss of function in a channel caused by a mutation may be compensated by high levels of expression. What one observes is strongly dependent on the expression level!

Mutagenesis studies have identified amino acids that affect the packing of the seal region of the channel, whereby the closed state is maintained. In *E. coli* MscL, this is defined by the region close to valine (Val) 21, with glycine (Gly) residues being critical in maintaining packing between the TM1 helices. The introduction of hydrophilic residues in this region causes the channel to open at low pressures (i.e., to become GOF) (Ou et al. 1998). In contrast, mutagenesis of the regions of the TM1 and TM2 helices that interact with the headgroups of the phospholipids at the periplasmic side of the membrane can cause loss of function (Yoshimura et al. 2004). Here, it is the replacement of hydrophobic residues with hydrophilic ones (Leu/Val to Asn) that leads to an increase in the pressure required to gate the channels and decreased function in hypoosmotic shock assays.

Overall, MscL is a well-understood channel and provides a clear structural transition from the closed to the open state.

3.4.1.2 MscS—Mechanosensitive Channel of Small Conductance

The MscS class of MS channels is very large and diverse with respect to size and properties (Levina et al. 1999; Li et al. 2002; Pivetti et al. 2003; Yoshimura et al. 2004; Li et al. 2007). The conductance values that have been reported are approximately 0.7 to 1.25 nanosiemens, and the channels usually show relatively extended open dwell times, lasting 10 to 50 times that of MscL, with approximately 200 ms being the mean open dwell time for *E. coli* MscS (Edwards et al. 2005). MscS channels also exhibit unusual selectivity in the solutes that are translocated, in contrast to MscL, which is truly nonspecific. MscS from *E. coli* is slightly anion selective (Martinac et al. 1987), MscK from *E. coli* is nonselective (Li et al. 2002), whereas MscS from *Methanococcus janaschii* has a slight preference for cations (Kloda and Martinac 2001). The structural basis for the ion selectivity is not understood, but recent work from the authors' laboratory indicates a probable role for portals in the cytoplasmic domain, because, in *E. coli* MscS, the inner surface of the portals has several basic residues that would repel cations. As discussed below, ions must flow through the portals from the cytoplasm to reach the pore. *E. coli* MscS is the only member of this class for which the structure has been demonstrated. The channel was shown to be a homoheptamer by both crystallography and by cross-linking studies (Bass et al. 2002; Miller et al. 2003b). Each subunit is 286 residues, and the crystal structure resolves the organization for amino acids 27 to 280 (Fig. 3.4). A highly complex organization was revealed, which has at its center the observation that each subunit is twisted around the axis of the pore (Fig. 3.4). The crystal structure probably does not reflect the "in-membrane" organization of the first two TM helices (TM1 and TM2)—it is more likely that they pack tightly against the pore generated from the TM3 helices. However, the whole structure can be seen to be organized so that the extreme carboxy terminus is located approximately 180° relative to the first defined residue of TM1 (Tyr27) (Fig. 3.4). This structural organization must be an important component of the gating mechanism and we

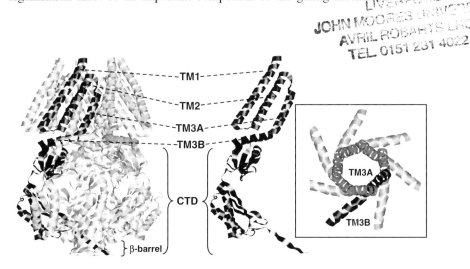

Fig. 3.4 The three-dimensional MscS structure. The *E. coli* MscS protein structure was determined by X-ray crystallography to a resolution of 3.9 Å in 2002 (Bass et al.). MscS is a homoheptamer with each subunit consisting of three TM spans: TM1 and TM2 spans forming a peripheral paddle-like structure and TM3 spans creating the central pore; followed by a large carboxy-terminal domain (*CTD*) terminating in a β-barrel structure (as indicated in a single subunit and the full heptameric molecule). TM3 spans have two specific parts: TM3A that lines the pore and TM3B, which connects the pore domain to the cytoplasmic domain after a distinct bend in the helix at Gly113. This kink causes TM3B to run tangential to the pore (*inset* shows amino acids 91–128 for each subunit, highlighting TM3A and TM3B). The figures were created using RasTop (http://sourceforge.net/projects/rastop/)

discuss below that its role is to provide resistance to the gating transition, closed to open, to ensure closure when operation is not required.

MscS subunits are multidomain, creating distinctive substructures—a tension sensor, a pore, and a carboxy-terminal vestibule through which solutes must pass *en route* to the external medium (Fig. 3.4). The amino terminus, residues 1 to 26, is located in the periplasm and is not highly conserved. Small deletions can be tolerated without perturbing the expression or assembly of the channel (Miller et al. 2003a). This region was not resolved in the crystal structure. The first visualized element of the protein consists of the initial two TM helices (TM1–TM2) that are organized in an antiparallel fashion as a mobile "paddle," attached to the pore-forming domain (TM3) by a short linker. This linker is not an inanimate connection and probably forms the periplasmic rim of the pore. The third TM helix, TM3, lines the pore and also connects the membrane domains to the cytoplasmic domain. TM3 comprises two elements. The amino-terminal half of TM3, TM3A, extends across the membrane, starting approximately at the junction between the two leaflets of the lipid bilayer. Seven TM3A helices create the pore, which is sealed by two rings of leucine residues (Leu105 and Leu109)—Leu109 lies very close to the cytoplasmic surface of the pore (Bass et al. 2002). Amino acid substitutions at these positions can render the channel easier to gate and confer on cells a GOF phenotype (Miller et al. 2003a). TM3B helices lie parallel to the membrane and tangential to the axis of the pore (Fig. 3.4). The path followed by this helix makes it possible for the TM3B helix to interact with the TM1–2 domain of another subunit, which may be the next or the next-but-one subunit, depending on the packing of the TM1–2 domain against the TM3A-generated pore domain in the closed and open structures. Such an arrangement may either increase the stability of the complex or create a mechanism for resetting the channel after gating.

The heptameric assembly of the carboxy-terminal domains creates a large hollow elongated "chamber" perforated by seven lateral portals that lie at the subunit interfaces. An axial portal is created by the carboxy termini of the seven subunits coming together into a β barrel. Both the external (cytoplasmic) and internal (continuous with the channel) surfaces of the vestibule are strongly hydrophilic, which is consistent with the role in containing solutes. Proteins will be excluded from the vestibule by the small size of the portals, which means that the composition of this compartment will reflect simply the ions of the cell cytoplasm. One can estimate the volume of the vestibule for a single channel is approximately 10^{-15} μl and that, given an intracellular concentration of ions in cells adapted to high osmolarity is approximately 1 M (600 mM K^+, 400 mM glutamate), one can guess that there are approximately 7 to 10 ions/vestibule, compared with approximately 10^{12} ions/cell. To lower the cytoplasmic osmolarity by 90% would require the movement of approximately 10^{11} ions/cell. Clearly, even with multiple channels, simply releasing the ions trapped in the vestibule will have a limited impact on cytoplasmic osmolarity. Consequently, ions must flow through the lateral portals and do so rapidly. The lateral portals are approximately 14 Å in diameter, which means that many solutes can potentially pass through with their water shell intact. The portal rim is constructed from parts of the upper-vestibule domain (the β or SM domain), the αβ domain that forms the lower half of the vestibule, and a short linker, which connects the SM domain to the αβ domain.

The effect of this organization is to create a "Chinese lantern" in which it is possible that structural changes could open and close the lateral portals (Edwards et al. 2004). The inner lining of the portal entry is strongly basic (Arg185 and Arg238), whereas the vestibule itself has rings of acidic residues that might bind cations. These features may attract anions into the vestibule and slightly restrict cation mobility, but this has not been tested experimentally.

MscS family members differ in their size and it is not known how closely the crystal structure of *E. coli* MscS relates to those of the larger proteins. Many of the proteins have extra amino-terminal membrane domains that must be packed around the core pore-forming TM1–3 membrane domains (Levina et al. 1999). The role of these extra domains is poorly understood, but, recently, GOF mutants have been described in these domains of MscK, which is the most extreme member of this family, in terms of size (Li et al. 2007). Extensions at the carboxy-terminal domains are also found in homologs in the genome databases. It is probable that these domains do not significantly affect the heptameric structure, because alkaline phosphatase (approximately 45 kDa) (Miller et al. 2003a) and green fluorescent protein (GFP) (approximately 27 kDa) have been fused to the *E. coli* MscS carboxy terminus without altering the structure.

The physical state of the crystallized *E. coli* MscS channel remains the subject of controversy—is it open or closed? The crystal structure shows a clear pore of 8-Å diameter, although simulations have suggested an even narrower pore (Bass et al. 2002). At first sight, the pore seemed to be open and this was the original view shared among the field. However, the lining of the exposed pore surface is extremely hydrophobic and thus, it has been proposed that the crystal structure represents the most closed form of the channel because of the formation of a vapor lock that prevents solute flow until the channel pore expands to create the conducting state (Anishkin and Sukharev 2004). Modeling of the packing of the TM3A helices indicates that their heptameric assembly brings them as close together as is feasible with retention of symmetry (Kim et al. 2004). Thus, further movement of the helices to provide a more "closed" state would have only a limited effect on the observed pore diameter. Consequently, there is an emerging consensus that the crystal structure represents a nonconducting state.

Rotation and tilting of the TM3A helices is associated with the transition to the open state (Edwards et al. 2005). The TM3A helices pack very tightly by virtue of a conserved surface created from alternating Ala and Gly residues (Bass et al. 2002). Our gating model proposes that the Ala residues form "knobs" that slide over the smooth "Gly" surface. Experiments with amino acid substitutions for these crucial residues have shown two phenomena consistent with the model. First, increasing the size of the "knobs" (Ala to Val substitution) or removing the Gly surface (Gly to Ala mutation) creates channels that gate at higher pressure, giving them a loss-of-function phenotype. Conversely, removing the "knob" (Ala to Gly mutation) creates a channel that gates more easily; GOF activity. Second, the severity of these classes of mutations, particularly those increasing the size of the "knob," is diminished if the change is made distant to the seal residues (Edwards et al. 2005). The data are consistent with the helices rotating and leaning away from the axis of the pore during the transition from the closed to the open state.

3.4.2   Sensing Osmotic Stress as a Mechanical Signal

There is no doubt that the major stresses that activate MS channels in bacterial cells are osmotic in origin (Mitchell and Moyle 1956; Britten and McClure 1962). Hypoosmotic stress provides the simplest example. When cells are transferred from a high to low osmolarity environment, an inrush of water presses the membrane against the wall. A limited expansion of the wall takes place, but it is likely that distortions in the membrane curvature occur and these result in activation of the channels. Patch-clamp studies on isolated membrane patches and on liposomes reconstituted solely with purified MS channel protein show that these channels gate in response to pressure applied across the lipid bilayer; detailed analysis has shown that the channels sense the tension in the bilayer (Sukharev et al. 1999; Sukharev 2002). There are also transient changes in the physiology of the organism that may trigger MS channels. Although these changes are less obviously similar to hypoosmotic shock, they may in fact be equivalent. Thus, during osmotic balancing, when cells exchange K^+ glutamate for compatible solutes, the release of the ions may take place through MS channels (Booth et al. 1988). On a simple level, one can consider that the channel proteins exist in a compressed state in the membrane and that the distortions introduced into membrane structure during hypoosmotic shock trigger the protein conformational changes. No specific sensors of membrane tension have been identified; to date, the evidence supports the idea that for MS channels to be effective transducers of hypoosmotic stress, the protein subunits must have only limited interaction with the phospholipid headgroups of the membrane (Booth et al. 2007b). Introduction of Asn residues into the ends of the TM1 helices in place of hydrophobic residues inhibits mechanotransduction, creating channels that require greater pressure to open in membrane patches, and, in cells, causes loss of function (Yoshimura et al. 2004; Nomura et al. 2006).

3.4.3   Biotechnological Applications of MS Channels

MS channels possess two characteristics that make them potential targets for novel antibacterial therapies. First, they open to form large holes in the membrane. Different classes of MS channels produce pores of different magnitudes, but early experiments suggested that MscL could allow passage of proteins (Ajouz et al. 1998; Berrier et al. 2000). Although the original evidence is often disputed, the data have inspired experiments to develop MscL as a delivery system for small proteins and peptides (Smisterova et al. 2005; van den Bogaart et al. 2006). The pharmaceutical industry has been interested for a long time in methods to attain slow or controlled release of drugs. This is particularly important for small protein therapeutics, where it is important to focus the activity of the protein at particular tissues. By using an MS channel to release the drug from a liposome, one may ultimately be able to maximize control over drug delivery. Second, MS channel pores allow the passage of a wide range of ions. Gating of MS channels could be used to perturb intracellular homeostasis by the introduction of ions or small molecules detrimental to cell survival (Jordan et al. 1999; Levina et al. 1999; McLaggan et al. 2002). One example of this would be induced uptake of H^+ if the cells were exposed to acidic conditions and the channels promoted to open, as seen in Fig. 3.2. In this scenario, the slightly alkaline physiological pH of the cytoplasm would be *rapidly* lowered by the influx of H^+ down their concentration gradient, leading to denaturation of proteins and resulting consequences. Depending on the application, strong acidic conditions could be used where feasible, for example to sterilize materials, to cause cell death, whereas mild acidic conditions could be used, for example, in foods, in which inhibition of bacterial growth is the biotechnical aim. Development of chemical or biochemical methods to activate channels could be used alone to reduce viability of contaminating bacteria by perturbing homeostasis or combined with acid pH or uptake of small toxic compounds, to kill pathogenic species. Molecules such as LPC and parabens have been shown to activate MS channels under certain conditions (Martinac et al. 1990; Nguyen et al. 2005), and recent studies describe a photocontrolled technique for activating MS channels in a regulated manner (Folgering et al. 2004). Additionally, using similar technology, MS channels could be used for targeted release of small therapeutic molecules placed in a liposome-like vehicle to achieve specific delivery of compounds in the body.

3.5   Concluding Remarks and Future Prospects

MS channels are major contributors to the maintenance of cellular integrity. Important insights have been gained into their structures, mechanisms of sensing membrane tension, and importance for cell physiology. In fact, the *E. coli* MscL and MscS channels are fairly well understood, although specific details are still required. In addition, major questions remain regarding other members of these families of channels. Clearly, we need to advance in-depth investigations into the gating systems of MS channels and explore chemical methods of activation to be able to create novel antibacterial strategies. However, we already possess the fundamental information and a great deal of critical knowledge to make this prospect successful.

Highly Recommended Readings

Anishkin A, Sukharev S (2004) Water dynamics and dewetting transition in the small mechanosensitive channel MscS. Biophys J 86:2883–2895.

Edwards MD et al. (2005) Pivotal role of the glycine-rich TM3 helix in gating the MscS mechanosensitive channel. Nat Struct Mol Biol 12:113–119.

Holtje J-V (1998) Growth of the stress-bearing and shape-maintaining murein sacculus of *Escherichia coli*. Microbiol Mol Biol Rev 62:181–203.

Miller S, Edwards MD, Ozdemir C, Booth IR (2003b) The closed structure of the MscS mechano-sensitive channel—Cross-linking of single cysteine mutants. J Biol Chem 278:32246–32250.

Perozo E, Kloda A, Cortes DM, Martinac B (2001) Site-directed spin-labeling analysis of recon-stituted MscL in the closed state. J Gen Physiol 118:193–206.

Perozo E, Rees DC (2003) Structure and mechanism in prokaryotic mechanosensitive channels. Current Opin Struct Biol 13:432–442.

van den Bogaart G, Krasnikov V, Poolman B (2007) Dual-color fluorescence-burst analysis to probe protein efflux through the mechanosensitive channel MscL. Biophys J 92:1233–1240.

Vollmer W, Holtje JV (2004) The architecture of the murein (peptidoglycan) in Gram-negative bacteria: vertical scaffold or horizontal layer(s)? J Bacteriol 186:5978–5987.

References

Ajouz B, Berrier C, Garrigues A, Besnard M, Ghazi A (1998) Release of thioredoxin via the mechanosensitive channel MscL during osmotic downshock of *Escherichia coli* cells. J Biol Chem 273:26670–26674.

Anishkin A, Sukharev S (2004) Water dynamics and dewetting transition in the small mechano-sensitive channel MscS. Biophys J 86:2883–2895.

Bakker EP, Booth IR, Dinnbier U, Epstein W, Gajewska A (1987) Evidence for multiple K+ export systems in *Escherichia coli*. J Bacteriol 169:3743–3749.

Bartlett JL, Levin G, Blount P (2004) An in vivo assay identifies changes in residue accessibility on mechanosensitive channel gating. Proc Natl Acad Sci USA 101:10161–10165.

Bass RB, Strop P, Barclay M, Rees DC (2002) Crystal structure of *Escherichia coli* MscS, a voltage-modulated and mechanosensitive channel. Science 298:1582–1587.

Batiza AF, Kuo MMC, Yoshimura K, Kung C (2002) Gating the bacterial mechanosensitive chan-nel MscL in vivo. Proc Natl Acad Sci USA 99:5643–5648.

Berrier C, Coulombe A, Szabo I, Zoratti M, Ghazi A (1992) Gadolinium ion inhibits loss of metabolites induced by osmotic shock and large stretch-activated channels in bacteria. Eur J Biochem 206:559–565.

Berrier C, Garrigues A, Richarme G, Ghazi A (2000) Elongation factor Tu and DnaK are trans-ferred from the cytoplasm to the periplasm of *Escherichia coli* during osmotic downshock presumably via the mechanosensitive channel MscL. J Bacteriol 182:248–251.

Blount P, Schroeder MJ, Kung C (1997) Mutations in a bacterial mechanosensitive channel change the cellular response to osmotic stress. J Biol Chem 272:32150–32157.

Blount P, Sukharev SI, Schroeder MJ, Nagle SK, Kung C (1996) Single residue substitutions that change the gating properties of a mechanosensitive channel in *Escherichia coli*. Proc Natl Acad Sci USA 93:11652–11657.

Booth IR (1985) Regulation of cytoplasmic pH in bacteria. Microbiological Reviews 49:359–378.

Booth IR, Cairney J, Sutherland L, Higgins CF (1988) Enteric bacteria and osmotic-stress—an integrated homeostatic system. J App Bacteriol 65:S35–S49.

Booth IR et al. (2007a) Physiological Analysis of Bacterial Mechanosensitive Channels. Sies M, Haeussinger D (eds) Methods in Enzymology 428:47–61.

Booth IR, Edwards MD, Black S, Schumann U, Miller S (2007b) Mechanosensitive channels in bacteria: signs of closure? Nature Reviews Microbiology 5(6):431–440.

Booth IR, et al. (2007c) Structure-function relations of MscS. Current Topics in Membranes 58, *in press.*

Britten RJ, McClure FT (1962) The amino acid pool in *Escherichia coli*. Microbiol Mol Biol Rev 26:292–335.

Buurman ET, McLaggan D, Naprstek J, Epstein W (2004) Multiple paths for aberrant transport of K+ in *Escherichia coli*. J Bacteriol 186:4238–4245.

Chang G, Spencer RH, Lee AT, Barclay MT, Rees DC (1998) Structure of the MscL homolog from *Mycobacterium tuberculosis*: A gated mechanosensitive ion channel. Science 282:2220–2226.

Denome SA, Elf PK, Henderson TA, Nelson DE, Young KD (1999) *Escherichia coli* mutants lacking all possible combinations of eight penicillin binding proteins: viability, characteristics, and implications for peptidoglycan synthesis. J Bacteriol 181:3981–3993.

Edwards MD et al. (2005) Pivotal role of the glycine-rich TM3 helix in gating the MscS mechanosensitive channel. Nat Struct Mol Biol 12:113–119.

Epstein W (1986) Osmoregulation by potassium-transport in *Escherichia-coli*. Fems Microbiology Reviews 39:73–78.

Epstein W, Davies M (1970) Potassium-dependent mutants of *Escherichia coli* K-12. J Bacteriol 101:836–843.

Folgering JH, Kuiper JM, de Vries AH, Engberts JB, Poolman B (2004) Lipid-mediated light activation of a mechanosensitive channel of large conductance. Langmuir 20:6985–6987.

Glauner B, Holtje JV, Schwarz U (1988) The composition of the murein of *Escherichia coli*. J Biol Chem 263:10088–10095.

Goodell EW (1985) Recycling of murein by *Escherichia coli*. J Bacteriol 163(1):305–310.

Hengge-Aronis R (2002a) Recent insights into the general stress response regulatory network in *Escherichia coli*. J Mol Microbiol Biotechnol 4:341–346.

Hengge-Aronis R (2002b) Signal transduction and regulatory mechanisms involved in control of the sigma(S) (RpoS) subunit of RNA polymerase. Microbiol Mol Biol Rev 66:373–395.

Holtje J-V (1998) Growth of the stress-bearing and shape-maintaining murein sacculus of *Escherichia coli*. Microbiol Mol Biol Rev 62:181–203.

Imhoff JF (1986) Osmoregulation and compatible solutes in eubacteria. FEMS Microbiology Reviews 39:57–66.

Isaac L, Ware GC (1974) The flexibility of bacterial cell walls. J Appl Bacteriol 37:335–339.

Jordan SL, Glover J, Malcolm L, Thomson-Carter FM, Booth IR, Park SF (1999) Augmentation of killing of *Escherichia coli* O157 by combinations of lactate, ethanol, and low-pH conditions. Applied and Environmental Microbiology 65:1308–1311.

Kim S, Chamberlain AK, Bowie JU (2004) Membrane channel structure of Helicobacter pylori vacuolating toxin: Role of multiple GXXXG motifs in cylindrical channels. Proc Natl Acad Sci USA 101:5988–5991.

Kloda A, Martinac B (2001) Structural and functional differences between two homologous mechanosensitive channels of *Methanococcus jannaschii*. EMBO J 20:1888–1896.

Koo SP, Booth IR (1994) Quantitative analysis of growth stimulation by glycine betaine in *Salmonella typhimurium*. Microbiology 140(Pt 3):617–621.

Kung C (2005) A possible unifying principle for mechanosensation. Nature 436:647–654.

Levina N, Totemeyer S, Stokes NR, Louis P, Jones MA, Booth IR (1999) Protection of *Escherichia coli* cells against extreme turgor by activation of MscS and MscL mechanosensitive channels: identification of genes required for MscS activity. EMBO J 18:1730–1737.

Li C, Edwards MD, Hocherl J, Roth J, Booth IR (2007) Identification of mutations that alter the gating of the *E. coli* mechanosensitive channel protein, MscK. Mol Microbiol 64(2):560–574.

Li Y, Moe PC, Chandrasekaran S, Booth IR, Blount P (2002) Ionic regulation of MscK, a mechanosensitive channel from *Escherichia coli*. EMBO J 21:5323–5330.

Mackinnon R (2000) Mechanism of ion conduction and selectivity in K^+ channels. J Gen Physiol 116:17 (abstract).

Martinac B, Adler J, Kung C (1990) Mechanosensitive ion channels of *E. coli* activated by amphipaths. Nature 348:261–263.

Martinac B, Buehner M, Delcour AH, Adler J, Kung C (1987) Pressure-sensitive ion channel in *Escherichia coli*. Proc Natl Acad Sci USA 84:2297–2301.

McLaggan D et al. (2002) Analysis of the *kefA2* mutation suggests that KefA is a cation-specific channel involved in osmotic adaptation in *Escherichia coli*. Mol Micro 43:521–536.

Mclaggan D, Naprstek J, Buurman ET, Epstein W (1994) Interdependence of K⁺ and glutamate accumulation during osmotic adaptation of *Escherichia-coli*. J Biol Chem 269:1911–1917.

Miller S, Bartlett W, Chandrasekaran S, Simpson S, Edwards M, Booth IR (2003a) Domain organization of the MscS mechanosensitive channel of *Escherichia coli*. EMBO J 22:36–46.

Miller S, Edwards MD, Ozdemir C, Booth IR (2003b) The closed structure of the MscS mechanosensitive channel—Cross-linking of single cysteine mutants. J Biol Chem 278:32246–32250.

Mitchell P, Moyle J (1956) Osmotic function and structure in bacteria. In: Spooner E, Stocker B (eds) Bacterial Anatomy. Cambridge University Press, Cambridge, pp 150–180.

Moe PC, Levin G, Blount P (2000) Correlating a protein structure with function of a bacterial mechanosensitive channel. J Biol Chem 275:31121–31127.

Nestorovich EM, Danelon C, Winterhalter M, Bezrukov SM (2002) Designed to penetrate: time-resolved interaction of single antibiotic molecules with bacterial pores. Proc Natl Acad Sci USA 99:9789–9794.

Nguyen T, Clare B, Guo W, Martinac B (2005) The effects of parabens on the mechanosensitive channels of *E. coli*. Eur Biophys J 34:389–395.

Nikaido H (1996) Outer membrane. In: Neidhardt F, Curtiss R III, Ingraham JL, Lin ECC, Low KB, Magasanik B, Reznikoff WS, Riley M, Schaechter M, Umbarger HE (ed) *Escherichia coli* and *Salmonella typhimurium* Cellular and Molecular Biology. ASM Press, Washington DC, pp 29–47.

Nikaido H, Vaara M (1985) Molecular basis of outer membrane permeability. Microbiological Reviews 19:1–32.

Nomura T, Sokabe M, Yoshimura K (2006) Lipid-protein interaction of the MscS mechanosensitive channel examined by scanning mutagenesis. Biophys J 91:2874–2881.

Ou X, Blount P, Hoffman RJ, Kung C (1998) One face of a transmembrane helix is crucial in mechanosensitive channel gating. Proc Natl Acad Sci USA 95:11471–11475.

Perozo E, Cortes DM, Sompornpisut P, Kloda A, Martinac B (2002a) Open channel structure of MscL and the gating mechanism of mechanosensitive channels. Nature 418:942–948.

Perozo E, Kloda A, Cortes DM, Martinac B (2001) Site-directed spin-labeling analysis of reconstituted MscL in the closed state. J Gen Physiol 118:193–206.

Perozo E, Kloda A, Cortes DM, Martinac B (2002b) Physical principles underlying the transduction of bilayer deformation forces during mechanosensitive channel gating. Nat Struct Biol 9:696–703.

Perozo E, Rees DC (2003) Structure and mechanism in prokaryotic mechanosensitive channels. Current Opin Struct Biol 13:432–442.

Pivetti CD et al. (2003) Two families of mechanosensitive channel proteins. Microbiol Mol Biol Rev 67:66–85.

Poolman B, Spitzer JJ, Wood JM (2004) Bacterial osmosensing: roles of membrane structure and electrostatics in lipid-protein and protein-protein interactions. Biochim Biophys Acta 1666:88–104.

Powl AM, Carney J, Marius P, East JM, Lee AG (2005) Lipid interactions with bacterial channels: fluorescence studies. Biochem Soc Trans 33:905–909.

Rhoads DB, Epstein W (1978) Cation transport in *Escherichia coli*. IX. Regulation of K⁺ transport. J Gen Physiol 72(3):283–295.

Roe AJ, Mclaggan D, Davidson I, O'Byrne C, Booth IR (1998) Perturbation of anion balance during inhibition of growth of *Escherichia coli* by weak acids. J Bacteriol 180:767–772.

Santos JM, Freire P, Vicente M, Arraiano CM (1999) The stationary-phase morphogene *bolA* from *Escherichia coli* is induced by stress during early stages of growth. Mol Microbiol 32:789–798.

Schleyer M, Schmid R, Bakker EP (1993) Transient, specific and extremely rapid release of osmolytes from growing cells of *Escherichia coli* K-12 exposed to hypoosmotic shock. Arch Microbiol 160:424–431.

Shapovalov G, Lester HA (2004) Gating transitions in bacterial ion channels measured at 3 μs resolution. J Gen Physiol 124:151–161.

Smisterova J, van Deemter M, van der Schaaf G, Meijberg W, Robillard G (2005) Channel protein-containing liposomes as delivery vehicles for the controlled release of drugs—optimization of the lipid composition. J Control Release 101:382–383.

Steinbacher S, Bass R, Strop P, Rees DC (2007) Structures of the prokaryotic mechanosensitive channels MscL and MscS. Current Topics in Membranes 58:1–24.

Stokes NR et al. (2003) A role for mechanosensitive channels in survival of stationary phase: Regulation of channel expression by RpoS. Proc Natl Acad Sci USA 100:15959–15964.

Strom AR, Kaasen I (1993) Trehalose metabolism in *Escherichia coli*; protection and stress regulation of gene expression. Mol Microbiol 8:205–210.

Strop P, Bass R, Rees DC (2003) Prokaryotic mechanosensitive channels. Adv Protein Chem 63:177–209.

Sukharev S (2000) Pulling the channel in many dimensions. Trends Microbiol 8:12–13.

Sukharev S (2002) Purification of the small mechanosensitive channel of *Escherichia coli* (MscS): the subunit structure, conduction, and gating characteristics in liposomes. Biophys J 83:290–298.

Sukharev SI, Blount P, Martinac B, Blattner FR, Kung C (1994a) A large-conductance mechanosensitive channel in *E. coli* encoded by *mscL* alone. Nature 368:265–268.

Sukharev S, Durell SR, Guy HR (2001) Structural models of the MscL gating mechanism. Biophys J 81(2):917–936.

Sukharev SI, Martinac B, Arshavsky VY, Kung C (1993) Two types of mechanosensitive channels in the *Escherichia coli* cell envelope: solubilization and functional reconstitution. Biophys J 65:177–183.

Sukharev SI, Sigurdson WJ, Kung C, Sachs F (1999) Energetic and spatial parameters for gating of the bacterial large conductance mechanosensitive channel, MscL. J Gen Physiol 113:525–540.

Szabo I, Petronilli V, Zoratti M (1992) A patch-clamp study of *Bacillus subtilis*. Biochim Biophys Acta 1112:29–38.

Szabo I, Petronilli V, Zoratti M (1993) A patch-clamp investigation of the *Streptococcus faecalis* cell membrane. J Membr Biol 131:203–218.

Templin MF, Ursinus A, Holtje JV (1999) A defect in cell wall recycling triggers autolysis during the stationary growth phase of *Escherichia coli*. EMBO J 18:4108–4117.

Tomasz A (1979) From penicillin-binding proteins to the lysis and death of bacteria: a 1979 view. Rev Infect Dis 1:434–467.

van den Bogaart G, Krasnikov V, Poolman B (2007) Dual-color fluorescence-burst analysis to probe protein efflux through the mechanosensitive channel MscL. Biophys J 92:1233–1240.

Vollmer W, Holtje JV (2001) Morphogenesis of *Escherichia coli*. Curr Opin Microbiol 4:625–633.

Vollmer W, Holtje JV (2004) The architecture of the murein (peptidoglycan) in Gram-negative bacteria: vertical scaffold or horizontal layer(s)? J Bacteriol 186:5978–5987.

Yoshimura K, Nomura T, Sokabe M (2004) Loss-of-function mutations at the rim of the funnel of mechanosensitive channel MscL. Biophys J 86:2113–2120.

4
Structural and Functional Flexibility of Bacterial Respiromes

David J. Richardson

Abstract Respiration is fundamental to the life of many bacterial species. It generally involves the transfer of electrons from low redox potential donors, such as NADH and succinate, through a range of integral membrane or membrane-associated electron transfer proteins, to higher potential electron acceptors. Free energy is released during this process that is used to drive the translocation of protons across the cytoplasmic membrane to generate a protonmotive force that can drive energy-consuming processes such as the synthesis of ATP, solute transport, and motility. There are a large number of different kinds of respiratory electron acceptors available in terrestrial and aquatic environments that bacteria can use as different electron acceptors. Some bacteria may be able to use one type of electron acceptor, but many species present an armory of respiratory enzymes that enable them to use

David J. Richardson
School of Biological Sciences, University of East Anglia, Norwich NR4 7TJ, UK
d.richardson@uea.ac.uk

W. El-Sharoud (ed.) *Bacterial Physiology: A Molecular Approach.*
© Springer-Verlag Berlin Heidelberg 2008

many chemically different kinds of electron acceptors. These include oxygen, elemental sulfur and sulfur oxyanions, organic sulfoxides, nitrogen oxyanions, nitrogen oxides, transition metal ions, radionuclides, and halogenated hydrocarbons. This respiratory diversity contributes to the ability of bacteria to colonize many of Earth's oxic, micro-oxic, and anoxic environments and underlies their contribution to the planet's key biogeochemical cycles, such as the carbon, nitrogen, and sulfur cycles. The emergence of the sequences of many hundreds of bacterial genomes during the last 10 years has provided much insight into the organization of respiratory systems in a range of bacteria, and reveals that many species display the capacity for a highly flexible respiratory system that enables them to use a range of different electron donors and acceptors in response to different environmental pressures. The proteins that make up a bacterium's respiratory system, and confer respiratory flexibility, define its "respirome." The organization and bioenergetics of a respirome can be illustrated through consideration of some well-studied paradigm Gram-negative organisms, such as the enteric bacterium *Escherichia coli*, the soil-denitrifying bacterium *Paracoccus denitrificans*, and insoluble metal oxide-reducing *Shewanella* species. These organisms can be used to describe the principles of electron transfer associated with the inner membrane, periplasmic compartment, and outer membrane, and to illustrate the different mechanisms by which bacteria can convert redox energy into protonmotive force. Their study provides paradigm models for understanding the respiratory systems of other organisms, for which genome sequences are available, but biochemical studies are less well advanced.

4.1   Introduction

Respiration is a catabolic process that is fundamental to all of the kingdoms of life (Nichols and Ferguson 2002). In human mitochondria, the ATP factories of our cells, electrons are extracted from organic carbon as it is catabolized through metabolic pathways, such as glycolysis and the tricarboxylic acid cycle. The electrons are then passed via freely diffusible carriers, such as NADH, or protein-bound carriers, such as $FADH_2$, into a multiprotein electron transfer pathway associated with the inner mitochondrial membrane (Fig. 4.1). Here the electrons migrate through a range of electron transferring redox cofactors, such as flavins, iron sulfur clusters, hemes, and copper centers, that are bound to integral membrane or membrane-associated protein complexes, and, ultimately, reduce oxygen to water (Fig. 4.1a). The electrons that enter the electron transport pathway have a low electrochemical potential, for example, the midpoint redox potential (at pH 7.0 this is termed $E^{0\prime}$) of the $NAD^+/NADH$ redox couple is approximately $-320\,mV$. This makes NADH a strong reductant. By contrast, the $E^{0\prime}$ of the O_2/H_2O couple is approximately $+820\,mV$, making it strongly oxidizing. Electrons, thus, flow "downhill" in energy

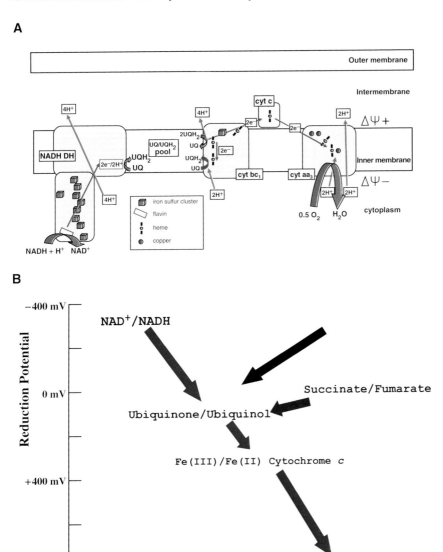

Fig. 4.1 Schemes for mitochondrial respiratory electron transport. **a** The topological organization of the electron transport system in the inner mitochondrial membrane. **b** The midpoint redox potentials of the key redox couples of the mitochondrial electron transfer system

terms from NADH to oxygen and the free energy (ΔG) released during this electron-transfer process is used to drive the translocation of protons across the inner mitochondrial membrane to generate a transmembrane proton electrochemical gradient or protonmotive force (Δp). This Δp has both a chemical (ΔpH) and electrical ($\Delta \psi$)

component. Because Δp has a value of approximately 150 to 200 mV, then a potential change (ΔE) of approximately this magnitude during transfer from donor to acceptor is required if that step is to be coupled to Δp generation (Fig. 4.1b). Overall, the transfer of two electrons from NADH to oxygen results in 10 positive charges being translocated across the membrane, and there are three key protonmotive steps that contribute to this: the NADH dehydrogenase (NADH:uniquinone oxidoreductase) that is a proton pump; the cytochrome bc_1 complex (ubiquinone: cytochrome c oxidoreductase) that moves positive charge across the membrane via a so-called Q-cycle that will be discussed later (see Sect. 4.3.1); and the cytochrome aa_3 oxidase that is a proton pump (Fig. 4.1).

This basic description of respiration is also fundamental to the Bacterial kingdom of life, with one key difference. The respiratory flexibility of the human mitochondrion is rather poor, because oxygen is the only significant electron acceptor and cytochrome aa_3 oxidase provides the only means of reducing this in an energy-conserving manner. However, in bacteria, a diverse range of electron acceptors can be used, including nitrogen oxides and oxyanions, sulfur oxyanions, elemental sulfur, organic sulfoxides, organic N-oxides, transition metal-containing minerals, radionuclides, and halogenated hydrocarbons (Richardson 2000). This respiratory diversity contributes to the ability of bacteria to colonize many of Earth's oxic, micro-oxic, and anoxic environments, and underlies their important contribution to the Earth's key biogeochemical cycles, such as the carbon, nitrogen, and sulfur cycles. The emergence of the sequences of many hundreds of bacterial genomes during the last 10 years has provided much insight into the organization of respiratory systems in a range of bacteria, and many species display the capacity for a highly flexible respiratory system that enables them to use a range of different electron donors and acceptors in response to different environmental pressures, so-called respiratory flexibility (Richardson 2000). Among some of the best-studied respiratory systems biochemically are those from as *Escherichia coli*, *Paracoccus denitrificans*, and *Shewanella* species, in which a number of the key respiratory proteins have been purified and for which a number of important molecular structures are emerging. The respiratory systems of these organisms have been studied for a number of years in the author's laboratory and will be used in this chapter to illustrate the different principles of respirome organization and energy conservation through which bacteria can convert redox energy into Δp. They, thus, provide paradigm systems for developing models for the respiratory systems of other organisms, including globally important pathogens, for which genome sequences are available but for which biochemical studies are less well advanced.

4.2   The Respirome of *E. coli*

E. coli is a facultative anaerobe that can switch between aerobic and anaerobic respiration in response to signals from the external environment, such as the oxygen tension. Many *E. coli* strains and a number of close relatives of *E. coli*, such as

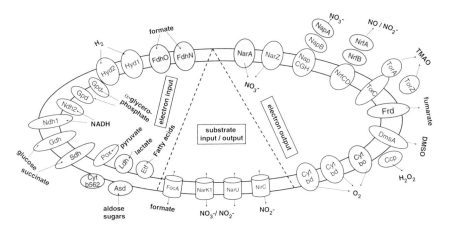

Fig. 4.2   The partial respirome of *E. coli*

members of the *Salmonella* genus, are pathogens, and this respiratory flexibility underpins the ability of these enteric pathogens to survive at a wide range of oxygen tensions within and outside of the host. A simple respiratory network for *E. coli* is shown in Fig. 4.2. The shaded proteins indicate those for which molecular structures have been resolved and illustrates that, although progress is being made toward a structural resolution of this respirome, there is much work still to be performed. The network is illustrated in two key sections; on the right-hand side of the figure, electron input into the respiratory systems is illustrated, whereas electron output is illustrated on the left-hand side. This diagram is not exhaustive of all of the electron input and output systems, but it shows some of the major systems that have been studied and it is immediately apparent that this network of electron transport proteins provides the organism with extensive respiratory flexibility with regard to exploiting a wide range of organic and inorganic electron donors to reduce a number of chemically distinct electron acceptors, including oxygen, nitrate, nitrite, dimethylsulfoxide (DMSO), trimethylamine *N*-oxide (TMAO), and fumarate. The regulation of the genes encoding the different respiratory enzymes is complex, and a description is beyond the scope of this chapter. However, the key point is that the whole respirome is not "on" at the same time, with the expression of the genes encoding the different complexes being regulated in response to many different environmental factors that include, for example, availability of respiratory substrates, such as oxygen, nitrate, and nitrite (Potter et al. 2001).

  The link between the electron donating enzymes and the electron-accepting enzymes in both mitochondria (Fig. 4.1) and *E. coli* (Fig. 4.2) is the quinone pool. Quinones are small freely diffusible, lipophilic, membrane-entrapped organic molecules that can carry two electrons and two protons when fully reduced (when they are then called quinols). Many of the quinones and quinols in a respiratory system are present as part of a pool in which the molecules of quinones and quinols exchange between free and bound species as they accept electrons from, or donate electrons to, the different electron transport proteins with which they interact. Different kinds of quinones have different electrochemical potentials. *E. coli* can synthesize more than one type of quinone. Ubiquinone ($E^{0'}$ UQ/UQH$_2$ ~+40 mV) predominates under aerobic conditions, and menaquinone ($E^{0'}$ MQ/MQH$_2$ ~−80 mV) predominates under anaerobic conditions, in which the cellular state is more reduced. Thus, under aerobic conditions, ubiquinol could be mediating electron transfer from the proton-pumping NADH dehydrogenase (Ndh1) to the proton-pumping cytochrome *bo* oxidase (cyt *bo*), whereas, under anaerobic conditions, menaquinol might be mediating electron transfer from formate dehydrogenase (Fdh-N) to nitrate (Nar, Nap) or nitrite (Nrf) (Fig. 4.2).

The importance of having both MQ and UQ can be illustrated when the succinate/fumarate redox couple is considered. The interconversion between fumarate and succinate is a reversible two-electron/two-proton process. The $E^{0'}$ of this redox couple is approximately 0 mV and it serves two functions in *E. coli*. Under aerobic conditions, succinate can act as an electron donor to the respiratory electron transport chain, being oxidized to fumarate in a reaction that relies on succinate dehydrogenase (Sdh) (Fig. 4.2) using the more oxidizing UQ as electron acceptor to pull the redox reaction in the oxidative direction. However, under anaerobic conditions, the reaction can go in the other direction, with fumarate now serving as a respiratory electron acceptor and being reduced to succinate. This requires a separate enzyme called fumarate reductase (Frd) (Fig. 4.2) that uses the more reducing menaquinol as electron donor to push the redox reaction in the reductive direction. The molecular structures of *E. coli* Frd and Sdh have both been resolved, and their comparison reveals considerable structural similarity, but the operating potentials of the redox centers have been "tuned" to these different physiological functions and also to reduce the release of reactive oxygen species in the aerobically active Sdh (Yankovskaya et al. 2003).

The consideration of the Sdh and Frd illustrate two biochemically distinct enzymes interacting with the same redox couple (i.e., the succinate/fumarate couple), but performing very different physiological functions. Analysis of the respirome reveals that there are a number of cases in which biochemically distinct enzymes apparently interface with the same redox couple and drive it in the same direction. One example is the oxygen/water couple. The cytochrome *bd* oxidase (cyt *bd*) and cyt *bo* both reduce oxygen to water. However, the former has a high affinity for oxygen and is expressed under micro-oxic conditions. In other cases, there are biochemically similar isozymes that catalyze the same reaction, for example, there are two isozymes of the membrane-bound nitrate reductase system, one constitutive (NarZ) and one expressed in anoxic environments when nitrate is present at high levels (NarA), and, likewise, of the membrane-bound formate dehydrogenase system (Fdh-N and Fdh-O), in which, again, one (Fdh-N) is highly induced by nitrate during anaerobic growth and the other can be found at low levels in aerobically grown cells (Potter et al. 2001; Richardson and Sawers 2002). Respiratory flexibility can also be provided by a single enzyme, as illustrated by the aldose sugar dehydrogenase that can extract electrons from a large number of monosaccharide, disaccharide, and polysaccharide aldose sugars by virtue of having an active site cofactor, pyrroloquinoline quinone (PQQ), in a highly solvent exposed cavity on the surface of the protein (Southall et al. 2006)

4.2.1 Protonmotive Coupling Mechanisms in the E. coli Respirome

Inspection of Fig. 4.2 reveals that the active site of the respiratory enzymes can be located in either the cytoplasm or the periplasm. This introduces the need for transporters to move substrates and products across the membrane (a few are indicated in the figure, but there are many more). The location of complex enzymes that bind redox cofactors in the periplasmic compartment also introduces the need to export these enzymes, and a system for exported folded proteins that can have redox cofactors preassembled in the cytoplasm called the TAT system has been described; this system plays an important role in many bacteria (Sargent 2007). However, in this chapter, we are focusing on the bioenergetics of respiratory systems and, in this respect, the location of the active sites of respiratory enzymes in different cellular compartments introduces some interesting considerations for mechanisms by which electron transfer from donor to acceptor can be coupled to the generation of Δp. Perhaps the most widely known protonmotive mechanism is the proton pump, in which the free energy in the redox reactions catalyzed by the enzyme is used to directly drive the translocation of protons across the cytoplasmic membrane. Examples in the E. coli respirome include the protonmotive NADH dehydrogenase (Ndh1) and the structurally defined cyt bo, the latter being an enzyme for which the proton channels have received a great deal of research attention, and key residues involved in proton movement can be resolved in the crystal structure (Abramson et al. 2000). This proton-pumping protonmotive mechanism is generally well described in biochemistry textbooks because it is present in the mitochondrial electron transport chain (see Sect. 4.1). When considering bacterial respiration, however, there are coupling mechanisms that can not be found in human mitochondria, and these will be considered in the following sections.

4.2.1.1 Protonmotive Oxidation of the Quinol Pool

First, we will consider the protonmotive redox loop and describe this with reference to two enzyme complexes, the membrane-bound nitrate reductase A (NarA) and the formate dehydrogenase (Fdh-N). NarA consists of three structural components: a catalytic NarG that binds a molybdenum ion as part of a complex Mo-bis-molybdopterin guanine dinucleotide cofactor (Mo-bis-MGD) and an iron sulfur cluster; NarH that binds four iron sulfur clusters and mediates electron transfer to NarG; and NarI, an integral membrane protein that binds two hemes and transfers electrons from the quinol pool to NarH (Fig 4.3). NarG and NarH are located in the cytoplasm at the membrane potential negative face of the cytoplasmic membrane (Fig 4.3). In NarI,

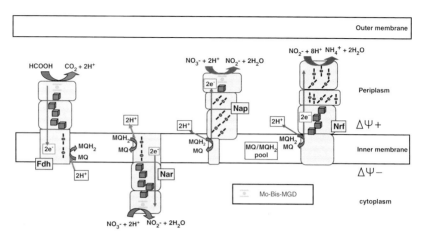

Fig. 4.3 Protonmotive Q-loops in *E. coli*

one of the two hemes is located at the periplasmic side of the protein and the other is at the cytoplasmic side. NarI receives electrons from quinol at the periplasmic side of the membrane and electrons move down an approximately 9-nm nanowire of eight redox centers and, ultimately, reduce nitrate at the membrane potential negative face of the cytoplasmic membrane (Fig. 4.3). This structurally defined nanowire comprises two hemes, five iron sulfur clusters, and the Mo-*bis*-MGD (Bertero et al. 2003). Because oxidation of quinol occurs at the periplasmic side of NarI, the protons are released into the periplasm (membrane potential positive side of the membrane) and the two electrons are moved from the heme on the periplasmic side to a heme on the cytoplasmic side (which has a higher $E^{0'}$). This transmembrane charge separation makes the Nar enzyme electrogenic (or protonmotive) in that a net of two positive charges is translocated across the membrane during transfer of two electrons to nitrate ($2\,q^+/2\,e^-$).

4.2.1.2 Protonmotive Reduction of the Quinone Pool

The process of electron transfer from quinol to nitrate via NarA presented a mechanism by which the free energy in the QH_2/nitrate couple (~400 mV) is conserved as Δp via an electrogenic redox loop mechanism. What about electron transfer into the quinone pool? *E. coli* produces formate under anaerobic conditions when organic carbon substrates are catabolized via pyruvate and the enzyme pyruvate formate-lyase. When nitrate reduction by NarA is coupled to electron input from formate via the nitrate inducible formate dehydrogenase (Fdh-N), two redox loops are brought together and the Fdh-N–NarA respiratory chain of *E. coli* emerges as a paradigm for a full protonmotive redox loop (Richardson and Sawers 2002) (Fig. 4.3). In Fdh-N, in contrast to NarGHI, the catalytic site is in the periplasm. However, similar to NarA, the electrons generated pass down an approximately 9-nm nanowire of redox centers that, in this case, connect the Mo-*bis*-MGD cofactor of Fdh-N, located in the periplasm, to a menaquinone reductase site at the cytoplasmic (membrane potential negative) face of the inner membrane. Similar to NarGHI, the wire comprises five iron sulfur clusters and two hemes. A large, approximately 350 mV, potential drop from formate to menaquinone allows efficient electron transfer against the Δp of approximately 200 mV, and the whole process serves, similar to NarA, to effectively translocate two positive charges across the membrane for every two electrons extracted from formate. Together then, the electron-carrying arms of the Fdh-N and NarA form a protonmotive redox loop that spans an electron-transfer distance of some 15 nm (Fig. 4.3), has a $\Delta E^{0\prime}$ of greater than 800 mV (−420 mV to + 420 mV), and has a coupling stoichiometry of $4H^+/2$ e⁻. When the whole electron transfer ladder from the Fdh-N Mo-*bis*-MGD via the quinone pool to the Nar Mo-*bis*-MGD is considered, it should be noted that the intermediary heme and iron sulfur cluster electron carriers are one-electron transfer centers. However, formate oxidation, quinone reduction, quinol oxidation, and nitrate reduction are two-electron reactions. Thus, the Mo-*bis*-MGD cofactors and Q reduction/QH_2 oxidation sites site at either end of the two coupled nanowires are crucial for coupling the one-electron/two-electron oxidoreductions.

In discussing the protonmotive redox loop coupling mechanisms, we will diverge transiently to examine the properties of an iron sulfur cluster electron nanowire. In the case of NarA and Fdh-N, iron sulfur clusters play a key role in the wire. Iron sulfur clusters are inorganic cofactors of iron and acid-labile sulfide that are normally bound to the proteins via cysteine or histidine residues, and they usually carry one electron at a time (Nichols and Ferguson 2001). Generally, if two redox cofactors are positioned within approximately 1.4 nm (or less) of each other, rapid electron transfer will take place, provided there is a sufficiently strong overall thermodynamic driving force (Moser et al. 1992) (Fig. 4.4). In the case of the formate/

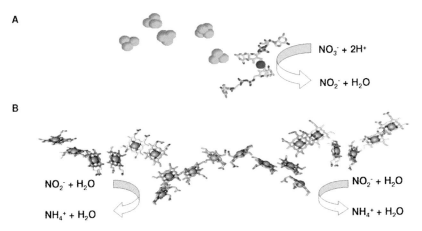

Fig. 4.4   Electron wires in *E. coli*. **a** The iron sulfur cluster wire of NarA. **b** The heme wire of NrfAB (taken from the model described in Clarke et al. 2007, figure courtesy of Dr T. Clarke)

menaquinol and menaquinol/nitrate redox couples, there is such a driving force and, thus, electrons will move rapidly through the wire, being "pushed" by formate and "pulled" by nitrate. The "wire"-like arrangement of the iron sulfur clusters (Fig. 4.4) can be seen in a number of different kinds of respiratory enzymes in *E. coli* that include hydrogenase, which is predicted to operate a redox loop mechanism similar to Fdh-N (Richardson and Sawers 2002), DMSO reductase (Cheng et al. 2005), and the seven iron sulfur cluster approximately 8-nm electron input arm of NADH dehydrogenase, for which a structure has recently been resolved from *Thermus thermophilus* (Sazanov and Hinchliffe 2006).

4.2.1.3   Nonprotonmotive Quinol Oxidation

Not all respiratory systems in *E. coli* are directly coupled to Δp generation. In addition to NarA, *E. coli* can synthesize a second type of respiratory nitrate reductase, called NapA. Unlike the NarA system, NapA is located in the periplasm (recall that, in NarA, the active site is in the cytoplasm). The periplasmic nitrate reductases are, similar to NarA, linked to quinol oxidation in respiratory electron transport chains, but they do not conserve the free energy in the QH_2/nitrate couple. This is because the electron delivery system, the quinol oxidizing proteins of the Nap system, is not electrogenic (Fig. 4.3). However, nitrate reduction via Nap can be coupled to energy conservation if the quinone reductase at the electron input side of the respiratory system generates a proton electrochemical gradient. This could, for example, be by a proton-pumping NADH dehydrogenase or the protonmotive formate dehydrogenase (Fig. 4.3). In fact, Nap forms part of a periplasmic enzymic system that can serve to reduce nitrate through to ammonium and which involves a second enzyme called NrfA (Figs. 4.2 and 4.3) (Potter et al. 2001). Quinol oxidation by the Nrf system is also not coupled to proton translocation and, therefore, the whole periplasmic nitrate reduction to ammonium respiratory systems relies on Δp generation at the level of electron input into the Q-pool, rather than at the level of electron output. The role of the Nap and Nrf systems is then to serve to turnover the QH_2 pool to ensure a continued supply of oxidized quinone for the quinone-reducing electron input components. In the case of nitrate reduction, however, why does *E. coli* make different types of enzymes that catalyze the same reaction with different coupling efficiencies? A clue comes from expression studies of *nap* in *E. coli* that show that the *nap* operon is induced at low nitrate concentrations, but repressed at higher nitrate concentrations that induce the *narA* operon (Potter et al. 2001). This suggests that Nap is a nitrate-scavenging system. Accordingly, in competition experiments in continuous cultures under nitrate-limited conditions, an *E. coli* strain expressing only *napA* out-competes a strain expressing only *narA*. This situation is reversed under carbon-limited conditions, in which the strain expressing *nap* is out-competed by a strain expressing *narA*. This reflects a higher affinity for intact cells for nitrate when the Nap system is expressed (Potter et al. 1999). In this context, it becomes significant that many of the pathogenic bacteria that may have to scavenge nitrate from the low levels present in many bodily fluids have the genetic information for Nap. Indeed, in some of these (e.g., *Hemophilus influenzae* and *Campylobacter jejuni*), Nap is the only nitrate reductase present (Potter et al. 2001).

In closing the discussion of the Nap and Nrf systems, it is notable that they are rich in proteins that bind multiple hemes covalently to the polypeptide chain, for example, the $NrfA_2B_2$ complex that reduces nitrite to ammonium can bind 20 such hemes (Clarke et al. 2007). From a structural and electron transfer perspective, the Nrf system then provides an illustration of a redox system in which electrons are transferred long distances through multiheme nanowires (Fig 4.4) that be compared with the use of chains of iron sulphur clusters to move electrons long distances discussed in Sect. 4.2.1.2. Multiheme electron wires are discussed further in Sect. 4.4 in the context of the respiration of metal oxides by *Shewanella* species.

4.2.1.4 Nonprotonmotive Quinone Reduction

In addition to protonmotive reduction of the Q-pool, there are also a number of examples in *E. coli* of nonprotonmotive reduction of the Q-pool, e.g., the Sdh, discussed earlier (page 102), where there is insufficient ΔE between the succinate/fumarate couple and the UQ/UQH_2 couple to drive an electrogenic reaction. A situation that is likewise true of electron input enzymes such as the electron transfer flavoprotein (ETF) quinone oxidoreductases that couple the β-oxidation of fatty acids to the respiratory chain (Fig. 4.2), and for which a structure of a homolog from porcine mitochondria has recently emerged (Zhang et al. 2006). A second NADH dehydrogenase (Ndh2) is also present in the respirome that is not protonmotive and, unlike Ndh1, dissipates the free energy in the NADH–UQ redox couple. In all of these cases, the generation of Δp is dependent on the electron output side of the respiratory chain. Nonprotonmotive electron input and output systems present opportunities for energy spillage during conditions of energy excess, when cellular reducing power needs to be dissipated (Calhoun et al. 1993). This redox-poising phenomenon will be discussed further with reference to *P. denitrificans* in Sect. 4.3.3.

4.3 The Modular Respiratory Network of *P. denitrificans*

A good example of respiratory flexibility can be found in *P. denitrificans*, a soil bacterium that is a member of the α-Proteobacteria and thought to be a close relative of the original progenitor of the mitochondrion. It is a denitrifying bacterium that can use nitrate as a respiratory electron acceptor and reduce it via nitrite, nitric oxide, and nitrous oxide to nitrogen gas (Fig. 4.5), with each intermediate also being a substrate for an energy-conserving electron transport system. This denitrification process is important and beneficial in water treatment processes for removal

108

D.J. Richardson

A

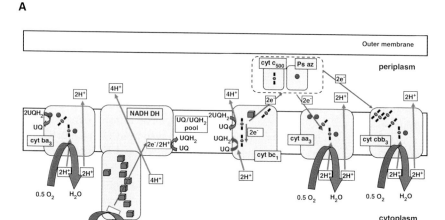

B

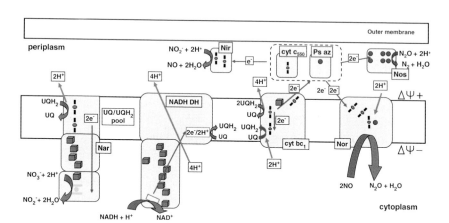

Fig. 4.5 The respiratory networks of *P. denitrificans*. **a** The aerobic network. **b** The anaerobic denitrification network. *Ps az*, pseudoazurin; *NADH DH*, NADH dehydrogenase

of nitrate and nitrite from wastewaters, but is detrimental in agriculture where it can lead to costly losses of fertilizers added to soil. Critically, however, another very important environmental impact of denitrifying bacteria is the emission of the gas nitrous oxide (N_2O). The most infamous greenhouse gases in the public eye are perhaps currently carbon dioxide and methane, and there is much international focus on reducing emissions of these by countries signed up to the Kyoto Protocol. However, nitrous oxide now ranks alongside carbon dioxide and methane as a cause for great concern. At present, nitrous oxide is perhaps best known to the public as laughing gas. However, throughout the 20th century, and continuing into the 21st century, nitrous oxide in the environment has increased by 50 parts per billion, and this atmospheric loading is increasing further by 0.25% each year. Although it only accounts for approximately 9% of total greenhouse gas emissions, it has a 300-fold greater global warming potential than carbon dioxide during the next 100 years, and an atmospheric lifetime of 150 years. This is most definitely not a laughing matter, and it is recognized in the Kyoto Protocol that is important to begin to predict the impact of environmental change and N_2O production and to mitigate these releases. Denitrifying bacteria are key global generators and consumers of nitrous oxide (Fig. 4.5). Much of the increase in production of nitrous oxide correlates to the increased application of nitrogenous fertilizers onto soils that began in the early 1900s. It is important that this does not remain the case. Thus, efforts to improve the prediction and management of nitrous oxide emissions will benefit from the better understanding of the factors that influence the net production of nitrous oxide by bacteria and this, in turn, requires an understanding of the structure and regulation of the respirome of a paradigm denitrifying bacterium.

Inspection of the *P. denitrificans* respirome (Figs. 4.5 and 4.6) reveals that a number of respiratory subnetworks with overlapping protein and quinol components can share the same energy-conserving membrane. These respiratory systems enable the organism to use oxygen, nitrogen oxyanions, and nitrogen oxides as electron acceptors depending on the physicochemical environment (Fig. 4.5). The genome sequence of this bacterium was completed in 2006 (http://genome.jgi-psf.

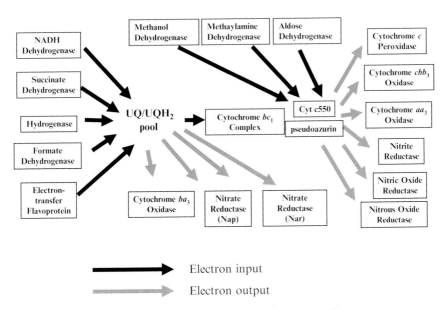

Fig. 4.6 The modular structure of the *P. denitrificans* respiratory network

org/draft_microbes/parde/parde.home.html). The bacterium is equipped with genes
that encode a number of biochemically distinct oxidases (Fig. 4.5). One of these,
the cytochrome aa_3 oxidase, operates at high oxygen tensions and terminates a
highly coupled electron transfer pathway that can be very similar to that described
for the mitochondrion in Sect. 4.1. However, at lower oxygen tensions, a high affinity cytochrome cbb_3 oxidase becomes more important, and a cytochrome c peroxidase is also expressed to enable the energy-conserving detoxification of the partially
reduced toxic oxygen species, hydrogen peroxide (H_2O_2). Under anoxic conditions,
the denitrification enzymes that are capable of reducing nitrogen oxyanions and
nitrogen oxides are synthesized. These can be coupled to the core electron-transport
pathway at the level of the ubiquinol pool or the cytochrome bc_1 complex and enable growth and metabolism of the organism in anoxic environments. Thus, by
modulating expression of different terminal oxidoreductases, which "lock" onto a
core electron-transfer system, it is possible for P. denitrificans to survive and proliferate at a range of oxygen tensions and adapt quickly in a rapidly changing environment (Richardson 2000). Understanding the factors that control the choice of
electron route through the respiratory networks of bacteria is topical challenge for
systems-based approaches to metabolic modeling. This is particularly true in periods of transition between aerobiosis and anaerobiosis or in rapidly fluctuating
environments, when subnetworks will often be simultaneously expressed. There
has been structural resolution of many of the key enzymes of the P. denitrificans
respirome (Ferguson and Richardson 2004), and there is a good understanding of
the transcription factors involved in regulating expression of the genes encoding the
respirome (van Spanning et al. 2006).

One of the features of the denitrification network is the branching electron
distribution at the level of the Q/QH$_2$ pool to either the nitrate reductases or to
the cytochrome bc_1 complex, from where they flow via dedicated cytochromes
c (heme proteins) or pseudoazurin (copper proteins) to the nitrite, nitric oxide,
and nitrous oxide reductases (Fig. 4.6). It is then possible to envisage the
organization of the P. denitrificans electron transport pathway as a modular one,
with each module connecting with each other and having one or more electron
input and output pathways. For example, modules would include: (1) the Q-pool
that has electron input through a number of primary dehydrogenases and
electron output via a number of quinol dehydrogenases and oxidases, including
the membrane-bound nitrate reductase system, the periplasmic nitrate reductase
system, the quinol oxidase, and the cytochrome bc_1 complex; (2) the cytochrome bc_1 complex that has a QH$_2$ electron input route and a number of electron output routes that include cytochrome c_{550} and the copper protein
pseudoazurin; and (3) the cytochrome c_{550} pool that has electron input at the
level of the cytochrome bc_1 complex and electron output to the nitrite, nitric
oxide and nitrous oxide reductases, and the cytochrome oxidases (Fig. 4.3).
This definition of modules provides for an experimental and modeling framework for from which to localize and quantify the effects of changes in pH, temperature, oxygen, and in the nature of the carbon and free energy source on
respiratory fluxes electron transfer. For example, the redox state of the Q/QH$_2$

pool or the cytochrome c_{550} pool could influence electron flux to redox enzymes that lie downstream of these modules (Otten et al. 2001; van Spanning et al. 2006).

4.3.1 Protonmotive Mechanisms in P. denitrificans: Q-Cycle Dependent Denitrification

In *P. denitrificans*, the process of denitrification under anaerobic conditions begins with reduction of nitrate to nitrite in the cytoplasm by a quinol-dependent Nar-type enzyme similar to that described for *E. coli* and which generates Δp via an electrogenic Q-loop mechanism (see Sect. 2.1.1). This is where the similarity with *E. coli* ends, because, thereafter, the reduction of nitrite proceeds via: (1) a periplasmic nitrite reductase containing two different kinds of heme cofactor (*c* heme and d_1 heme; cd_1Nir) that is distinct from the *E. coli* Nrf nitrite reductase system described in Sect. 4.2.1.3; (2) an integral membrane nitric oxide reductase is not found in *E. coli*; and (3) a periplasmic copper-containing nitrous oxide reductase also not found in *E. coli*. In some denitrifying bacteria, the cd_1Nir nitrite reductase is substituted for by a copper enzyme (Zumft 1997). These three enzymes all couple into the Q/QH_2 pool of respiratory network at the level of the cytochrome bc_1 complex module. This complex is also not present in *E. coli*, and consideration of the mechanism of coupling in denitrification allows for the introduction of the protonmotive Q-cycle associated with the cytochrome bc_1 complex (Figs. 4.1 and 4.5). In this cycle, a total of four electrons are extracted from two quinol molecules at the periplasmic face of the cytoplasmic membrane. Two of these electrons move out of the cytochrome bc_1 complex module into the cytochrome c_{550} (or pseudoazurin) module. The remaining two electrons move back across the membrane through two stacked hemes of the cytochrome *b* subunit of the complex to the cytoplasmic face of the membrane, where they reduce a molecule of Q to QH_2. The consequence of this is that there is a net oxidation of one molecule of quinol and a net movement of two positive charges across the membrane to give a $2q^+/2\,e^-$ ratio that is identical to that of the protonmotive Q-loop described for Nar (see Sect. 2.1.1). In the context of denitrification, this means that reductions of nitrite, nitric oxide, and nitrous oxide are bioenergetically equivalent to the reduction of nitrate by the Nar membrane-bound nitrate reductase system. Piecing the bioenergetics of the entire denitrification process together, it can then be seen that reduction of $2NO_3^-$ to N_2 requires 10 electrons and results in the movement of 30 positive charges across the membrane when these originate from NADH via the protonmotive NADH dehydrogenase (van Spanning et al. 2006) (Fig 4.5b). In oxygen respiration, however, 10 electrons flowing from NADH to oxygen via the cytochrome bc_1 complex and cytochrome aa_3 oxidase can be coupled to the movement of 50 positive charges across the membrane (Fig 4.5a), and this illustrates why oxygen is usually the preferred electron acceptor if it is available. Critically, however, because turnover of the cytochrome bc_1 complex is dependent on the availability of both Q and QH_2 (Fig. 4.1), it is highly dependent on the redox state of the Q/QH_2 pool. Thus, turnover can be compromised when the Q-pool is either over reduced or over oxidized. This can lead to a process called aerobic nitrate respiration, which will be discussed in Sect. 4.3.3.

4.3.2   Comparison of Protonmotive and Nonprotonmotive Quinol-Dependent Nitrite Reduction in P. denitrificans and E. coli

In considering the *P. denitrificans* cytochrome cd_1 nitrite reductase and the Q-cycle, it is informative to return to the *E. coli* NrfA nitrite reductase (see Sect. 4.2.2.3). NrfA is an ammonium-genic respiratory nitrite reductase, whereas cytochrome cd_1 nitrite reductase is nitric oxide-genic. Why does this cytochrome cd_1 nitrite reductase release nitric oxide when NrfA does not? Nitrite reduction to ammonium is thought to proceed via bound NO and NH_2OH intermediates (Einsle et al. 2002). Both cd_1Nir and NrfA are competent NH_2OH reductases, but only NrfA is an NO reductase (Angove et al. 2002; Richter et al. 2002). The reason lies in the operating potential window of the enzymes. Although the $E^{0\prime}$ of the nitrite/nitric oxide and NH_2OH/ammonium couples are of high potential and, therefore, accessible by both enzymes, the $E^{0\prime}$ NO/NH_2OH couple is approximately $-100\,mV$. The cytochrome cd_1 enzyme operates at some $300\,mV$ higher than this and, therefore, cannot access this couple. However, NrfA needs to be able to access this couple and, therefore, is tuned to operate in this low potential domain. This also has the bioenergetic consequence that NrfA loses a coupling site compared with cytochrome cd_1 because it is never found as a cytochrome bc_1 complex-dependent enzyme in any bacterium, presumably because it is thermodynamically unable to operate at sufficiently high potentials to be able to couple to this protonmotive Q-cycle complex.

Why might NrfA "sacrifice" a coupling site? Nitric oxide is a potent cytotoxin and, therefore, to prevent autocytotoxicity, it makes some biochemical sense to tightly bind nitric oxide when it forms as part of nitrite reduction and further reduce it to ammonium rather than release it. In some cases, there is also evidence to suggest this respiratory system may serve a role in detoxifying exogenous nitric oxide. Cytotoxic nitric oxide is part of the innate response of hosts to the infective invasion of pathogens, for example, through production from arginine by the inducible nitric oxide synthase. In addition to enzymatic routes for nitric oxide synthesis, there are also chemical routes, for example, nitrite in swallowed saliva is reduced to nitric oxide by the intragastric acidity, and this has been suggested to be important for the intragastric clearance of ingested microorganisms. To balance the usefulness and toxicity of this ubiquitous molecule, many microorganisms have mechanisms for both its metabolism and detoxification. Enteric food-borne pathogens, such as *E. coli*, will encounter nitric oxide in a wide range of the oxic, micro-oxic, and anoxic environments that they encounter both within and outside of their animal hosts, and have a number of enzymes implicated in its removal, which include the cytoplasmic NADH-dependent enzymes flavohemoglobin and flavorubredoxin and, notably in the context of this chapter, the cytochrome c nitrite reductase, NrfA (Fig. 4.2). The role of NrfA in nitric oxide detoxification emerged with the observation that an *E. coli nrf* strain has much greater sensitivity to nitric oxide than the parent strain (Poock et al. 2002). The *nrf* genes encode a multiprotein complex, for which a role in coupling quinol oxidation to nitrite reduction during anoxic or micro-oxic growth in the presence of nitrate and nitrite is well established. In *E. coli*, quinol oxidation by NrfD is followed by electron transfer, via NrfC and NrfB, to NrfA, a pentaheme-containing cytochrome c nitrite reductase. NrfA homologs are present in a wide range of bacteria, and in vitro studies have shown that their substrates include nitric oxide and hydroxylamine in addition to nitrite. This led to the proposal that direct nitric oxide reduction by NrfA, rather than nonspecific reduction by another Nrf component, is responsible for providing *E. coli* with resistance toward nitric oxide under anaerobic conditions. The nonprotonmotive nature of the MQH_2–NrfA electron transport system means that this will not be subject to thermodynamic backpressure of the protonmotive force, and this may be beneficial for a detoxification system.

In contrast to NrfA, the consequence of the *P. denitrificans* cytochrome cd_1 nitrite reductase releasing nitric oxide is that the denitrification process then produces the potent nitric oxide cytotoxin as a free intermediate (Fig 4.5). Therefore, it has to have a "built-in" detoxification system, the nitric oxide reductase (Nor), that catalyzes the reaction $2NO + 2e^- + 2H^+ \rightarrow N_2O + H_2O$ (Watmough et al. 1999). It is an integral membrane protein, and analysis of its primary structure has established that it is a relative, and perhaps progenitor, of the cytochrome aa_3 oxidase that is widespread in bacteria and also present in human mitochondria (Figs. 4.1 and 4.5). There are multiple Nor branches. In one group, which includes the enzyme of *P. denitrificans*, the catalytic subunit, NorB, is predicted to comprise 12 transmembrane helices and form a complex with a membrane-anchored monoheme c-type cytochrome (NorC). This cytochrome mediates electron transfer between the protonmotive cytochrome bc_1 complex and NorB, and these types of Nor are known as cNOR (Fig. 4.6). In a second class of Nor, the NorC protein is not present and the NorB protein has a C-terminal extension which folds to give two extra transmembrane helices. This class of Nor is a quinol-oxidizing enzyme and is known as the qNOR class and is often found in pathogenic bacteria, for example, in *Neisseria meningitides* (Rock et al. 2007). Thus, similar to oxidases (e.g., Fig. 4.5), there are cytochrome c-dependent and quinol-dependent Nor systems. This would lead to a cytochrome bc_1 complex-independent route for electron transport and, thus, the cNORs and qNORs would have different coupling efficiencies, resembling the situation described earlier for the membrane-bound and periplasmic nitrate reductases (section 4.2.1). There will then be a small price to pay in energy coupling during denitrification, with the $q^+/10e$ ratio being 28 and 8 with NADH and succinate as electron donor, compared with 30 and 10 when the cyt bc_1-linked NorC-dependent system operates in *P. denitrificans* (van Spanning et al. 2006). Crucially, the main function of a Nor systems is probably to ensure detoxification of nitric oxide. Thus, Nor mutants of *P. denitrificans* fail to grow anaerobically because the nitric oxide that accumulates kills the culture (Butland et al. 2001). In the light of this, it is pertinent to note that Nor enzymes are present in many important pathogens, including those that do not denitrify, where it may play an important role in the detoxification of nitric oxide produced by the hosts' innate immune systems (Rock et al. 2007) and also in many soil bacteria and phototrophs, for example, *Rhodobacter capsulatus* (Richardson 2000), in which it may equip the bacterium with a detoxification system to protect against nitric oxide produced by other organisms in the environment.

4.3.3 Aerobic Nitrate Respiration

P. denitrificans, similar to *E. coli,* can synthesize both the membrane bound (Nar) and periplasmic (Nap) types of nitrate reductase (Berks et al. 1995; Richardson 2000). The Nar system is predominantly expressed under anaerobic denitrifying growth conditions, whereas the Nap system is predominantly expressed under aerobic growth conditions. Consideration of the bioenergetic properties of each system offers a physiological rationale. In the case of Nar, we have already seen that when quinol is oxidized at the periplasmic face of the cytoplasmic membrane by the NarI subunit, protons are ejected into the periplasm whereas the electrons flow back across the membrane, such that the enzyme is electrogenic (see Sect. 4.2.2.1). We have also seen that ,in the case of Nap, quinol is also oxidized at the periplasmic face of the cytoplasmic membrane by NapC, but the electrons also flow into the periplasm and, thus, the free energy in the QH_2/NO_3^- redox couple is not conserved as Δp and is, therefore, dissipated. In *E. coli* Nap, we have seen that Nap is expressed anaerobically and operates in nitrate-limited anaerobic respiration. Similarly, there are denitrifying bacteria, for example, *Bradyrhizobium japonicum* (van Spanning et al. 2006), in which Nap catalyzes the first step in denitrification. When considered in isolation, the energy coupling of Nar and Nap seem markedly different: $q^+/2e^- = 6$ (Nar) and 4 (Nap) with NADH as the electron donor and 2 (Nar) and 0 (Nap) with succinate as the electron donor. When considered in the context of the entire denitrification pathway, we have already seen that the total number of positive charges translocated for denitrification of $2NO_3^-$ to N_2 with NADH as the reductant is 30 for a total of 10 electrons transferred when Nar catalyzes the first step (see Sect. 4.3.2). This falls to 26 when Nap catalyzes the first step and, thus, the energetic loss of using Nap rather than Nar is only 13%. The difference is, however, much more acute when succinate is the electron donor. We have seen already that the succinate to UQ electron transfer is nonprotonmotive in contrast to the protonmotive reduction by the proton-pumping NADH dehydrogenase. Thus, the reduction $2NO_3^-$ to N_2 with succinate as the reductant results in 10 positive charges translocated across the membrane, for a total of 10 electrons transferred when Nar catalyzes the first step and only 6 positive charges translocated when Nap catalyzes the first step, representing an energetic loss of 40% (van Spanning et al. 2007).

However, because, in *P. denitrificans*, *nap* is expressed aerobically and not anaerobically, then, unlike in *E. coli* and *B. japonicum*, it must have a role in aerobic rather than anaerobic respiratory metabolism. Under aerobic conditions, two electrons proceeding from the quinol pool, via the cytochrome bc_1 complex, to the cytochrome aa_3 oxidase can be coupled to the translocation of six positive charges across the membrane (Fig. 4.5b), whereas there will be no charge translocation when two electrons proceed to Nap. Therefore, any diversion of electrons from oxygen to nitrate results in energy dissipation. What then is the role for an aerobically active, non-oxygen consuming, respiratory electron transport chain that dissipates, rather than conserves, the free energy potentially available between UQH_2 and the terminal electron acceptor?

Nap is synthesized at the highest levels during aerobic chemoheterotrophic growth on highly reduced carbon substrates. If oxidation of this substrate to the level at which it can be assimilated results in the release of more reductant than is needed for the generation of the ATP required for the metabolism of the carbon, then a means of disposing of the excess reductant must be available (Richardson 2000). In principle, excess reductant can be removed via respiratory electron-transport pathways, but efficiently coupled pathways, such as the cytochrome bc_1 complex and cytochrome aa_3 oxidase-dependent pathways (Fig. 4.5) will not turn over at high rates in the presence of a large Δp. Furthermore, we saw earlier that the turnover of this complex is dependent on the redox state of the Q/QH_2 pool (section 4.3.1). Cellular over reduction will ultimately result in over reduction of the Q/QH_2 pool as a consequence of the high NADH/NAD⁺ ratio driving turnover of the Q-dependent NADH dehydrogenase. The resulting increase in the QH_2/Q ratio will then limit turnover of the cytochrome bc_1 complex because of restricted availability of Q as a substrate for the cytoplasmic Q-reductase (Fig. 4.5). Consequently, the maximum rate of growth substrate use is only possible if electrons are disposed of by relatively uncoupled cytochrome bc_1 complex-independent pathways, such as those provided by Nap.

The redox components that comprise the Nap electron-transport system have rather low redox potentials and the system depends only on QH_2. Thus, electron flow through the Nap system may be slow when the QH_2/Q ratio is low and protonmotive cytochrome bc_1-dependent electron transport to the proton pumping oxidase is favored, with six positive charges being moved across the membrane per two electrons flowing through this route. However, at high QH_2/Q ratios, turnover of the cytochrome bc_1 complex module becomes limited by Q availability, and there is a stronger thermodynamic driving force for electron transport through the poorly coupled Nap system, in which no positive charges pass across the membrane when the two electrons are used for this "aerobic" nitrate respiration. Re-oxidation of QH_2 via Nap will serve to repoise the Q pool, lowering the QH_2/Q ratio so that electron flux switches back to the more highly coupled cytochrome bc_1-dependent oxidase systems. Consequently, only a small fraction of the total electron flux through the respiratory chain need actually pass through the Nap pathway to maintain the poise of the electron transport network and maximize flux through the more highly coupled pathways. This has been illustrated in continuous cultures studies of *P. denitrificans* in malate- or butyrate-limited chemostat cultures. Butyrate is a highly reduced carbon substrate, the anabolism of which generates excess reductant. Expression of *nap* is at its highest during butyrate metabolism (Ellington et al. 2002), and the rate of aerobic nitrate respiration by Nap in aerobic butyrate-limited chemostat cultures is some 20% of the rate anaerobic nitrate respiration by Nar in aerobic butyrate-limited cultures (Sears et. al. 1997). In contrast, during metabolism of substrates that are more oxidized, such as malate of succinate, *nap* expression is very low (Ellington et al. 2002), and, accordingly, in aerobic malate-limited cultures, that rate of aerobic nitrate respiration is only approximately 1% compared with that of anaerobic malate-limited cultures (Sears et al. 1997).

4.3.4   C-1 Metabolism and Reverse Electron Transport

In addition to catabolising complex organic compounds, *P. denitrificans* is able to grow on the C-1 compounds methanol and methylamine (Baker et al. 1998). These substrates are oxidized, by the PQQ and tryptophan tryptophyl quinone enzymes methanol dehydrogenase and methylamine dehydrogenase, to formaldehyde. Electrons from the oxidation reaction are donated to the electron transport chain, via copper containing proteins, at the level of the periplasmic cytochrome *c* pool (Fig. 6) and then move to the respiratory electron acceptors e.g. via the cytochrome aa_3 oxidase or cbb_3 oxidase. These are proton-translocating and so generate the protonmotive force needed for growth. The formaldehyde generated is subsequently oxidized, via formate, to carbon dioxide. This reaction directly yields cytoplasmic NADH that can be used in the reductive reactions of CO_2 fixation into organic carbon. In principle NADH can also be generated by bacteria by a process known as reverse electron transport. Here there is electron bifurcation from an electron donor either in the periplasmic cytochrome *c* module or quinol pool module such that, in addition to proceeding to terminal oxidases via proton motive electron transfer complexes, some electrons move back through a cytochrome bc_1 complex – Q pool - NADH dehydrogenase network in an 'uphill' fashion in which electron transfer is driven by, and so consumes, the protonmotive force. This is important in respiratory autotrophic metabolism in which some bacteria can fix carbon dioxide into organic carbon using NADH generated from such reverse electron transfer from simple inorganic electron donors. These include the globally important nitrifying bacteria that oxidize ammonia to nitrite at the periplasmic side of the membrane, via a quinol dependent ammonium mono-oxygenase and a cytochrome *c* dependent hydroxylamine oxidoreductase. Some of the electrons from this process are used to reduce oxygen to water via the protonmotive cytochrome bc_1 complex and cytochrome aa_3 oxidase (Ferguson et al. 2006). Some are used in the ammonia monoxygenase reaction, but around 5% of the electrons extracted are driven uphill to reduce NAD^+ using the protonmotive force generated from the electrons that were transferred to oxygen. In the case of *P. denitrificans* examples of reverse electron transfer occur when hydrogen or thiosulfate are used as electron donors (energy sources) for autotrophic growth (Baker et al. 1998), providing further examples of its respiratory flexibility.

4.4 Respiratory Electron Transport Across Two Membranes in *Shewanella*

Some of the most abundant respiratory substrates in many of Earth's anoxic environments are insoluble minerals, particularly those of Fe(III), such as hematite and goethite. Fe(III) mineral oxide reduction is one of the most widespread respiratory processes in anoxic zones and has wide environmental significance, influencing several biogeochemical cycles (Lovley 1997; Lovley and Coates 1997). Unlike the other terminal electron acceptors discussed thus far in this chapter (e.g., oxygen and nitrate), mineral oxides cannot freely diffuse into cells. If a Gram-negative bacterium is to be able to use Fe(III) or Mn(IV) minerals as respiratory electron acceptors, it faces the problem of moving electrons generated from cellular metabolism across two cell membranes and the intervening periplasm to the site of reduction of the insoluble extracellular species. The free energy released in this process must also be conserved as a proton-electrochemical gradient across the inner membrane if the process is to be competent in supporting growth and maintenance of the organism. In the case of the members of the Fe(III)-respiring genus *Shewanella*, it is emerging that this problem may be solved by using a number of tetraheme and decaheme *c*-type cytochromes to form a multiheme electron "nanowire" between the inner and the outer membranes (Fig. 4.7).

4.4.1 The Shewanella Cytochrome C "ome"

Shewanella species are renowned for their incredible respiratory versatility and are able to use more than 20 terminal electron acceptors for respiration. The *Shewanella oneidensis* MR-1 genome encodes up to 42 putative *c*-type cytochromes (Heidelberg et al. 2002), many of which are predicted to be multiheme cytochromes because of the presence of multiple CXXCH motifs within their primary sequences. Heme

A

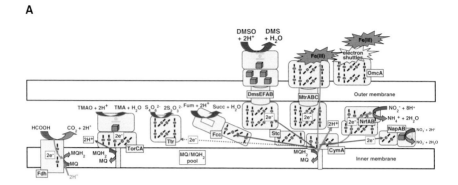

B

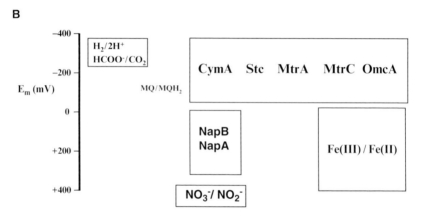

Fig. 4.7 Electron transport at two membranes in *Shewanella*. **a** A scheme for electron transport to the outside of the cell. **b** The redox potentials of the electron transport components

groups are covalently attached to the cysteines in the mature protein, and the histidine becomes a ligand to the heme iron. These 42 multiheme *c*-type cytochromes form a cytochrome c "ome" that has attracted a great deal of interest recently, and the functions and structures of the number of the cytochromes have emerged. These include the tetraheme quinol dehydrogenase CymA (Myers and Myers 1992; Schwalb et al. 2003); the "metal respiration" system OmcA–MtrCAB, which includes two outer membrane decaheme cytochromes (OmcA and MtrC), a periplasmic decaheme cytochrome (MtrA), and a putative outer membrane β-barrel protein (MtrB) (Pitts et al. 2003; Myers and Myers 2001; Beliaev et al. 2001; Richardson 2000; Beliaev and Saffarini 1998); tetraheme cytochrome domains of the structurally defined flavocytochrome *c* Frds (Fcc) (Bamford et al. 1999; Leys et al. 1999; Taylor et al. 1999); structurally defined octaheme tetrathionate reductases (Mowat et al. 2004); multiheme-dependent TMAO reductases (Tor) (Czjzek et al. 1998); multiheme nitrite reductases of the Nrf type described above (section 4.2.1.3) for

E. coli; and a number of small periplasmic cytochromes, such as the structurally defined small tetraheme cytochrome *c* (Stc) (Leys et al. 2002; Gordon et al. 2000) (Fig. 4.7). All of these multiheme cytochromes have in common that, when characterized, they operate in a rather low potential domain (0 to −350 mV). The large number of *c*-type cytochromes produced by *Shewanella* species and their extensive respiratory flexibility has led to the genome sequence of approximately 10 different species emerging.

It is notable that, in many *Shewanella* species, the respirome is distributed on both the inner and outer membranes. The outer membrane cytochromes OmcA and MtrC are thought to be the terminal contact point of the "multiheme electron conducting wire" connecting the bacterial inner membrane to the insoluble terminal electron-accepting minerals of Fe(III) and Mn(IV). This view is supported by the experimental observation that mutation of OmcA or MtrC has no effect on the ability of *S. oneidensis* MR-1 to reduce soluble respiratory substrates such as nitrate, nitrite, or anthraquinone-2,6-disulfonate, however, a Δ*mtrC* strain of *S. oneidensis* MR-1 had a decreased ability to reduce insoluble Fe(III)-oxides (Belieav et al. 2001).

When we discussed the *E. coli* respiratory system, we paused to consider electron wires made up of iron sulfur clusters (see Sect. 4.2.1.2). The *Shewanella* respiratory chains provide examples of electrons being moved long distance through electron wires made up of hemes. At present, the structure of the outer membrane decaheme cytochromes is not known; however, primary structure analysis suggests that it, and periplasmic decaheme MtrA, comprise two approximately 150-amino acid pentaheme modules. These modules may be structurally organized in a similar way to the *E. coli* pentaheme NrfB protein that is involved in electron transfer to the nitrite reductase NrfA, and for which a structure-based model for a 20-heme NrfAB complex has been proposed (Fig. 4.4) (Clarke et al. 2007; Bamford et al. 2002). The NrfB protein is essentially a small electron wire with a 4-nm chain of five hemes (Fig 4.4b). A heme group is covalently attached to an approximately 150-amino acid polypeptide chain approximately every 20 to 25 amino acids via thiol ester linkages to the cysteines of five classical CXXCH attachment sites (Fig. 4.4b), so that all five of the minimum heme distances between neighboring hemes is greater than 0.6 nm. All of the hemes are sufficiently solvent exposed to introduce the possibility of electron exchange between each heme and an external electron acceptor/donor (Clarke et al. 2007). Packing many similar such multiheme nanowires together in the periplasm or on the cell surface could, in principle, provide for rapid electron transport across the periplasm and multiple sites of electron transfer to an amorphous substrate, such as mineral iron.

From a physiological and thermodynamic view, the contribution from individual hemes of the multiheme cytochromes to the redox activity is perhaps less important to consider than the observation that the multiheme chain as a whole operates in a low redox potential domain. However, the low redox potentials of the Fe(III) respiration system has bioenergetic consequences that limit the number of protonmotive coupling sites that can be associated with it. Depending on growth conditions, electrons could enter the MQH_2 pool via the activity of primary dehydrogenases, such as the proton-pumping NADH dehydrogenase, or electrogenic formate dehydrogenase or hydrogenase. The redox potential of the NAD^+/NADH, CO_2/$HCOO^-$, or $2H^+$/H_2 couples in the cell will lie in the −300 mV to −400 mV domain and that of MQ/MQH_2 will be approximately −50 mV to −100 mV. This thermodynamic energy gap (~300 mV) is sufficient to allow for generation of Δp (~200 mV) through coupling electron transport from donor to acceptor to either proton translocation or an electrogenic redox loop, as already described for Fdh-N in Sect. 4.2.1.1 (Fig. 4.3). However, there is not such strong thermodynamic force of free energy change available for driving electron transfer from the quinol pool through the periplasmic and outer membrane low redox potential heme pool of CymA, Stc, MtrA, OmcA, and MtrC in a protonmotive manner. Mechanistically, electrons need to be drawn through this system by a strong oxidant; for example, a Fe(III) mineral oxide. There is no Δp coupling site in this part of the electron transport process, thus, no matter how oxidizing the Fe(III) species that terminates the electron transport chain at the outside of the cell, the free energy in the MQH_2/Fe(III) couple will be dissipated. Thus, Fe(III) respiration is constrained to only be coupled to Δp generation at the level of electron input into the MQ-pool (Fig. 4.7). This makes it bioenergetically equivalent to the bacterial periplasmic nitrate reductase and nitrite systems of *E. coli* described in Sect. 4.2, that are also present in many *Shewanella* species (Fig. 4.7). It is notable that the MtrC outer membrane decaheme cytochrome may provide for a range of extracellular respiratory processes; in addition to Fe(III) oxide and Mn(IV) oxide respiration, it has also been implicated in the respiration-linked extracellular formation of uranium (IV) oxide species when using soluble uranium (VI) as a respiratory substrate. This process may have applications in removal of the radionuclide from contaminated waters (Marshall et al. 2006).

The respiratory flexibility of *Shewanella* species also includes the ability to use DMSO as a respiratory electron acceptor. In *E. coli*, the active site of the DMSO reductase, a Mo-*bis*-MGD enzyme, is located in the periplasmic compartment (Fig. 4.2). There is no coupling site between the quinol pool and the terminal reductase site, and, therefore, the DMSO respiring system can only be coupled at the level of electron input into the Q-pool, similar to the periplasmic nitrate and nitrite reductase systems (Fig. 4.2). In *Shewanella* species, genes encoding homologs of the DMSO reductase catalytic subunit cluster with genes that encode homologs of the decaheme cytochromes associated with mineral iron oxide respiration. Rather surprisingly, the catalytic subunit has been localized to the outside of the cell and, therefore, it seems that Shewanellae are configured to respire insoluble forms of DMSO that are abundant in oceans and they must use similar electron transfer proteins to get electrons to the outside of the cell as those used for Fe(III) and Mn(IV) respiration (Gralnick et al. 2006). Despite the different location of the active sites, the energy conservation associated with electron transfer to periplasmic DMSO reductase and extracellular DMSO reductases is predicted to be the same, with protonmotive steps being at the level of electron input into the Q-pool in both cases.

4.4.2 Bioinformatics on Bacterial Respiromes—"XX" a Cautionary Tale

In the multiheme c-type cytochromes involved in Fe(III) respiration (Fig. 4.1), the heme cofactor is covalently attached via two cysteine residues organized in a heme c binding motif. Bioinformatics with such c-type cytochromes, including those of *Shewanella* species, relies on the conventional wisdom that if you count the number of CXXCH motifs you can determine the number of bound hemes without having to purify the protein. However, many *Shewanella* species have a gene that encodes novel multiheme cytochrome, MccA, that contains seven conventional heme c binding motifs (CXXCH), but additionally has several single cysteine residues and a conserved CH signature. Mass spectrometric analysis of purified MccA from *Wolinella succinogenes* suggests that two of the single cysteine residues are actually part of an unprecedented CX15CH sequence involved in heme c binding. Thus, the MccA binds eight not seven hemes. A heme lyase gene that encodes a protein involved in attaching heme to the polypeptide chain was found to be specifically required for the maturation of MccA in *W. succinogenes*. Heme lyase-encoding genes are also present in the vicinity of the *mccA* genes of *Shewanella* species, and are upregulated when thiosulphate is being used as an electron donor (Beliaev et al. 2005), suggesting a dedicated cytochrome c maturation pathway for a multiheme system involved in some way in thiosulfate respiration. The results necessitate reconsideration of computer-based prediction of putative heme c binding motifs in bacterial proteomes, and highlights the need for careful biochemical analysis of the other *Shewanella* multiheme cytochromes of the sort described for MtrC that do have conserved Cys residues other than those of CXXCH motifs, to be certain of the cofactor content (Hartshorne et al. 2007).

4.5 Electron Transport Across One Membrane— Gram-Positive Bacteria

Thus far, this chapter has focused on Gram-negative bacteria and we have considered electron transfer on the cytoplasmic and periplasmic faces of the inner membrane, electron transfer across the periplasm, and electron transfer on the outer face of the outer membrane. In closing, we will turn briefly to consider Gram-positive bacteria. By definition, Gram-positive bacteria differ from Gram-negative bacteria in having a cytoplasmic membrane, but not an outer membrane or *bi*-membrane-confined periplasmic compartment. Despite having a fundamentally different cellular structure, Gram-positive respiratory systems follow the basic bioenergetic rules already described for the Gram-negative bacteria, and many of the respiratory systems described for Gram-negative bacteria have homologous counterparts in Gram-positive bacteria. Thus, for example, there are Gram-positive denitrifying bacteria, that, similar to *P. denitrificans*, reduce nitrate to dinitrogen, such as

Bacillus azotoformans (Suharti and DeVries 2005), in which the complete denitrification electron transport system is tightly associated with the cytoplasmic membrane. Gram-positive bacteria are very important in many environments, and space does not permit an exhaustive review of them but we will briefly consider respiratory flexibility in actinomycetes that include soil streptomycetes and in pathogenic mycobacteria.

Dealing first with mycobacteria, the pathogen *Mycobacterium tuberculosis* can grow in its human host and also persist in a nongrowing latent state. In terms of respiration, *M. tuberculosis* will experience a range of different environments during the infective process and must be equipped to adapt to these environments. We saw this also with enteric pathogens, such as *E. coli* (see Sect. 4.2). In many ways, however, the organization of the mycobacterial respiratory electron transport has modular properties that resemble those of *P. denitrificans* (see Sect. 4.3). It has a branched aerobic respiratory chain in which electrons can flow from either proton-pumping NADH dehydrogenase (Ndh1), nonproton-pumping NADH dehydrogenase (Ndh2), or Sdh complexes into a MQ/MQH_2 pool. From there, they can proceed directly to an MQH_2-dependent cyt *bd* or, via a cytochrome bc_1 complex, to a cytochrome aa_3 oxidase. This branching at the level of the Q-pool to cyt bc_1 complex-dependent and quinol-dependent oxidases is reminiscent of that already described for *P. denitrificans* (see Sect. 4.3). Interestingly, however, there is a suggestion that cytochrome bc_1 and cytochrome aa_3 oxidase form a supercomplex in mycobacteria that negates the need for an equivalent of a periplasmic water-soluble electron carrier, such as the cytochrome c_{550} or pseudoazurin present in *P. denitrificans*, and this may reflect the absence of a true periplasm in mycobacteria, so that electron transport is tightly associated with the membrane (Magehee et al. 2006). Respiratory super complexes may also emerge to be widespread in a number of species of Gram-positive bacteria.

Similar to *E. coli*, the respiratory branch of mycobacteria that terminates in the *bd*-type oxidase has been shown to be important for microaerobic respiration. Thus, mycobacteria are equipped with highly coupled low-affinity oxidase systems for life at high oxygen tensions and less well-coupled high-affinity oxidase systems for life in low oxygen-tension environments. In addition, mycobacteria also have a membrane-bound MQH_2-dependent nitrate reductase with an active site in the cytoplasm, similar to the protonmotive Nar enzymes of both *E. coli* and *P. denitrificans*, which equips them for survival in the absence of oxygen (Matsoso et al. 2005; Wayne et al. 2001). Interestingly, *Mycobacterium bovis* Bacille Calmette-Guèrin (BCG) mutants deficient in nitrate reductase activity are compromised for virulence in immunodeficient mice (Weber et al. 2000). Gene expression studies during mouse lung infection has provided evidence for the transition from highly coupled (almost mitochondrial-like; Fig. 4.1) respiratory electron transfer (Ndh1)-MQ–cytochrome bc_1–cytochrome aa_3; $q^+/2e^- = 10$) to poorly coupled aerobic (Ndh2–MQ–cyt *bd*; $q^+/2e^- = 2$) or anaerobic (Ndh2–MQ–Nar; $q^+/2e^- = 2$) systems during transition from acute infection (associated with exponential bacterial growth) to chronic infection (associated with cessation of growth) (Shi et al. 2005).

Turning to consider the streptomycetes, soil rhizosphere bacteria that can form dense mycelial structures, a similar respiratory flexibility to the closely related mycobacteria is apparent. For example, the important antibiotic-producing organism, *Streptomyces coelicolor* has the capacity for oxygen respiration, but additionally has the genetic information for three isozymes of the protonmotive membrane-bound Nar type of nitrate reductase (van Keulen et al. 2005). Conditions have not yet been established whereby these nitrate reductases can sustain growth, but they may be important for sustaining metabolism in anoxic environments. Similar to the mycobacteria, *S. coelicolor* also has high- and low-affinity oxidase systems. Transition to the high-affinity cyt *bd* system is regulated in response to the metabolic state of the cell through a DNA binding protein called Rex that is a sensor of the redox state of the NADH/NAD$^+$ and that is widespread among Gram-positive bacteria, including *Bacillus*, *Staphylococcus*, *Listeria*, and *Streptococcus* pathogens that also exhibit various degrees of respiratory flexibility (Brekasis and Paget 2003).

Finally, one group of Gram-positive bacteria that were for a long time not thought to exhibit respiratory flexibility, or indeed any respiration at all, are the lactic acid bacteria (LAB). In this group of organisms, energy conservation was long thought to be confined to fermentation of sugars, and LAB were thought to be incapable of respiration, not least because they did not have the complete biosynthetic machinery to make heme that is central to aerobic respiration. However, some LAB, such as *Lactococcus lactis*, have a complete menaquinone biosynthetic machinery and the genes encoding a cyt *bd* respiratory oxidase (Wegmann et al. 2007). Thus, it seems that *L. lactis* can assemble a functional protonmotive aerobic respiratory system if heme is present in the growth medium (Duwat et al. 2001), and, in light of the genome sequences available, we can infer this to be most likely a NADH–MQ–cyt *bd* electron transfer chain. There is no evidence for a protonmotive NADH dehydrogenase in the *L. lactis* genome sequence and, thus, the protonmotive step will most likely be at the level of electron output from the MQH_2 pool through the protonmotive cyt *bd*, which, by analogy to *E. coli* cyt *bd*, may have a q$^+$/e$^-$ ratio of 1. Thus, this will not be a highly coupled pathway, but, nevertheless, its operation does seem to increase growth yield (Duwat et al. 2001). Although the LAB respiratory system can hardly be called flexible, it does illustrate that even in very well-characterized industrially important bacteria, the genome sequences helped to uncover hitherto unexpected respiratory functions.

4.6   Future Perspectives

In this chapter, we have reviewed the topological and associated bioenergetic organization of a range of respiratory systems that illustrate principles of respiratory energy conservation and respiratory flexibility that are widespread among bacteria. This respiratory diversity has shaped many of Earth's environments, and its importance is reflected by a current interest in understanding the implications of respiratory flexibility for biotechnology and medical microbiology. Our mechanistic understanding of key enzymes involved in bacterial respiration at a molecular level has increased exponentially during the last 10 years, and the complete structural definition of a model bacterial respirome, such as that of *E. coli*, can be seen as achievable target for the next 10 years. Mapping what we know regarding the organization and regulation of bacterial respiromes from well-characterized organisms onto less well-characterized organisms using bioinformatics approaches is also proving informative, and a better understanding of the operation of well-characterized branched respiratory systems, such as that of *P. denitrificans*, through collaborations in systems biology projects between mathematicians and biochemists, should prove fruitful for a wider understanding among less well-characterized organisms. The importance of studying respiratory flexibility is perhaps nowhere better illustrated than in pathogens that cause globally important diseases. Almost all bacterial pathogens present a degree of respiratory flexibility. In many cases, this has not been studied biochemically, but is apparent from bioinformatics analysis of genome sequences. However, from the limited studies available, it is clear that studying the nature, degree, and regulation of this respiratory flexibility in pathogens will make a significant contribution to understanding the survival of these bacteria both inside and outside of the host organism.

Acknowledgments I thank the UK BBSRC and US DOE for supporting my own research work covered in this chapter. I also thank Tom Clarke and Seth Hartshorne for their contributions to the text and figures. I am very grateful to Lars Bakken, Nils Blüthgen, Mark Blyth, Vince Moulton, Stuart Ferguson, Hans Westerhof and Rob van Spanning for many fruitful discussions on the developing field of Paracoccus denitrificans systems biology. Finally, thanks to Andrea, Rosemary, and Alexander.

Highly Recommended Readings

Baker SC, Ferguson SJ, Ludwig B, Page MD, Richter OM, van Spanning RJ. (1998) Molecular genetics of the genus *Paracoccus*: metabolically versatile bacteria with bioenergetic flexibility. Microbiol Mol Biol Rev 62:1046–1078
Ferguson SJ, Richardson DJ (2004) Electron transport systems for reduction of nitrogen oxides and nitrogen oxyanions. Davide Zannoni (Ed.) Respiration in Archaea & Bacteria. Vol. 2: Diversity of Prokaryotic Respiratory Systems, (2004) pp. 169–206 Kluwer Academic Publishers, The Netherlands
Lovley DR (1997) Microbial Fe(III) reduction in subsurface environments. FEMS Microbiol Rev 20:305–313
Nicholls DG, Ferguson SJ (2002) Bioenergetics 3. London: Academic Press
Potter L, Angove H, Richardson DJ, Cole JA (2001) Nitrate reduction in the periplasm of Gram negative bacteria. Adv Microbiol 45:52–102
Richardson DJ (2000) Bacterial respiration: a flexible process for a changing environment. Microbiol 146:551–571
van Spanning RJ, Richardson DJ, Ferguson SJ (2006) The regulation of denitrification. In "The Nitrogen Cycle" Elsevier Press
Zumft WG (1997) Cell biology and molecular basis of denitrification. Microbiol Mol Biol Rev 61:533–616

References

Abramson J, Riistama S, Larsson G, Jasaitis A, Svensson-Ek M, Laakkonen L, Puustinen A, Iwata S, Wikstrom M (2000) The structure of the ubiquinol oxidase from *Escherichia coli* and its ubiquinone binding site. Nat Struct Biol 7:910–917

Angrove H, Cole JA, Richardson DJ, Butt JN (2002) Protein film voltammetry reveals distinctive fingerprints of nitrite and hydroxylamine reduction by a cytochrome c nitrite reductase. J Biol Chem 277:23374–23381

Baker SC, Ferguson SJ, Ludwig B, Page MD, Richter OM, van Spanning RJ (1998) Molecular genetics of the genus *Paracoccus*: metabolically versatile bacteria with bioenergetic flexibility. Microbiol Mol Biol Rev 62:1046–1078

Beliaev AS, Klingeman DM, Klappenbach JA, Wu L, Romine MF, Tiedje JA, Nealson KH, Fredrickson JK, Zhou J (2005) Global transcriptome analysis of *Shewanella oneidensis* MR-1 exposed to different terminal electron acceptors. J Bacteriol 187:7138–7145

Bamford V, Dobbin PS, Richardson DJ, Hemmings AM (1999) The high resolution X-ray structure of the open form of a flavocytochrome c_3. Nat Struct Biol 6:1104–1107

Bamford V, Angrove H, Seward H, Butt JN, Cole JA, Thomson AJ, Hemmings AH, Richardson DJ (2002) The structure and spectroscopy of the cytochrome c nitrite reductase of *Escherichia coli*. Biochem 41:2921–2931

Beliaev AS, Saffarini DA, McLaughlin JL, Hunnicutt D (2001) MtrC, an outer membrane decahaem c cytochrome required for metal reduction in Shewanella putrefaciens MR-1. Mol Microbiol 39:722–730

Beliaev AS, Saffarini DA (1998) *Shewanella putrefaciens* mtrB encodes an outer membrane protein required for Fe(III) and Mn(IV) reduction. J Bacteriol 180:6292–6297

Berks BC, Ferguson SJ, Moir JW, Richardson DJ (1995). Enzymes and associated electron transport systems that catalyse the respiratory reduction of nitrogen oxides and oxyanions. Biochim Biophys Acta 1232:97–173

Bertero MG, Rothery RA, Palak M, Hou C, Lim D, Blasco F, Weiner JH, Strynadka NC (2003) Insights into the respiratory electron transfer pathway from the structure of nitrate reductase A. Nat Struct Biol 10:681–687

Blattner FR, Plunkett G III, Bloch CA, et al. (1997). The complete genome sequence of *Escherichia coli* K-12. Science 277:1453–1474

Brekasis D, Paget MS (2003) A novel sensor of NADH/NAD+ redox poise in *Streptomyces coelicolor* A3. EMBO J. 22:4856–4865

Butland G. Spiro S, Watmough N, Richardson DJ (2001) Two conserved glutamates in the bacterial nitric oxide reductase are essential for activity but not assembly of the enzyme. J Bacteriol 183:189–199

Calhoun MW, Oden KL, Gennis RB, Demattos MJT, Neijssel OM (1993) Energetic efficiency of *Escherichia coli*—Effects of mutations in components of the aerobic respiratory-chain J Bacteriol 17:3020–3025

Cheng VW, Rothery RA, Bertero MG, Strynadka NC, Weiner JH (2005) Investigation of the environment surrounding iron-sulfur cluster 4 of *Escherichia coli* dimethylsulfoxide reductase. Biochemistry 44:8068–8077

Clarke TA, Cole JA, Richardson DJ, Hemmings AM (2007) The crystal structure of the pentahaem c-type cytochrome NrfB and characterisation of its solution-state interaction with the pentahaem nitrite reductase. NrfA. Biochem JBiochem J. 406:19–30

Czjzek M, Dos Santos JP, Pommier J, Giordano G, Mejean V, Haser R (1998) Crystal structure of oxidized trimethylamine N-oxide reductase from *Shewanella massilia* at 2.5 A resolution. J Mol Biol 284:435–447

Duwat P, Cochu A, Ehrlich SD, Boyle MD (2001) Respiration capacity of the fermenting bacteria *Lactococcus lactis* and its positive effects on growth and survival. J Bacteriol 179:4473–4479

Einsle O, Messerschmidt A, Huber R, Kroneck PM, Neese F (2002) Mechanism of the six-electron reduction of nitrite to ammonia by cytochrome c nitrite reductase. J Am Chem Soc 124:11737–11745

Ellington MJK, Bhakoo KK, Sawers G, Richardson DJ, Ferguson SJ (2002) Hierarchy of carbon source selection in *Paracoccus pantotrophus*: Strict correlation between reduction state of the carbon substrate and aerobic expression of the *nap* operon. J Bacteriol 184: 4767–4774

Ferguson SJ, Richardson DJ (2004) Electron transport systems for reduction of nitrogen oxides and nitrogen oxyanions. Davide Zannoni (Ed.) Respiration in Archaea & Bacteria. Vol. 2: Diversity of Prokaryotic Respiratory Systems, (2004) pp. 169–206 Kluwer Academic Publishers, The Netherlands

Gordon EHJ, Pike AD, Hill AE, Cuthbertson PM, Chapman SK, Reid GA (2000) Identification and characterization of a novel cytochrome c(3) from Shewanella frigidimarina that is involved in Fe(III) respiration. Biochem J 349:153–158

Gralnick JA, Vali H, Lies DP, Newman DK (2006) Extracellular respiration of dimethyl sulfoxide by *Shewanella oneidensis* strain MR-1. Proc Natl Acad Sci USA 103:4669–4674

Hartshorne RS, Kern M, Meyer B, Clarke TA, Karas M, Richardson DJ, Simon J (2007) A dedicated haem lyase is required for the maturation of a novel bacterial cytochrome c with unconventional covalent haem binding. Mol Microbiol 64:1049–1060

Heidelberg JF, Paulsen IT, Nelson KE, Gaidos EJ, Nelson WC, Read TD, Eisen JA, Seshadri R, Ward N, Methe B, Clayton RA, Meyer T, Tsapin A, Scott J, Beanan M, Brinkac L, Daugherty S, DeBoy RT, Dodson RJ, Durkin AS, Haft DH, Kolonay JF, Madupu R, Peterson JD, Umayam LA, White O, Wolf AM, Vamathevan J, Weidman J, Impraim M, Lee K, Berry K, Lee C, Mueller J, Khouri H, Gill J, Utterback TR, McDonald LA, Feldblyum TV, Smith HO, Venter JC, Nealson KH, Fraser CM (2002) Genome sequence of the dissimilatory metal ion-reducing bacterium *Shewanella oneidensis*. Nat Biotechnol 20:1118–11123

Leys D, Tsapin AS, Nealson KH, Meyer TE, Cusanovich MA, Van Beeumen JJ (1999) Structure and mechanism of the flavocytochrome c fumarate reductase of *Shewanella putrefaciens* MR-1. Nat Struct Biol 6:1113–1117

Leys D, Meyer TE, Tsapin AS, Nealson KH, Cusanovich MA, Van Beeumen JJ (2002) Crystal structures at atomic resolution reveal the novel concept of "electron-harvesting" as a role for the small tetraheme cytochrome c. J Biol Chem 277:35703–25711

Lovley DR (1997) Microbial Fe(III) reduction in subsurface environments. FEMS Microbiol Rev 20:305–313

Lovley DR, Coates JD (1997) Bioremediation of metal contamination. Curr Opin Biotechnol 8:285–289

Myers CR, Myers JM (1992) Localization of cytochromes to the outer membrane of anaerobically grown Shewanella putrefaciens MR-1. J Bacteriol 174:3429–3438

Matsoso LG, Kana BD, Crellin PK, Lea-Smith DJ, Pelosi A, Powell D, Dawes SS, Rubin H, Coppel RL, Mizrahi V (2005) Function of the cytochrome bc_1-aa_3 branch of the respiratory network in mycobacteria and network adaptation occurring in response to its disruption. J Bacteriol 187:6300–6308

Megehee JA, Hosler JP, Lundrigan MD (2006) Evidence for a cytochrome bcc-aa3 interaction in the respiratory chain of *Mycobacterium smegmatis*. Microbiology. 152:823–829

Moser CC, Keske JM, Warncke K, Farid RS, Dutton PL (1992) Nature of biological electron transfer. Nature 355:796–802

Mowat CG, Rothery E, Miles CS, McIver L, Doherty MK, Drewette K, Taylor P, Walkinshaw MD, Chapman SK, Reid GA (2004) Octaheme tetrathionate reductase is a respiratory enzyme with novel heme ligation. Nat Struct Mol Biol 11:1023–1024

Myers JM, Myers CR (2001) Role for outer membrane cytochromes OmcA and OmcB of Shewanella putrefaciens MR-1 in reduction of manganese dioxide. Appl Environ Microbiol 67:260–269

Nicholls DG, Ferguson SJ (2002) Bioenergetics 3. London: Academic Press

Marshall MJ, Beliaev AS, Dohnalkova AC, Kennedy DW, Shi L, Wang Z, Boyanov MI, Lai B, Kemner KM, McLean JS, Reed SB, Culley DE, Bailey VL, Simonson CJ, Saffarini DA, Romine MF, Zachara JM, Fredrickson JK (2006) c-Type cytochrome-dependent formation of U(IV) nanoparticles by *Shewanella oneidensis*. PLoS Biol 2006 4:e268

Otten MF, Stork DM, Reijnders WN, Westerhoff HV, Van Spanning RJ (2001) Regulation of expression of terminal oxidases in *Paracoccus denitrificans*. Eur J Biochem 268:2486–2497

Pitts KE, Dobbin PS, Reyes-Ramirez F, Thomson AJ, Richardson DJ, Seward HE (2003) Characterization of the Shewanella oneidensis MR-1 decaheme cytochrome MtrA: expression in Escherichia coli confers the ability to reduce soluble Fe(III) chelates. J Biol Chem 278:27758–27765

Poock SR, Leach, ER, Moir JWB, Cole JA, Richardson DJ (2002) The respiratory detoxification of nitric oxide by *Escherichia coli*. J Biol Chem 277:23664–23669

Potter LC, Millington P, Griffiths L, Thomas GH, Cole JA. (1999) Competition between *Escherichia coli* strains expressing either a periplasmic or a membrane-bound nitrate reductase: does Nap confer a selective advantage during nitrate-limited growth? Biochem J 344:77–84

Potter L, Angove H, Richardson DJ, Cole JA (2001) Nitrate reduction in the periplasm of Gram negative bacteria. Adv Microbiol 45:52–102

Richardson DJ (2000) Bacterial respiration: a flexible process for a changing environment. Microbiol 146:551–571

Richardson DJ, Sawers G (2002) PMF through the redox loop. Science 295:1842–1843

Richter CD, Allen JW, Higham CW, Koppenhofer A, Zajicek RS, Watmough NJ, Ferguson SJ. (2002) Cytochrome cd_1, reductive activation and kinetic analysis of a multifunctional respiratory enzyme. J Biol Chem 277:3093–3100

Rock JD, Thomson MJ, Read RC, Moir JW (2007). Regulation of denitrification genes in *Neisseria meningitidis* by nitric oxide and the repressor NsrR. J Bacteriol 189:1138–1144

Sargent F (2007) Constructing the wonders of the bacterial world: biosynthesis of complex enzymes. Microbiol 153:633–651

Sazanov LA, Hinchliffe P (2006) Structure of the hydrophilic domain of respiratory complex I from Thermus thermophilus. Science 311:1430–1436

Sears HJ, Spiro S, Richardson DJ (1997) Effect of carbon substrate and aeration on nitrate reduction and expression of the periplasmic and membrane bound nitrate reductases in carbon limited cultures of *Paracoccus denitrificans* Pd1222. Microbiol 143:3767–3774

Shi L, Sohaskey CD, Kana BD, Dawes S, North RJ, Mizrahi V, Gennaro ML (2005) Changes in energy metabolism of *Mycobacterium tuberculosis* in mouse lung and under in vitro conditions affecting aerobic respiration. Proc Natl Acad Sci USA 102:15629–15634

Schwalb C, Chapman SK, Reid GA (2003) The tetraheme cytochrome CymA is required for anaerobic respiration with dimethyl sulfoxide and nitrite in *Shewanella oneidensis*. Biochemistry 42:9491–4947

Southall S, Joel J, Richardson DJ, Oubrie A (2006). Structural and kinetic characterisation of a novel aldose sugar dehydrogenase from *Escherichia coli* (2006) J Biol Chem 281:30650–30659

Suharti, de Vries S (2005) Membrane-bound denitrification in the Gram-positive bacterium *Bacillus azotoformans*. Biochem Soc Trans 33:130–133

Taylor P, Pealing SL, Reid GA, Chapman SK, Walkinshaw MD (1999) Structural and mechanistic mapping of a unique fumarate reductase. Nat Struct Biol 6:1108–1112

van Keulen G, Alderson J, White J, Sawers RG (2005) Nitrate respiration in the actinomycete *Streptomyces coelicolor*. Biochem Soc Trans 33:210–212

van Spanning RJ, Richardson DJ, Ferguson SJ (2006) The regulation of denitrification. In "The Nitrogen Cycle" Elsevier Press

Watmough NJ, Butland G, Cheesman MR, Moir JW, Richardson DJ, Spiro S (1998) Nitric oxide in bacteria: synthesis and consumption. Biochim Biophys Acta 1411:456–474

Wayne LG, Sohaskey CD (2001) Nonreplicating persistence of *Mycobacterium tuberculosis*. Ann Rev Microbiol 55:139–163

Weber I, Fritz C, Ruttkowski S, Kreft A, Bange FC. (2000) Anaerobic nitrate reductase (*narGHJI*) activity of *Mycobacterium bovis* BCG in vitro and its contribution to virulence in immunodeficient mice. Mol Microbiol. 35:1017–1025

Wegmann U, O'Connell-Motherway M, Zomer A, Buist G, Shearman C, Canchaya C, Ventura M, Goesmann A, Gasson MJ, Kuipers OP, van Sinderen D, Kok J. (2007) Complete genome sequence of the prototype lactic acid bacterium *Lactococcus lactis* subsp. cremoris MG1363. J Bacteriol 189:3256–3270

Yankovskaya V, Horsefield R, Tornroth S, Luna-Chavez C, Miyoshi H, Leger C, Byrne B, Cecchini G, Iwata S (2003) Architecture of succinate dehydrogenase and reactive oxygen species generation. Science 299:700–704

Zhang J, Frerman FE, Kim JJ (2006) Structure of electron transfer flavoprotein-ubiquinone oxidoreductase and electron transfer to the mitochondrial ubiquinone pool. Proc Natl Acad Sci USA 103:16212–16217

Zumft WG (1997) Cell biology and molecular basis of denitrification. Microbiol Mol Biol Rev 61:533–616

5
Protein Secretion in Bacterial Cells

Christos Stathopoulos(✉), Yihfen T. Yen, Casey Tsang, and Todd Cameron

Abstract Approximately 20% of the proteins synthesized in a bacterial cell are transported outside of the cytoplasm. These proteins are localized either within the different compartments of the cell envelope or are released into the extracellular growth medium. Bacteria have evolved multiple pathways to secrete proteins across their cell envelopes. In this chapter, we examine similarities and differences among the several types of protein secretion pathways found in bacteria, as well as in archaea. Because the applied aspects of bacterial protein secretion play a vital role in biological and pharmaceutical industries, a more comprehensive understanding of the mechanisms underlying these secretion systems is essential to the advancement of present technologies.

5.1   Introduction

To survive and multiply, bacteria must be able to interact with their external environment. They need to obtain nutrients from their surroundings, communicate with other bacterial cells, and, in the case of pathogenic or symbiotic bacteria,

Christos Stathopoulos
Biological Sciences Department, California State
Polytechnic University, USA
stathopoulos@csupomona.edu

W. El-Sharoud (ed.) *Bacterial Physiology: A Molecular Approach.*
© Springer-Verlag Berlin Heidelberg 2008

interact with host cells and tissues. Secretion of proteins into the external environment is essential for all of these processes. To reach their final destinations, secreted proteins must pass through one or more membranes, and bacterial cells have evolved a variety of mechanisms for this purpose.

"Protein secretion" refers to the transport of a protein from the bacterial cytoplasm to the outside of the cell, whereas "protein export" indicates the transport of a protein from the cytoplasm to the periplasm and the outer membrane (OM) in Gram-negative bacteria. In Gram-positive bacteria and archaea, a secreted protein needs to be transported across only one membrane (known as the cytoplasmic membrane [CM]), whereas, in Gram-negative bacteria, both cytoplasmic and OMs and the periplasm have to be crossed. Because of the complexity of the Gram-negative cell envelope, multiple mechanisms of protein secretion (known as "secretion pathways" or "secretion systems") have evolved (Table 5.1 and Fig. 5.1). Protein secretion in Gram-positive bacteria is less understood, with fewer pathways being identified in these organisms compared with their Gram-negative counterparts. Protein secretion in archaea represents an entirely new area of research, which, with the help of molecular tools, attracts growing attention. This chapter will discuss protein secretion in Gram-negative and Gram-positive bacteria, as well as archaea. Studies of protein secretion in bacteria, such as the cyanobacteria, the green sulfur bacteria, and the spirochetes, are in their infancy and will not be covered here.

5.2 Protein Secretion in Gram-Negative Bacteria

5.2.1 Two-step Secretion Pathways

In Gram-negative bacteria, secretion involving the Sec translocase or the twin-arginine translocase (Tat) takes place in two discrete steps: first, one of these translocases (or translocons) moves a protein from the cytoplasm to the periplasm; next, a separate complex translocates the protein across the OM. Whereas the Tat translocon specializes in transporting proteins that have been already folded within the cytoplasm, Sec translocates unfolded or partially folded polypeptides that fold into functional proteins only after reaching the periplasm or extracellular space. Additionally, the signal-recognition particle (SRP), using parts of the Sec translocon, is mainly for insertion of proteins into the CM.

5.2.1.1 Export Across the CM

5.2.1.1.1   Sec Translocase

The Sec translocase, the most thoroughly studied of the CM translocons, is a large CM protein complex comprised of a channel and associated accessory proteins. Several key components of the translocase are universally conserved throughout

Table 5.1 Secretion pathways in Gram-negative bacteria

Characteristics	I	II	III	IV	V (autotransporter)	Two-partner secretion	VI	Chaperone/usher
					Secretion pathways			
Secretion apparatus	3 components	Numerous components	Numerous components	Numerous components	1 component	2 components	Unknown	2 components
Sec-dependent substrate secretion	No	Yes	No	No; one exception	Yes	Yes	No	Yes
Substrate localization	Cell surface; extracellular	Cell surface; extracellular	Within host cells; extracellular	Within host cells; extracellular	Cell surface; extracellular	Cell surface; extracellular	Extracellular	Cell surface
Substrate function	Vary	Vary	Mainly virulence	Vary	Mainly virulence	Mainly virulence	Mainly virulence	Mainly virulence

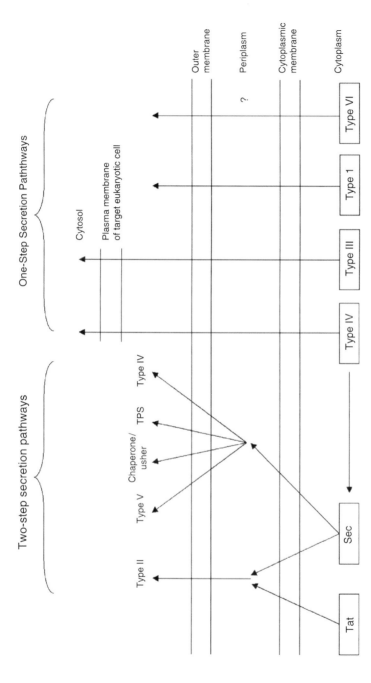

Fig. 5.1 Overview of the protein secretion systems of Gram-negative bacteria. Two-step secretion pathways require Sec or Tat translocases for secretion across the bacterial inner (cytoplasmic) membrane. One-step secretion pathways are Sec (Tat) independent. Type IV can be either a two-step (Sec-dependent) or a one-step mechanism. Type III and two-step Type IV pathways secrete effector molecules across three membranes and directly into the cytosol of the target eukaryotic cell. The newly discovered Type VI is probably a one-step secretion pathway; however, its secretion apparatus and mechanism of secretion remain currently unknown

evolution (Tam et al. 2005). In Gram-negative bacteria, the majority of the proteins transported across the CM are routed through the Sec translocase (de Keyzer et al. 2003). Among the Sec-dependent proteins are virulence factors such as extracellular toxins, pilus and nonpilus adhesins, invasins, and proteases, all of which make research on the Sec translocase particularly relevant in the field of medical microbiology (Stathopoulos et al. 2000).

In *Escherichia coli*, the pathway through Sec begins as a cytoplasmic chaperone, SecB, recognizes a presecretory polypeptide with the correct N-terminal signal sequence. A typical Sec-recognizable signal sequence is 18 to 30 amino acids in length, and is composed of three distinct domains: a positively charged N domain, a hydrophobic H domain, and a C domain containing the cleavage site. The role of SecB as a chaperone is to maintain the presecretory polypeptide in a conformation loose enough for secretion and to deliver the protein to SecA, an ATPase bound at the SecYEG membrane pore (Stathopoulos et al. 2000). Translocation is initiated with the binding of ATP to SecA. This causes a significant conformational change in SecA that releases SecB and leads to the insertion of approximately 2.5 kDa of the signal sequence and mature protein sequence into the translocation channel of SecYEG (Driessen et al. 1998; Stathopoulos et al. 2000). Accessory proteins SecDF, YajC, and YidC that bind to the SecYEG pore play various roles in the regulation of translocation (de Keyzer et al. 2003) or CM protein insertions (Dalbey and Chen 2004).

Once the presecretory protein is inserted into the translocation channel, the conformational change previously induced in SecA is reversed on ATP hydrolysis (Stathopoulos et al. 2000). This frees SecA to rebind and translocate another approximately 2.5-kDa section. Thus, to transport a full length of polypeptide through the Sec translocon, repeated ATP hydrolysis by SecA is required, at least during the initial stages. During the later stages, movement of the polypeptide through the translocation channel seems to be driven mainly by the protonmotive force (PMF) (Vrontou and Economou 2004) (see Chap. 4 for more details on PMF). Once fully translocated across the CM, the polypeptide remains in the periplasm, anchored to the CM by the signal sequence (Stathopoulos et al. 2000). Translocation by Sec is generally characterized as a posttranslational process; but this might actually occur before the protein is fully synthesized, especially with larger proteins (Newman and Stathopoulos 2004).

After translocation, the signal sequence of a presecretory protein is cleaved by one of the three known types of integral membrane peptidases (Stathopoulos et al. 2000). The majority of proteins secreted in *E. coli* via this pathway are processed by signal peptidase I. In contrast, the signal sequences of precursor lipoproteins are cleaved by signal peptidase II, whereas the N termini of the type IV pilus substrates are processed by prepilin peptidase (Stathopoulos et al. 2000; Newman and Stathopoulos 2004). Once the signal sequence has been cleaved, the mature polypeptide is released into the periplasm and *en route* for its final destination, whereas the signal peptide is further degraded by protease IV (Stathopoulos et al. 2000).

5.2.1.1.2 Signal-Recognition Particle

The *E. coli* SRP is composed of a 4.5 S RNA segment and a protein named Ffh (for *fifty-four homolog*). Similar to the SecB chaperone, the SRP acts to target presecretory polypeptides to the Sec translocon (de Keyzer et al. 2003). However, SecB and SRP are functionally distinct, in that SRP-mediated targeting is a co-translational rather than a posttranslational process (Newman and Stathopoulos 2004). Furthermore, the SRP route is specialized for CM proteins (Luirink and Sinning 2004).

The 4.5 S RNA segment and the Ffh protein of an *E. coli* SRP are homologous to the eukaryotic 7 S RNA segment and SRP 54-kDa subunit, respectively (Dalbey and Chen 2004). The *E. coli* SRP receptor, FtsY, is also homologous to the eukaryotic SRα (Dalbey and Chen 2004). Despite the homology, the *E. coli* SRP has not shown signs of being capable of arresting translation like its eukaryotic counterpart (Dalbey and Chen 2004). SRP-mediated CM translocation begins as Ffh and the 4.5 S RNA segment bind to a nascent polypeptide while it is being translated by the ribosome (Luirink and Sinning 2004). The signal sequence is immediately targeted to the membrane-bound FtsY in a GTP-dependent manner (Dalbey and Chen 2004; Luirink and Sinning 2004). Once in association with FtsY, GTP is hydrolyzed, resulting in the release of the SRP. It is unknown exactly how the nascent polypeptide is then transferred to the SecYEG complex, but, once this occurs, the hydrophobic domains are incorporated into the membrane using both SecY and YidC. ATPase SecA and the PMF are both required for translocation of hydrophilic domains. The PMF serves mainly as the driving force for smaller hydrophilic regions of less than 40 residues, with SecA acting only on domains greater than 60 residues in length (Dalbey and Chen 2004). After translocation, signal sequences may be cleaved from CM proteins, or simply be incorporated into the membrane protein structure. Any misfolded or misassembled membrane proteins are removed from the membrane and degraded in the cytoplasm by the ATP-dependent protease FtsH (Dalbey and Chen 2004).

The SRP-recognizable signal sequence is very similar to the tripartite signal sequence recognized by SecB. However, the central hydrophobic section of an SRP-recognizable signal sequence contains additional hydrophobic residues, which play an important distinguishing role in determining a protein's routing through the SRP pathway. Additional features of a signal sequence, such as being able to form a helical structure and possessing more than usually basic N-terminal regions, also contribute to SRP pathway usage (Luirink and Sinning 2004).

5.2.1.1.3 Twin-Arginine Translocase

The Tat export system is named after the two consecutive arginine residues located in the signal sequence of its substrate. The system is a CM protein channel complex comprised of four different components. What distinguishes this pathway from other CM export systems is its ability to transport proteins that have completely folded within the cytoplasm (Berks et al. 2000). In *E. coli*, Tat is largely used for the export of redox cofactor-containing proteins that are involved in anaerobic respiration chains; however, a variety of proteins, including some virulence factors, have also been shown to be Tat dependent (Voulhoux et al. 2001). Interestingly, Tat is found to be responsible for the integration of a small number of proteins into the CM, a task mainly carried out by YidC (Hatzixanthis et al. 2003).

The Tat transport machinery in *E. coli* consists of the CM proteins TatA, TatB, TatC, and TatE (Berks et al. 2000). In other bacteria, the Tat exporter usually possesses only one homolog each of TatA, TatB, and TatC (Palmer and Berks 2007). TatE, a relatively ineffective substitute of TatA, seems to have arisen from a gene duplication event in *E. coli*. A signal sequence recognition complex is formed out of TatBC units, whereas the membrane pore is formed by a TatA homo-oligomer. Unlike Sec, the Tat export process has not been thoroughly studied; of what is known, once a Tat signal sequence has been released from its chaperones, it binds to the TatBC complex in an energy-independent process. The TatA complex then binds to the TatBC complex, and, using the PMF, transports the protein in an energy intensive process. After translocation, the signal sequence is cleaved by leader peptidase, and the protein is freed into the periplasm for further export (Palmer and Berks 2007).

The Tat system recognizes proteins with a signal sequence with the following characteristics: a longer positively charged N-terminal region than in Sec, a conserved (S/T)-R-R-x-F-L-K motif, a less hydrophobic H region, and a positively charged C region (Berks et al. 2000; Voulhoux et al. 2001). Although the arginines in the conserved motif are vital for Tat export, the other residues are each found in more than 50% of Tat signal sequences. It is probable that the conserved Tat motif of the translocated substrate is recognized by TatC, the most highly conserved element of the Tat translocon, during transmembrane translocation (Berks et al. 2000).

Identified Tat substrates have diameters ranging from 20 to 70 Å, and the majority of the substrates are folded proteins containing cofactors. This raises a question regarding how the Tat translocation channel can accommodate such large substrates but at the same time prevent the leakage of ions (Berks et al. 2000). It turns out that the TatA homo-oligomer forms a ring-shaped complex with a capped channel. These TatA rings have been isolated with channel diameters of up to 70 Å (Gohlke et al. 2005), which is wide enough to allow the passage of large-size substrates (Palmer and Berks 2007). Because of the inherent difficulty of translocating its substrates, Tat is understandably a slow export system. In the closely related plant thylakoid ΔpH-dependent system, a single transporter moves approximately one 150-residue protein in 1 min, whereas an *E. coli* Sec translocase can transport roughly 10,000 amino acids in the same time (Berks et al. 2000). For these reasons, it is not surprising that there are only approximately 35 known candidate proteins targeted to the Tat system in *E. coli*, whereas more than 450 secreted proteins go through the Sec system (Palmer and Berks 2007).

5.2.1.2   Translocation Across the OM

5.2.1.2.1   Type V

The type V or autotransporter (AT) secretion system is uniquely compact, comprised of only one polypeptide that harbors both the substrate and secretion components: an N-terminal Sec-dependent signal sequence, an internal passenger domain, and a C-terminal translocator domain. Whereas the N-terminal domain enables the AT precursor to be targeted to the Sec translocase for transport across the CM, the C-terminal translocator domain mediates translocation of the AT across the OM. Together, the two terminal domains govern the secretion of the internal passenger domain across the cell envelope and to the cell surface or the extracellular environment (Henderson et al. 2004; Newman and Stathopoulos 2004).

An AT's N-terminal signal sequence is approximately 25 to 30 residues long. Several AT precursor proteins were found to have extra-long signal sequences. It has been proposed that such an N-terminal extension may have a function in preventing premature folding of an AT protein (Szabady et al. 2005). After CM translocation, the C-terminal domain of an AT protein inserts into the OM and forms a β-barrel with a central hydrophilic channel. Some AT β-barrels seem to be monomeric, whereas others are trimeric (Oomen et al. 2004; Cotter et al. 2005; Meng et al. 2006). The β-barrel mediates the translocation of the passenger domain across the OM (Henderson et al. 2004; Newman and Stathopoulos 2004). It is still unclear whether the passenger goes through the β-barrel pore to reach the cell surface, or requires accessory proteins such as the Omp85 complex to mediate its OM translocation (Voulhoux et al. 2003; Kostakioti et al. 2005). After export to the cell surface, the passenger domain is thought to fold into its native conformation with the aid of an intramolecular chaperone located between the passenger and the translocator of an AT (Oliver et al. 2003). After folding, some ATs remain attached to the β-barrel and are exposed on the bacterial cell surface, whereas some are released into the external environment by autoproteolysis or proteolysis facilitated by OM proteases (Jacob-Dubuisson et al. 2004; Newman and Stathopoulos 2004). Unlike many other secretion systems that use ATP or the PMF to power the process, secretion of an AT does not seem to require any external energy source (Thanassi et al. 2005).

The substrates of type V secretion are large proteins, approximately 100kDa in size, with some exceptions. The majority of these are virulence proteins associated with bacterial pathogenicity (Henderson et al. 2004). The first identified members of the AT family were proteases from *Neisseria gonorrhoeae*, capable of cleaving human immunoglobulin (Ig) A1 (Pohlner et al. 1987). Other examples include the Tsh hemagglutinin/protease of avian pathogenic *E. coli*, the VacA cytotoxin of *Helicobacter pylori*, and the Hap adhesin/protease of *Haemophilus influenzae* (Henderson et al. 2004). Rather than directly contributing to bacterial virulence, several AT proteins were recently found to have an indirect role, being responsible for processing and maturation of other virulence proteins. These include NalP (van Ulsen et al. 2003) from *Neisseria meningitidis*, and SphB1 from *Bordetella pertussis* (Coutte et al. 2001).

5.2.1.2.2   Two-Partner Secretion

The two-partner secretion (TPS) system shares some similarities with the AT pathway. Both pathways mainly secrete proteins associated with bacterial virulence. Unlike the AT system, which packs the substrate and secretion components into one polypeptide, these two components are separated in two different polypeptides in the TPS system. In this system, the secreted substrate is termed TpsA, whereas the Omp85-like translocator named TpsB is capable of forming a channel in the OM, allowing TpsA to exit the cell. Both proteins are encoded by genes located on the same operon, and each component harbors its own N-terminal signal sequence (Jacob-Dubuisson et al. 2004; Newman and Stathopoulos 2004).

Translocation of both TpsA and TpsB into the periplasm uses the Sec translocase (Jacob-Dubuisson et al. 2004); thereafter, TpsB inserts into the OM and adopts a β-barrel structure. The barrel is formed by the C terminus of the TpsB polypeptide (Meli et al. 2006), leaving the remaining one third of the polypeptide localized in the periplasm. Similarly to Omp85, the N-terminal domain of TpsB contains POTRA, or *polypeptide transport-associated*, repeats (Sanchez-Pulido et al. 2003; Gentle et al. 2005). Not only does the N terminus influence the TpsB channel activity, it also participates in the recognition of the TpsA substrate (Hodak et al. 2006; Meli et al. 2006). The interaction between TpsA and TpsB has been shown to occur between the N-terminal regions of TpsA and the N-terminal domain of TpsB (Hodak et al. 2006). Such interaction is also needed to open up the gated TpsB channel, allowing the secretion of TpsA into the extracellular environment in the order of N to C terminus. Because of the general large size of a TpsA protein, its translocation across the CM and OM is probably coupled. TpsA tends to stay in the periplasm briefly and is very likely to be transported across the OM in an unfolded manner. After it reaches the cell surface, TpsA is thought to fold gradually into its native conformation. Maturation of several TpsA proteins is attained only after further proteolytic processing by extracellular proteases. Subsequently, part of TpsA remains attached to the cell surface, whereas the rest is released into the external medium (Jacob-Dubuisson et al. 2004).

Substrates of the TPS system are mostly large proteins of sizes greater than 300 kDa (Jacob-Dubuisson et al. 2004). Most TpsA proteins are virulence factors. The most characterized members include the FHA adhesin of *B. pertussis* (Lambert-Buisine et al. 1998), ShlA hemolysin (Hly) of *Serratia marcescens* (Braun et al. 1992), and HMW1/HMW2 adhesins of non-typeable *H. influenzae* (Grass and St Geme 2000).

5.2.1.2.3   Chaperone/Usher Pathway

The chaperone/usher (CU) pathway is another secretion system involved in the virulence of several Gram-negative bacterial pathogens. It is responsible for the secretion of pilus subunits and the assembly of these into rod-like organelles protruding on the bacterial surface (Thanassi and Hultgren 2000). The CU system is comprised of two major components: the periplasmic chaperone and the OM usher. The chaperone acts to keep each substrate/subunit in a secretion competent state. Moreover, it targets the substrate to the usher pore, through which secretion to the surface takes place. A pilus is, thus, polymerized by repeatedly adding the substrates onto the growing organelle from its base (Sauer et al. 2004).

After being released into the periplasm from the Sec translocase, the pilus subunit is targeted to a periplasmic chaperone. Specific regions in both termini of a subunit are needed for chaperone recognition (Soto et al. 1998). Interaction between a chaperone and a pilus subunit is essential for the chaperone to block premature subunit–subunit interactions in the periplasm and to facilitate correct folding of the subunit. Structural features of both the chaperone and the substrate play an important role in their interaction. The chaperone protein consists of two domains, each of which is formed by an Ig-like fold containing seven β-strands. In contrast, each pilus subunit harbors a domain that interacts with the chaperone, which contains an Ig fold that lacks a β-strand. The missing β-strand forms a groove in the domain of the subunit and exposes its hydrophobic core. During chaperon–subunit interaction, the chaperone donates one of the β-strands to the subunit. The strand goes into the groove of the subunit and completes its Ig fold, in a process called the donor strand complementation. Such a complementation covers the hydrophobic core of the subunit, which stabilizes the structure and prevents premature pilus assembly (Sauer et al. 2004).

Once the chaperone–subunit complex is formed, the chaperone delivers the subunit to the OM usher. The usher is a dimeric channel protein with an internal channel diameter of 2 to 3 nm (Li et al. 2004). Interaction between the chaperone–subunit complex and the usher dissociates the subunit from the complex and enables it to bind to a previously assembled subunit in the elongating pilus. The same structural complementation that occurred between the chaperone and the subunit is now used to join the two subunits together. At the usher, the dissociated chaperone leaves the groove in the previously bound subunit unoccupied and exposed. An N-terminal extension of a subunit at the end of the elongating pilus then fits into the groove of the newly arriving substrate, in a process called the donor strand exchange (Sauer et al. 2004). Such an assembly process is thought to take place at the periplasmic side of the usher before OM translocation. It is also suggested that the usher organizes the order of subunit incorporation in a tip to base fashion, that is, the adhesin subunit is incorporated before the subunits that form the pilus rod. After assembly, the secreted pilus coils into its final conformation and remains bound to the cell surface (Thanassi and Hultgren 2000). There is no evidence that CU secretion requires external energy (Thanassi et al. 2005).

Two well-characterized CU systems are type 1 and P pili, which are commonly found in the *Enterobacteriaceae*. Each pilus is composed of several different proteins that form a main pilus rod with an adhesin subunit attached to its tip, which confers adhesive property on the pilus (Kuehn et al. 1992). This property enables bacteria possessing the type 1 and P pili to colonize in the urinary tract through cell adherence (Roberts et al. 1994). The CU system is also responsible for secreting subunits that form the hemagglutinating pilus in the respiratory pathogen, *H. influenzae*, and the capsular F1 antigen in the bubonic plague-causing bacterium, *Yersinia pestis* (Sauer et al. 2004).

5.2.1.2.4   Type II

The Type II secretion pathway is responsible for the secretion of various hydrolytic enzymes and toxins in Gram-negative bacteria. In most cases, the type II secretion apparatus involves 12 to 15 different protein components, designated by letters A through O and S. These proteins are the products of genes that tend to cluster in the same operon (Johnson et al. 2006). The substrates of type II secretion fold into their functional conformation in the periplasm, before reaching the secretion channel. The substrates are secreted across the OM through the only integral OM protein component in the apparatus, protein D, which belongs to the secretin family (Nouwen et al. 2000; Thanassi 2002). In many species, such secretion is pilus mediated (Sandkvist 2001; Johnson et al. 2006). The rest of the components in the apparatus are CM and periplasm localized, with the exception of lipoprotein S, which is a periplasmic protein associated with the OM (Sandkvist 2001).

Secretion of cholera toxin from *Vibrio cholerae* has become a model for the type II pathway. The toxin consists of six protein subunits. Individual subunits pass through the Sec translocase to reach the periplasm, where they fold into their native conformation aided by a disulfide isomerase DsbA (Sandkvist 2001). Folded subunits are then assembled and targeted to the type II secretion apparatus for translocation across the OM, through the stable β-barrel formed by the C-terminal domain of OM protein D (Nouwen et al. 1999; Nouwen et al. 2000; Thanassi 2002). The N-terminal domain of protein D, in contrast, is thought to be essential for interacting with other secretion components, recognizing specific substrates, and gating the channel formed by the C-terminal domain (Nouwen et al. 1999). In addition to protein D, substrate secretion also requires a pilus-like structure formed by other components in the apparatus, named proteins G, H, I, J, and K. It has been proposed that the pilus may act as a piston that pushes the substrates through the pore by motions of extension and retraction, thereby facilitating the secretion of substrates across the OM. Secretion by the type II pathway is an energy dependent process requiring ATP and the PMF (Sandkvist 2001; Johnson et al. 2006).

The first characterized substrate of this pathway was the starch-hydrolyzing lipoprotein, pullulanase (PulA) of *Klebsiella oxytoca* (Pugsley et al. 1997). The best-studied virulence protein secreted by this system, however, is probably the cholera toxin of *V. cholerae* (Merritt and Hol 1995). Other substrates of the type II secretion pathway include proteases, cellulases, pectinases, phospholipases, lipases, and toxins (Sandkvist 2001; Johnson et al. 2006).

5.2.2   One-Step Secretion Pathways

5.2.2.1   Type I

The type I secretion system delivers protein substrates to the cell surface or external environment of Gram-negative bacteria. It uses a secretion apparatus comprised of merely three components: a CM-localized ATPase that powers secretion, an OM channel that performs protein export, and a membrane-fusion protein that connects the ATPase and the OM channel while forming a conduit for the substrate's periplasmic passage. Substrates for this system are mostly acidic proteins with isoelectric points (pIs) of approximately 4. These proteins harbor tandem repeats in their sequences that identify them as adhesins of various molecules or ions. Depending on the bacterial species, the substrates may have diverse sizes and functions (Holland et al. 2005).

The mechanism of type I secretion was largely elucidated through the study of the hemolysin (Hly) transporter in pathogenic *E. coli*. In this system, the newly synthesized 110-kDa substrate HlyA is maintained in an unfolded, secretion-competent state by a cytoplasmic chaperone. Unlike those secreted by Sec-dependent secretion systems, a type I secretion substrate such as HlyA carries an uncleaveable secretion signal of approximately 50 residues in the C terminus. This signal allows HlyA to be recognized by the HlyB ATPase (Schindel et al. 2001). The cytoplasmic membrane-localized HlyB exists as a monomeric protein that belongs to the large ABC-binding cassette protein family (Holland et al. 2005). The binding between HlyA and HlyB promotes a conformational change in HlyB. The conformational change, in turn, promotes the binding of HlyB to ATP, thereby releasing the HlyA substrate into the secretion pathway. HlyA also interacts with HlyD, the periplasmic conduit that bridges the CM HlyB and the OM channel, TolC. Interactions among HlyA, HlyB, and HlyD recruit TolC to the site (Koronakis et al. 2004). TolC is a trimeric OM protein with long α-helical periplasmic extensions (Koronakis et al. 2000). It is proposed that HlyD exerts a force on these α helices, which subsequently forces the TolC channel in the OM to open up in an iris-like twisting fashion, thus, allowing the passage of HlyA (Koronakis et al. 2000; Koronakis et al. 2004). After translocation across the OM, the HlyA substrate reaches the cell surface and begins to fold into a functional protein. Secreted HlyA is capable of forming a pore that can insert itself into the plasma membrane of the eukaryotic host cell, resulting in cell lysis (Schindel et al. 2001).

Substrates of the type I system are secreted at a rapid rate and in an unfolded conformation (Holland et al. 2005). Although many substrates contribute to bacterial virulence, as seen in HlyA, several are also involved in structural maintenance or cellular metabolism. Examples of type I substrates include surface-layer proteins (RsaA of *Caulobacter crescentus* and SapA of *Campylobacter fetus*), metal-binding proteases (PrtA, B, C, and G of *Erwinia chrysanthemi*), toxins (RtxA of *V. cholerae* and Colicin V of *E. coli*), adhesins (LapA of *Pseudomonas* species), and lipases (LipA of *S. marcescens* and *Pseudomonas fluorescens*) (Thompson et al. 1998; Delepelaire 2004).

5.2.2.2   Type III

The type III secretion apparatus is called an injectisome. It is a large complex comprised of more than 20 different proteins and takes the shape of a syringe with a needle. The "syringe" portion spans the cell envelope and is composed of multiple-ring structures that stack on top of one another, forming a cylindrical channel. Protruding above the cell surface is a "needle," through which, secreted substrates can be delivered directly into the eukaryotic host cell (Cornelis 2002; Galan and Wolf-Watz 2006). Type III secretion is notorious for being the major virulence determinant in several disease-causing bacteria, including mammalian pathogens (*Yersinia* and *Salmonella* species) and plant pathogens (*Pseudomonas syringae* and *Erwinia* species) (He et al. 2004; Galan and Wolf-Watz 2006).

In addition to the secretion apparatus, the type III secretion system also includes numerous secretion substrates (effectors and translocators) and many other proteins that regulate secretion (regulators and chaperones). Genes encoding these various components of the type III system are usually clustered. The gene cluster can be chromosome localized in some bacteria, or plasmid localized in others. Several organisms are also found to possess more than one type III system. Expressing these numerous gene products, indeed, consumes much cellular energy. The system may, thus, not be fully expressed or secreting its substrates until the bacterium comes in contact with its eukaryotic host cell. Once in contact, the cues of when to release the substrates are then transduced from the environmental stimuli, through the regulatory proteins, to the secretion system. The secretion substrates are synthesized in the cytoplasm with one or two noncleavable, Sec-independent signal sequences located in the N terminus. The substrates are kept in a secretion-competent conformation by cytoplasmic chaperones, which may perhaps even assist in targeting the proteins to the entrance/base of the secretion channel. It is not clear how substrates are recognized by or transported through the secretion channel. Because an ATPase is discovered in the cytoplasmic side of the CM of all type III apparatuses, it is assumed that the substrates are pumped out of the channel using the energy from ATP hydrolysis (He et al. 2004).

Two types of secretion substrates are released from the cell through the type III secretion channel. The translocators have the ability to disrupt or form pores in the eukaryotic host cell membrane, allowing the "needle" to dock onto and penetrate the host cell (He et al. 2004; Galan and Wolf-Watz 2006). The effectors are then delivered through the needle complex into the host's cytoplasmic compartment, where they can exert their functions (He et al. 2004). In *Y. pestis*, the effector proteins (Yops) are able to inhibit phagocytosis and host immune responses (Cornelis 2002). In enteric pathogen *Salmonella*, type III effectors (Sips and Sops) are responsible for bacterial invasion and survival within the macrophages or intestinal epithelial cells (Schlumberger and Hardt 2006). Interestingly, type III secretion can also be found in bacteria that form a symbiotic relationship with insects, nematodes, or plants. Here, the effectors allow the bacteria to interact with their hosts (He et al. 2004; Galan and Wolf-Watz 2006).

5.2.2.3   Type IV

The type IV secretion system is related to the bacterial conjugation system, designated for DNA transfer. However, in pathogenic bacteria such as *B. pertussis* and

H. pylori, this system is used for virulence protein secretion. In intracellular pathogens such as *Legionella pneumophila* and *Bartonella henselae*, the system is essential for bacterial replication within their eukaryotic host cells (Backert and Meyer 2006). Similarly to the type III secretion apparatus, the type IV transporter is a large complex comprised of a surface-exposed pilus attached to a channel that spans the cell envelope. Furthermore, substrates are usually delivered directly into eukaryotic host cells via direct cell-to-cell contact.

Much of what is learned about the type IV pathway comes from the study of the VirB system in *Agrobacterium tumefaciens*. This plant pathogen delivers into its host cell a processed plasmid via a secretion channel complex formed by 11 proteins. Genes encoding these proteins are located on the same operon. Two of the proteins, VirB4 and VirB11, are ATPases localized in the CM. Although the energy required to drive substrate secretion comes from ATP hydrolysis by VirB11 and an accessory protein, VirB4 plays a structural stabilization role. In fact, both ATP and the PMF are required to power substrate translocation. The DNA substrate to be transferred is processed in the bacterial cytoplasm and coupled to a protein, which carries a signal sequence in its C terminus. The secretion signal enables the protein–DNA substrate to be targeted to VirB6, an integral protein located in the CM. The substrate is then transferred through the secretion channel. Thereafter, it can pass through an existing pilus attached to the secretion apparatus, as described in the "channel" model. Alternatively, it can be "pushed" forward by a growing pilus, as proposed in the "piston" model (Christie 2004). In addition to the DNA–protein substrate, several other effector proteins are also translocated. These protein substrates seem to assist in the nuclear targeting and genome integration of the transferred DNA. Once integrated into the host genome, the bacterial DNA promotes uninhibited plant cell division that eventually leads to the formation of a crown gall tumor (Garcia-Rodriguez et al. 2006).

The export of the pertussis toxin from *B. pertussis*, the organism that causes whooping cough, is an exception to the typical type IV secretion route. In fact, its secretion can be parallel to that of the cholera toxin by type II secretion. The pertussis toxin is composed of five different proteins, each of which carries a typical Sec-dependent cleavable signal sequence in the N terminus and is, thus, transported into the periplasmic compartment through the Sec complex. Assembly of these five substrates into an active toxin takes place in the periplasm. Thereafter, the toxin is secreted to the external environment via the type IV secretion channel. One of the subunits then mediates internalization of the toxin into the host cell. Within the cytoplasm, this protein interrupts the signaling system of the host cell and also interferes with its communication with other immune cells (Backert and Meyer 2006).

5.2.2.4 Type VI

This is the most recently described protein secretion pathway, whose function and secretion mechanism remain largely unknown. The system has been identified in *Edwardsiella tarda* (Rao et al. 2004), *V. cholerae* (Pukatzki et al. 2006), *Pseudomonas aeruginosa (Mougous et al. 2006)*, and enteroaggregative *E. coli* (Dudley et al. 2006) by genetic screening, microarrays, and proteomic analysis. Although the actual secretion apparatus or components have not yet been found, several protein substrates have already been characterized. It was experimentally demonstrated that several of these substrates could not be secreted by other known protein secretion systems (Rao et al. 2004; Pukatzki et al. 2006). The substrates from different bacterial species are not homologous to one another at the protein sequence level. However, all of them seem to be small-sized proteins (less than 20 kDa) lacking an N-terminal signal sequence, and are secreted to the external environment (Dudley et al. 2006). More significantly, all of these proteins have been implicated in bacterial virulence to their hosts (Dudley et al. 2006). Therefore, type VI secretion seems to be another pathway used by pathogenic bacteria for the delivery of virulence determinants.

5.3   Protein Secretion in Gram-Positive Bacteria

In Gram-positive bacteria, proteins can be sorted to at least four different final destinations: the cytoplasm, the CM, the cell wall, and the extracellular growth medium. By far, the largest number of secreted proteins in Gram-positive bacteria is secreted via the Sec translocase. However, similar to the Gram-negative bacteria, most Gram-positive bacteria also have SRP and Tat secretion pathways. In addition, at least two other secretion systems specific to Gram-positive bacteria have been evolved to secrete virulence factors or components of surface appendages.

5.3.1   Sec, SRP, and Tat-Dependent Secretion

The Sec translocase of *Bacillus subtilis* consists of SecA (motor), SecYEG complex (pore), SecDF, and, possibly, YrbF (a homolog of *E. coli* YajC), SpoIIIJ (a homolog of *E. coli* YidC), and YqjG (a homolog of mitochondrial Oxa1) (Yamane et al. 2004). A SecB homolog has not been identified for *B. subtilis*, but it has been suggested that the general chaperone, CsaA, may play a similar role by targeting presecreted proteins to the Sec translocase. This view is supported by the finding that the *B. subtilis* CsaA has binding affinity for preproteins and SecA (Muller et al. 2000). However, a CsaA homolog has not been found in some other Gram-positive species, e.g., *Staphylococcus aureus*. Unlike *E. coli* K12, which expresses SecD and SecF, *B. subtilis* expresses a single SecDF fusion protein, which compensates for the functions of the two proteins in *E. coli* (Bolhuis et al. 1998).

The Sec-dependent signal peptides of *B. subtilis* are, on average, longer and more hydrophobic than those of *E. coli*. One of the most remarkable features of the *B. subtilis* Sec-dependent secretion is the presence of five type I signal peptidases (SipS, SipT, SipU, SipV, and SipW). SipS and SipT seem to be responsible for processing the majority of secreted preproteins, whereas the other three proteases play a minor role in protein secretion (Tjalsma et al. 2004). SipW is homologous to the signal peptidases found in sporulating Gram-positive bacteria, archaea, and the endoplasmic reticulum (ER) of the eukaryotic cell (ER-type of signal peptidases). Moreover, SipW has been reported to be required for the processing of only one secreted protein, the spore-associated protein, TasA. All other signal peptidases of *B. subtilis* are of prokaryotic type (P-type). Although possessing multiple type I signal peptidases, *B. subtilis* contains only one type II signal peptidase (LspA).

The SRP of Gram-positive bacteria is similar in composition to that of the Gram-negative bacteria. It consists of a small cytoplasmic RNA (scRNA), the Ffh protein, a histone-like protein (HBsu), and the receptor protein, FtsY (Yamane et al. 2004). The scRNA consists of approximately 270 nucleotides and its predicted structure is strikingly similar to that of the human SRP 7 S RNA, although it lacks domain III (Althoff et al. 1994). In contrast, the SRP 4.5 S RNA of *E. coli* is approximately 120 nucleotides long and is predicted to fold into a single hairpin corresponding to domain IV of the human counterpart. When *E. coli* 4.5 S RNA, a truncated form of scRNA of *B. subtilis* corresponding to domain IV or human 7 S RNA, is expressed in *B. subtilis* lacking scRNA, all three RNAs compensate functionally for the absence of scRNA. These studies provided strong evidence that the hairpin structure of domain IV, which is evolutionarily well conserved in all three SRP RNAs, plays a key role in SRP-mediated protein secretion. Similarly to *E. coli*, SRP is essential for viability and growth in *B. subtilis*. However, this is not the case for all Gram-positive bacteria. For example, the SRP is not essential in *Streptococcus mutans*. In this organism, the SRP is merely required for growth under stressful conditions, such as low pH or high salt. This finding suggests that SRP is more important in the secretion of some bacteria than in other bacteria, and that, in *S. mutans*, the role of SRP may be to secrete proteins that protect the cell against environmental insults (Sibbald et al. 2006).

Most Gram-positive bacterial species seem to contain at least one set of Tat genes in their genomes. For example, *B. subtilis* contains two *tatC* genes (*tatCd* and *tatCy*) and three *tatA* genes (*tatAd, tatAy*, and *tatAc*). These *tat* genes encode two distinct Tat translocases, each possessing its own substrates. In contrast, a few other species, such as *S. epidermidis, S. aureus*, and *Mycoplasma* species, do not have Tat genes and seem to lack a Tat pathway (Dilks et al. 2003). *Streptomyces coelicolor* and *Mycobacterium tuberculosis*, with 145 and 31 predicted Tat substrates, respectively, are the Gram-positive bacteria (both of the Actinobacteria phylum) that seem to secrete the largest number of proteins via the Tat system. Nonetheless, both species contain only one set of Tat genes, which indicates that the number of Tat systems does not usually correlate with the number of secretion substrates (Dilks et al. 2003; Digiuseppe Champion and Cox 2007). The Tat translocase of most Gram-positive bacteria is made of only TatA and TatC, and lacks TatB. It has been hypothesized that TatA may compensate for the absence of TatB in these organisms. Studies in *B. subtilis* have indicated that not all precursor proteins with twin arginine residues in their predicted signal sequence require the Tat translocase for secretion (Jongbloed et al. 2002). This finding implies that the Tat-dependent signal sequences in some Gram-positive organisms may be different from those identified in the *E. coli* Tat substrates.

5.3.2 ESX-1 (Snm) Pathway

The ESX-1 secretion pathway was first described for *Mycobacterium tuberculosis* and is also referred to as the "secretion in mycobacteria" (Snm) or ESAT-6 pathway (Berthet et al. 1998). At least two small virulence proteins, ESAT-6 and CFP-10, are known to be secreted via this pathway in the mycobacterial cell (Digiuseppe Champion and Cox 2007). Homologs of ESAT-6 have been identified in other Gram-positive bacteria, including *B. subtilis*, *B. anthracis*, and *S. aureus* (Pallen et al. 2003). In *S. aureus*, two proteins, EsxA and EsxB, have been found to be secreted via the ESAT-6 pathway, which seems to involve at least six other proteins, all part of the same gene cluster. At least four of these six proteins are predicted to be localized in the CM. Studies with the ESAT-6 protein have shown that it has a C-terminal signal sequence that promotes secretion via this pathway (Champion et al. 2006).

5.3.3 Pseudopilin-Export (Com) Pathway

Studies with *B. subtilis* have resulted in the identification of four proteins, ComGC, ComGD, ComGE, and ComGG, which contain N-terminal pseudopilin-like signal peptides and are thought to participate in the formation of a pilus-like structure on the cell wall (Sibbald et al. 2006). Secretion of these four proteins requires a pseudopilin-specific signal peptidase (ComC), an integral membrane protein (ComGB), and an ATPase (ComGA) that is located at the cytoplasmic side of the membrane. The mechanism of secretion of these proteins and the function of the Com pilus-like structure in *B. subtilis* remain unclear.

5.4 Protein Secretion in Archaea

Archaea constitute the third domain of life, with bacteria and eukarya forming the other two domains. These prokaryotic organisms are predominantly isolated from environments that are characterized by extreme conditions, such as high or low temperatures, high salinity, and high or low pH values. To withstand hostile environmental conditions, these organisms have developed unique cell envelope and cell surface structures that have not been seen in prokaryotic or eukaryotic cells. Membranes of the archaea are composed of glycerol-ether lipids that contain isoprenoid side chains, whereas the cell walls are formed by surface-layer proteins that are directly anchored to the CM. With the exception of *Ignicoccus* species, which have an OM and a periplasm (Rachel et al. 2002), all other known archaeal cells are surrounded by a single membrane (CM).

Annotation of more than 25 archaeal genomes has provided great insights into the presence and distribution of known secretion systems in this domain of life. Components of the Sec and SRP pathways were found to be present in all archaeal genomes, establishing the belief that these secretion systems are conserved in all three domains of life. Moreover, components of the Tat system were also identified in several archaeal genomes.

The archaeal Sec translocase shows similarities and differences to its counterparts in bacterial and eukaryotic organisms. Structural analysis of the Sec61 protein-conducting channel of *Methanocaldococcus jannaschii* has revealed that it is more similar to its eukaryotic counterparts than to the bacterial SecYEG complex (Van den Berg et al. 2004). Moreover, archaea possess a homolog of the eukaryotic oligosyltransferase that is absent in the bacterial Sec translocase. However, homologs of the eukaryotic pore-associated subunits (such as Sec62/63, Sec71/72, and TRAM) have not been identified in archaea, and many archaea possess homologs of the bacterial subunits YidC, SecD, and SecF, which are not found in eukarya, with the exception of organelles such as mitochondria and chloroplasts (Pohlschroder et al. 2005). Equally interesting is the observation that archaea lack SecA, the translocation ATPase of the bacterial Sec system. This finding raises questions regarding the energetics of the Sec-mediated secretion in archaea. It has been suggested that archaea may use an ATPase that is completely different from SecA to provide the energy for secretion, or, alternatively, the secretion process may entirely rely on the ion gradient that exists across the cytoplasm membrane.

To date, two signal peptidases have been identified in archaea. These are the type I signal peptidase (SPI), responsible for the cleavage of secretory signal peptides from the majority of secreted proteins, and the prepilin peptidase-like signal peptidase (TFPP), which processes signal peptides from prepilin-like protein. Archaeal SPI more closely resembles its eukaryotic counterpart. In contrast, TFPP has a catalytic mechanism that is similar to that of the corresponding bacterial enzymes, but with archaeal specificities. In addition, a peptidase needed for the degradation of signal peptides after their removal from the secreted proteins has also been found (Ng et al. 2007). However, a homolog of the bacterial signal peptidase II (SPII), responsible for the removal of signal sequences from secreted lipoproteins, has not been identified yet in any archaeal genomes. It is likely that, because of the unusual nature of archaeal lipids, an archaeal SPII equivalent may have unique properties and has escaped detection by the current bioinformatical methods.

The archaeal SRP consists of a 7S RNA, proteins SRP54 and SRP19, and a protein receptor, FtsY. The archaeal SRP most resembles that of the eukaryotes, although the archaeal SRP receptor is more similar to its bacterial counterpart (Maeshima et al. 2001). It remains to be seen whether there are additional accessory proteins associated with the SRP, and whether translational arrest also takes place in archaea.

Examination of the Tat translocase in archaea has revealed that Tat components and substrates are present in several Crenarchaeota and Euryarchaeota. However, all archaea lack a TatB homolog. Some archaea contain more than one copy of the Tat genes. Furthermore, although the number of predicted Tat substrates varies from one organism to another, most archaea seem to contain fewer genes that could encode Tat secreted proteins compared with *E. coli* K12 (Pohlschroder et al. 2005). Interestingly, some archaea, such as the extreme halophile *Halobacterium salinarum* NRC-1 strain and the haloalkaliphile *Natronomonas pharaonis*, show strong preference for the Tat pathway over the more universal Sec pathway to use as a major secretion route in the cell (Dilks et al. 2003; Pohlschroder et al. 2005). This unexpected shift in protein secretion routing, which has been seen only for halophilic archaea to date, may be an adaptation to the high-salt environment. Proteins adapted to high-salt conditions might fold faster in the cytoplasm, and, as a result, halophilic organisms might have chosen the Tat pathway to secrete these proteins that could not otherwise be exported by the Sec translocase. Future studies will be needed to determine the validity of this hypothesis (Dilks et al. 2003).

Archaeal genomes contain several putative homologs of the Gram-negative bacterial type I, II, and IV secretion system components, although complete secretion systems have not been found in these organisms. On the contrary, homologs of components of the type III and V secretion systems have not been reported. It is possible that archaeal cells may use adapted type I-, II-, or IV-like secretion systems to target subunits of their cell surface structures (surface layer, flagella, and pili) to the extracellular space. Currently, it is unclear how the subunits of these structures are secreted or assembled in vivo. It is apparent that future research and a better understanding of the biogenesis of the archaeal surface structures may lead to the discovery of novel secretion pathways that have not been identified yet.

5.5 Bacterial Protein Secretion and Biotechnological Applications

During the last few decades, bacterial protein secretion systems have been exploited for various biotechnological applications, including vaccine development, drug design, bioremediation, and enzyme/drug production. These applications typically use bacterial hosts, most commonly, *E. coli*, for the expression of recombinant proteins (Georgiou and Segatori 2005). Through the use of one of several bacterial protein secretion pathways, vaccine antigens can be delivered to the external environment of microbes and be presented to immunize hosts. Currently, pathways type I, type II, type III, type V, and CU have all been used to display heterologous epitopes in the making of live recombinant vaccines (Georgiou et al. 1997; Chen et al. 2006; Zhu et al. 2006). Heterologous protein display is also a powerful tool in drug design. Combined with the tool of fluorescence-activated cell sorting (FACS),

the technology enables rapid selection of potential drug candidates by simultaneously screening a repertoire of microbes expressing various ligand-binding peptides
on their surface (Georgiou et al. 1997). On the industrial front, recombinant bacteria expressing enzymes or chelating peptides have been developed to serve as self-
regenerating cleaning agents for pollutant removal (Georgiou et al. 1997;
Jacob-Dubuisson et al. 2004).

Overproduction of desired substrates through manipulation of bacterial protein
secretion systems also has many uses in commercial and pharmaceutical industries.
Although the Gram-negative bacterium *E. coli* is the most popular host in laboratories
(Georgiou and Segatori 2005), Gram-positive bacteria *Bacillus* species are more
frequently used in large-scale production of industrial enzymes, because they
lack lipopolysaccharides (LPS)-associated safety complications. Although native
Bacillus proteins, such as detergent proteases and amylases, are common industrial products, *Bacillus* has also been the host for human insulin and epidermal
growth factor production in the pharmaceutical industry (Westers et al. 2004). In
comparison to Gram-positive and Gram-negative bacteria, the technology of
recombinant protein expression in archaea still lags behind, mainly because of
inadequate understanding of these systems. Potentially, archaea can become the
most practical hosts used for protein expression in the industrial setting, because
many of these organisms can tolerate extreme temperature, pH, or pressure variation
(Park et al. 2004).

Applications using protein display or expression require the host systems to
deliver substantial levels of quality proteins. In *E. coli*, optimization of protein
export levels usually involves tweaking the signal peptides (Economou 2002; Lee
et al. 2006). Recently, there is evidence that phage display involving SRP-mediated
translocation seems to be more effective than the SecB route (Steiner et al. 2006).
Whether this applies for all substrates remains to be investigated. In addition to the
Sec system, Tat translocation has been explored as a new method of exporting
active, folded protein substrates (Lee et al. 2006). Although the translocation
efficiency of the Tat system is not as high as Sec (Georgiou and Segatori 2005), Tat
serves as a natural "quality control" checkpoint for correctly folded substrates. This
property has been exploited in the development of a screening assay that can
discriminate between correctly folded and aggregated β-peptides of Alzheimer's
amyloid (Fisher et al. 2006).

Despite recent advances, many applications involving protein expression and
delivery still face significant challenges. Achieving the desired levels of display or
yield of the functional target protein is still the primary struggle, but this is hindered
by many factors, including protein misfolding and proteolytic degradation
(Economou 2002; Lee et al. 2006). One way to circumvent such obstacles is by
coexpressing protein substrates with the necessary chaperones that can assist in
folding and targeting (Westers et al. 2004; Georgiou and Segatori 2005). Another
way is to express protein substrates in protease-deficient mutant strains (Westers
et al. 2004). There is no doubt that a better understanding of protein secretion
mechanisms is needed to overcome present challenges.

Highly Recommended Readings

de Keyzer J, van der Does C, Driessen AJ (2003) The bacterial translocase: a dynamic protein channel complex. Cell Mol Life Sci 60: 2034–2052

Palmer T, Berks RBC (2007) The Tat protein export pathway. In: Ehrmann M (ed) The periplasm. ASM Press, Washington D.C., pp 16–29

Kostakioti M, Newman CL, Thanassi DG, Stathopoulos C (2005) Mechanisms of protein export across the bacterial outer membrane. J Bacteriol 187: 4306–4314

Tjalsma H, Antelmann H, Jongbloed JD, Braun PG, Darmon E, Dorenbos R, Dubois JY, Westers H, Zanen G, Quax WJ, Kuipers OP, Bron S, Hecker M, van Dijl JM (2004) Proteomics of protein secretion by Bacillus subtilis: separating the "secrets" of the secretome. Microbiol Mol Biol Rev 68: 207–233

Digiuseppe Champion PA, Cox JS (2007) Protein secretion systems in Mycobacteria. Cell Microbiol 9: 1376–1384

Pohlschroder M, Gimenez MI, Jarrell KF (2005) Protein transport in Archaea: Sec and twin arginine translocation pathways. Curr Opin Microbiol 8: 713–719

Georgiou G, Segatori L (2005) Preparative expression of secreted proteins in bacteria: status report and future prospects. Curr Opin Biotechnol 16: 538–545

References

Althoff S, Selinger D, Wise JA (1994) Molecular evolution of SRP cycle components: functional implications. Nucleic Acids Res 22: 1933–1947

Backert S, Meyer TF (2006) Type IV secretion systems and their effectors in bacterial pathogenesis. Curr Opin Microbiol 9: 207–217

Berks BC, Sargent F, Palmer T (2000) The Tat protein export pathway. Mol Microbiol 35: 260–274

Berthet FX, Rasmussen PB, Rosenkrands I, Andersen P, Gicquel B (1998) A Mycobacterium tuberculosis operon encoding ESAT-6 and a novel low-molecular-mass culture filtrate protein (CFP-10). Microbiology 144 (Pt 11): 3195–3203

Bolhuis A, Broekhuizen CP, Sorokin A, van Roosmalen ML, Venema G, Bron S, Quax WJ, van Dijl JM (1998) SecDF of Bacillus subtilis, a molecular Siamese twin required for the efficient secretion of proteins. J Biol Chem 273: 21217–21224

Braun V, Hobbie S, Ondraczek R (1992) Serratia marcescens forms a new type of cytolysin. FEMS Microbiol Lett 79: 299–305

Champion PA, Stanley SA, Champion MM, Brown EJ, Cox JS (2006) C-terminal signal sequence promotes virulence factor secretion in Mycobacterium tuberculosis. Science 313: 1632–1636

Chen LM, Briones G, Donis RO, Galan JE (2006) Optimization of the delivery of heterologous proteins by the Salmonella enterica serovar Typhimurium type III secretion system for vaccine development. Infect Immun 74: 5826–5833

Christie PJ (2004) Type IV secretion: the Agrobacterium VirB/D4 and related conjugation systems. Biochim Biophys Acta 1694: 219–234

Cornelis GR (2002) Yersinia type III secretion: send in the effectors. J Cell Biol 158: 401–408

Cotter SE, Surana NK, St Geme JW 3rd (2005) Trimeric autotransporters: a distinct subfamily of autotransporter proteins. Trends Microbiol 13: 199–205

Coutte L, Antoine R, Drobecq H, Locht C, Jacob-Duisson F (2001) Subtilisin-like autotransporter serves as maturation protease in a bacterial secretion pathway. Embo J 20: 5040–5048

Dalbey RE, Chen M (2004) Sec-translocase mediated membrane protein biogenesis. Biochim Biophys Acta 1694: 37–53 de Keyzer J, van der Does C, Driessen AJ (2003) The bacterial translocase: a dynamic protein channel complex. Cell Mol Life Sci 60: 2034–2052

Delepelaire P (2004) Type I secretion in gram-negative bacteria. Biochim Biophys Acta 1694: 149–161

Digiuseppe Champion PA, Cox JS (2007) Protein secretion systems in Mycobacteria. Cell Microbiol 9: 1376–1384

Dilks K, Rose RW, Hartmann E, Pohlschroder M (2003) Prokaryotic utilization of the twin-arginine translocation pathway: a genomic survey. J Bacteriol 185: 1478–1483

Driessen AJ, Fekkes P, van der Wolk JP (1998) The Sec system. Curr Opin Microbiol 1: 216–222

Dudley EG, Thomson NR, Parkhill J, Morin NP, Nataro JP (2006) Proteomic and microarray characterization of the AggR regulon identifies a pheU pathogenicity island in enteroaggregative Escherichia coli. Mol Microbiol 61: 1267–1282

Economou A (2002) Bacterial secretome: the assembly manual and operating instructions (Review). Mol Membr Biol 19: 159–169

Fisher AC, Kim W, DeLisa MP (2006) Genetic selection for protein solubility enabled by the folding quality control feature of the twin-arginine translocation pathway. Protein Sci 15: 449–458

Galan JE, Wolf-Watz H (2006) Protein delivery into eukaryotic cells by type III secretion machines. Nature 444: 567–573

Garcia-Rodriguez FM, Schrammeijer B, Hooykaas PJ (2006) The Agrobacterium VirE3 effector protein: a potential plant transcriptional activator. Nucleic Acids Res 34: 6496–6504

Gentle IE, Burri L, Lithgow T (2005) Molecular architecture and function of the Omp85 family of proteins. Mol Microbiol 58: 1216–1225

Georgiou G, Segatori L (2005) Preparative expression of secreted proteins in bacteria: status report and future prospects. Curr Opin Biotechnol 16: 538–545

Georgiou G, Stathopoulos C, Daugherty PS, Nayak AR, Iverson BL, Curtiss R 3rd (1997) Display of heterologous proteins on the surface of microorganisms: from the screening of combinatorial libraries to live recombinant vaccines. Nat Biotechnol 15: 29–34

Gohlke U, Pullan L, McDevitt CA, Porcelli I, de Leeuw E, Palmer T, Saibil HR, Berks BC (2005) The TatA component of the twin-arginine protein transport system forms channel complexes of variable diameter. Proc Natl Acad Sci USA 102: 10482–10486

Grass S, St Geme JW 3rd (2000) Maturation and secretion of the non-typable Haemophilus influenzae HMW1 adhesin: roles of the N-terminal and C-terminal domains. Mol Microbiol 36: 55–67

Hatzixanthis K, Palmer T, Sargent F (2003) A subset of bacterial inner membrane proteins integrated by the twin-arginine translocase. Mol Microbiol 49: 1377–1390

He SY, Nomura K, Whittam TS (2004) Type III protein secretion mechanism in mammalian and plant pathogens. Biochim Biophys Acta 1694: 181–206

Henderson IR, Navarro-Garcia F, Desvaux M, Fernandez RC, Ala'Aldeen D (2004) Type V protein secretion pathway: the autotransporter story. Microbiol Mol Biol Rev 68: 692–744

Hodak H, Clantin B, Willery E, Villeret V, Locht C, Jacob-Dubuisson F (2006) Secretion signal of the filamentous haemagglutinin, a model two-partner secretion substrate. Mol Microbiol 61: 368–382

Holland IB, Schmitt L, Young J (2005) Type 1 protein secretion in bacteria, the ABC-transporter dependent pathway (review). Mol Membr Biol 22: 29–39

Jacob-Dubuisson F, Fernandez R, Coutte L (2004) Protein secretion through autotransporter and two-partner pathways. Biochim Biophys Acta 1694: 235–257

Johnson TL, Abendroth J, Hol WG, Sandkvist M (2006) Type II secretion: from structure to function. FEMS Microbiol Lett 255: 175–186

Jongbloed JD, Antelmann H, Hecker M, Nijland R, Bron S, Airaksinen U, Pries F, Quax WJ, van Dijl JM, Braun PG (2002) Selective contribution of the twin-arginine translocation pathway to protein secretion in Bacillus subtilis. J Biol Chem 277: 44068–44078

Koronakis V, Eswaran J, Hughes C (2004) Structure and function of TolC: the bacterial exit duct for proteins and drugs. Annu Rev Biochem 73: 467–489

Koronakis V, Sharff A, Koronakis E, Luisi B, Hughes C (2000) Crystal structure of the bacterial membrane protein TolC central to multidrug efflux and protein export. Nature 405: 914–919

Kostakioti M, Newman CL, Thanassi DG, Stathopoulos C (2005) Mechanisms of protein export across the bacterial outer membrane. J Bacteriol 187: 4306–4314

Kuehn MJ, Heuser J, Normark S, Hultgren SJ (1992) P pili in uropathogenic E. coli are composite fibres with distinct fibrillar adhesive tips. Nature 356: 252–255

Lambert-Buisine C, Willery E, Locht C, Jacob-Dubuisson F (1998) N-terminal characterization of the Bordetella pertussis filamentous haemagglutinin. Mol Microbiol 28: 1283–1293

Lee PA, Tullman-Ercek D, Georgiou G (2006) The bacterial twin-arginine translocation pathway. Annu Rev Microbiol 60: 373–395

Li H, Qian L, Chen Z, Thibault D, Liu G, Liu T, Thanassi DG (2004) The outer membrane usher forms a twin-pore secretion complex. J Mol Biol 344: 1397–1407

Luirink J, Sinning I (2004) SRP-mediated protein targeting: structure and function revisited. Biochim Biophys Acta 1694: 17–35

Maeshima H, Okuno E, Aimi T, Morinaga T, Itoh T (2001) An archaeal protein homologous to mammalian SRP54 and bacterial Ffh recognizes a highly conserved region of SRP RNA. FEBS Lett 507: 336–340

Meli AC, Hodak H, Clantin B, Locht C, Molle G, Jacob-Dubuisson F, Saint N (2006) Channel properties of TpsB transporter FhaC point to two functional domains with a C-terminal protein-conducting pore. J Biol Chem 281: 158–166

Meng G, Surana NK, St Geme JW 3rd, Waksman G (2006) Structure of the outer membrane translocator domain of the Haemophilus influenzae Hia trimeric autotransporter. Embo J 25: 2297–2304

Merritt EA, Hol WG (1995) AB5 toxins. Curr Opin Struct Biol 5: 165–171

Mougous JD, Cuff ME, Raunser S, Shen A, Zhou M, Gifford CA, Goodman AL, Joachimiak G, Ordonez CL, Lory S, Walz T, Joachimiak A, Mekalanos JJ (2006) A virulence locus of Pseudomonas aeruginosa encodes a protein secretion apparatus. Science 312: 1526–1530

Muller JP, Ozegowski J, Vettermann S, Swaving J, Van Wely KH, Driessen AJ (2000) Interaction of Bacillus subtilis CsaA with SecA and precursor proteins. Biochem J 348 Pt 2: 367–373

Newman CL, Stathopoulos C (2004) Autotransporter and two-partner secretion: delivery of large-size virulence factors by gram-negative bacterial pathogens. Crit Rev Microbiol 30: 275–286

Ng SY, Chaban B, VanDyke DJ, Jarrell KF (2007) Archaeal signal peptidases. Microbiology 153: 305–314

Nouwen N, Ranson N, Saibil H, Wolpensinger B, Engel A, Ghazi A, Pugsley AP (1999) Secretin PulD: association with pilot PulS, structure, and ion-conducting channel formation. Proc Natl Acad Sci USA 96: 8173–8177

Nouwen N, Stahlberg H, Pugsley AP, Engel A (2000) Domain structure of secretin PulD revealed by limited proteolysis and electron microscopy. Embo J 19: 2229–2236

Oliver DC, Huang G, Nodel E, Pleasance S, Fernandez RC (2003) A conserved region within the Bordetella pertussis autotransporter BrkA is necessary for folding of its passenger domain. Mol Microbiol 47: 1367–1383

Oomen CJ, van Ulsen P, van Gelder P, Feijen M, Tommassen J, Gros P (2004) Structure of the translocator domain of a bacterial autotransporter. Embo J 23: 1257–1266

Pallen MJ, Chaudhuri RR, Henderson IR (2003) Genomic analysis of secretion systems. Curr Opin Microbiol 6: 519–527

Palmer T, Berks RBC (2007) The Tat protein export pathway. In: Ehrmann M (ed) The periplasm. ASM Press, Washington D.C., pp 16–29

Park HS, Kayser KJ, Kwak JH, Kilbane JJ 2nd (2004) Heterologous gene expression in Thermus thermophilus: beta-galactosidase, dibenzothiophene monooxygenase, PNB carboxy esterase, 2-aminobiphenyl-2,3-diol dioxygenase, and chloramphenicol acetyl transferase. J Ind Microbiol Biotechnol 31: 189–197

Pohlschroder M, Gimenez MI, Jarrell KF (2005) Protein transport in Archaea: Sec and twin arginine translocation pathways. Curr Opin Microbiol 8: 713–719

Pugsley AP, Francetic O, Possot OM, Sauvonnet N, Hardie KR (1997) Recent progress and future directions in studies of the main terminal branch of the general secretory pathway in Gram-negative bacteria—a review. Gene 192: 13–19

Pukatzki S, Ma AT, Sturtevant D, Krastins B, Sarracino D, Nelson WC, Heidelberg JF, Mekalanos JJ (2006) Identification of a conserved bacterial protein secretion system in *Vibrio cholerae* using the Dictyostelium host model system. Proc Natl Acad Sci USA 103: 1528–1533

Rachel R, Wyschkony I, Riehl S, Huber H (2002) The ultrastructure of Ignicoccus: evidence for a novel outer membrane and for intracellular vesicle budding in an archaeon. Archaea 1: 9–18

Rao PS, Yamada Y, Tan YP, Leung KY (2004) Use of proteomics to identify novel virulence determinants that are required for Edwardsiella tarda pathogenesis. Mol Microbiol 53: 573–586

Roberts JA, Marklund BI, Ilver D, Haslam D, Kaack MB, Baskin G, Louis M, Mollby R, Winberg J, Normark S (1994) The Gal(alpha 1–4)Gal-specific tip adhesin of Escherichia coli P-fimbriae is needed for pyelonephritis to occur in the normal urinary tract. Proc Natl Acad Sci USA 91: 11889–11893

Sanchez-Pulido L, Devos D, Genevrois S, Vicente M, Valencia A (2003) POTRA: a conserved domain in the FtsQ family and a class of beta-barrel outer membrane proteins. Trends Biochem Sci 28: 523–526

Sandkvist M (2001) Type II secretion and pathogenesis. Infect Immun 69: 3523–3535

Sauer FG, Remaut H, Hultgren SJ, Waksman G (2004) Fiber assembly by the chaperone-usher pathway. Biochim Biophys Acta 1694: 259–267

Schindel C, Zitzer A, Schulte B, Gerhards A, Stanley P, Hughes C, Koronakis V, Bhakdi S, Palmer M (2001) Interaction of Escherichia coli hemolysin with biological membranes. A study using cysteine scanning mutagenesis. Eur J Biochem 268: 800–808

Schlumberger MC, Hardt WD (2006) Salmonella type III secretion effectors: pulling the host cell's strings. Curr Opin Microbiol 9: 46–54

Sibbald MJ, Ziebandt AK, Engelmann S, Hecker M, de Jong A, Harmsen HJ, Raangs GC, Stokroos I, Arends JP, Dubois JY, van Dijl JM (2006) Mapping the pathways to staphylococcal pathogenesis by comparative secretomics. Microbiol Mol Biol Rev 70: 755–788

Soto GE, Dodson KW, Ogg D, Liu C, Heuser J, Knight S, Kihlberg J, Jones CH, Hultgren SJ (1998) Periplasmic chaperone recognition motif of subunits mediates quaternary interactions in the pilus. Embo J 17: 6155–6167

Stathopoulos C, Hendrixson DR, Thanassi DG, Hultgren SJ, St Geme JW 3rd, Curtiss R 3rd (2000) Secretion of virulence determinants by the general secretory pathway in gram-negative pathogens: an evolving story. Microbes Infect 2: 1061–1072

Steiner D, Forrer P, Stumpp MT, Pluckthun A (2006) Signal sequences directing cotranslational translocation expand the range of proteins amenable to phage display. Nat Biotechnol 24: 823–831

Szabady RL, Peterson JH, Skillman KM, Bernstein HD (2005) An unusual signal peptide facilitates late steps in the biogenesis of a bacterial autotransporter. Proc Natl Acad Sci U S A 102: 221–226

Tam PC, Maillard AP, Chan KK, Duong F (2005) Investigating the SecY plug movement at the SecYEG translocation channel. Embo J 24: 3380–3388

Thanassi DG (2002) Ushers and secretins: channels for the secretion of folded proteins across the bacterial outer membrane. J Mol Microbiol Biotechnol 4: 11–20

Thanassi DG, Hultgren SJ (2000) Assembly of complex organelles: pilus biogenesis in gram-negative bacteria as a model system. Methods 20: 111–126

Thanassi DG, Stathopoulos C, Karkal A, Li H (2005) Protein secretion in the absence of ATP: the autotransporter, two-partner secretion and chaperone/usher pathways of gram-negative bacteria (review). Mol Membr Biol 22: 63–72

Thompson SA, Shedd OL, Ray KC, Beins MH, Jorgensen JP, Blaser MJ (1998) *Campylobacter fetus* surface layer proteins are transported by a type I secretion system. J Bacteriol 180: 6450–6458

Tjalsma H, Antelmann H, Jongbloed JD, Braun PG, Darmon E, Dorenbos R, Dubois JY, Westers H, Zanen G, Quax WJ, Kuipers OP, Bron S, Hecker M, van Dijl JM (2004) Proteomics of protein secretion by Bacillus subtilis: separating the "secrets" of the secretome. Microbiol Mol Biol Rev 68: 207–233

Van den Berg B, Clemons WM Jr, Collinson I, Modis Y, Hartmann E, Harrison SC, Rapoport TA (2004) X-ray structure of a protein-conducting channel. Nature 427: 36–44

van Ulsen P, van Alphen L, ten Hove J, Fransen F, van der Ley P, Tommassen J (2003) A Neisserial autotransporter NalP modulating the processing of other autotransporters. Mol Microbiol 50: 1017–1030

Voulhoux R, Ball G, Ize B, Vasil ML, Lazdunski A, Wu LF, Filloux A (2001) Involvement of the twin-arginine translocation system in protein secretion via the type II pathway. Embo J 20: 6735–6741

Voulhoux R, Bos MP, Geurtsen J, Mols M, Tommassen J (2003) Role of a highly conserved bacterial protein in outer membrane protein assembly. Science 299: 262–265

Vrontou E, Economou A (2004) Structure and function of SecA, the preprotein translocase nanomotor. Biochim Biophys Acta 1694: 67–80

Westers L, Westers H, Quax WJ (2004) Bacillus subtilis as cell factory for pharmaceutical proteins: a biotechnological approach to optimize the host organism. Biochim Biophys Acta 1694: 299–310

Yamane K, Bunai K, Kakeshita H (2004) Protein traffic for secretion and related machinery of Bacillus subtilis. Biosci Biotechnol Biochem 68: 2007–2023

Zhu C, Ruiz-Perez F, Yang Z, Mao Y, Hackethal VL, Greco KM, Choy W, Davis K, Butterton JR, Boedeker EC (2006) Delivery of heterologous protein antigens via hemolysin or autotransporter systems by an attenuated ler mutant of rabbit enteropathogenic Escherichia coli. Vaccine 24: 3821–3831

6
Regulation of Transcription in Bacteria by DNA Supercoiling

Charles J. Dorman

Abstract Understanding mechanisms of gene regulation is a major goal of modern molecular biology and much work has focused on the central roles of DNA binding proteins in controlling the key events in gene expression. This chapter takes a different approach by considering the contribution of the genetic material itself to gene regulatory processes. DNA is often regarded as a passive partner in the gene regulatory relationship, a mere substrate on which the proteins act. Here, we will examine evidence that the conformation of DNA has a significant influence on the gene expression process, at least at the level of transcription. The focus of this review is on transcription regulation in bacteria by DNA supercoiling, with an emphasis on the Gram-negative organism, *Escherichia coli*, and its close relatives, not least because much of the relevant groundbreaking work has been conducted in these microbes.

Charles J Dorman
Department of Microbiology, Moyne Institute of Preventive Medicine, Trinity College
Dublin, Dublin2, Ireland
cjdorman@tcd.ie

W. El-Sharoud (ed.) *Bacterial Physiology: A Molecular Approach.*
© Springer-Verlag Berlin Heidelberg 2008

6.1 Negative Supercoiling

The double-stranded DNA in most cells is maintained in an underwound state. That is to say, it has a deficit of helical turns. This places the molecule under torsional stress, causing it to adopt a conformation that allows it to relieve this stress. The result is usually the plectonemic writhing that we associate with supercoiled DNA in bacteria such as *Escherichia coli* (Drlica 1992; Drlica et al. 1999; Sinden 1994). The pathway of the DNA within the writhed molecule has a negative sign, hence, the term negative supercoiling. It is important to note that DNA can also be over-wound, resulting in the addition of helical turns (Fig 6.1). This also leads to a writhed structure, but one with a positive sign. DNA in this state is said to be posi-tively supercoiled. Although negative supercoiling is the rule in bacteria such as *E. coli*, positive supercoils can arise as a result of certain DNA-based reactions, as we shall see below (Sect. 6.4). In some bacteria that inhabit extremely stressful environments, such as those at a constant high temperature in deep-sea locations, positive supercoiling, with its associated tightening of the genomic DNA, may be the normal situation, perhaps as a way of imposing stability on the DNA duplex in conditions that would otherwise be structurally disruptive (Hsieh and Plank 2006; Kikuchi 1990; Kikuchi and Asai 1984).

Topology is the branch of mathematics that deals with the shapes of objects, and the structure of DNA can be described conveniently in topological terms. Double-stranded DNA has a linking number, L, which is a statement of the number of times one strand in the duplex crosses the path of the other. When the DNA is fully

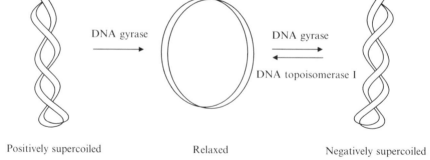

Positively supercoiled   Relaxed   Negatively supercoiled

Fig. 6.1 Positive and negative writhing of DNA. The circular ribbon represents a covalently closed DNA molecule and this is shown in a relaxed conformation in the center. In this state, the DNA lies in the plane of the page. Overwinding the DNA duplex causes it to adopt a positively writhed conformation, equivalent to positive supercoiling, as shown on the left. Underwinding the DNA causes it to adopt negative writhe, as shown on the right. The positively supercoiled mole-cule is a substrate for DNA gyrase that relaxes it using an identical enzymatic process to that used to introduce negative supercoils (as shown on the right). DNA topoisomerase I relaxes the nega-tively supercoiled DNA molecule

relaxed, it has a characteristic linking number, L_0. Integer increases or decreases in L result in departures from L_0 in a positive or negative direction, respectively. Linking number changes occur after breakage of one or both DNA strands followed by swiveling or rotation of the cut end or ends and then religation. The addition or subtraction of links has consequences for the other DNA topological parameters, twist (T) and writhe (W). Twist is a measurement of the number of turns of the DNA strands around the helical axis, whereas writhe describes the number of times the duplex axis winds around itself. Linking number, twist, and writhe are interdependent, and their relationship is described by the equation: $L = T + W$.

This interrelationship means that changes in linking number are expressed as compensatory changes in the other parameters thus: $\Delta L = \Delta T + \Delta W$.

The linking number change can be partitioned between changes in twist and writhe or it can become manifest exclusively in changes to one parameter or the other. Negative changes in L that result in untwisting of the duplex have the potential to break hydrogen bonds in the DNA and, thus, assist biological reactions, such as transcription initiation, that rely on the formation of single-stranded regions. This represents one of the main ways that DNA supercoiling can influence gene expression. However, transcription can be modulated at a number of levels by underwinding of the DNA template. The writhing that accompanies reductions in linking number can assist or impede long-range interactions along the DNA, and this can have consequences for the interaction of bound regulatory proteins with RNA polymerase. These interactions can influence the recruitment of polymerase to promoters and can, therefore, affect the formation of closed transcription complexes. The conversion of these to open complexes can be modulated, in turn, by the duplex unwinding reactions referred to above. DNA superhelicity also has the potential to influence transcript elongation and termination. Thus, every step in the transcription process can be affected by the superhelicity of the DNA.

6.2 DNA Topoisomerases

The negative superhelical state of the DNA arises, at least in part, because of the action of DNA gyrase (Wang 1996; Westerhoff et al. 1988). This is an ATP-dependent topoisomerase, an enzyme that changes the linking number of the DNA substrate in steps of two (Table 6.1). Its ATP dependence links gyrase activity to the [ATP]/[ADP] ratio and, hence, to the energy charge of the cell (van Workum et al. 1996). This is thought to provide a useful connection between DNA superhelicity and bacterial metabolism. The activity of gyrase is opposed by DNA topoisomerase I, a type I enzyme that alters the linking number of negatively supercoiled DNA molecules in steps of one (Table 6.1). The opposing activities of gyrase and topoisomerase I are thought to maintain negative superhelicity within physiologically tolerable limits (Fig. 6.1). The two enzymes regulate supercoiling homeostatically, and the expression of the genes that encode them reflects this balance. The promoter of the *topA* gene (topoisomerase I) is activated by DNA negative superhelicity and becomes repressed as the DNA is relaxed (Tse-Dinh 1985; Tse-Dinh

Table 6.1 The topoisomerases of *E. coli*

Protein name	Type[+]	Gene	Molecular mass (kDa)	Needs ATP?	Needs Mg^{2+}?	Remarks
Topoisomerase I	I	topA	197	No	Yes	Binds to cleavage site via a 5′-phosphotyrosine bond; DNA swivelase making a transient cut in one strand of the DNA duplex, relaxes negatively supercoiled DNA
Topoisomerase III	I	topB	73.2	No	Yes	Relaxes negatively supercoiled DNA; decatenase activity; catenase activity (with RecQ)
DNA gyrase	II	gyrA gyrB	105 (GyrA) 95 (GyrB)	Yes	Yes	Negative supercoiling relaxes negative or positively supercoiled DNA
Topoisomerase IV	II	parC parE	175 (ParC) 170 (ParE)	Yes	Yes	Decatenase activity; relaxes negative supercoils; cannot introduce negative supercoils

*Type I enzymes alter the linking number of DNA in steps of one (Δ Lk =1); type II enzymes do this in steps of 2 (ΔLk=2).

and Beran 1988; Weinstein-Fischer et al. 2000; Weinstein-Fischer and Altuvia 2007). In contrast, the promoters of the *gyrA* and *gyrB* genes (A and B gyrase subunits, respectively) are activated by DNA relaxation and inhibited by negative supercoiling (Menzel and Gellert 1983, 1987). *E. coli* also expresses two other topoisomerases. These are topoisomerases III and IV. Topoisomerase III is a type I enzyme and is responsible for decatenation reactions through its ability to make single-stranded breaks in DNA. Topoisomerase IV is a type II enzyme related to gyrase but lacking the ability to introduce negative supercoiling (Kato et al. 1990; Drlica and Zhao 1997) (Table 6.1). Its primary role is to decatenate daughter chromosomes at the end of each round of chromosome replication. Mutants with defects in the genes (*parC* and *parE*) that encode topoisomerase IV have an increased tendency to produce anucleate daughter cells at cell division.

6.3 DNA Supercoiling and Nucleoid Organization

The negatively supercoiled state of bacterial DNA contributes to the organization of the genetic material in the nucleoid. This structure consists of the chromosome and its associated proteins. Left to its own devices, the *E. coli* chromosome would adopt a globular structure with a radius approximately five times too great to fit in the cell (Trun and Marko 1998). The organization of the chromosome into up to 400 independent looped domains and the imposition on these domains of plectonemic writhing by negative supercoiling plays an important part in compacting the nucleoid (Deng et al. 2005; Postow et al. 2004; Stein et al. 2005). The heterologous family of nucleoid-associated proteins imparts further organization (Dorman and Deighan 2003; Dorman 2004; Hardy and Cozzarelli 2005). The most important of these, from an organizational perspective, are H-NS, Fis, and HU, with H-NS and HU being leading candidates for the role of domain boundary elements, whereas Fis is thought to associate with the apices of the supercoiled looped domains. Frequently, these proteins cooperate with DNA supercoiling to modulate transcription (Dorman and Deighan 2003).

6.4 Supercoiling Alters Transcription and Vice Versa

The capacity for DNA supercoiling to influence transcription is perhaps intuitively obvious given the effect of underwinding of the DNA duplex on the stability of hydrogen bonds between its two strands. A great deal of work has established that transcription in bacteria is indeed responsive to changes in superhelicity of the DNA template. Much of this comes from investigations of individual promoters, and other evidence has come from whole genome studies using DNA microarray methods. In the latter studies, global transcription is monitored after inhibition of topoisomerases by mutation or treatment with inhibitors. When this procedure was carried out in *E. coli*, transcription of 106 genes increased after DNA relaxation whereas transcription of 200 genes decreased. The genes that were affected carried out a diverse range of functions in the cell and were located in all parts of the genome (Peter et al. 2004).

When discussing the well-established effects of DNA supercoiling on transcription, it is important to keep in mind the fact that the transcription process can, in turn, influence DNA supercoiling (Fig. 6.2). Early work with the cloning vector pBR322 suggested a link between transcription and DNA supercoiling in which the reading of the gene coding for the tetracycline resistance protein affected the topology of the plasmid (Pruss and Drlica 1986). It was subsequently suggested that the movement of RNA polymerase along the DNA template, with the associated strand separation, results in the creation of a domain of overwound (or positively super-coiled) DNA ahead and an underwound (or negatively supercoiled) domain behind the moving transcription complex (Liu and Wang 1987) (Fig. 6.2). This leads quickly to a situation in which the polymerase will become jammed unless the associated topological complex is resolved. It is thought that the viscous drag in the cytoplasm, together with the coupling of transcription and translation in bacteria, creates a barrier to the rotation of the transcription complex around the DNA. The DNA may not be free to rotate either because of anchoring at looped domain barriers, or the physical connection of the coupled transcription and translation complex to

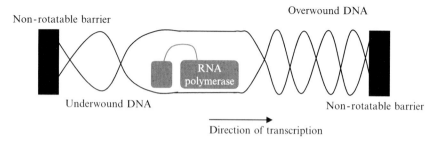

Direction of transcription

Fig. 6.2 The Twin-domain supercoil model. A segment of double-stranded DNA is tethered at each end to a nonrotatable support. RNA polymerase tracks along one strand within a single-stranded bubble. The polymerase cannot rotate, and its forward movement creates a domain of overwound or positively supercoiled DNA ahead and a domain of underwound or negatively supercoiled DNA behind. The underwound DNA can be relaxed by DNA topoisomerase I, whereas the overwound DNA can be relaxed by DNA gyrase

the cell membrane in the case of genes coding for secreted or membrane-associated proteins. The resolution of this apparent impasse depends on the activities of the topoisomerases of the cell. DNA topoisomerase I can relax the negatively super-coiled domain behind the transcription complex while DNA gyrase relaxes the positively supercoiled domain that lies ahead (Liu and Wang 1987; Massé and Drolet 1999a; Mielke et al. 2004). This "Twin-supercoiled domain" model has received experimental support and is generally accepted as a useful description of the impact of the process of transcription on DNA supercoiling (Chen and Wu 2003; Chen et al. 1992; Leng and McMacken 2002; Leng et al. 2004; Pruss and Drlica 1989; Rhee et al. 1999; Wang and Dröge 1997; Wu et al. 1988).

In this way, the process of transcription generates positive supercoiling. The movement of DNA polymerase during DNA replication can also create positive supercoiling. This overwinding of the DNA duplex poses a problem for the cell because it can interfere with polymerase movement and because the associated tightening of the DNA can impede strand separation required for gene expression. The creation of local domains of negative supercoiling also has the potential to affect gene expression.

The phenomenon of promoter coupling is an interesting by-product of the ability of transcription to influence DNA superhelicity (Fig. 6.3). Here, supercoiling changes generated during the transcription of one gene may affect the performance of the promoter of another gene (Chen and Wu 2003; Wu et al. 1988). This can occur when the genes are organized divergently, convergently, or in tandem on the chromosome (or other replicon). These promoter–promoter interactions need not be confined to pairs of genes but can spread along the DNA to create promoter relays when the DNA supercoiling signal is telegraphed over several kilobases of the chromosome. The *leuABCD–leuO–ilvIH* region of the *Salmonella* chromosome represents one well-characterized example of such a relay (Chen et al. 1992; Fang and Wu 1998; Hanafi and Bossi 2000; Lilley and Higgins 1991; Opel and Hatfield 2001; Opel et al. 2001). Observations of this type reveal previously hidden subtleties

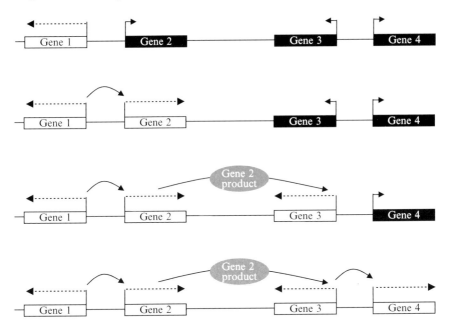

Fig. 6.3 Local DNA supercoiling and a promoter relay. The promoter of gene 1 is active, creating a microdomain of negatively supercoiled DNA that extends back to the promoter of divergently transcribed gene 2. The product of the second gene is a transcription factor that activates the promoter of gene 3. The creation of a microdomain of negatively supercoiled DNA by the gene 3 promoter activates the upstream and divergently oriented promoter of gene 4. This scenario is based on the promoter relay that operates in the *leuABCD–ilvIH* locus of the *Salmonella enterica* chromosome

in the relationships among bacterial genes that share a common DNA template and show how genes have the ability to modulate the expression of their neighbors through transient modification of the structure of the genetic material itself.

6.5   The Homeostatic Model of DNA Supercoiling

Evidence that changes in DNA supercoiling can affect the expression of genes in bacteria has come from a number of sources. The homeostatic model of DNA supercoiling management has at its heart a role for supercoiling in modulating the activities of the *gyrA*, *gyrB*, and *topA* genes in ways that allow linking number changes to feed back onto topoisomerase gene transcription. This was discovered in studies in which the *topA* gene was inactivated by mutation and a shift in global supercoiling levels to more negatively supercoiled values was detected. This results in a change in the expression of genes with DNA supercoiling-sensitive promoters. Mutants deficient in the *topA* locus grow poorly, but normal growth is restored if they acquire compensatory mutations that reduce the activity of DNA gyrase (DiNardo et al. 1981; Pruss et al. 1982). Compensation can also arise because of amplification of the number of copies of the *parC* and *parE* genes on the chromosome (Kato et al. 1990) (Table 6.1). This phenomenon has been referred to in the past as involving *toc*, from the locus for *topoisomerase one compensation* (Dorman et al. 1989; Raji et al. 1985). The *toc* event permits compensation by providing an increased level of topoisomerase IV, with its DNA relaxing activity, to replace that lost through *topA* inactivation (Free and Dorman 1994; Kato et al. 1990; McNairn et al. 1995). Topoisomerase III can also suppress the effects associated with a *topA* mutation when the *topB* gene is expressed in multicopy (Brocolli et al. 2000).

R-loop formation is an important consequence of *topA* inactivation (Phoenix et al. 1997). These arise when transcripts base-pair with the coding DNA strand, causing the noncoding strand to be displaced as a single-stranded bubble. Their formation is promoted by negative supercoiling, and it has been proposed that a major in vivo role of topoisomerase I lies in the suppression of R-loop formation (Brocolli et al. 2000). R-loop generation in *topA* mutants can be suppressed by overexpression of topoisomerase III or RNase H. The former is thought to relax the DNA template and suppress R-loop creation, whereas RNase H degrades the RNA component of those loops that do occur (Brocolli et al. 2000; Massé and Drolet 1999b; Drolet et al. 1995).

Investigators often use multicopy plasmids such as the cloning vectors pUC18, pBR322, or pACYC184 as reporters of global supercoiling trends (Kelly et al. 2006; Ó Cróinín et al. 2006; Peter et al. 2004). Linking number changes in the plasmids can be measured electrophoretically in agarose gels containing chloroquine, a DNA intercalating agent that separates the different topological isomers (or topoisomers) of the plasmid. This technique allows one to compare a wild-type bacterium with a *topA* mutant. Similarly, wild-type bacteria that have been treated with a DNA gyrase-inhibiting antibiotic can be compared with an untreated control, allowing one to detect the effect of gyrase inhibition on plasmid linking number. More recently, the effect on the gene expression profile of the cell of altering global DNA supercoiling levels by treatments of this type has been examined using DNA microarray methods. As one might expect, alterations in DNA supercoiling result in widespread changes in the bacterial transcriptome (Cheung et al. 2003; Gmuender et al. 2001; Peter et al. 2004).

6.6   Environmental Modulation of DNA Supercoiling

Experiments with antibiotics or with mutants with deficiencies in topoisomerase expression established that changing the linking number of DNA had the potential to influence gene expression. Detailed investigations using reductionist molecular biology methods showed that these effects were usually caused by the influence of negative supercoiling on promoter function (Borowiec and Gralla 1985; Bowater et al. 1994; Free and Dorman 1994). However, for linking number changes to be meaningful in a truly physiological setting, it was necessary to demonstrate that they could occur in response to signals encountered by bacteria in the normal course of their lives (Dorman and Ní Bhriain 1992).

Early work showed that environmental signals could result in a linking number difference in reporter plasmids isolated from bacteria undergoing stress. It was shown that bacterial DNA underwent a linking number change when the culture was shifted from an aerobic to an anaerobic environment (Dixon et al. 1988; Dorman et al. 1988; Ní Bhriain et al. 1989; Yamamoto and Droffner 1985). Similarly, osmotic upshift altered the linking number of reporter plasmids in Gram-negative (Higgins et al. 1988; McClellan et al. 1990) and Gram-positive bacteria (Sheehan et al. 1992). More recently, DNA microarray analysis has shown that osmotic stress alters the expression of very many genes around the *E. coli* chromosome (Church et al. 2003). The increases in negative supercoiling that accompany the onset of osmotic upshift play a role in selecting for *topA* compensatory mutations in *topA* mutants. If the *topA* mutants avoid osmotic stress, compensation is not required for the bacteria to have normal rates of growth (Dorman et al. 1989). Other environmental parameters that were found to affect DNA supercoiling levels include temperature (Dorman et al. 1990; Goldstein and Drlica 1984; Friedman et al. 1995) and pH (Karem and Foster 1993).

The link between the environmental stimuli and the linking number of the DNA consists of DNA gyrase and the [ATP]/[ADP] ratio of the cell (Hsieh et al. 1991a, 1991b; van Workum et al. 1996). Gyrase activity requires ATP, is inhibited by ADP, and the ratio of these molecules is a measure of the metabolic flux of the cell. Metabolic activity is, in turn, responsive to the environment. One might predict that direct manipulation of metabolic pathways can alter the linking number of the DNA in the cell. Consistent with this view is the finding that DNA linking number is affected by the carbon source in the bacterial growth medium (Balke and Gralla 1987) and that it varies throughout the growth of a bacterial batch culture (Dorman et al. 1988). This leads to a model in which the dynamic nature of the level of negative supercoiling of bacterial DNA has the potential to influence the gene expression program of the cell as the organism interacts with its environment in time and space (Dorman 1995).

How does this work? Every gene in the cell has the potential to respond to changes in DNA supercoiling. This is because the free energy imparted to the DNA by reductions in linking number can result in alterations in DNA twist or writhe or both that affect promoter function. Such a general influence offers the possibility of coordinating the responses of very large numbers of genes to changes in environmental parameters. However, the control offered by DNA linking number change alone is very crude and can be regarded only as an underlying or background influence within the global gene expression program of the cell. Refinement of the transcriptional response to environmental changes involves other regulatory influences. One of these is provided by the nucleoid-associated proteins, discussed above as contributors to the organization of the bacterial nucleoid (see Sect. 6.3). It is abundantly clear that most nucleoid-associated proteins also play key roles in the regulation of transcription (Dorman and Deighan 2003). Further and more targeted control is exerted by the conventional transcription factors of the cell. These are the many sequence-specific repressors and activators of transcription that control the activity of RNA polymerase. This leads to a hierarchical view of the control of bacterial transcription, in which DNA supercoiling is at the apex by virtue of its potential to influence most promoters, followed by the nucleoid-associated proteins and then the transcription factors. With its responsiveness to environmental stress, one can envisage DNA supercoiling as a candidate for an early and primitive transcription regulator during the evolution of bacterial gene regulatory circuits (Dorman 2002). How is its contribution to gene regulation integrated with that of the nucleoid-associated proteins?

6.7 Nucleoid-Associated Proteins and DNA Supercoiling

By some estimates, *E. coli* and its close relatives have at least 12 distinct types of nucleoid-associated proteins (Azam and Ishihama 1999). These are roughly equivalent in function to the histone proteins from eukaryotic chromatin, although they do not resemble these in amino acid sequence. One clue to their role, and to the existence of a chromatin-like substance in bacteria, comes from measurements of the superhelical density of bacterial DNA. This parameter, known as σ, is equivalent to the average number of superhelical turns per helical turn of the DNA (Sinden 1994). When it is measured for DNA in vivo, the value of -0.025 is found to be half that obtained when the measurement is made for purified DNA in vitro, -0.05 (Bliska and Cozzarelli 1987). Why is this? The explanation is that the approximately half of the DNA in the cell is in complexes with protein, and the wrapping of the DNA around these proteins removes approximately half of the supercoils. When the DNA is purified, the purification process removes the proteins and the fully supercoiled nature of the DNA is revealed. Thus, one can consider the operational or effective level of supercoiling, one that has the power to influence biological activity, in contrast to the absolute value, which is only seen when the DNA is removed from the cells and bound protein (Blot et al. 2000). These proteins are the nucleoid-associated proteins. The most prominent members of the group are H-NS, Fis, HU, and integration host factor (IHF) (Dorman and Deighan, 2003). The negatively supercoiled looped domain structure of the bacterial chromosome has already been described, as has the possibility that H-NS and HU may be involved in forming domain boundaries. IHF resembles a more specific version of the HU protein (Swinger and Rice 2004). Both are composed of heterologous subunits and have an AB structure. Homomeric versions of each exist and these have biological activities that are different from those of the heteromers (Claret and Rouvière-Yaniv 1997; Mangan et al. 2006). Transcriptomic analysis using DNA microarrays have shown that IHF has widespread effects on transcription in *E. coli* and *Salmonella* (Arfin et al. 2000; Mangan et al. 2006). It is also associated with the control of transcription through modulation of DNA supercoiling. This is achieved through its ability to bind and bend DNA within regions of the genome that are prone to the formation of single-stranded regions when negatively supercoiled and then to transmit DNA twist to nearby transcription promoters. This mechanism is discussed in Sect. 6.9.

The Fis protein shares this twist-transmission property with IHF (see Sect. 6.9) and also makes a number of other contributions to the control of DNA supercoiling and, hence, transcription. Similar to IHF, Fis has been shown by transcriptomic analysis to affect the expression of a very large number of genes in *E. coli* and *Salmonella*, and some of these effects are known to involve a role for DNA supercoiling (Blot et al. 2006; Kelly et al. 2004). The Fis protein was discovered originally as a cofactor in the site-specific recombination reactions that are catalyzed by the serine invertase family of site-specific recombinases in Gram-negative bacteria and their bacteriophage (Sanders and Johnson 2004). Subsequently, the Fis protein was shown to be a transcriptional regulator capable of acting as an activator or repressor, depending on the location of its binding site relative to the promoter. One of the promoters that it represses is that of its own gene, *fis*, allowing the protein to feed back negatively onto its own expression (Pratt et al. 1997). The role of Fis as a transcription activator has been studied intensively at the promoters of genes coding for stable RNA (transfer RNA [tRNA] and ribosomal RNA) in *E. coli* (Schneider et al. 2003). The protein binds as a dimer to one or three sites located in an upstream activator sequence (UAS) (Fig. 6.4). Its role is partly to act as a conventional transcription factor that contacts RNA polymerase and partly as a buffering agent to maintain the local level of negative DNA supercoiling at values that are optimal for promoter function (McLeod et al. 2002; Rochman et al. 2002; 2004; Travers et al. 2001) (Fig. 6.4). The occupation of the Fis binding sites is most

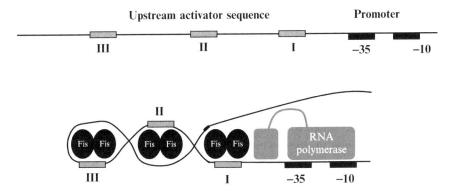

Fig. 6.4 Activation of a stable RNA gene promoter by Fis and DNA supercoiling. A typical stable RNA gene promoter is shown in a linear conformation (*top*). The positions of the −35 and −10 hexamers are shown, together with the three Fis binding sites that make up the UAS. The lower portion of the figure shows the promoter occupied by RNA polymerase and each of the Fis binding sites occupied by a Fis dimer. The Fis dimer at site I can interact with RNA polymerase, making the type of protein–protein contact that is characteristic of a conventional transcription activator. In addition, the three Fis dimers stabilize a writhed structure immediately upstream of the promoter, preserving a microdomain of negatively supercoiled DNA that can assist promoter function. Finally, the writhing of the DNA promotes an additional DNA contact over the back of the polymerase, stabilizing its interaction with the promoter

likely at the early stages of exponential growth in *E. coli* cultures, a period when the intracellular concentration of the protein is highest (Appleman et al., 1998). It is attractive to interpret this as a mechanism to boost the supply of components for the translational machinery of the cell as the culture leaves lag phase and enters the physiologically demanding exponential phase of growth.

The proposal that Fis can preserve the topological state of negatively supercoiled DNA has received support from experiments in which the protein was shown to bind preferentially to DNA of particular superhelical density and to protect the DNA from further supercoiling by DNA gyrase or relaxation by topoisomerase I (Schneider et al. 1997). Fis can make a more general contribution to the control of DNA supercoiling in the cell through its role as a repressor of the genes (*gyrA* and *gyrB*) that code for DNA gyrase and its ability to regulate the gene (*topA*) that encodes topoisomerase I (Schneider et al. 1999; Keane and Dorman 2003; Weinstein-Fischer and Altuvia 2007; Weinstein-Fischer et al. 2000).

The *fis* gene is itself regulated by DNA topology because its promoter requires negative superhelicity to function (Schneider et al. 2000). This integrates its expression nicely into that part of the global regulatory circuitry of the cell that relies on changes in DNA negative superhelicity to affect gene expression.

6.8   DNA Supercoiling and the Stringent Response

The stringent response refers to the process by which a bacterium adjusts to a buildup of uncharged tRNA molecules that result from a depletion in the resources needed to sustain protein synthesis (Jain et al. 2006). When the stringent response is activated, those genes that code for ribosome components, other translation factors, and the Fis protein become repressed at the level of transcription. This is because stringently regulated genes have promoters that are difficult to use under nutrient-depleted growth conditions. The feature that distinguishes them from other promoters is a G+C-rich discriminator sequence that is usually located between the Pribnow (or −10) box and the transcription start site (position +1) (Pemberton et al., 2000; Travers, 1980). This DNA element is characterized by its G+C content rather than any specific base sequence and it is not thought to bind a regulatory protein. Instead, its role is to raise the activation energy of the promoter by increasing the number of hydrogen bonds that must be broken for an open transcription complex to be created (Figueroa-Bossi et al. 1998). In fast-growing bacteria, the energy charge is sufficient to maintain DNA supercoiling through the action of gyrase at levels that are sufficient to overcome the discriminator. As DNA relaxes with the onset of starvation conditions, the transcription machinery can no longer open the G+C-rich promoters. Other regulatory signals also influence this process, notably the stringent response second messengers, guanosine tetraphosphate (ppGpp) and pentaphosphate (ppGppp) that modify RNA polymerase (Ohlsen and Gralla, 1992). In this way, DNA supercoiling plays a direct role in controlling the expression of the protein synthesis machinery of the cell (Travers and Muskhelishvili, 2005). Its responsiveness to growth phase provides a useful link to cellular physiology and its control over the expression of the Fis protein forms a valuable subroutine in the program because Fis serves to sustain the activity of promoters involved in the expression of ribosome components and other translation factors (Schneider et al., 2000). Fis also feeds back onto the supercoiling machinery through its transcriptional regulation of topoisomerase genes (see Sect. 6.7). The result is an integrated and homeostatically regulated molecular machine in which the physiological state of the cell sets DNA superhelical density and this, in turn, modulates the gene expression program that sustains the cell.

6.9 Stress-Induced Duplex Destabilization

The ability of negative superhelicity to drive structural transitions in the DNA duplex is known as stress-induced duplex destabilization (SIDD) (Kowalski et al. 1988; Hatfield and Benham 2002). The ability of supercoiling-induced destabilization of the DNA duplex to influence transcription initiation has been discussed above (Sect. 6.4). The free energy of supercoiling can also drive other types of structural transition. These include the formation of Z DNA, with its ability to disturb the transcription of nearby genes (Benham 1990; Rahmouni and Wells 1989, 1992), and the triple-stranded form known as H-DNA (Htun and Dahlberg 1989) that has been proposed as a factor in driving phase variation in pathogenic bacteria (Belland 1991). Negative supercoiling of DNA can also aid the extrusion of cruciform structures, in which inverted repeat sequences form a four-way junction with single-stranded loops at the ends of two opposing arms (Lilley 1980).

The discovery that the binding of a protein to a region of DNA that is prone to SIDD can displace the tendency toward the formation of a single-stranded bubble to another location represents an important advance. The probability that a portion of the DNA within a specific domain will become single stranded on the introduction of a given degree of negative superhelicity can be predicted *in silico* (Benham 1990; Fye and Benham 1999; Sheridan et al. 2001; Sun et al. 1995). Analyses of topologically closed systems, such as plasmids, reveal a hierarchy of SIDD-prone sequences whose members can be ranked according to the probability that they will become single stranded in response to the torsional stress of negative supercoiling under a given set of environmental conditions. The insight that protein binding can suppress bubble formation at a SIDD-prone site raises the possibility that the free energy can be displaced to a second site that had previously been less preferred for bubble formation. If this second site is a transcription promoter, the result is a mechanism of transcription activation in which DNA twisting is transmitted from the suppressed SIDD-prone element to the promoter to facilitate open complex formation. This has been examined experimentally for bacterial promoters and very clear examples have now been described. In one of these, the promoter of the *ilvGMEDA* operon is activated by the binding of IHF to an upstream sequence within a DNA element, with a high probability of forming a bubble in response to negative supercoiling. The binding of IHF suppresses bubble formation within the SIDD site, and the resulting displacement of DNA twist activates the *ilvGMEDA* promoter (Parekh et al. 1996; Sheridan et al. 1998). This type of effect is not restricted to IHF. The Fis protein performs an analogous task in *E. coli* at the promoter of *leuV*, a gene that encodes a leucine tRNA (Opel et al. 2004). These examples show that the influence of DNA supercoiling can be moved from place to place along the chromosome through the binding of proteins capable of suppressing torsional stress-induced bubble formation. Taken together with the promoter coupling mechanism, this shows that the genetic material itself acts in a manner analogous to a telegraph, linking regulatory events and shuttling regulatory signals between them. This suggests that a model of transcriptional regulation in bacteria that concentrates chiefly on the roles of conventional transcription factors that recruit and activate RNA polymerase by direct protein–protein interaction is incomplete. It is also necessary to appreciate the subtleties of the regulatory influence of the structure of DNA itself.

6.10   DNA Supercoiling and Bacterial Virulence

A natural corollary of the discovery that environmental stresses such as temperature, osmolarity, pH, and anaerobiosis influence DNA supercoiling is that genes involved in bacterial infection are likely to be supercoiling sensitive (Dorman 1991; 1995). This is because these same environmental signals are known to modulate the expression of many bacterial virulence genes in a variety of pathogens (Dorman 2004b; Hromockyj et al. 1992; Marceau 2005; Rhen and Dorman 2005; Rohde et al. 1999; Slamti et al. 2007; Sue et al. 2004). The sensitivity of virulence gene transcription to changes in DNA supercoiling has been demonstrated in several bacterial species using mutants with deficiencies in topoisomerase expression or treatments with topoisomerase-inhibiting antibiotics (Bang et al. 2002; Beltrametti et al. 1999; Dorman, 1991; Dorman et al. 1990; Falconi et al. 1998; Fournier and Klier 2004; Galán and Curtiss 1990; Graeff-Wohlleben et al. 1995; Marshall et al. 2000; Ó Cróinín et al. 2006; Parsot and Mekalanos 1991; Rohde et al. 1994; 1999; Tobe et al. 1995).

The genes that encode the type III secretion system and its effector proteins in *Shigella flexneri* are regulated by temperature, osmolarity, and pH (Maurelli et al. 1984; Mitobe et al. 2005; Porter and Dorman 1994), all of which are known to affect the global level of negative DNA supercoiling. These virulence genes are organized in a complicated regulatory cascade in which an AraC-like transcription factor known as VirF activates the transcription of a second regulatory gene, called *virB*, whose product, in turn, displaces the H-NS repressor protein from the promoters in the cascade, resulting in transcription activation (Dorman 2004b; Turner and Dorman 2007) (Fig. 6.5). Activation of the *virF* promoter also involves displacement of the H-NS repressor but, here, protein removal is accomplished by a restructuring of the DNA in the regulatory region in response to temperature (Prosseda et al. 2004). DNA supercoiling is needed

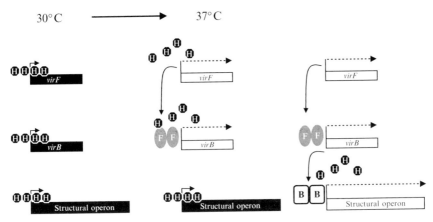

Fig. 6.5 The regulatory cascade controlling the expression of virulence genes in *Shigella flexneri*. The virulence genes are in a repressed state when the bacterium is growing at 30°C because of the presence of repression complexes in which the H-NS protein binds to each of the promoters. This binding is affected by the degree of supercoiling of the promoter DNA. When the temperature is increased to 37°C, the repression complex at the *virF* regulatory gene is disrupted because of a change in the local DNA curvature that causes displacement of H-NS. The VirF protein cooperates with the negatively supercoiled DNA at the *virB* promoter to activate transcription of this second regulatory gene. The VirB anti-repressor, in turn, remodels the DNA at the promoters of the structural genes to displace H-NS and upregulate their transcription

for the H-NS-mediated repression of the *virF* promoter (Falconi et al. 1998). The sensitivity of the *virB* intermediate regulatory gene to DNA supercoiling has been demonstrated by exploiting insights gained from the Twin-domain super-coiling model. It has been shown that this gene can be transcriptionally activated under normally nonpermissive growth conditions if an inducible promoter is placed upstream oriented away from *virB*. When this upstream promoter is activated, the resulting minidomain of negative supercoiling switches on the *virB* promoter (Tobe et al. 1995).

The virulence genes of *S. flexneri* that encode its type III secretion apparatus are located on a large virulence plasmid. *Salmonella enterica* serovar Typhimurium has two type III secretion systems that are involved in virulence, and each of these is encoded by a pathogenicity island located in the chromosome. These islands, SPI1 and SPI2, have an A+T base content in their DNA that is significantly higher than that of the rest of the genome and it is thought that the islands have been acquired by lateral gene transfer from an unidentified source (Groisman and Ochman 1997; Ochman et al. 2000). SPI1 is required for invasion of mammalian epithelial cells (Hardt et al. 1998; Lostroh and Lee 2001; Mills et al. 1995; Wood et al. 1996), whereas SPI2 is required for intracellular survival (Cirillo et al. 1998; Hensel 2000; Hensel et al. 1998; Ochman et al. 1996; Waterman and Holden 2003). The promoters within the islands are sensitive to DNA supercoiling changes (Galán and Curtiss 1990; Marshall et al. 2000; Ó Cróinín et al. 2006). Their activities are also influenced by several nucleoid-associated proteins, including H-NS and Fis, both of which are frequently found to affect the expression of genes that respond to changes in DNA superhelicity (Kelly et al. 2004; Lucchini et al. 2006; Navarre et al. 2006; Schechter et al. 2003; Wilson et al. 2001). Expression of SPI1 genes is required before invasion, and SPI2 gene expression is needed in the intracellular environment. SPI1 gene transcription is enhanced by growth conditions (a temperature of 37°C, anaerobiosis, and high osmolarity) that reduce the linking number of reporter plasmids, and these are conditions that are assumed to occur in the gut at the surface of the epithelium. In contrast, expression of SPI2 genes is optimal in conditions that favor the relaxation of supercoiled DNA. Experiments in which the superhelicity of a reporter plasmid was monitored in bacteria growing in epithelial cells and macrophage showed relaxation of the plasmid topoisomers in the macrophage (Marshall et al. 2000; Ó Cróinín et al. 2006). This activation also required the Fis DNA binding protein, suggesting that it plays a role in the maintenance of an appropriate DNA topology at SPI2 gene promoters during intracellular growth of *Salmonella* (Ó Cróinín et al. 2006). Thus, virulence genes, including those that have been acquired by horizontal transfer, use the same range of local and global regulatory devices as housekeeping genes in the ancestral component of the genome. This observation raises some interesting questions regarding the evolution of bacteria, and, in particular, the evolution of their gene regulatory circuits.

6.11   DNA Supercoiling and Gene Regulation from an Evolutionary Perspective

The dual role played by DNA supercoiling in imposing structure on the nucleoid and in influencing transcription has been discussed. Its ability to act both locally, as in the Twin-domain model, and globally makes it an ideal mechanism for exerting regulatory influences through the complement of genes in the cell. This proposal is made more compelling by the observation that negative supercoiling of bacterial DNA responds to environmental stimuli, including those encountered by pathogens during infection. Several authors have placed DNA supercoiling within a regulatory hierarchy in which its general effects on gene expression are tempered and refined by regulatory influences that are more specific (Dorman 1991, 1995, 2002; Hatfield and Benham 2002; Travers and Muskhelishvili 2005). This confers on DNA superhelicity a background role in the regulatory affairs of the cell, an agent that sets the scene against which the more specific players act. It is still there today and it can influence genes acquired by horizontal transfer as soon as they are established in the cell, either as insertions in the chromosome or as autonomously replicating plasmids. This provides an immediate, although crude, means of influencing the transcription of the newcomers that can act in combination with nucleoid-associated proteins such as H-NS and Fis (Dorman 2007; Lucchini et al. 2006; Navarre et al. 2006). These global regulators may serve to "tame" incoming sequences until specific regulatory mechanisms have evolved that permit them to be expressed in ways that benefit the cell. Horizontal genetic transfer is a very important driver of bacterial evolution and has implications for symbiosis, pathogenesis, and the emergence and spread of resistance to antimicrobials, such as antibiotics. Because horizontal transfer with a minimal impact on recipient fitness is so important for the emergence of new forms of bacteria, this is an aspect of DNA topological control of transcription that is likely to receive much attention in the near future.

Acknowledgments Research in the author's laboratory is supported by grants from Science Foundation Ireland, the Wellcome Trust and the Health Research Board.

Highly Recommended Readings

Azam AT, Ishihama A (1999) Twelve species of the nucleoid-associated protein from *Escherichia coli*: sequence recognition specificity and DNA binding affinity. J Biol Chem 274:33105–33113

Balke VL, Gralla JD (1987) Changes in linking number of supercoiled DNA accompany growth transitions in *Escherichia coli*. J Bacteriol 169:4499–4506

Benham CJ (1990) Theoretical analysis of transitions between B- and Z-conformations in torsionally stressed DNA. Nature 286:637–638

Blot N, Mavathur R, Geertz M, Travers A, Muskhelishvili G (2006) Homeostatic regulation of supercoiling sensitivity coordinates transcription of the bacterial genome. EMBO 7:710–715

Dorman CJ (1991) DNA supercoiling and the environmental regulation of virulence gene expression in bacterial pathogens. Infection and Immunity 59:745–749

Dorman CJ (1995) DNA topology and global control of bacterial gene expression. Microbiology 141:1271–1280

Dorman CJ (2002) DNA topology and the regulation of bacterial gene expression. In: (DA Hodgson & CM Thomas, eds) Switches, Signals, Regulons and Cascades: Regulation of Bacterial Gene Expression Soc Gen Microbiol Symp 60:41–56

Dorman CJ (2004a) Nucleoid organization of bacterial chromosomes. In: (W Lennarz and MD Lane, eds) Encyclopedia of Biological Chemistry, Elsevier Science, San Diego, CA 3:115–118

Dorman CJ (2007) H-NS, genome sentinel. Nat Rev Microbiol 5:157–161

Sinden RR (1994) DNA structure and function. Academic Press, San Diego

References

Appleman JA, Ross W, Salomon J, Gourse RL (1998) Activation of *Escherichia coli* rRNA transcription by FIS during a growth cycle. J Bacteriol 180:1525–1532

Arfin SM, Long AD, Ito ET, Tolleri L, Riehle MM, Paegle ES, Hatfield GW (2000) Global gene expression profiling in *Escherichia coli* K12. The effects of integration host factor. J Biol Chem 275:29672–29684

Azam AT, Ishihama A (1999) Twelve species of the nucleoid-associated protein from *Escherichia coli*: sequence recognition specificity and DNA binding affinity. J Biol Chem 274:33105–33113

Balke VL, Gralla JD (1987) Changes in linking number of supercoiled DNA accompany growth transitions in *Escherichia coli*. J Bacteriol 169:4499–4506

Bang IS, Audia JP, Park YK, Foster JW (2002) Autoinduction of the *ompR* response regulator by acid shock and control of the *Salmonella enterica* acid tolerance response. Mol Microbiol 44:1235–1250

Belland RJ (1991) H-NDA formation by the coding region repeat elements of neisserial *opa* genes. Mol Microbiol 5:2351–2360

Beltrametti F, Kresse AU, Guzman CA (1999) Transcriptional regulation of the *esp* genes of enterohemorrhagic *Escherichia coli*. J Bacteriol 181:3409–3418

Benham CJ (1990) Theoretical analysis of transitions between B- and Z-conformations in torsionally stressed DNA. Nature 286:637–638

Bliska JB, Cozzarelli NR (1987) The use of site-specific recombination as a probe of DNA structure and metabolism in vivo. J Mol Biol 194:205–218

Blot N, Mavathur R, Geertz M, Travers A, Muskhelishvili G (2006) Homeostatic regulation of supercoiling sensitivity coordinates transcription of the bacterial genome. EMBO 7:710–715

Borowiec JA, Gralla JD (1985) Supercoiling response of the *lac* ps promoter in vitro. J Mol Biol 184:587–598

Bowater RP, Chen D, Lilley DM (1994) Modulation of *tyrT* promoter activity by template supercoiling in vivo. EMBO J 13:5647–5655

Brocolli S, Phoenix P, Drolet M. (2000) Isolation of the *topB* gene encoding DNA topoisomerase III as a multicopy suppressor of *topA* null mutations in *Escherichia coli*. Mol Microbiol 35:58–68

Chen CC, Wu HY (2003) Transcription-driven DNA supercoiling and gene expression control. Front Biosci 8:d430–439

Chen D, Bowater R, Dorman CJ, Lilley DMJ (1992) Activity of the *leu-500* promoter depends on the transcription and translation of an adjacent gene. Proc Natl Acad Sci USA 89:8784–8788

Cheung KJ, Badarinarayana V, Selinger DW, Janse D, Church GM (2003) A microarray-based antibiotic screen identifies a regulatory role for supercoiling in the osmotic stress response of *Escherichia coli*. Genome Res 13:206–215

Cirillo D, Valdivia R, Monack D, Falkow S (1998) Macrophage-dependent induction of the *Salmonella* pathogenicity island 2 type III secretion system and its role in intracellular survival. Mol Microbiol 30:175–188

Claret L, Rouvière-Yaniv J (1997) Variation in HU composition during growth of Escherichia coli: the heterodimer is required for long term survival. J Mol Biol 273:93–104

Deng S, Stein RA, Higgins NP (2005) Organization of supercoil domains and their reorganization by transcription. Mol Microbiol 57:1511–1521

DiNardo S, Voelkel KA, Sternglanz R, Reynolds AE, Wright A (1981) *Escherichia coli* DNA topoisomerase I mutants have compensatory mutations in DNA gyrase genes. Cell 31:43–51

Dixon RA, Henderson NC, Austin S (1988) DNA supercoiling and aerobic regulation of transcription from the *Klebsiella pneumoniae nifLA* promoter. Nucleic Acids Res 16:9933–9946

Dorman CJ, N Ní Bhriain, CF Higgins (1990) DNA supercoiling and the environmental regulation of virulence gene expression in *Shigella flexneri*. Nature 344:789–792

Dorman CJ (1991) DNA supercoiling and the environmental regulation of virulence gene expression in bacterial pathogens. Infect Immun 59:745–749

Dorman CJ (1995) DNA topology and global control of bacterial gene expression. Microbiology 141:1271–1280

Dorman CJ (2002) DNA topology and the regulation of bacterial gene expression. In: (DA Hodgson, Thomas CM, eds) Switches, Signals, Regulons and Cascades: Regulation of Bacterial Gene Expression Soc Gen Microbiol Symp 60:41–56

Dorman CJ (2004a) Nucleoid organization of bacterial chromosomes. In: (W Lennarz, Lane MD, eds) Encyclopedia of Biological Chemistry, Elsevier Science, San Diego, CA 3:115–118

Dorman CJ (2004b) Virulence gene regulation in *Shigella*. In *Escherichia coli* and *Salmonella*: Cellular and Molecular Biology. (Curtiss R III, Ingraham JL, Kaper JB, Maloy S, Neidhardt FC, Riley MM, Squires CJ, Wanner BL, Bock A, eds) 3rd edition, *EcoSal* online. Posted 29th December, 2004 at http://www.ecosal.org. American Society for Microbiology, Washington, D.C.

Dorman CJ (2007) H-NS, genome sentinel. Nat Rev Microbiol 5:157–161

Dorman CJ, Deighan P (2003) Regulation of gene expression by histone-like proteins in bacteria. Curr Opin Genet Dev 13:179–184

Dorman CJ, Ní Bhriain N (1992) Global regulation of gene expression during environmental adaptation: implications for bacterial pathogens. In: (CE Hormaeche, Penn CW, Smyth CJ, eds). Molecular Biology of Bacterial Infection Cambridge: Cambridge University Press, pp. 193–230

Dorman CJ, Lynch AS, Ní Bhriain N, Higgins CF (1989) DNA supercoiling in *Escherichia coli*: *topA* mutations can be suppressed by DNA amplifications involving the *tolC* locus. Mol Microbiol 3:531–540

Dorman CJ, Ní Bhriain N, Higgins CF (1990) DNA supercoiling and the environmental regulation of virulence gene expression in *Shigella flexneri*. Nature 344:789–792

Dorman CJ, Barr GC, Ní Bhriain N, Higgins CF (1988) DNA supercoiling and the anaerobic and growth-phase regulation of *tonB* gene expression. J Bacteriol 170:2816–2826

Drlica K (1992) Control of bacterial DNA supercoiling. Mol Microbiol 6:425–433

Drlica K, Zhao X (1997) DNA gyrase, topoisomerase IV, and the 4-quinolones. Microbiol Mol Biol Rev 61:377–392

Drlica K, Wu E-D, Chen C-R, Wang J-Y, Zhao X, Qiu L, Malik M, Kayman S, Friedman SM (1999) Prokaryotic DNA topology and gene expression. In: (Baumberg S, ed) Prokaryotic Gene Expression, Oxford: Oxford University Press, pp 141–168

Drolet M, Phoenix P, Menzel R, Massé E, Liu L, Crouch RJ (1995) Overexpression of RNase H partially complements the growth defect of an *Escherichia coli* Δ*topA* mutant: R-loop formation is a major problem in the absence of DNA topoisomerase I. Proc Natl Acad Sci USA 92:3526–3530

Falconi M, Colonna B, Prosseda G, Micheli G, Gualerzi CO (1998) Thermoregulation of *Shigella* and *Escherichia coli* EIEC pathogenicity. A temperature-dependent structural transition of DNA modulates accessibility of *virF* promoter to transcriptional repressor H-NS. EMBO J 17:7033–7043

Fang M, Wu H-Y (1998) A promoter relay mechanism for sequential gene activation. J Bacteriol 180:626–633

Figueroa-Bossi N, Guerin M, Rahmouni R, Leng M, Bossi L (1998) The supercoiling response of a bacterial tRNA promoter parallels its responsiveness to stringent control. EMBO J 17:2359–2367

Fournier B, Klier A (2004) Protein A gene expression is regulated by DNA supercoiling which is modified by the ArlS-ArlR two-component system of *Staphylococcus aureus*. Microbiol 150:3807–3819

Free A, Dorman CJ (1994) *Escherichia coli tyrT* gene transcription is sensitive to DNA supercoiling in its native chromosomal context: effect of DNA topoisomerase IV over-expression on *tyrT* promoter function. Mol Microbiol 14:151–161

Friedman SM, Malik M, Drlica K (1995) DNA supercoiling in a thermotolerant mutant of *Escherichia coli*. Mol Gen Genet 248:417–422

Fye RM, Benham CJ (1999) Exact method for numerically analyzing a model of local denaturation in superhelically stressed DNA. Phys Rev 59:3408–3426

Galán JE, Curtiss R III (1990) Expression of *Salmonella typhimurium* genes required for invasion is regulated by DNA supercoiling. Infect Immun 58:1879–1865

Gmuender H, Kuratli K, Di Padova K, Gray CP, Keck W, Evers S (2001) Gene expression changes triggered by exposure of *Haemophilus influenzae* to novobiocin or ciprofloxacin: combined transcription and translation analysis. Genome Res 11:28–42

Goldstein E, Drlica K (1984) Regulation of bacterial DNA supercoiling: plasmid linking numbers vary with growth temperature. Proc Natl Acad Sci USA 81:4046–4050

Graeff-Wohlleben H, Deppisch H, Gross R (1995) Global regulatory mechanisms affect virulence gene expression in *Bordetella pertussis*. Mol Gen Genet 247:86–94

Groisman EA, Ochman H (1997) How *Salmonella* became a pathogen. Trends Microbiol 5:343–349

174 C.J. Dorman

Hanafi EE, Bossi L (2000) Activation and silencing of *leu-500* promoter by transcription-induced DNA supercoiling in the *Salmonella* chromosome. Mol Microbiol 37:583–594

Hardt W-D, Urlaub H Gálan JE (1998) A substrate of the centisome 63 type III protein secretion system of *Salmonella typhimurium* is encoded by a cryptic bacteriophage. Proc Natl Acad Sci USA 95:2574–2579

Hardy CD, Cozzarelli NR (2005) A genetic selection for supercoiling mutants of *Escherichia coli* reveals proteins implicated in chromosome structure. Mol Microbiol 57:1636–1652

Hatfield GM, Benham CJ (2002) DNA topology-mediated control of global gene expression in *Escherichia coli*. Annu Rev Microbiol 36:175–203

Hensel M (2000) *Salmonella* pathogenicity island 2. Mol Microbiol 36:1015–1023

Hensel M, Shea JE, Waterman SR, Mundy R, Nikolaus T, Banks G, Vazquez-Torres A, Gleeson C, Fang FC, Holden DW (1998) Genes encoding putative effector proteins of the type III secretion system of *Salmonella* pathogenicity island 2 are required for bacterial virulence and proliferation in macrophages. Mol Microbiol 30:163–174

Higgins CF, Dorman CJ, Stirling DA, Waddell L, Booth IR, May G, Bremer E (1988) A physiological role for DNA supercoiling in the osmotic regulation of gene expression in *S. typhimurium* and *E.coli*. Cell 52:569–584

Hromockyj AE, Tucker SC Maurelli AT (1992) Temperature regulation of *Shigella* virulence: identification of the repressor gene *virR*, an analogue of *hns*, and partial complementation by tyrosyl transfer RNA (tRNA₁(Tyr)). Mol Microbiol 6:2113–2124

Hsieh TS, Plank JL (2006) Reverse gyrase functions as a DNA renaturase: annealing of complementary single-stranded circles and positive supercoiling of a bubble substrate. J Biol Chem 281:5640–5647

Hsieh L-S, Burger RM, Drlica K (1991a) Bacterial DNA supercoiling and [ATP]/[ADP] ratio: changes associated with a transition to anaerobic growth. J Mol Biol 219:443–450

Hsieh L-S, Rouvière-Yaniv J, Drlica K (1991b) Bacterial DNA supercoiling and [ATP]/[ADP] ratio: changes associated with salt shock. J Bacteriol 173:3914–3917

Htun H, Dahlberg HJ (1989) Topology formation of triple-stranded H-DNA. Science 243:1571–1576

Jain V, Kumar M, Chatterji D (2006) ppGpp: stringent response and survival. J Microbiol 44:1–10

Karem K, Foster J (1993) The influence of DNA topology on the environmental regulation of a pH-regulated locus in *Salmonella typhimurium*. Mol Microbiol 10:75–86

Kato J-I, Nishimura Y, Imamura R, Niki H, Hiraga S, Suzuki H (1990) New topoisomerase essential for chromosome segregation in *E. coli*. Cell 63:393–404

Keane OM, Dorman CJ (2003) The *gyr* genes of *Salmonella enterica* serovar Typhimurium are repressed by the factor for inversion stimulation, Fis. Mol Genet Genomics 270:56–65

Kelly A, Goldberg MD, Carroll RK, Danino V, Hinton JCD, Dorman CJ (2004) A global role for Fis in the transcriptional control of metabolic and type III secretion genes of *Salmonella enterica* serovar Typhimurium. Microbiology 150:2037–2053

Kelly A, Conway C, Ó Cróinín T, Smith SGJ, Dorman CJ (2006) DNA supercoiling and the Lrp protein determine the directionality of *fim* switch DNA inversion in *Escherichia coli* K-12. J Bacteriol 188:5356–5363

Kikuchi, A (1990) Reverse gyrase. In: (Cozzarelli N, Wang JC, eds) DNA Topology and its Biological Effects, Cold Spring Harbor Laboratory Press, Cold Spring Harbor NY, pp 285–298

Kikuchi A, Asai K (1984) Reverse gyrase: a topoisomerase which introduces positive superhelical turns into DNA. Nature 309:677–681

Kowalski D, Natale DA, Eddy MJ (1988) Stable DNA unwinding, not breathing accounts for single-strand-specific nuclease sensitivity of specific A+T-rich sequences. Proc Natl Acad Sci USA 85:9464–9468

Leng F, McMacken R (2002) Potent stimulation of transcription-coupled DNA supercoiling by sequence-specific DNA-binding proteins. Proc Natl Acad Sci USA 99:9139–9144

Leng F, Amado L, McMacken R (2004) Coupling DNA supercoiling to transcription in defined protein systems. J Biol Chem 279:47564–47571

Lilley DMJ (1980) The inverted repeat as a recognizable structural feature in supercoiled DNA molecules. Prof Natl Acad Sci USA 77:6468–6472

Lilley DMJ, Higgins CF (1991) Local DNA topology and gene expression: the case of the *leu-500* promoter. Mol Microbiol 5:779–783

Liu LF, Wang JC (1987) Supercoiling of the DNA template during transcription. Proc Natl Acad Sci USA 84:7024–7027

Lostroh CP, Lee CA (2001) The *Salmonella* pathogenicity island-1 type III secretion system. Microbes Infect 3:1281–1291

Lucchini S, Rowley G, Goldberg MD, Hurd D, Harrison M, Hinton JC (2006) H-NS mediates the silencing of laterally acquired genes in bacteria. PLoS Pathog 2:e81

Mangan MW, S Lucchini, V Danino, T Ó Cróinín, JCD Hinton, CJ Dorman (2006) The integration host factor (IHF) integrates stationary phase and virulence gene expression in *Salmonella enterica* serovar Typhimurium. Mol Microbiol 59:1831–1847

Marceau M (2005) Transcriptional regulation in *Yersinia*: an update. Curr Issues Mol Biol 7:151–177

Marshall DG, Bowe F, Hale C, Dougan G, Dorman CJ (2000) DNA topology and adaptation of *Salmonella typhimurium* to an intracellular environment. Philos Trans R Soc Lond B Biol Sci 355:565–574

Massé E, Drolet M (1999a) Relaxation of transcription-induced negative supercoiling is an essential function of *Escherichia coli* DNA topoisomerase I. J Biol Chem 274:16654–16658

Massé E, Drolet M (1999b) *Escherichia coli* topoisomerase I inhibits R-loop formation by relaxing transcription-induced negative supercoiling. J Biol Chem 274:16659–16664

Maurelli AT, Blackmon B, Curtiss R III (1984) Temperature-dependent expression of virulence genes in *Shigella* species. Infect Immun 43:195–201

McClellan JA, Boublikova P, Palecek E, Lilley DMJ (1990) Superhelical torsion in cellular DNA responds directly to environmental and genetic factors. Proc Natl Acad Sci USA 87:8373–8377

McLeod SM, Aiyar SE, Gourse RL, Johnson RC (2002) The C-terminal domains of the RNA polymerase alpha subunits: contact site with Fis and localization during co-activation with CRP at the *Escherichia coli prop* P2 promoter. J Mol Biol 316:517–529

McNairn E, Ní Bhriain N, Dorman CJ (1995) Over-expression of the *Shigella flexneri* genes coding for DNA topoisomerase IV compensates for loss of DNA topoisomerase I: Effect on virulence gene expression. Mol Microbiol 15:507–517

Menzel R, Gellert M (1983) Regulation of the genes for *E. coli* DNA gyrase: homeostatic control of DNA supercoiling. Cell 34:105–113

Menzel R, Gellert M (1987) Fusions of the *Escherichia coli gyrA* and *gyrB* control regions to the galactokinase gene are inducible by coumermycin treatment. J Bacteriol 169:1272–1278

Mielke SP, Fink WH, Krishnan VV, Gronbech-Jensen N, Benham CJ (2004) Transcription-driven twin supercoiling of a DNA loop: a Brownian dynamics study. J Chem Phys 121:8104–8112

Mills DM, Bajaj V, Lee CA (1995) A 40-kilobase chromosomal fragment encoding *Salmonella typhimurium* invasion genes is absent from the corresponding region of the *Escherichia coli* K-12 chromosome. Mol Microbiol 15:749–759

Mitobe J, Arakawa E, Watanabe H (2005) A sensor of the two-component system CpxA affects expression of the type III secretion system through posttranscriptional processing of InvE. J Bacteriol 187:107–113

Muskhelishvili G, Travers A (2003) Transcription factor as a topological homeostat. Front Biosci 8:d279–285

Navarre WW, Porwollik S, Wang Y, McClelland M, Rosen H, Libby SJ, Fang FC (2006) Selective silencing of foreign DNA with low GC content by the H-NS protein in *Salmonella*. Science 313:236–238

Ní Bhriain N, Dorman CJ, Higgins CF (1989) An overlap between osmotic and anaerobic stress responses: a potential role for DNA supercoiling. Mol Microbiol 3:933–942

Ochman H, Lawrence JG, Groisman EA (2000) Lateral gene transfer and the nature of bacterial innovation. Nature 405:299–304

Ó Cróinín T, Carroll RK, Kelly A, Dorman CJ (2006) Roles for DNA supercoiling and the FIS protein in modulating expression of virulence genes during intracellular growth of *Salmonella enterica* serovar Typhimurium. Mol Microbiol 62:869–882

Ohlsen KL, Gralla JD (1992) Interrelated effects of DNA supercoiling, ppGpp, and low salt in on melting within the *Escherichia coli* ribosomal RNA *rrnB* P1 promoter. Mol Microbiol 6:2243–2251

Opel ML, Hatfield GW (2001) DNA supercoiling-dependent transcriptional coupling between the divergently transcribed promoters of the *ilvYC* operon of *Escherichia coli* is proportional to promoter strengths and transcript lengths. Mol Microbiol 39:191–198

Opel ML, Arfin SM, Hatfield GW (2001) The effects of DNA supercoiling on the expression of operons of the *ilv* regulon of *Escherichia coli* suggest a physiological rationale for divergently transcribed operons. Mol Microbiol 39:1109–1115

Opel ML, Aeling KA, Holmes WM, Johnson RC, Benham CJ, Hatfield GW (2004) Activation of transcription initiation from a stable RNA promoter by a Fis protein-mediated DNA structural transmission mechanism. Mol Microbiol 53:665–674

Parekh BS, Sheridan SD, Hatfield GW (1996) Effects of integration host factor and DNA super-coiling on transcription from the *ilvPG* promoter of *Escherichia coli*. J Biol Chem 271:20258–20264

Parsot C, Mekalanos JJ (1992) Structural analysis of the *acfA* and *acfD* genes of *Vibrio cholerae*, effects of DNA topology and transcriptional activators on expression. J Bacteriol 174:5211–5218

Pemberton IK, Muskhelishvili G, Travers AA, Buckle M (2000) The GC-rich discriminator region of the *tyrT* promoter antagonizes the formation of stable pre-initiation complexes. J Mol Biol 299:859–864

Peter BJ, Arsuaga J, Breier AM, Khodursky AB, Brown PO, Cozzarelli NR. 2004. Genomic tran-scriptional response to loss of chromosomal supercoiling in *Escherichia coli*. Genome Biol 5: R87

Phoenix P, Raymond M-A, Massé E, Drolet M (1997) Roles of DNA topoisomerases in the regula-tion of R-loop formation in vitro. J Biol Chem 272:1473–1479

Porter ME, Dorman CJ (1994) A role for H-NS in the thermo-osmotic regulation of virulence gene expression in *Shigella flexneri*. J Bacteriol 176:4187–4191

Postow L, Hardy CD, Arsuaga J, Cozzarelli NR (2004) Topological domain structure of the *Escherichia coli* chromosome. Genes Dev 18:1766–1779

Pratt TS, Steiner T, Feldman LS, Walker KA, Osuna R (1997) Deletion analysis of the *fis* promoter region in *Escherichia coli*: antagonistic effects of integration host factor and Fis. J Bacteriol 179:6367–6377

Prosseda G, Falconi M, Giangrossi M, Gualerzi CO, Micheli G, Colonna B (2004) The *virF* promoter in *Shigella*: more than just a curved DNA stretch. Mol Microbiol 51:523–537

Pruss GJ, Drlica K (1986) Topoisomerase I mutants: the gene on pBR322 that encodes resistance to tetracycline affects plasmid DNA supercoiling. Proc Natl Acad Sci USA 83:8952–8956

Pruss GJ, Drlica K (1989) DNA supercoiling and prokaryotic transcription. Cell 56:521–523

Pruss GJ, Manes SH, Drlica K (1982) *Escherichia coli* DNA topoisomerase mutants: increased supercoiling is corrected by mutations near gyrase genes. Cell 31:35–42

Raji A, Zabel DJ, Laufer, CS, Depew RE (1985).Genetic analysis of mutations that compensate for loss of *Escherichia coli* DNA topoisomerase I. J Bacteriol 162:1173–1179

Rahmouni AR, Wells RD (1989) Stabilization of Z DNA in vitro by localized supercoiling. Science 246:358–363

Rahmouni AR, Wells RD (1992) Direct evidence for the effect of transcription on local DNA supercoiling in vivo. J Mol Biol 223:131–144

Rhee KY, Opel M, Ito E, Hung S, Arfin SM, Hatfield GW (1999) Transcriptional coupling between the divergent promoters of a prototypic LysR-type regulatory system, the *ilvYC* operon of *Escherichia coli*. Proc Natl Acad Sci USA 96:14294–14299

Rhen M, Dorman CJ (2005) Hierarchical gene regulators adapt *Salmonella enterica* to its host milieus. Int J Med Microbiol 295:487–502

Rochman M, Aviv M, Glaser G, Muskhelishvili G (2002) Promoter protection by a transcription factor acting as a local topological homeostat. EMBO Rep 3:355–360

Rochman M, Blot N, Dyachenko M, Glaser G, Travers A, Muskhelishvili G (2004) Buffering of stable RNA promoter activity against DNA relaxation requires a far upstream sequence. Mol Microbiol 53:143–152

Rohde JR, Fox JM, Minnich SA (1994) Thermoregulation in *Yersinia enterocolitica* is coincident with changes in DNA supercoiling. Mol Microbiol 12:187–199

Rohde JR, Luan XS, Rohde H, Fox JM, Minnich SA (1999) The *Yersinia enterocolitica* pYV virulence plasmid contains multiple intrinsic DNA bends which melt at 37 degrees C. J Bacteriol 181:4198–4204

Sanders ER, Johnson RC (2004) Stepwise dissection of the Hin-catalyzed recombination reaction from synapsis to resolution. J Mol Biol 340:753–766

Schechter LM, Jain S, Akbar S, Lee CA (2003) The small nucleoid-binding proteins H-NS, HU, and Fis affect *hilA* expression in *Salmonella enterica* serovar Typhimurium. Infect Immun 71:5432–5435

Schneider DA, Ross W, Gourse RL (2003) Control of rRNA expression in *Escherichia coli*. Curr Opin Microbiol 6:151–156

Schneider R, Travers A, Muskhelishvili G (1997) FIS modulates growth phase-dependent topological transitions of DNA in *Escherichia coli*. Mol Microbiol 26:519–530

Schneider R, Travers A, Kutateladze T, Muskhelishvili G (1999) A DNA architectural protein couples cellular physiology and DNA topology in *Escherichia coli*. Mol Microbiol 34:953–964

Schneider R, Travers A, Muskhelishvili G (2000) The expression of the *Escherichia coli fis* gene is strongly dependent on the superhelical density of DNA. Mol Microbiol 38:167–175

Sheehan BJ, Foster TJ, Dorman CJ, Park S, Stewart GSAB (1992) Osmotic and growth phase-dependent regulation of the *eta* gene of *Staphylococcus aureus*: a role for DNA supercoiling. Mol Gen Genet 232:49–57

Sheridan SD, Benham CJ, Hatfield GW (1998) Activation of gene expression by a novel DNA structural transmission mechanism that requires supercoiling-induced DNA duplex destabilization in an upstream activating sequence. J Biol Chem 273:21298–21308

Sheridan SD, Opel ML, Hatfield GW (2001) Activation and repression of transcription initiation by a distant DNA structural transition. Mol Microbiol 40:684–690

Sinden RR (1994) DNA structure and function. Academic Press, San Diego, CA

Slamti L, Livny J, Waldor MK (2007) Global gene expression and phenotypic analysis of a *Vibrio cholerae rpoH* deletion mutant. J Bacteriol 189:351–362

Stein RA, Deng S, Higgins NP (2005) Measuring chromosome dynamics on different time scales using resolvases with varying half-lives. Mol Microbiol 56:1049–1061

Sun HZ, Mezei M, Fye R, Benham CJ (1995) Monte Carlo analysis of conformational transitions in superhelical DNA. J Chem Phys 103:8653–8665

Sue D, Fink D, Wiedmann M, Boor KJ (2004) Sigma B-dependent gene induction and expression in *Listeria monocytogenes* during osmotic and acid stress conditions simulating the intestinal environment. Microbiology 150:3843–3855

Swinger KK, Rice PA (2004) IHF and HU: flexible architects of bent DNA. Curr Opin Struct Biol 14:28–35

Tobe T, Yoshikawa M, Sasakawa C (1995) Thermoregulation of *virB* transcription in *Shigella flexneri* by sensing of changes in local DNA superhelicity. J Bacteriol 177:1094–1097

Travers AA (1980) Promoter sequence for stringent control of bacterial ribonucleic acid synthesis. J Bacteriol 141:973–976

Travers A, Muskhelishvili G (2005) DNA supercoiling—a global transcriptional regulator for enterobacterial growth? Nature Rev Microbiol 3:157–169

Trun NJ, Marko JF (1998) Architecture of the bacterial chromosome. ASM Press 64:276–283

Tse-Dinh Y-C (1985) Regulation of the *Escherichia coli* topoisomerase I gene by DNA supercoiling. Nucleic Acids Res 13:4751–4763

Tse-Dinh Y-C, Beran RK (1988) Multiple promoters for transcription of DNA topoisomerase I gene and their regulation by DNA supercoiling. J Mol Biol 202:735–742

Turner EC, Dorman CJ (2007) H-NS antagonism in *Shigella flexneri* by VirB, a virulence gene transcription regulator that is closely related to plasmid partition factors. J Bacteriol 189:3403–3413

van Workum M, van Dooren SJ, Oldenburg N, Molenaar D, Jensen PR, Snoep JL, Westerhoff HV (1996) DNA supercoiling depends on the phosphorylation potential in *Escherichia coli*. Mol Microbiol 20:351–360

Waterman SR, Holden DW (2003) Functions and effectors of the *Salmonella* pathogenicity island 2 type III secretion system. Cell Microbiol 5:501–551

Wang JC (1996) DNA topoisomerases. Annu Rev Biochem 65:635–692

Wang Z, Dröge P (1997) Long-range effects in a supercoiled DNA domain generated by transcription in vitro. J Mol Biol 271:499–510

Weinstein-Fischer D, Altuvia S (2007) Differential regulation of *Escherichia coli* topoisomerase I by Fis. Mol Microbiol 63:1131–1144

Weinstein-Fischer D, Elgrably-Weiss M, Altuvia S (2000) *Escherichia coli* response to hydrogen peroxide: a role for DNA supercoiling, topoisomerase I and FIS. Mol Microbiol 35:1413–1420

Westerhoff HV, Odea MH, Maxwell A, Gellert (1988) DNA supercoiling by DNA gyrase: a static head model. Cell 12:157–181

Wu HY, Shyy SH, Wang JC, Liu LF (1988) Transcription generates positively and negatively supercoiled domains in the template. Cell 53:433–440

Yamamoto N, Droffner M (1985) Mechanisms determining aerobic or anaerobic growth in the facultative anaerobe *Salmonella typhimurium*. Proc Natl Acad Sci USA 82:2077–2081

7
Quorum Sensing

Simon Swift(✉), Maria C. Rowe, and Malavika Kamath

7.1 Quorum Sensing, Bacterial Signals and Autoinducers.. 180
7.2 Signals and Signal Production 182
    7.2.1 Acyl Homoserine Lactones.. 183
    7.2.2 Posttranslationally Modified Peptides... 185
    7.2.3 AI-2: The LuxS Signal... 185
    7.2.4 Is Signal Generation a Regulatory Step? 187
    7.2.5 How Does the Signal Exit the Cell?... 187
7.3 Signal Perception and Response Regulation.. 187
    7.3.1 LuxR Receptors for Acyl-HSLs... 188
    7.3.2 LuxN-Type Receptors for Acyl-HSLs 189
    7.3.3 The Response to AI-2 190
    7.3.4 The Response to Modified Peptide Signals 190
    7.3.5 The Quorum Response and the QS Regulon 191
7.4 Studying QS.. 192
    7.4.1 A Brief History of the Discovery of QS in *Erwinia*.......................... 192
    7.4.2 A Brief History of the Study of QS in *Aeromonas* 194
    7.4.3 QS in *E. coli*.. 196
7.5 Regulation of Microbial Physiology by QS... 197
    7.5.1 QS and Secondary Metabolism... 198
    7.5.2 QS and Virulence 204
    7.5.3 Biofilms... 212
7.6 QS in the Real World 214
    7.6.1 The Effect of Other Species... 214
7.7 New Perspectives and Applications... 216
Highly Recommended Readings.. 217
References... 218

Abstract Bacteria use small molecule signals to communicate with each other. Intercellular signalling at high population cell densities is termed quorum sensing and explains many aspects of bacterial physiology observed in single species cultures entering stationary phase in the laboratory. Quorum sensing is used by diverse species to control a multitude of phenotypic traits that often include virulence factors

Simon Swift
Department of Molecular Medicine and Pathology, University of Auckland, Auckland,
New Zealand
s.swift@auckland.ac.nz

(e.g., exoenzymes) and secondary metabolites (e.g., antibiotics and biosurfactants). In this review, diversity in the biochemistry and molecular biology of signal production, signal sensing, and signal response are introduced. The elucidation of the roles of quorum sensing in bacterial virulence and in biofilm formation will be used to illustrate experimental approaches commonly used. The understanding of quorum sensing obtained in vitro will be considered in the light of studies describing the activities of bacteria in the real situations of infection and biofilm formation. The relevance of quorum sensing to the activities of bacteria in real situations is discussed, taking into account the role of (1) other bacterial species; (2) the host; and (3) changes in other, nonsignalling, parameters within the environment.

7.1 Quorum Sensing, Bacterial Signals and Autoinducers

Bacteria are able to sense changes within the environment that they inhabit. On perception of change, bacteria are able to respond by altering their phenotype to provide the activities best suited to success in the new environment. The expression of a modified phenotype often relies on new gene expression. In quorum sensing (QS), the environmental parameter being sensed is the number or density of other bacteria, particularly of the same species, also present. The study of QS in numerous species has led to the concept of the *quorate* population, which we can define as a population of bacteria that is above a threshold number or density, and that is able to coordinate gene expression and, thus, its phenotypic activities (Fuqua et al. 1994).

QS relies on the production and release of small molecule signals by the bacterium into its environment. These signals have also been termed "autoinducers" and bacterial "pheromones." Put simply, the population grows and more signal is produced until a threshold concentration is reached that the bacterium perceives, and responds to, by activating (or sometimes repressing) gene expression. The key properties of a QS system are, therefore:

1. The small molecule signal
2. The signal synthase
3. The signal receptor
4. The signal response regulator
5. The genes regulated (the QS regulon)

A good example is the control of bioluminescence in symbiotic populations of *Vibrio fischeri* within the light organ of the Hawaiian squid, where only above a certain number of bacteria will be able to produce enough bioluminescence to be visible and assist the squid's hunting (reviewed by Nyholm and McFall-Ngai 2004). The *lux* genes are contained within divergent transcripts. The *luxR* gene transcript encodes a protein housing the signal receptor and the signal response regulator. The transcript of the remaining *lux* genes *luxICDABE* of the *lux* operon is activated by LuxR in the presence of the signal, an acylated homoserine lactone *N*-3-oxohexanoyl-L-homoserine lactone (3-oxo-C6-HSL) (Eberhard et al. 1981; Engebrecht et al.

1983). The signal is produced by LuxI, encoded by the first gene of the *lux* operon. At low population density, the low level of transcription of the *lux* operon is insufficient to activate LuxR. As the population grows in the laboratory flask or within the light organ of the Hawaiian squid, the levels of signal reach a threshold level that activates LuxR. The LuxR/3-oxo-C6-HSL complex activates the transcription from the promoter of the *lux* operon resulting in the following:

1. The expression of more LuxI, so more signal is produced and, hence positive feedback occurs. The term "autoinducer" is used by some to describe QS signals because of this positive feedback, whereby the signal induces the production of more signal.
2. The expression of the *luxAB* genes that encode the luciferase, *luxCDE* genes that encode the enzymes that produce substrate for the luciferase and, hence, bioluminescence (light).

The *lux* system has been a paradigm for "autoinduction" (Nealson et al. 1970; Nealson 1977) and QS for many years and the system is now described in great detail. Recent studies have uncovered a greater complexity (Nyholm and McFall-Ngai 2004; Milton 2006; Visick and Ruby 2006). One of the most exciting discoveries is that, in addition to the *lux* operon genes, the QS regulon also contains genes encoding activities involved in the initiation and maintenance of the symbiosis with the squid (Lupp et al. 2003; Lupp and Ruby 2005). Indeed, the ability of bacteria to be able to regulate many genes encoded at different sites on the chromosome with the same system to allow coordination of expression with high cell density is one of the most important features of QS. This is best illustrated in the examples of pathogenic bacteria, in which the regulation of virulence factors, e.g., by *Pseudomonas aeruginosa* (Wagner et al. 2007) or *Staphylococcus aureus* (Kong et al. 2006), occurs via QS. A population of significant size can produce sufficient toxins and exoenzymes to overcome a host, whereas lower numbers of bacteria would simply not do enough damage and only induce inflammatory responses that would contain the nascent infection.

The examples mentioned above are based initially on laboratory observations in the culture flask, and sometimes do not wholly reflect the situation in real life. In more detailed study, it has been demonstrated that the quorum response may be activated by small numbers of bacteria within a small, enclosed space, e.g., intracellular *S. aureus* in the endosome (Qazi et al. 2001), and that, in some cases, QS may act as a diffusion sensor rather than a sensor of population size (Redfield 2002). Moreover, in considering QS in the wider environment, it has been demonstrated that other organisms (both prokaryotic and eukaryotic) can perceive, respond, and even interfere with the QS activities of a given species in vivo (see Sect. 7.6.1).

For the purposes of this chapter, it will be assumed that the change in the population parameter is perceived by the bacterium and that the response is a change in gene expression. The nature of signalling mechanisms will be examined first, and then the effect these have on the bacterial phenotype. Figure 7.1 summarises the processes and overall principles.

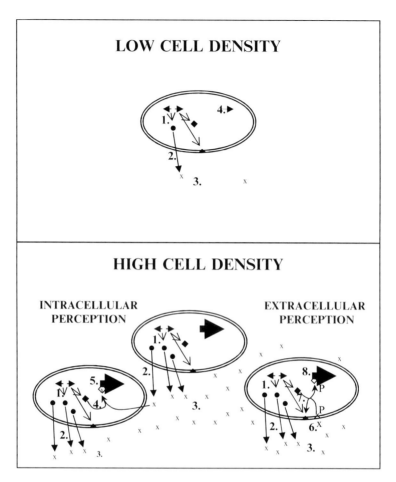

Fig. 7.1 Quorum sensing. At *low cell density*, genes for the biosynthesis, perception, and response to the signal are expressed at a basal level (*1*). The signal (x) is made and exits the cell (*2*), but does not accumulate (*3*), and the quorum response is not activated (*4*). At *high cell density*, the genes for the biosynthesis, perception, and response to the signal are expressed (*1*). The signal is made and exits the cell (*2*), where it accumulates (*3*), and is perceived (*4, 6*). If perception is intracellular, there is direct activation of the response regulator (*4*), e.g., LuxR type, and the quorum response is activated (*5*). If perception is extracellular, there is activation of a sensor kinase (*6*), phosphotransfer to activate the response regulator (*7*), and the quorum response is initiated (*8*)

7.2 Signals and Signal Production

The chemistry of QS signals classifies most Gram-negative bacteria as users of *N*-acyl homoserine lactones, whereas most Gram-positive bacteria use peptide-based signals. The Autoinducer (AI) 2 signal family (derivatives of the LuxS product, 4,5-dihydroxyl-pentanedione [DPD]) are proposed to be universal QS signals that act across a great number of bacterial species (Schauder et al. 2001). Moreover, the discovery of the widespread nature of bacteria-to-bacteria signalling has stimulated research that has highlighted the presence of many other potential signal chemistries, including unsaturated fatty acids (Wang et al. 2004), fatty acyl methyl esters (Flavier et al. 1997), quinolones (Pesci et al. 1999), cyclic dipeptides (Holden et al. 1999), and indole (Wang et al. 2001). See Fig. 7.2 for some signal structures.

The small molecule signal defines QS; it is released from the bacterial cell and allows communication with other (bacterial) cells within the population. One significant area for discussion regarding QS has focussed on what makes a small molecule found in spent culture supernatants a QS signal? The argument is most intensive around the area of signalling in *Escherichia coli*, because, despite numerous claims of QS roles for various components of culture supernatants, none really satisfy this requirement for QS: that the cellular response extends beyond the physiological changes required to metabolise or detoxify the molecule (Winzer et al. 2002a, b).

7.2.1 Acyl Homoserine Lactones

Signal generation for acyl homoserine lactones (acyl-HSLs) seems simply to be the coupling of amino acid and fatty acid biosynthesis. Proteins homologous to LuxI represent the major family of acyl-HSL synthases (see Lerat and Moran 2004). However, a second type of acyl-HSL synthase (LuxM family) has been found in *Vibrio* species (reviewed by Milton 2006). The primary molecular substrates for this reaction have been determined as *S*-adenosyl methionine (SAM) and acylated acyl carrier protein (ACP) in a number of independent studies for members of the LuxI family (reviewed in Fuqua and Eberhard 1999) and AinS, a LuxM type synthase from *V. fischeri* (Hanzelka et al. 1999).

X-ray crystallography of LuxI type proteins from *Pantoea* (*Erwinia*) *stewartii* (EsaI; 3-oxo-C6-HSL synthase) (Watson et al. 2002) and *P. aeruginosa* (LasI; *N*-[3-oxododecanoyl]-L-homoserine lactone [3-oxo-C12-HSL] synthase) (Gould et al. 2004) has been used to explain biochemical and mutational studies of LuxI-type proteins (Hanzelka et al. 1997; Parsek et al. 1997). It is thought that acyl-ACP binds to the enzyme first, which is followed by a conformational rearrangement in the N-terminal region of the protein that precedes SAM binding within an N-terminal pocket containing the conserved residues arginine 23, phenylalanine 27, and tryptophan 33. *N*-Acetylation of SAM then occurs, followed by lactonisation and the release of acyl-HSL, holo-ACP, and 5′-methylthioadenosine. The core catalytic fold of EsaI and LasI shares features essential for phosphopantetheine binding and *N*-acylation that are found in the GNAT family of *N*-acetyltransferases and also in LuxM-type acyl-HSL synthases (Watson et al. 2002; Gould et al. 2004).

ACP binds to the acyl-HSL synthase at a surface-exposed binding site including residues lysine 150 and arginine 154. Acyl-ACP binding places the acyl group into a hydrophobic pocket (EsaI) or tunnel (LasI). The pocket in EsaI is much smaller than that in LasI, and favours short chain acyl-ACPs (Watson et al. 2002), whereas the tunnel in LasI can accommodate longer acyl-ACPs (Gould et al. 2004). Both

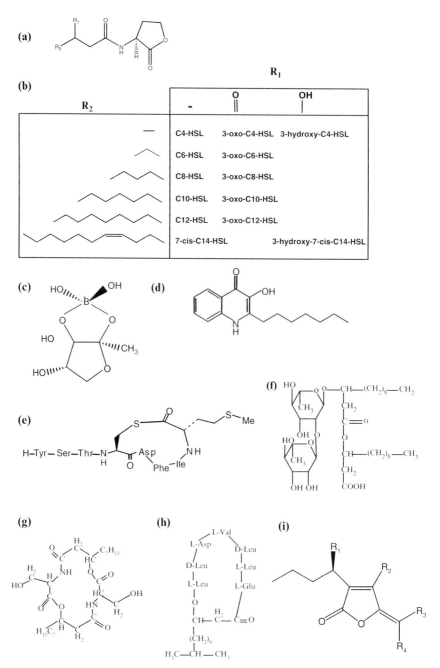

Fig. 7.2 Molecular structures of signals and biosurfactants. **a** The basic structure of the acyl-HSL signal molecule family. **b** The variable groups R1 and R2 are shown for a number of acyl-HSLs that have been referred to in this review and in the wider literature. The nomenclature adopted in Dunny and Winans (1999) is used; **c** *V. harveyi* AI-2; **d** the pseudomonas quinolone signal (PQS), 2-heptyl-3-hydroxy-4-quinolone; **e** the *S. aureus* group I cyclic peptide thiolactone; **f** rhamnolipid 1; **g** serrawettin W2; **h** surfactin; **i** halogenated furanones (R1 may be H, OAc, or OH; R2, Br or H; R3, Br or H; and R4, Br or I)

Esal and Lasl are LuxI-type proteins that produce 3-oxo-acyl-HSLs and possess either a serine or a threonine residue at position 140. Acyl-HSL synthases (e.g., AhyI, RhlI, SwrI) possessing either alanine or glycine residues at position 140 produce acyl-HSLs lacking C3-substitutions (Watson et al. 2002). The side-chain of the amino acid at position 140 protrudes into the acyl-chain pocket and mutation of EsaI to valine at 140 reduces enzyme activity, presumably by reducing access to the pocket. Mutation of EsaI to alanine at 140 shifts the preference of the enzyme to acyl-ACP substrates without a C3-substitution (Watson et al. 2002).

Advances in understanding the mechanisms of synthesis and acyl side chain specificity will be of benefit in designing novel antipathogenic drugs that may prevent activation of virulence gene expression by inhibiting acyl-HSL synthesis (see Sect. 7.7).

7.2.2 Posttranslationally Modified Peptides

For peptide signals, the ribosomal synthesis of a precursor propeptide is followed by processing, which often introduces other chemical groups such as lipid moieties, as with the ComX pheromone of *Bacillus subtilis* (Okada et al. 2005) or intramolecular bonds such as thiolactone, in the staphylococcal autoinducing peptide (AIP) (Ji et al. 1997; McDowell et al. 2001). Then, a cleavage of the processed precursor occurs to release the mature peptide.

7.2.3 AI-2: The LuxS Signal

To date, the only QS system shared by both Gram-positive and Gram-negative organisms involves the production of AI-2 via LuxS (Surette et al. 1999; Xavier and Bassler 2005). In *Vibrio harveyi*, the regulation of bioluminescence is under the control of parallel QS systems (reviewed by Milton 2006). System 1 involves an acyl-HSL synthesised by a LuxM synthase, and the LuxN receptor kinase sensor (Bassler et al. 1993). In System 2, the signal synthase is LuxS (Surette et al. 1999) and the signal (AI-2) is a furanosyl borate diester (3A-methyl-5,6-dihydro-furo [2,3-D][1,3,2] dioxaborole-2,2,6,6A-tetrol; abbreviated as S-THMF-borate) as identified from X-ray crystallography of the ligand-bound receptor, LuxP (Bassler et al. 1994; Chen et al. 2002). The *luxS* gene is conserved in many bacterial species, and molecules activating an AI-2 biosensor are found in spent supernatants from diverse bacterial species, including both Gram-positive and Gram-negative bacteria, and leading to the suggestion that AI-2 may be a universal signal for interspecies communication (Schauder et al. 2001).

AI-2 is formed as a metabolic byproduct of the activated methyl cycle (AMC). The AMC recycles SAM, which acts as the main methyl donor in eubacterial, archaebacterial, and eukaryotic cells (reviewed by Vendeville et al. 2005). After methyl donation, SAM is converted to a toxic metabolite *S*-adenosyl-L-homocysteine (SAH). Detoxification of SAH in *V. harveyi*, *E. coli*, and many other bacteria is a

two-step process, involving first Pfs enzyme (5′-methylthioadenosine/S-adenosylho-
mocysteine nucleosidase) to generate S-ribosyl homocysteine (SRH), which acts as
the substrate for LuxS. SRH is converted to adenine, homocysteine (which is
converted to methionine and then SAM), and DPD, the precursor for AI-2 (Vendeville
et al. 2005; Winzer et al. 2002a). Some bacteria and eukaryotes are able to replace
this two-step reaction with a single enzyme, SAH hydrolase, which converts SAH to
homocysteine without producing AI-2 (Winzer et al. 2002b) (Fig. 7.3).

The DPD precursor is a highly unstable molecule that may spontaneously inter-
convert to a number of related structures depending on the environment (Waters and
Bassler 2005), including the form that is stabilised by forming a complex with boron
(Chen et al. 2002) in AI-2 signalling in *V. harveyi* system 2. The putative AI-2 signals
of other bacteria, e.g., *E. coli*, may be formed via different routes depending on the

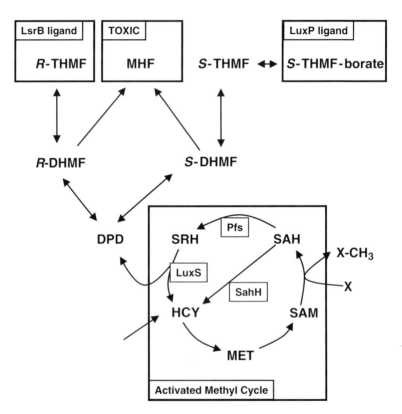

Fig. 7.3 LuxS, AI-2, and the AMC. The AMC fed by metabolic homocysteine (*HCY*) through
methionine (*MET*), SAM, a methyl group (CH₃) donor, SAH recycling to HCY via the one step
process through SahH or in AI-2-producing bacteria via SRH and the Pfs and LuxS activities.
LuxS also releases DPD, which spontaneously cyclises to *R*-2,4-dihydroxy-2-methyldihydro-3-
furanone (R-*DHMF*) or *S*-DHMF. *R*-DHMF hydration forms the LsrB ligand (2*R*,4*S*)-2-methyl-
2,3,3,4-tetrahydroxytetrahydrofuran (R-*THMF*). *R*-DHMF or *S*-DHMF hydrolysis forms the toxic
metabolite 4-hydroxy-5-methyl-3(2*H*)-furanone (*MHF*). *S*-DHMF hydration forms *S*-THMF,
which forms a diester with boric acid to form the LuxP ligand

cyclisation product of DPD (Fig. 7.3). It is hypothesised that alternate forms of AI-2 may be more active within a specific niche or may reflect the variation in the function of AI-2, such as QS versus metabolic roles (Vendeville et al. 2005).

7.2.4 Is Signal Generation a Regulatory Step?

In many cases, the expression of signal synthase forms part of the quorum response, providing positive feedback that allows a rapid induction of the high cell density phenotype (e.g., *V. fischeri*). For some signals, substrate availability may coordinate signal production with nutrition, although there is little evidence to suggest that this is a widespread strategy.

7.2.5 How Does the Signal Exit the Cell?

In the case of acyl-HSL molecules with short acyl chains, the freely diffusible nature of these molecules has been demonstrated (Kaplan and Greenberg 1985; Pearson et al. 1999). Acyl-HSLs with longer acyl chains do not seem to escape the cell membranes as easily, and 3-oxo-C12-HSL, for example, is actively pumped from the *P. aeruginosa* cell (Pearson et al. 1999). Peptide signals commonly undergo active export, with ATP-binding cassette (ABC) transporters commonly used (e.g., for CSP [competence-stimulating peptide, *Streptococcus pneumoniae*], CSF [competence- and sporulation-stimulating factor, also termed the Phr pheromones, which is Sec dependent; Simonen and Palva, 1993; *B. subtilis*], Nisin [*Lactococcus lactis*], and possibly ComX [*B. subtilis*]) (see Kleerebezem et al. 1997; Lazazzera et al. 1999; Michiels et al. 2001 for reviews). Note the PhrA signals controlling sporulation in *Bacillus* are thought to be part of an export–import circuit in which signals are exported from the bacterial cell, undergo processing, and are then reimported via the oligopeptide permease (Opp) system. It is thought that only the producer cell is affected and that these pheromones are not a population-wide signal (Perego 1997).

7.3 Signal Perception and Response Regulation

In QS, the environmental parameter the bacterium perceives is the level of signal external to the cell. Perception of the signal can be accomplished by surface-exposed membrane receptors or intracellular receptors (see Fig. 7.1). For the major classes of signal acyl-HSLs, AI-2 and posttranslationally modified peptides, examples of both internal and external sensing are apparent.

The response to signal perception is intracellular, most commonly affecting activation or repression of gene expression. In the simplest case, the signal diffuses into a cell and acts as a ligand for a protein influencing the initiation of transcription. For extracellular perception, signal transduction via phosphotransfer to proteins affecting transcription occurs.

7.3.1 LuxR Receptors for Acyl-HSLs

Perception of acyl-HSLs by LuxR family response regulators is intracellular. The LuxR-type acyl-HSL receptors can be described as an N-terminal acyl-HSL binding domain and a C-terminal transcriptional regulatory domain that contains a helix-turn-helix (HTH) DNA binding motif. Interaction with DNA is as a dimer, recognising a sequence of dyad symmetry located within the regulatory region of target genes. The recognition sequence, a *lux* or *lux*-type box, is approximately 20 bp in length (reviewed by Lerat and Moran 2004; Nasser and Reverchon 2007).

The majority of LuxR-type proteins studied in detail to date are transcriptional activators, when bound to their coactivating acyl-HSL ligand. TraR (*Agrobacterium tumefaciens*), LuxR (*V. fischeri*), and LasR and RhlR (both *P. aeruginosa*) bind to their recognition sequences as dimers, or higher-order multimers in the case of CarR (*Erwinia carotovora* subsp. *carotovora* [Ecc]) (Welch et al. 2000), and favour the recruitment of RNA polymerase at the target promoter (Lamb et al. 2003; Schuster et al. 2004; Urbanowski et al. 2004; Zhu and Winans 2001). The LuxR-type proteins bind their acyl-HSL ligands in a 1:1 stoichiometric ratio (Schuster et al. 2004; Welch et al. 2000; Zhu and Winans 2001). In the case of *A. tumefaciens*, TraR perceives the *N*-(3-oxooctanoyl)-L-homoserine lactone (3-oxo-C8-HSL) signal as a monomer on the inner face of the inner cytoplasmic membrane (Qin et al. 2000). Holo-TraR dimerises and is cytoplasmic, where it acts as a transcriptional activator for the quorum response (Qin et al. 2000).

Not all LuxR-type proteins act as transcriptional activators. Genetic, in vitro DNA binding assays and phylogenetic studies have identified EsaR (*Pantoea* [*Erwinia*] *stewartii*; Minogue et al. 2002), YpsR (*Yersinia paratuberculosis*; Atkinson et al. 1999), SpnR (*Serratia marcescens*; Horng et al. 2002), ExpR (*Erwinia chrysanthemi*; Nasser et al. 1998), and VirR (Ecc; Burr et al. 2006) as a group of LuxR-type proteins that act as repressors in the absence of their derepressing cognate acyl-HSL (Lerat and Moran 2004).

X-ray crystallography has revealed that LuxR-type proteins interact with their ligand at an acyl-HSL binding cavity (Vannini et al. 2002; Yao et al. 2006; Zhang et al. 2002). The highly conserved residues at position 57 (tryptophan) and 70 (aspartate) are important in the stabilisation of acyl-HSL binding. Mutations in TraR in this region have identified the tyrosine at position 53 as being important in discriminating in favour of the 3-oxo substituted ligand (Chai and Winans 2004). Other mutations in the acyl-HSL cavity of LuxR-type proteins have affected chain length specificity (Chai and Winans 2004; Finney et al. 2002; Kiratisin et al. 2002; Lamb et al. 2003; Pappas and Winans 2003; White and Winans 2005). Studies with various analogues of the acyl-HSL signal have identified a number of agonistic and antagonistic structures. The most important conserved feature of the signal that affects its activity as a ligand is chain length, but various alterations of the lactone ring head group have also been shown to have profound effects (reviewed by Nasser and Reverchon 2007).

Using studies of LuxR and TraR as evidence (Choi and Greenberg 1991; Luo and Farrand 1999; Sitnikov et al. 1996) it is thought that ligand binding induces a conformational change in the LuxR-type transcriptional activators that permits dimerisation and unmasks the DNA binding domain. After DNA binding, there is also evidence to suggest that interaction with the C-terminal domain of the alpha subunit of RNA polymerase contributes to the recruitment of RNA polymerase, and to the initiation of transcription (Finney et al. 2002; Johnson et al. 2002; Qin et al. 2000; Stevens et al. 1994; Stevens et al. 1999). In studies of repressor LuxR-type proteins, it seems that the apo-protein binds DNA and blocks access to the promoter. The presence of the appropriate ligand releases the repression, and it is hypothesised that ligand binding induces conformational changes that interfere with DNA binding (Castang et al. 2006; Frezza et al. 2006).

7.3.2 LuxN-Type Receptors for Acyl-HSLs

The investigation of the control of bioluminescence in *V. harveyi* and *V. fischeri* has identified not only a second acyl-HSL synthase family, but also a membrane receptor family. In *V. harveyi*, *N*-(3-hydroxybutanoyl)-L-homoserine lactone (3-hydroxy-C4-HSL) is produced by LuxM (Bassler et al. 1993; Cao and Meighen 1989). In the absence of signal, the sensor kinase LuxN autophosphorylates and relays phosphates to LuxU (Freeman and Bassler 1999a; Henke and Bassler 2004), a regulator also phosphorylated by two other sensor kinases, LuxQ and CqsS (Henke and Bassler 2004). LuxQ is the sensor kinase perceiving AI-2 (Bassler et al. 1994). Phospho-LuxU phosphorylates LuxO, which activates the expression of a collection of small regulatory RNAs (sRNAs) at σ^{54}-dependent promoters (Freeman and Bassler 1999b). In the presence of the RNA chaperone, Hfq, the sRNAs destabilise the mRNA encoding LuxR (Lenz et al. 2004; Tu and Bassler 2007). LuxR here is a transcriptional activator for the *luxCDABEGH* operon and other genes involved in virulence, but not an acyl-HSL receptor and not homologous to *V. fischeri* LuxR (Swartzman et al. 1992). In the absence of functional LuxR, there is no bioluminescence.

At high cell density, 3-hydroxy-C4-HSL is produced (and also the ligands activating LuxQ and CqsS), inducing LuxN, LuxQ, and CqsS phosphatase activities that dephosphorylate LuxU and lead to the inactivation of LuxO, allowing LuxR to be expressed and transcriptional activation to occur (Freeman and Bassler 1999b; Freeman et al. 2000; Henke and Bassler 2004).

Homologues of LuxM and N have been found in *V. fischeri* (AinS and R) and *V. anguillarum* (VanM and N), where they contribute to the regulation of gene expression through a phosphotransfer pathway involving LuxU and LuxO-type proteins (Croxatto et al. 2004; Gilson et al. 1995; Kuo et al. 1994; Milton et al. 2001).

7.3.3   The Response to AI-2

In *V. harveyi*, AI-2 (S-THMF-borate) (Chen et al. 2002) is bound by the periplasmic protein LuxP (Neiditch et al. 2005), which then activates the dephosphorylase activity of LuxQ, leading to inactivation of LuxO and the expression of LuxR. LuxP and Q homologues exist in other vibrio species, where they are involved in the control of virulence factor expression (*V. cholerae*, *V. anguillarum*, and *V. vulnificus*) bioluminescence and symbiosis factors (*V. fischeri*) (see Milton 2006 for a review).

Molecules able to activate LuxQ are produced by many other bacteria via LuxS and are also termed AI-2. There is debate regarding whether these molecules are actually QS signals, or whether they are simply waste products of the AMC (Sun et al. 2004; Vendeville et al. 2005; Winzer et al. 2002a, b). Certainly, *luxS* mutations have profound phenotypic effects (DeLisa et al. 2001), but these may be caused by the toxic effects of disrupting the AMC. One question is whether these other bacteria possess AI-2 receptors and signal transduction mechanisms to affect gene expression. Although there are homologues of LuxP, Q, U, and O; they are only found together in *Vibrio* species (Sun et al. 2004). Unlike AI-2 signalling in these *Vibrio* species where a phosphorylation cascade is initiated when extracellular threshold levels of AI-2 are reached, AI-2 signalling within *E. coli*, *Salmonella*, and other organisms depends on the active uptake of DPD. In *Salmonella*, the cyclic derivative of DPD, (2R,4 S)-2-methyl-2,3,3,4-tetrahydroxytetrhydrofuran (R-THMF), binds to a homologue of the periplasmic binding protein LsrB (Taga et al. 2001 2003). LsrB is part of an ABC transporter encoded by the *lsrACDBFGE* operon. The putative ATPase of the ABC transporter, a sugar binding protein, a membrane channel, and other proteins encoded by the *lsr* operon show similarity to proteins encoded by the b1513 operon in *E. coli*. The repressor LsrR regulates the *lsr* operon. Downstream of *lsrR* is a gene encoding an AI-2 kinase. Phosphorylation of AI-2 is proposed to occur after import to allow sequestration within the cytoplasm. Phosphorylation of AI-2 causes LsrR to relieve its repression of the *lsr* operon, allowing further AI-2 import (Taga et al. 2003; Xavier et al. 2007) (see Fig. 7.3).

7.3.4   The Response to Modified Peptide Signals

Two-component signal transduction systems predominate in Gram-positive bacteria. The majority of peptide signals are perceived by sensor kinase proteins, which generally activate transcriptional activators of the quorum response (see Kleerebezem et al. 1997 for a review). The posttranslationally modified peptide (e.g., AIP in *S. aureus*, ComX in *B. subtilis*) binds to the surface-exposed transmembrane receptor histidine kinase (e.g., AgrC in *S. aureus*; ComP in *B. subtilis*), promoting autophosphorylation. Phosphotransfer to the response regulator (e.g., AgrA in *S. aureus*; ComA in *B. subtilis*) initiates expression of the quorum response. In *S. aureus*, two promoters are activated in the QS regulon: the *agr* promoter P2 activates expression of RNAII, which encodes *agrBDCA*; and the *agr* promoter P3, which encodes the regulatory RNA, RNAIII

(Benito et al. 2000). Phospho-AgrA, and AgrA, to a lesser extent, bind to consensus DNA sequences for the LytR family of response regulators (Koenig et al. 2004). Phospho-AgrA binds to the P2 site with approximately a 10-fold greater affinity than to the P3 site. In vitro electrophoretic mobility shift assays with wild-type and mutant P2 and P3 sequences demonstrated that a deviation of two bases in P3 away from the consensus LytR sequence was responsible for the differential binding (Koenig et al. 2004). It is proposed that AgrA first activates the P2 promoter, where autoinduction initiates positive feedback that increases AgrA concentrations to activate transcription at P3. AgrA activates transcription from P2 and P3 in concert with another global regulator, SarA, that has been shown to bind *agr* promoter DNA (Dunman et al. 2001; Heinrichs et al. 1996). SarA and AgrA DNA binding footprints overlap on P2, and the details of how these two regulators interact to control expression are subject to speculation (Koenig et al. 2004). Phospho-ComA directly activates transcription at a number of promoters, and a palindromic consensus binding site sequence has been identified (Lazazzera et al. 1999). In addition to the directly activated genes of the ComX quorum response in *B. subtilis*, there is indirect activation of more than 100 genes through the activity of competence transcription factor, ComK (Berka et al. 2002; Turgay et al. 1998), and the expression of an additional 89 genes is indirectly affected through ComK-independent mechanisms (Comella and Grossman 2005).

In Gram-positive bacteria, the exception to the two component signal transduction system rule are the Phr pheromones of *B. subtilis* that enter the cell through the OPP (Perego et al. 1991; Rudner et al. 1991). Phr pheromones are perceived internally by Rap phosphatases (which they inhibit) and, thereby, influence the quorum response by affecting the level of phosphorylated transcriptional activators (see Perego 1998 for a review).

7.3.5 The Quorum Response and the QS Regulon

In the first studies of QS, the phenotypic traits under investigation were known, and it was their regulation that was under investigation (e.g., the control of bioluminescence in *V. fischeri*). In later studies, the signalling mechanism was identified first and then the extent of the regulon was determined. A strategy of mutation of the signalling genes and observation of high cell density phenotypic traits was developed to identify regulated genes and contributions to whole phenotypes, e.g., to biofilm formation or virulence (see Swift 2003 for a review). The analysis of signalling mutants, using both proteomic and transcriptomic approaches, is now being applied to further describe the quorum response (e.g., Comella and Grossman 2005; Nouwens et al. 2003; Wagner et al. 2007).

It is now clear that the quorum response is comprised of directly controlled genes (the QS regulon) and indirectly controlled genes. Direct control of transcription by QS activates, or, in some cases, derepresses, gene expression. The model is simple: signal accumulates, acts to stimulate DNA binding by a transcriptional activator (or reduces DNA binding by a repressor), and new gene expression occurs at genes that are at least open for transcription (i.e., not being repressed by a second mechanism). In *Agrobacterium tumefaciens*, a process of anti-activation can occur, in which TraM can form a stable complex with TraR/3-oxo-C8-HSL and which can even disrupt the TraR–DNA complex to ensure that activation of the quorum response occurs at the correct time (Hwang et al. 1999; Qin et al. 2007).

Regulatory proteins and sRNAs are also part of the QS regulon and these mediate indirect QS effects on the quorum response. The contribution of these secondary regulators of the quorum response is especially apparent in DNA microarray studies analysing the bacterial transcriptome (e.g., Comella and Grossman 2005; Dunman et al. 2001; Wagner et al. 2007).

QS is not the only factor controlling gene expression, and other inputs are essential in controlling what we define as the quorum response. Some genes will not be expressed unless cell density and another environmental parameter are satisfied. The clearest illustration of this came from a comparison of what happens when the cognate signal is added exogenously to *V. fischeri*, *Erwinia carotovora*, and *P. aeruginosa*. In *V. fischeri*, the expression of bioluminescence is advanced, and expression may occur at low cell density (Nealson 1977). The same is true for carbapenem biosynthesis in *E. carotovora* (Chan et al. 1995), but it is not possible to advance, for example, exoenzyme production by *P. aeruginosa* (Jones et al. 1993), without first making mutations in additional regulators (Diggle et al. 2002).

7.4   Studying QS

The early studies of QS developed from investigations of particular phenotypic traits, i.e., bioluminescence of *V. fischeri* (Nealson et al. 1970), carbapenem biosynthesis by *Erwinia carotovora* (Bainton et al. 1992a, b), elastase production by *P. aeruginosa* (Passador et al. 1993), and conjugation in *Agrobacterium tumefaciens* (Piper et al. 1993), and their regulation. As the importance of this novel regulatory mechanism became apparent, similar systems were sought in other species (Swift et al. 1993) and there was renewed interest in other signalling systems including the regulation of conjugation in *Enterococcus faecalis* (Dunny et al. 1978) and the production of antibiotics by *Streptomyces* (Hara and Beppu 1982). Initially, this was particularly fruitful, with reporter strains used to demonstrate signal production and screen for signal synthase clones or null mutants. More challenging has been the search for true QS signalling systems in bacteria such as *E. coli*, in which many molecules have been identified from culture supernatants that influence gene expression.

7.4.1   A Brief History of the Discovery of QS in Erwinia

In the early 1990s, the regulation of bioluminescence by the LuxR, LuxI, 3-oxo-C6-HSL autoinducer system in *V. fischeri* had been well documented (see Meighen 1991 for a review) and was regarded as being interesting, but somewhat esoteric, and not relevant to all bacteriology. An important event in changing this view and developing the field of QS was the discovery of 3-oxo-C6-HSL-mediated gene regulation in a terrestrial bacterium, Ecc (Bainton et al. 1992a, b). Reflection on the process of discovery will highlight important aspects of QS and will also allow comparison with how we study QS today.

Carbapenem antibiotics are beta-lactams that act by inhibiting bacterial cell wall synthesis by inactivating penicillin-binding proteins (PBPs). Carbapenems are important broad-spectrum antibiotics because they are stable to most clinically relevant beta-lactamases. Production of carbapenems for clinical use is by total synthesis and is associated with high costs. The discovery that some species of bacteria produced simple carbapenems opened the possibility of a fermentative route for the production of cheaper intermediates for the synthesis of carbapenem antibiotics (Bonfiglio et al. 2002; Bycroft et al. 1988; Sleeman et al. 2004).

One approach to study carbapenem biosynthesis in Ecc was to identify the intermediates that accumulated when the biosynthetic pathway was disrupted by mutations. A simple strategy involved transposon mutagenesis of Ecc strain ATCC39048, screening for mutants that did not produce carbapenem (using a carbapenem-sensitive reporter strain) and then cross-feeding to look for those mutants that could restore antibiotic biosynthesis (Bainton et al. 1992a, b). Two classes of mutant were obtained: class 1 were thought to be blocked early in the carbapenem biosynthetic pathway, and class 2 were presumably blocked later, and assumed to accumulate an intermediate that could be used by class 1 to complete the synthesis. The "intermediate" was purified using high-performance liquid chromatography (HPLC) of organic solvent extracts of culture supernatants from class 2 mutants and identified using nuclear magnetic resonance (NMR) and mass spectrometry (MS) to be 3-oxo-C6-HSL (Bainton et al. 1992a, b).

3-oxo-C6-HSL is the V. fischeri autoinducer, not an intermediate in carbapenem biosynthesis, which suggested that it might be involved in the regulation of carbapenem synthesis. A biosensor plasmid for 3-oxo-C6-HSL was constructed that contained luxR and the divergent luxICDABE promoter from V. fischeri, but no luxI. Instead, the promoter was fused to the luxCDABE genes of Photorhabdus luminescens, which gave a good bioluminescence signal at 37°C in the presence of 3-oxo-C6-HSL and other acyl-HSLs of similar structure (Swift et al. 1993; Winson et al. 1998). E. coli does not produce acyl-HSLs, and E. coli transformed with the biosensor plasmid will only bioluminesce in the presence of exogenous acyl-HSLs or complementing luxI homologues (Bainton et al. 1992a, b; Swift et al. 1993).

The QS signal synthase gene from Ecc, termed carI, was cloned (Swift et al. 1993). The inability to produce 3-oxo-C6-HSL by class 1 mutants was complemented by transformation with the carI gene, spent Ecc supernatants, and also synthetic 3-oxo-C6-HSL (Bainton et al. 1992b; Jones et al. 1993; McGowan et al. 1995). The disruption of carI (and a homologous gene, expI, in a second, noncarbapenem-producing Ecc strain, SCC3193) had additional effects on the phenotype of Ecc, also known as a plant pathogen causing soft rot; a substantial reduction in the synthesis of exoenzymes, and a concomitant reduction in virulence (Jones et al. 1993; Pirhonen et al. 1993). Hence, QS controls the expression numerous genes at different locations within the genome, a regulon. A convergently transcribed (to carI/expI) luxR homologue was discovered and termed expR and, later, eccR (Andersson et al. 2000; Bell et al. 2004). In addition, the cloning and sequencing of the carbapenem biosynthetic pathway identified a second *luxR* homologue, *carR* (McGowan et al. 1995). Disruption of *carR* resulted in the loss of carbapenem biosynthesis (McGowan et al. 1995, 1996, 1997), but did not affect exoenzyme production. Disruption of *eccR* also did not affect exoenzyme production (Andersson et al. 2000). Publication of the genome sequence of a closely related plant pathogen *Erwinia carotovora* subsp. *atroseptica* (Eca) strain SCRI1043 (Bell et al. 2004) revealed the presence of a novel *luxR* homologue that proved to be the missing QS regulator of exoenzymes, termed VirR (Burr et al. 2006). Disruption of *virR* in Eca gave no obvious phenotype, however, a double mutant of *virR* and *expI* (the *carI* homologue of Eca) restored wild-type levels of exoenzyme production, suggesting that VirR negatively regulates gene expression in the presence of 3-oxo-C6-HSL. The *virR* gene was found in other species of *Erwinia*, including Ecc strain ATCC 39048, in which *virR* and *virR/carI* mutations had similar phenotypes to their Eca counterparts (Burr et al. 2006).

To understand the mechanism of VirR action, it is necessary to understand some of the non-QS regulators of virulence in *Erwinia* species. RsmA was identified in phytopathogenic Ecc strain 71 (which does not possess the *car* operon) (Chatterjee et al. 1995; Cui et al. 1995). By analogy with CsrA in *E. coli*, it is thought that RsmA acts by binding to specific messenger RNAs (mRNAs) and targeting them for degradation (reviewed by Majdalani et al. 2005). Mutants of *rsmA* are hypervirulent and overexpress exoenzymes (Chatterjee et al. 1995; Cui et al. 1995). In the absence of QS signal, the VirA homologue in Ecc71 activates *rsmA* expression, thus, negatively regulating the level of exoenzyme mRNAs. In the presence of QS signal, affinity for the *rsmA* promoter, and presumably transcription, is substantially reduced (Cui et al. 2006).

In summary, the dissection of QS in *Erwinia* species began with the finding that 3-oxo-C6-HSL production was required for the activation of carbapenem antibiotic production. The mutant unable to produce the QS signal was also deficient in exoenzyme production and substantially less virulent. The role of QS in the control of phytopathogenicity led to the study of QS in numerous strains of Ecc and other *Erwinia* species. QS controls virulence determinants in these closely related strains, but not always through identical mechanisms. Nevertheless, it is now clear (after 15 years of research) that, in Ecc strain ATCC39048, there are three LuxR-type proteins that respond to signal production via one LuxI-type protein to control the expression of dedicated QS regulons in concert with numerous other regulatory mechanisms that allow integration of a range of environmental parameters to control gene expression (reviewed by Barnard and Salmond 2007).

7.4.2   A Brief History of the Study of QS in Aeromonas

The revelation that acyl-HSL-mediated QS existed beyond the genus *Vibrio* stimulated a period of discovery in which biosensor complementation was used as a first step in elucidation of QS-controlled phenotype in numerous Gram-negative bacteria, including species of *Erwinia, Vibrio, Serratia, Burkholderia, Pseudomonas, Rhizobium, Rhodobacter, Yersinia, Chromobacterium*, and *Aeromonas* (see Swift et al. 2001 for a review). The research in this area was enabled by the development of numerous biosensors for acyl-HSLs with different chain lengths based on the preferences of different LuxR-type proteins (reviewed by Steindler and Venturi 2007). Reporters were either constructed on plasmids for transformation into *E. coli* (e.g., Shaw et al. 1997; Winson et al. 1998) or by mutation of *luxI* type genes in bacteria with an easily measurable QS-controlled phenotype, such as pigment production (McClean et al. 1997). The identification of specific acyl-HSL signals was further facilitated by the development of a low-technology thin-layer chromatography method (Shaw et al. 1997) and improvements in other separation, molecular identification, and chemical synthesis methodologies (Fekete et al. 2007; Ortori et al. 2007).

Aeromonas hydrophila presents itself as a case in which the discovery process went according to plan (Swift et al. 1997, 1999). The bacteria produced a positive T-streak on agar with the *Chromobacterium violaceum* CV026 biosensor, and concentrated extracts were shown by thin layer chromatography to contain *N*-butanoyl-L-homoserine lactone (C4-HSL) and *N*-hexanoyl-L-homoserine lactone (C6-HSL). MS of the HPLC-purified material confirmed this. The *luxI* homologue, *ahyI*, was cloned by complementation in *E. coli*. Chromosomal inactivation of *ahyI* gene resulted in a substantial reduction in extracellular protease activity, which was recovered by the addition of exogenous signal. Protease was originally hypothesised as a regulated characteristic because of its known stationary phase induction. This effect on the protease activity was observable on agar plates, in quantitative assays of supernatant proteins and by polyacrylamide gel electrophoresis analysis of secreted proteins. Importantly, antagonism of the QS response with long-chain acyl-HSLs inhibited protease production, providing early evidence that essential virulence factors could be inhibited by interfering with QS (Swift et al. 1999). Nevertheless, these strains were not significantly impaired in virulence using a gnotobiotic *Artemia franciscana* (brine shrimp) model (Defoirdt et al. 2005).

The choice of strain can be important in virulence studies, and laboratory adapted strains, e.g., AH-1N used in the above studies can give nonrepresentative results (see Fux et al. 2005). An *ahyR* mutation in a different strain of *A. hydrophila*, J1, is severely attenuated in fish (*Xiphophophorus helleri Hecke*) infection, with the LD50 increasing from 1.8×10^5 to more than 10^9 bacteria. The main virulent determinants, including proteases, amylase, DNase, haemolysin, and S layer, could not be detected in the *ahyR* mutant in vitro (Bi et al. 2007). AhyR acts as both a negative and a positive regulator of *ahyI* and, hence, C4-HSL production, in a growth phase-dependent manner (Kirke et al. 2004). As such, AhyR seems to be involved in the timing of C4-HSL production, and it is not known whether the control of virulence factor expression is direct or indirect.

Subsequently, mutations in *ahyI* were shown to be deficient in producing the characteristic mushroom-like stacks seen in normal *A. hydrophila* biofilms grown in a flow-through model (Lynch et al. 2002), and *ahyR* mutations were shown to form biofilms with a significant increase in surface coverage (Lynch et al. 2002).

The individual genes from within the biofilm QS regulon responsible for these effects have not yet been identified.

7.4.3 QS in E. coli

Salmonella and *E. coli* do not have a *luxI* gene or any acyl-HSL synthase and, therefore, do not synthesise acyl-HSLs. *E. coli* and *Salmonella* do possess a LuxR-type protein, SdiA, which is acyl-HSL responsive and regulates genes contributing to the adhesion to host tissues and the resistance to complement killing (Ahmer 2004; Michael et al. 2001; Smith and Ahmer 2003). The biological role of SdiA, and the detection of acyl-HSLs presumably produced by other bacterial species, is yet to be defined.

E. coli and *Salmonella* are paradigm species of bacterial life (Neidhardt et al. 1996), and it is hypothesised that they must surely produce a QS signal. For this reason, the culture supernatants of *E. coli* and *Salmonella* have been extensively interrogated for the presence of potential QS signals, and many candidates have been proposed (reviewed by Ahmer 2004; Walters and Sperandio 2006; Winzer et al. 2002a).

The role of AI-2 as a QS molecule in *E. coli* and *Salmonella* is controversial. In the true sense of the word, a cell-to-cell signalling molecule is a small diffusible molecule that has a function in cell-to-cell communication (Winzer et al. 2002a). Within conditioned media, a large number of bacterial products can be found and may have the potential to serve as cell-to-cell signals within a QS system. The presence of bacterial products, e.g., fermentation metabolites and medium degradation products all provide a milieu that, when added to culture of low cell density, will trigger a variety of responses unrelated to cell-to-cell signalling (Winzer et al. 2002a). Conversely, a true signalling molecule is produced during specific stages of growth, under certain physiological conditions or in response to environmental change. The molecule accumulates extracellularly and is recognised by a specific receptor. Threshold concentrations of the molecule generate a concerted response in which the cellular response extends beyond physiological changes required to metabolise or detoxify the signalling molecule (Winzer et al. 2002a).

7.4.3.1 Is AI-2 an *E. coli or Salmonella* QS Signal?

Studies comparing *luxS* mutants (unable to produce AI-2) with wild-type *E. coli* found that, based on DNA microarray analysis, greater than 400 genes were either upregulated or downregulated in the *luxS* mutant when compared with the parent strain (Sperandio et al. 2001), concluding that AI-2 signalling was a global regulatory system in *E. coli*. This study neglected the fact that LuxS was vital for the AMC and the production of a feedback mechanism within the cycle (via SAH). The *luxS* and *pfs* genes are located adjacent to other genes involved in metabolic reactions linked to the AMC, further suggesting a role in metabolism rather than QS (Vendeville et al. 2005). There is strong argument for the LuxS protein as a metabolic enzyme involved primarily in the detoxification of SAH, and AI-2 is a byproduct of this process (Sperandio et al. 2003; Vendeville et al. 2005; Winzer et al. 2002b).

The regulation of type III secreted virulence determinants, flagella, and motility genes has been linked to AI-2 signalling in enterohaemorrhagic *E. coli* (EHEC) (Sperandio et al. 2002a, b). Careful study of this regulation has demonstrated a regulatory role for another extracellular product, AI-3, that is not produced by a *luxS* mutant. AI-3 is the activating signal for virulence gene transcription and is not dependent on LuxS for synthesis. It is proposed that the pleiotropic effects of a *luxS* mutation on AMC and amino acid metabolism affects the availability of synthesis precursors for AI-3 (Walters et al. 2006). AI-3 is a chemically distinct molecule from AI-2 in that it binds C-18 HPLC columns and can only be eluted with methanol.

AI-2 and AI-3 activity may be differentiated by two assays. AI-2 produces bioluminescence in *V. harveyi*, whereas AI-3 shows no activity and AI-3 is able to activate transcription of virulence genes in EHEC in which AI-2 has no effect (Surette et al. 1999; Walters et al. 2006). The catecholamine neurotransmitters epinephrine and norepinephrine can replace AI-3 as a signal in the regulation of virulence genes in EHEC, where these effects may also be blocked by adrenergic receptor antagonists, suggesting that AI-3 may be structurally similar to epinephrine and norepinephrine (Clarke et al. 2006) and have a role in host–bacteria communication (Lyte and Ernst 1992).

A membrane sensor kinase, QseC, is activated by AI-3, epinephrine, and norepinephrine, suggesting a role in intraspecies, interspecies, and interkingdom signalling (Clarke et al. 2006). QseC is part of a two-component system as a sensor kinase activating response regulator QseB to activate transcription of the flagella regulon for swimming motility in EHEC (Clarke et al. 2006). Amino acid sequence analysis shows that QseC is conserved in other enteric bacteria that have also been shown to respond to catecholamines, e.g., *Shigella*, *Salmonella*, and *Yersinia* (Freestone et al. 1999; Kendall and Sperandio 2007). AI-3-, epinephrine-, and norepinephrine-activated QseC also activates another response regulator, QseA, which is one of many activators of the expression of the genes encoded on the locus of enterocyte effacement (LEE) locus of enteropathogenic *E. coli* (EPEC) and EHEC, and, therefore, central to the regulation of enterovirulence in these pathogens (Sharp and Sperandio 2007).

In *E. coli*, it is clear that extracellular products can affect gene expression and, hence, bacterial phenotype. In the case of AI-2, and also other molecules such as indole (Wang et al. 2001), it is likely that the phenotypic changes observed were consequences of experiments that disrupted normal metabolism. The evidence supporting a role for AI-3 as a signal molecule is stronger, although whether that is as a true QS signal or possibly as an amplifier of host signals, e.g., catecholamines produced by damaged tissue, is an issue yet to be resolved.

7.5 Regulation of Microbial Physiology by QS

QS controls gene expression and defines a high cell density phenotype. Research studying the various components of the high cell density phenotype has identified some common traits, regulated by QS in its various evolutionary forms. That is to say that whether bacteria use acyl-HSLs, modified peptides, activators, or repressors to actuate their QS control, there are a number of traits that seem to be commonly regulated by QS. To illustrate this, the example of the regulation of biosurfactant secondary metabolites will be discussed. In many cases, QS coordinates the activation (or repression) of transcription from numerous promoters at sites on the bacterial genome: the QS regulon. It is clear here that the coordinated combination of expressed gene products is necessary for the bacterial population to display phenotypes that are more complex. The roles of QS in the control of virulence and biofilm formation will be discussed as examples.

7.5.1 QS and Secondary Metabolism

A secondary metabolite is a compound that is not necessary for growth or maintenance of cellular functions but is synthesised, often for the protection of a cell, during the stationary phase of the growth.

Microbial biosurfactants are surface-active molecules produced by a wide variety of microorganisms, including bacteria, yeasts, and filamentous fungi. The surfactant properties of these molecules may be attributed to their amphipathic nature in that they are composed of both hydrophobic and hydrophilic moieties. This enables them to effectively reduce surface and interfacial tensions, dissolve hydrophobic compounds, and alter the hydrophobicity of the microbial cell surface. The phylogenetic diversity of organisms that produce biosurfactants is reflected in their varied chemical structures and surface properties. All known microbial biosurfactants are classified as low molecular weight, high molecular weight, or particulate biosurfactants (Desai and Banat 1997). The hydrophilic component is usually an amino acid, polypeptide, monosaccharide, disaccharide, or polysaccharide, and the hydrophobic component is usually a saturated or unsaturated fatty acid. Low molecular weight biosurfactants are glycolipids; lipopeptides or lipoproteins; and fatty acids, phospholipids and neutral lipids (Rosenberg and Ron 1999).

The synthesis and regulation of biosurfactant production is directed by specific environmental signals and is often a cell density-dependent phenomenon. The diversity of chemical structures and physicochemical properties of biosurfactants indicates that they are synthesised by microorganism for a variety of purposes. These include:

1. Enhancing the bioavailability of hydrophobic substrates by forming micelles/emulsions.
2. To facilitate the surface translocation of swarming bacteria by overcoming surface tension.
3. Attachment and detachment of bacteria from hydrophobic substrates by influencing cell surface properties.

Growth of bacteria on hydrophobic substrates such as polyaromatic hydrocarbons (PAHs) stimulates the bacterial synthesis of biosurfactants, so as to facilitate the use of these compounds as a source of carbon. Because growth on such substrates is limited to the interface between water and oil, the release of biosurfactants enhances bacterial growth by partitioning at the hydrophobic–hydrophilic interface. This increases the surface area over which the bacteria can grow. It is mostly at this interface that bacteria can proceed to degrade the compound, with the help of their surface-associated oxygenase enzymes that oxidise the highly reduced ring structures that characterise hydrophobic xenobiotics. The key, therefore, to more efficient, accelerated growth on hydrocarbons is increased contact between cells and hydrocarbon, and this is afforded by the biosurfactants. From a bioremediation perspective, this is crucial because the initial ring cleavage is the rate-limiting step in biodegradation, and microbial biosurfactants can overcome this limitation.

A second environmental advantage of biosurfactant production is swarming migration in bacteria. Bacterial swarming is a flagella-driven movement accompanied by the production of extracellular slime, including biosurfactants. Swarming may be considered as a means to colonise new niches that are more nutritionally endowed (reviewed by Daniels et al. 2003). It is cell density dependent, with specific nutritional and surface associated signals that lead to differentiation of cells into the swarmer state. Biosurfactants function as wetting agents by reducing the surface tension, thus, facilitating the smooth movement of these cells (Daniels et al. 2003). Mutants deficient in biosurfactant production are unable to spread over a solid surface such as an agar plate (Daniels et al. 2003).

Biosurfactants are known to alter the surface properties of the secreting cell, which may, in turn, influence the interaction between the cell and the hydrocarbon. Cell surface properties arise from the unique chemical structure of the cell surface. For example, the Gram-negative bacterium, *P. aeruginosa*, has an outer membrane containing lipopolysaccharides (LPS). The variable O-Antigen of the LPS extends into the surrounding environment and consists of 15 to 20 repeating monomers of a three- to five-sugar subunit. The structure of this O-Antigen contributes to cell surface hydrophilicity. The interaction between the surfactant and the bacterial cell is thought to occur in two ways:

1. Formation of micelles that coat the hydrophobic compound and, thus, allow its uptake into the cell.
2. Altering the cell surface hydrophobicity by the release of LPS.

In the second instance, the biosurfactant may interact with the cell surface in two ways to cause changes to its hydrophobicity. The biosurfactant rhamnolipid (reviewed later in detail) directly removes the LPS through its solubilization or indirectly through the complexation of magnesium cations that are crucial for maintaining strong LPS–LPS interactions in the outer membrane (Al-Tahhan et al. 2000).

In either case, the loss of the LPS from the outer membrane results in high adherence to hydrocarbons and enhanced degradation of the hydrophobic compound. Therefore, biosurfactants interact with the secreting cells to determine the outcome of the cells' interaction with its environment.

7.5.1.1 Rhamnolipid Production by *P. aeruginosa*

Rhamnolipids are classified as low molecular weight glycolipids, composed of disaccharides acylated with long-chain fatty acids or hydroxy fatty acids. They are mainly produced during growth on hydrocarbons or carbohydrates. Their synthesis occurs at late exponential or stationary phase and is usually associated with nitrogen limitation (Ochsner et al. 1994). Rhamnolipids consist of one or two molecules of the sugar rhamnose linked to one to two molecules of β-hydroxydecanoic acid. Various types of rhamnolipids have been identified depending on the combinations of rhamnose and decanoate. The rhamnolipids principally detected in culture supernatants include rhamnolipid 1 (L-rhamnosyl-L-rhamnosyl-β-hydrocyde-canoyl-β-hydroxydecanoate) and rhamnolipid 2 (L-rhamnosyl-β-hydrocydecanoyl-β-hydroxydecanoate) (see Fig. 7.2).

Rhamnolipid is produced during the stationary phase of growth, and biosynthesis occurs via a series of glycosyl transfer reactions catalysed at each step by specific rhamnosyltransferases (Ochsner and Reiser 1995). The nucleotide-linked sugar thymidine diphosphate-rhamnose (TDP-rhamnose) is the donor and β-hydroxyde-canoyl-β-hydroxydecanoate is the acceptor. Rhamnosyltransferase 1 (catalysing the first step in rhamnolipid synthesis) is encoded by the *rhlAB* operon (Ochsner et al. 1994), which is able to restore rhamnolipid activity in mutant strains. The operon encodes two proteins, RhlA and RhlB, which encode the fully functional enzyme (Sullivan 1998). The amino acid sequence of RhlA has revealed a putative signal peptide at the N terminus, and there are at least two putative membrane-spanning domains in the RhlB protein (Ochsner et al. 1994). This suggests that RhlA is in the periplasm and RhlB is in the cytoplasmic membrane. In studies involving heterologous host expression (*E. coli* and other Pseudomonads) of rhamnosyltransferase (Ochsner et al. 1994), enzyme activity was observed when just the *rhlB* gene was induced, indicating that the RhlB protein is the functional enzyme. However, levels of rhamnolipids in the supernatant of induced *Pseudomonas* cultures was significantly higher when both *rhlA* and *rhlB* genes were expressed, indicating the involvement of RhlA protein in the activity of RhlB. It has been suggested (Ochsner et al. 1994) that RhlA may be involved in the synthesis or transport of precursor substrates for rhamnosyltransferase or in stabilization of RhlB in the cytoplasmic membrane.

Biosurfactant production in *P. aeruginosa* is tightly regulated and under the control of a QS system (Sullivan 1998). The LasR (LuxR-type protein)/LasI (3-oxo-C12-HSL) synthase pair regulate the expression of a large regulon that includes virulence factors such as the elastase gene, *lasB*, and a second signalling pair RhlR/RhlI (C4-HSL synthase) (Pearson et al. 1997; Pesci et al. 1997). In the case of rhamnolipid synthesis, the transcription on *rhlAB* is under the control of RhlR and the signal molecule C4-HSL. *rhlR* and *rhlI* are located immediately downstream of the *rhlAB* (Winson et al. 1995; Pearson et al. 1997). The provision of RhlR and C4-HSL is not sufficient for RhlAB expression, because further levels of regulation exist to silence this part of the quorum response in *P. aeruginosa* (Medina et al. 2003a). Experiments in *P. aeruginosa* and *E. coli* have shown that transcription from the *rhlAB* promoter does not occur in logarithmic growth, even when the presence of RhlR and C4-HSL is verified (Medina et al. 2003a). Additional levels of negative regulation are a common feature of genes encoding elements of the quorum response in *P. aeruginosa* (e.g., RpoS [Whiteley et al. 2000]; RsmA [Pessi et al. 2001]; DksA [Branny et al. 2001]; QscR [Chugani et al. 2001]; and MvaT [Diggle et al. 2002]). In the case of RhlAB expression, there is a requirement for the stationary phase sigma factor, RpoS, as well as RhlR and C4-HSL. In addition, RhlR also binds to the *rhlAB* promoter in the absence of C4-HSL as a repressor (Medina et al. 2003b).

P. aeruginosa is a bacterium that can adapt to relatively diverse environments and situations. The ability to produce rhamnolipid has been demonstrated to be an advantage to *P. aeruginosa* cells colonising various environments by promoting surface motility (Caiazza et al. 2005), maintaining biofilm channels (Davey et al. 2003; Lequette and Greenberg 2005), and rapidly killing neutrophils attracted to sites of *P. aeruginosa* infection (Jensen et al. 2007).

7.5.1.2 Serrawettin Production by *Serratia*

Serratia species, including *Serratia liquefaciens* MG1, reclassified recently as *S. marcescens* (Rice et al. 2005), and *S. marcescens* produce the extracellular biosurfactant serrawettin. *Serratia* are opportunistic pathogens capable of colonizing a wide variety of surfaces in water, soil, and the digestive tracts of humans and animals (Grimont and Grimont 1978). Serrawettin is a high molecular weight lipodepsipentapeptide, and three types have been identified that include W1, W2, and W3 (Lindum et al. 1998; Li et al. 2005) (Fig.7.2). Serrawettins are synthesised nonribosomally, sharing a basic structure of five amino acids in a cyclic arrangement, with variation arising from the number (one or two) and composition of substituent acyl chains (Li et al. 2005; Lindum et al. 1998). W1, W2, and W3 are produced by *S. marcescens* strains 274 (ATCC 1388), NS 25, and NS 45, respectively. Serrawettin W2 has also been reported as a wetting agent produced by *S. marcescens* (*liquefaciens*) strain MG1 (Lindum et al. 1998). One physiological role of serrawettin is to facilitate the movement of cells over various surfaces, enhancing the flagellum-dependent (swarming motility) and flagellum-independent (sliding motility) expansion of bacterial colonies on the surface of agar media (Eberl et al. 1999; Henrichsen 1972; Matsuyama et al. 1992). The antibiotic properties of serrawettin have also been described (Wasserman et al. 1961).

Swarming behaviour in *Serratia* species is initiated by two genetic switches that control the differentiation into the swarmer phenotype (Daniels et al. 2003). The first is the *flhDC* master operon that regulates expression of the flagellar regulon (Eberl et al. 1996). The second switch encodes a QS system (Givskov et al. 1998). In *S. marcescens* (*liquefaciens*), this system is composed of two genes, *swrR* and *swrI*. *swrR* is a *luxR* homologue that is the transcriptional activator of the target genes (Daniels et al. 2003) and *swrI*, a *luxI* homologue, catalyses the formation of C4-HSL and C6-HSL in a molar ratio of 10:1 (Eberl et al. 1996). The *swrA* gene is part of the QS regulon in *S. marcescens* (*liquefaciens*), discovered as 1 of 19 promoterless Tn*5 luxAB* inserts in the *S. marcescens* (*liquefaciens*) MG1 *swrI* mutant that respond to the exogenous addition of C4-HSL (Lindum et al. 1998). Serrawettin biosynthesis requires the *swrA* gene product, SwrA, which is a multido-main enzyme homologous to a number of other nonribosomal peptide synthetases and with general homology to the protein superfamily of nonribosomal peptide synthetases (NRPS) (Lindum et al. 1998). Serrawettin production is associated with the formation of swarmer colonies because it effectively reduces surface tension and facilitates the spreading of cells over a given surface, i.e., an agar plate. The swarming activity of *S. marcescens* (*liquefaciens*) strains with mutations in *swrI* and *swrA* is recovered by the addition of exogenous surfactants, including serrawettin W2 (Lindum et al. 1998).

S. marcescens (*liquefaciens*) MG1 has been reidentified as a nonpigmented (not producing the characteristic red pigment prodigiosin) *S. marcescens*. QS has been studied in three additional strains of *S. marcescens*. In ATCC 39006 (pigmented) and strain 12 (non-pigmented), homologues to SwrR and SwrI have been identi-fied, and the main acyl-HSL is conserved as C4-HSL (Thomson et al. 2000; Coulthurst et al. 2006). In strain 12, the regulation of surfactant and swarming motility by QS is apparent, with SwrR or SmaR activating transcription of *swrA* in the presence of C4-HSL. For the nonflagellated strain SS-1, a different QS cassette is present; here the SpnR/SpnI (3-oxo-C6-HSL synthase) pair also control sur-factant production, which allows for sliding motility, but by derepression of SpnR in the presence of acyl-HSL (Horng et al. 2002).

QS in the various strains of *Serratia* studied have identified overlapping QS regulons encoding various secondary metabolites (e.g., prodigiosin and carbap-enem) and exoenzymes co-regulated by other regulators, such as RsmA in a situa-tion reminiscent of *Erwinia* (reviewed by van Houdt et al. 2007). Strains of *Serratia* have also been identified that neither produce acyl-HSLs nor carry a *luxI* homo-logue, but do express these same phenotypes (van Houdt et al. 2005). In *S. marces-cens* SS-1, the *spnRI* genes are located within a Tn*3* family transposon termed Tn*TIR* (Wei et al. 2006) that supports an important role for horizontal gene transfer in the spread of QS genes, as proposed by Lerat and Moran (2004). Indeed, phage-mediated transfer of QS genes and QS-regulated genes between strains of *Serratia* has been demonstrated in the laboratory (Coulthurst et al. 2006). The evidence sup-porting the probability of horizontal transfer of QS genes and QS-regulated genes, at least between members of the genus *Serratia*, is an unlooked-for outcome of research investigating secondary metabolism in *Serratia*.

7.5.1.3 Surfactin—The Biosurfactant of *Bacillus*

The biosurfactant surfactin is produced by a Gram-positive bacterium, *Bacillus subtilis*. Surfactin is a low molecular weight cyclic lipopeptide composed of seven amino acids linked to a fatty acid moiety (Fig. 7.2) (Rosenberg and Ron 1999). The peptide moiety is synthesised by a nonribosomal peptide synthetase, which consists of a large multienzyme complex that links amino acids by ester and amide bonds

(Menkhaus et al. 1993). An acyl transferase is also required for the initial step involving the transfer of a hydroxy fatty acid to the first amino acid in the peptide (Menkhaus et al. 1993). Surfactin production is regulated by a cell density-responsive mechanism that is based on the accumulation of a posttranslationally modified peptide pheromone, ComX (Magnuson et al. 1994).

Biosynthesis of surfactin requires the products of the *srf* operon, which encodes the three subunits of the surfactin synthetase enzyme catalysing the nonribosomal peptide synthesis. The *srf* operon consists of four open reading frames (ORF) (Fabret et al. 1995), each of which specifies one domain within the peptide synthetase itself. For example, the ORF designated SrfAA (SrfORF1) contains the amino acid-activating domains for the addition of glutamate, leucine and D-leucine, the first three amino acids in the arrangement of surfactin. The second ORF, SrfAB (SrfORF2), contains the amino acid-activating domains for the addition of valine, aspartate, and D-leucine. The third ORF, SrfAC, contains the amino acid-activating domains for leucine (Galli et al. 1994) and also encodes a thioesterase responsible for peptide termination. The last ORF, SrfAD, is not required for surfactin synthesis and shows sequence homology to a thioesterase type II motif. Another gene found several kilobases downstream of the *srf* operon is also essential for surfactin synthesis; it is designated *sfp* (Nakano et al. 1992). Sfp is a phosphopantetheinyl transferase required for the activation of the surfactin synthetase via posttranslational modifications (Lambalot et al. 1996; Quadri et al. 1998).

srf expression is also associated with the onset of the competent state in *B. subtilis*. Competence is the ability to uptake foreign DNA (e.g., from lysed cells) from the external environment. The gene *comS* encodes ComS, a competence regulatory protein, which is responsible for the development of competence in *B. subtilis* via prevention of the degradation of the competence transcription factor ComK (Berka et al. 2002; D'Souza et al. 1994; Turgay et al. 1998). *comS* is embedded, out of frame, within the *srf* operon, and its expression is dependent on the expression of the *srf* operon because *comS* and *srf* share a common promoter. It is suggested that a possible role for the co-regulation of these two genes is to coincide the onset of the competent state with the synthesis of a lytic agent (surfactin) (Cosby et al. 1998).

QS regulation uses the *comQ, comX, comP, comA* signalling cassette, which is similar to many signalling cassettes found in Gram-positive bacteria (Kleerebezem et al. 1997). ComX serves as the QS regulatory signal molecule, the equivalent of the acyl-HSL in Gram-negative bacteria. It is a farnesylated decapeptide produced from pre-ComX by the action of ComQ and secreted by *B. subtilis* to accumulate extracellularly (Ansaldi et al. 2002; Bacon-Schneider et al. 2002; Magnuson et al. 1994; Tortosa et al. 2001). ComX interacts with a two-component signal transduction system on the cell membrane. On perception of ComX, the membrane-bound histidine kinase ComP donates a phosphate group to the response regulator ComA, which is, thus, activated and stimulates the transcription of the Com regulon, which has been defined using a DNA microarray approach (Comella and Grossman 2005) and which includes the *srf* operon (Solomon et al. 1996).

Another signal peptide, CSF or PhrC, also influences the expression of the *srf* operon. It is an export–import regulatory peptide of the same family as the sporulation regulator, PhrA (Core and Perego 2003; Cosby et al. 1998). CSF participates in the regulation of *srf* expression by inhibiting the activity of RapC, a known negative regulator of *srf* expression. RapC is a phosphatase that inhibits *srf* expression by dephosphorylating ComA and rendering it unable to bind to the *srf* promoter (Solomon et al. 1996). CSF is involved in maintaining ComA in the phosphorylated state and, thus, positively regulates *srf* transcription at low cell density.

From the above, it could be concluded that biosurfactant synthesis is under the control of QS in many Gram-negative and Gram-positive organisms. The synthesis of biosurfactants is associated with the onset of other phenotypes and is part of a coordinated high cell density response by bacteria that clearly transcends the molecular details of how that cell density dependent expression is achieved.

7.5.2 QS and Virulence

The roles of population size, evasion of host defences, and QS are entwined in the control of pathogenicity of at least two important pathogens, *S. aureus* and *P. aeruginosa*. Both organisms are common in our environment and are responsible for a wide range of infections (see Lowy et al. 1998; Lyczak et al. 2000 for reviews). Often these infections are hospital acquired and are difficult to treat because of antibiotic resistance (Bonomo and Szabo 2006; Schito 2006). The pathogenesis of both species relies on the coordinated expression of multiple virulence factors, a process in which QS, via acyl-HSLs for *P. aeruginosa* and via modified peptides for *S. aureus*, has a central role. Allied with this is the capacity of both organisms to form infection-related biofilms (Kirisits and Parsek 2006; Kong et al. 2006). A biofilm is a persistent mode of growth at a surface within a polymeric matrix exhibiting a resistant physiology. The bacterial cells within a biofilm are at high cell densities, and cell-to-cell signalling has been shown to play a central regulatory role in the development of a mature, resistant biofilm (see Sect. 7.5.3 below).

7.5.2.1 QS Is Essential for the Full Virulence of *P. aeruginosa*

P. aeruginosa uses a multilayered hierarchical QS cascade that links Las signalling (LasR/LasI/3-oxo-C12-HSL), Rhl-signalling (RhlR/RhlI/C4-HSL), 4-quinolone signalling (PQS), and genetically unlinked LuxR-type regulators, QscR and VqsR, to integrate the regulation of virulence determinants and the development of persistent biofilms with survival under environmental stress (Diggle et al. 2007; Juhas et al. 2004; Lequette et al. 2006; Pearson et al. 1997; Pesci et al. 1997, 1999). The quorum response of *P. aeruginosa* is extensive (Lequette et al. 2006; Schuster et al. 2003; Wagner et al. 2003 2004 2007) and provides for the coordinated activation of major virulence determinants. The quorum response can be subdivided into genes (1) that are induced only by 3-oxo-C12-HSL, (2) that are induced only by C4-HSL, (3) that are induced either by C4-HSL or 3-oxo-C12-HSL, and (4) that are only induced by C4-HSL and 3-oxo-C12-HSL (Whiteley et al. 1999), and the quorum response is summarised in Fig. 7.4.

More importantly, it has been possible to show that acyl-HSL signalling is essential for the development of full virulence by *P. aeruginosa* during an infection. The effect of specific mutations in *rhlI*, *lasR*, and *lasI* has been investigated in murine models of acute pulmonary infections and burn wound infections (Pearson et al. 2000; Rumbaugh et al. 1999; Tang et al. 1996). In the burn wound model, *lasR*, *lasI*, and *rhlI* mutants are significantly less virulent than the parent *P. aeruginosa* strain, PAO1 (Rumbaugh et al. 1999). After 48 hours, the wild-type strain shows an average mortality of 94% compared with mutants of *lasR* (28% mortality), *lasI* (47%), *rhlI* (47%), and *lasI, rhlI* double mutant (7%). The virulence of the mutants was restored by complementation with plasmids expressing LasI, RhlI, or LasI and RhlI.

The virulence of *P. aeruginosa* is linked to the production of exoproducts that degrade tissue and allow the spread of bacteria to deeper tissue. To assess the spread of *P. aeruginosa* within the burned skin, bacterial counts were made at the site of

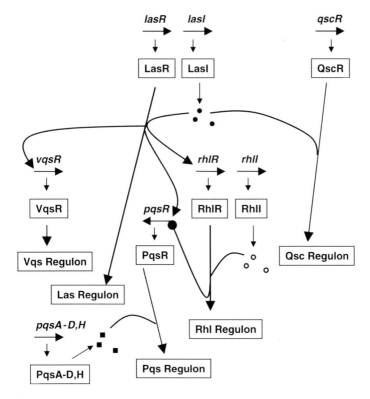

Fig. 7.4 QS in *P. aeruginosa*. A summary of the complex hierarchy of signalling in *P. aeruginosa* showing the primary genes (*italics*) and proteins (*boxed*) involved. The LasI signal 3-oxo-C12-HSL is shown as a *filled circle*. The RhlI signal C4-HSL is shown as an *open circle*. The *Pseudomonas* quinolone signal (*Pqs*) is shown as a *filled box*. The *pqsABCD* operon encodes the enzymes responsible for the biosynthesis of 2-heptyl-4-quinolone (*HHQ*) from anthranilic acid. PqsH catalyses the conversion of HHQ to PQS (Diggle et al. 2006; not discussed in the main text). Gene regulation by activation is shown by a *thick-lined arrow*. Gene regulation via repression is shown by a *thick line ending with a circle*

inoculation and at a site 15-mm distant. Single *rhlI* and *lasI* mutations had no significant effect on the spread of the bacteria, mutants with defects in *lasR* or both *rhlI* and *lasI* showed no spread to the distant site until after 16 hours from inoculation. (Rumbaugh et al. 1999). These data suggest that although there is some redundancy in the control of the important virulence factors via *las* and *rhl* signalling, QS is necessary for the optimal coordination of virulence factor expression for pathogenicity.

A similar situation is apparent in the pulmonary infection model. Of the mice inoculated with the parental strain, 55% developed confluent pneumonia throughout the lungs, with a mortality rate of 21% of the inoculated animals. In contrast, only 10% of mice inoculated with a *rhlI*, *lasI* double mutant developed pneumonia, and this was much less severe than that seen with the parent strain. Full virulence could be restored to the double mutant by complementation of the *rhlI*, *lasI* mutations with plasmid-borne copies of *rhlI* and *lasI* (75). In agreement with a role for signalling in pulmonary infection, Tang et al. (1996) demonstrated that, although a *lasR* mutant could colonise the murine lung, it was avirulent, being unable to achieve high cell densities, cause pneumonia, or penetrate into deeper tissues.

The various QS signals of *P. aeruginosa* coordinate the expression of many individual phenotypic traits to present a bacterial population best able to survive within the confines of an infection. The use of different signals and response regulators provides for flexibility in the timing of the deployment of individual gene groupings and the integration of this transcriptional activation with other signals from the bacterial environment. Moreover, the signals of *P. aeruginosa* provide more than QS capabilities influencing immune responses, vasodilation, and other bacterial species (see Sect. 7.6).

7.5.2.2   QS Is Essential for the Full Virulence of *S. aureus*

S. aureus is an opportunistic pathogen deploying a range of adhesins, evasins, and aggressins (Lowy et al. 1998). The collection of genes expressed during an infection that is required for the establishment and progression of disease have been termed the "virulon." The controlled expression of the virulon during an infection is central to the development of disease. The regulation of expression relies on the response to changing conditions resulting from penetration into host tissues and the resultant changes that occur because of bacterial and immune activities. The study of staphylococcal virulence has helped develop the concept of "antipathogenic" drugs (Williams 2002); these compounds do not kill the bacteria, but simply inhibit the expression of destructive virulence factors.

7.5.2.2.1   Agr, QS, and RNAIII

The virulon of *S. aureus* can be classified as surface factors (involved in adhesion and immune evasion, e.g., protein A) and secreted factors (toxins and enzymes involved in damaging the host, e.g., α haemolysin, toxic shock syndrome toxin

[TSST], and proteases). A pleiotropic transposon mutant that was downregulated for secreted factors and upregulated for surface factors was first described in 1986 (Kong et al. 2006; Lyon and Novick 2004; Recsei et al. 1986) and has since been characterised in great detail. The mutation is in the accessory gene regulator region in *agrA*. The control of gene regulation through *agr* is in response to increasing bacterial cell density.

During the initial, low population-density stages of a staphylococcal infection, the expression of surface proteins binding extracellular matrix molecules, e.g., fibronectin, collagen, and fibrinogen, and to the Fc region of immunoglobulin, i.e., Protein A, is favoured. This is thought to promote evasion of host defences and the successful colonisation of host tissues. *S. aureus* challenges the host immune system by eliciting a regional inflammation and subsequent abscess formation (Clarke and Foster 2006). Inside the effectively closed system of the abscess, bacterial population density increases and secreted enzymes and toxins are induced that efficiently destroy white blood cells and liberate nutrients from tissue.

The Agr locus consists of two divergent operons, P2 and P3 (Kong et al. 2006; Lyon and Novick 2004) (Fig. 7.5). The P2 operon comprises the *agrBDCA* signalling cassette. P3 encodes the RNAIII molecule that acts as an intracellular signal controlling the transcription of genes within the Agr regulon. AgrD encodes a small peptide that is cleaved and processed in a process that involves AgrB, and which results in the secretion of a thiolactone peptide or AIP. AIP perception is by the

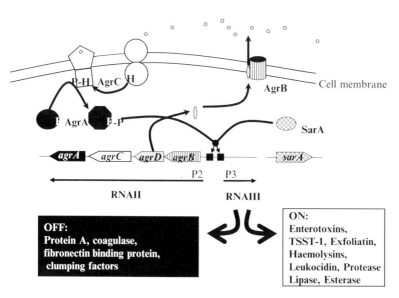

Fig. 7.5 QS in *S. aureus*. The *agr* signalling cassette encodes AgrB, which participates in the export and processing of the AIP (*grey circle*) from the precursor encoded by *agrD*. AgrC perceives the threshold concentration of AIP, which induces autophosphorylation on a histidine (*H*) residue and phosphotransfer to an aspartate (*E*) residue on AgrA. Phospho-AgrA activates transcription of the P2 (*agrBDCA*) and P3 (RNAIII) operons in concert with SarA. RNAIII mediates the quorum response

sensor kinase, AgrC. Molecular events proceed after perception to induce AgrC autophosphorylation on a conserved histidine residue and the subsequent phospho-transfer involving an aspartate residue on AgrA. A simple model in which AgrA-P activates transcription in concert with SarA is now proposed (Dunman et al. 2001; Heinrichs et al. 1996) to activate transcription of P2 (a positive feedback, autoinducing loop) and P3 (producing RNAIII) operons.

The intracellular signal RNAIII then activates transcription of the exoenzyme and toxin genes while downregulating transcription of the surface-associated factors. Unlike other pleiotropic regulators, AgrA does not seem to directly influence the initiation of transcription at other promoters. RNAIII is a regulatory RNA. The mechanisms of action are not wholly understood. It acts to increase or decrease the transcription of many genes, most likely by affecting the availability of other regulatory proteins. RNAIII is also able to increase translation of α-haemolysin by interaction with the *hla* mRNA and blocking translation of protein A by interaction with the initiation codon of the *spa* mRNA (Johansson and Cossart 2003; Majdalani et al. 2005). The study of *S. aureus* in vitro has shown Agr to be the dominant regulator.

7.5.2.2.2 Antipathogenic Drugs

Different groups of staphylococci produce AIPs with different structure/activity profiles (Chan et al. 2004). AIP variants are recognised by group-specific AgrC sensor kinases. The AIPs from different groups do not cross-activate each other and can inhibit the activation of the virulon in vitro. A consensus inhibitory sequence has been synthesised and shown to reduce the virulence of *S. aureus* infections (see Table 7.1). The antagonistic peptide inhibits the production of damage causing virulence factors, and, thus, is antipathogenic. This illustrates the concept of QS blockade (Chan et al. 2004; Williams 2002), which is discussed further in Sect. 7.6.1.

7.5.2.2.3 Agr Is Not the Only Regulator of the *S. aureus* Virulon

Additional studies have identified mutations that affect the expression of the *S. aureus* virulon. The genes involved may act in concert with *agr* or RNAIII to effect control and include the Sar family of regulators (Arvidson and Tegmark 2001), and the two component systems comprising SaeRS, SrrAB, and ArlSR (Liang et al. 2005; Novick and Jiang 2003; Yarwood et al. 2001).

The first member of the Sar family to be identified was SarA (Staphylococcal accessory regulator) (Cheung et al. 1992). SarA is required for the activation of the *agr* P3 promoter to produce RNAIII and, thus, activate the expression of secreted virulence factors (Dunman et al. 2001; Heinrichs et al. 1996). Members of the Sar family normally act as repressors. SarA also exerts control independently of Agr via repression of *spa* (protein A) and *sarT* (a repressor of *hla*) expression (Gao and Stewart 2004; Oscarsson et al. 2006). Rot (repressor of toxins) is a member of the Sar family that is repressed by RNAIII (Said-Salim et al. 2003). Changes in

Table 7.1 The study of *S. aureus* virulence *in vivo*

Model	Study reference	Strains tested	Conclusion
Arthritis (mouse)	Abdelnour et al. (1993)	8325–4 (parent) WA250 (*agrA*::Tn551)	*agrA* mutant is much less able to cause disease
Endocarditis (rabbit)	Cheung et al. (1994)	*sar-l agrA*::Tn551 mutants of strains RN6390 and RN450	*sarA/agrA* mutant reduced infectivity and intravegetation bacterial numbers. Comparable rates of bacteraemia. Inoculum size important—higher inocula of sar-, agr-, and sar-/agr- gave disease, but showed reduced rates of attachment to aortic valve vegetations
Endophthalmitis (rabbit)	Booth et al. (1995)	ISP479 (parent) ISP546 (*agrA*::Tn551)	Mutation in *agrA* greatly reduced the severity of disease but did not eliminate the ability to colonise and grow in vitreous. Some disease was apparent, but much reduced in severity
Osteomyelitis (rabbit)	Gillaspy et al. (1995)	UAMS-1 (osteomyelitis isolate) UAMS-4 (*agrA*::Tn551 transductant)	Mutation in *agr* reduced the incidence and severity of disease but did not eliminate the ability to colonise bone and cause disease
Skin abscess (mouse)	Mayville et al. (1999)	*agrA*::Tn551 and wild type ± QS inhibitor	Mutant is less able to cause disease. QS inhibitor reduces signs of disease
Endophthalmitis (rat)	Giese et al. (1999)	RN6390 parent and *agrA/sarA* mutant	Both strains induced clinical signs of inflammation and inflammatory cell infiltration. The effects in the rats injected with the mutant strain were less severe
Cystic fibrosis lung infection	Goerke et al. (2000)	Isolates from infected sputa	*agr* (RNAIII) not expressed. Other regulation of protein A (also repressed) and haemolysins (variable)

(continued)

Table 7.1 (continued)

Model	Study reference	Strains tested	Conclusion
Device-related infection (guinea pig) exudates	Goerke et al. (2001)	Isolates from infected sputa (agr+ and agr−), Newman, RN6390, and *agrA*, *sarA*, and *sae* mutants	In vitro, some correlation between RNAIII and *hla* expression. RNAIII expression low in vivo; *hla* remained high. RNAIII not controlling expression in vivo. Expression in vivo at lower cell densities, as density increases, both RNAIII and *hla* expression decreased. Agr, Sar, and Sae mutants established infection at similar level to parent. Hla expression in vitro much reduced for all strains. Only reduced for *sae* mutants in vivo
Subcutaneous polyethylene chambers (rabbit) and in rabbit serum	Yarwood et al. (2002)	TSST-producing strains, RN4282, and *agrA*::Tn551 derivative	RNAII and RNAIII repressed in vivo, and did not correlate with in vitro controlled genes, i.e., toxins expressed
Arthritis and osteomyelitis (mouse)	Blevins et al. (2003)	UAMS-1 and *agrA*, *sarA*, and *agrA/sarA* mutants	The virulence of single mutants was attenuated, and the double mutant more so
Systemic infections (intra-peritoneal and intravenous routes), mouse	Benton et al. (2004)	Tn assay to identify mutants unable to survive	*agr* mutants not identified. *saeS*, and *arlS* were identified

environmental conditions affect the expression of *sar* genes. A feature of many is the presence of three promoters that respond to different stimuli, e.g., P*sar*A3 is part of the alternative sigma factor σB regulon. σB responds to stress (Deora et al. 1997).

The SaeRS two-component system was discovered as a transposon mutant with a pleiotropic effect on secreted virulence factors, but without having any effect on RNAIII (Giraudo et al. 1994, 1997). The SrrAB two-component system downregulates Agr induction in low-oxygen conditions (Pragman et al. 2004). The autolysin regulated locus (ArlSR) two-component system downregulates secreted and surface-exposed virulence factors (Liang et al. 2005) (see Fig. 7.6).

7.5.2.2.4 The Study of *S. aureus* Virulence and QS In Vivo

The central role of *agr* has been demonstrated in vitro, but how important is it in vivo? The role of various regulatory loci has been studied or implied in a variety of disease models summarised in Table 7.1. The main conclusions are:

1. Parent *S. aureus* strains are more virulent than their isogenic *agrA* and *sarA* mutants.
2. *agrA* mutants are still able to cause disease.
3. RNAIII expression is substantially repressed in the infecting bacteria while expression of the virulon remains.

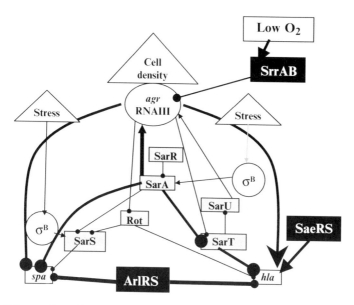

Fig. 7.6 Other regulatory input influencing the staphylococcal quorum response. Members of the Sar family are in *grey boxes*. Additional regulators thought to have greater influence in vivo are shown as *black boxes*. The major regulatory pathways are shown as *thick lines*. *Lines ending with arrows* indicate activation. *Lines ending with circles* indicate repression

4. *saeRS* and *arlRS* are expressed at elevated levels by infecting bacteria.
5. Tagged transposon mutagenesis studies have identified mutants with reduced in vivo survival in *sae* and *arl* (but not *agr* or *sar*).
6. SaeRS and ArlRS have roles that are more significant than *agr* and *sar* in the regulation of the virulon in disease, and the evidence presented does not demonstrate a good correlation between the situation in vitro and the situation in disease.

In conclusion, it is clear that QS in *S. aureus* is part of the complex network that controls virulence, and that interfering with QS can upset the timing and extent of virulence gene expression that may, ultimately, lead to a less virulent infection. QS is not, however, absolutely essential for the expression of damage-causing virulence determinants in an infection.

7.5.3 Biofilms

A biofilm is a complex aggregation of microorganisms often characterised by surface attachment, structural heterogeneity, and an extracellular matrix of polymeric substances. The development of cells in an organised manner within a biofilm can be enhanced by bacteria-to-bacteria communication. Roles for QS have been identified in early development and the final dispersal phases (Parsek and Greenberg 2005; Stanley and Lazazzera 2004). Although biofilms commonly exist in nature as multispecies aggregations, most laboratory studies have used single-species biofilms. However, single-species biofilms may be found in clinical situations growing at surfaces within sterile sites of the body.

The role of QS in biofilm formation has been demonstrated for many bacterial species, with mutation and signal antagonism impairing biofilm formation by bacteria using acyl-HSL, modified peptide, and AI-2 signalling systems. The first studies demonstrated a general role for QS in biofilm formation, maturation, or dispersal primarily using microscopy techniques to visualise the gross impact of mutations or signal antagonists on biofilm development through time. The application of transcriptomics and mutations of QS controlled genes has begun to elucidate the roles of individual components of the quorum response to the biofilm phenotype, and their relative role in different environments.

In *Streptococcus pneumoniae*, signalling via the CSP promotes biofilm formation. Mutants lacking the CSP receptor are compromised in their ability to form biofilms and are less virulent in pneumonia models of infection, where biofilm growth in tissue is an advantage. Conversely, where planktonic growth is an advantage in a sepsis model of infection, the CSP receptor mutants were more virulent (Oggioni et al. 2006).

Modified peptide signalling is also important in biofilm formation by *S. aureus*, a major cause of nosocomial infections involving biofilms on implanted devices and in tissue (Eggiman et al. 2004; Lowy et al. 1998). Primary attachment of *S. aureus* involves MSCRAMM proteins (negatively regulated by QS) binding to surfaces coated with host proteins (e.g., vitronectin, fibronectin) (Clarke and Foster

2006). The formation of microcolonies increases bacterial cell population density and *agr*-mediated QS within the microcolony activates new gene expression (Pratten et al. 2001; Yarwood et al. 2004). SarA also activates biofilm activities (Beenken et al. 2003; Pratten et al. 2001; Valle et al. 2003). Intercellular interactions are promoted because of polysaccharide intercellular adhesion (PIA) (Beenken et al. 2004; Kristian et al. 2004) and alpha haemolysin (Caiazza and O'Toole 2003). Detachment of individual cells, small groups of cells, and larger emboli may involve *agr*–QS-activated enzyme activities because detaching cells express green fluorescent protein from a recombinant P3 promoter construct (Yarwood et al. 2004). The process continues with regrowth of biofilm areas (Yarwood et al. 2004) and the continued detachment of biofilm biomass as increasingly larger clumps (Fux et al. 2004). QS-regulated activities that promote detachment have also been identified for *S. marcescens* (*liquefaciens*) MG1, but only when the biofilm is grown in low levels of nutrients (Rice et al. 2005). This emphasises the importance of other environmental parameters in addition to bacterial cell density in the control of biofilm phenotypes (reviewed by Horswill et al. 2007).

In *B. subtilis*, the architecture of the biofilm accommodates specialised endospore-containing fruiting bodies. QS-regulated surfactant mediates the formation of pillar structures here by reducing surface tension (Branda et al. 2001). The QS-regulated rhamnolipid surfactant has been shown to be involved in the maintenance of water channels within *P. aeruginosa* biofilms and also the mushroom-like biofilm pillar structures (Davey et al. 2003; Lequette and Greenberg 2005). Las and Rhl QS signalling, possibly through VqsR, QscR, and PQS signalling, have all been implicated in the activation of genes required for the formation of a *P. aeruginosa* biofilm, including both architectural features and the activation of genes involved in the detoxification of waste products within the biofilm (De Kievit et al. 2001; Hassett et al. 1999). The biofilm growth of *P. aeruginosa* is especially important in the development of chronic infections of the cystic fibrosis lung (Parsek and Greenberg 2005).

In *Pseudomonas putida*, QS controls the production of the putisolvin surfactants. The regulation of putisolvins is similar to the regulation of rhamnolipid production by *P. aeruginosa*; however, the effects are different. *P. putida* mutants unable to produce the putisolvin produce much thicker biofilms, leading to the suggestion that, in this case, the surfactant aids colonisation of the *P. putida* by breaking up existing biofilms, thus, making the space available for colonisation by *P. putida* (Dubern et al. 2006). QS regulation of other traits enabling better competition within multispecies biofilms have been described. In *S. marcescens* (*liquefaciens*) MG1, acyl-HSL-mediated QS activates resistance to protozoan grazing of biofilm cells, through a mechanism that involves bacterial cell filamentation (Queck et al. 2006).

Other extracellular bacterial products have been implicated in the control of biofilm formation. The role of AI-2- and LuxS-mediated signals have been proposed for a number of bacteria, although, again, some of these effects may be caused by effects on central metabolism rather than through QS signalling (Doherty et al. 2006). In *Vibrio cholerae*, the unidentified CAI-1 activates the sensor kinase, CqsA, which acts via phosphorylation of LuxO to activate HapR, a transcriptional activator of polysaccharide biosynthesis that promotes biofilm formation and favours colonisation (Liu et al. 2007).

In conclusion, biofilms are high cell density "cities of microbes" and the ideal environment for QS. Cell density-regulated determinants control biofilm formation, maturation, and maintenance of the structural architecture and the dispersal from biofilms to varying degrees for diverse species in different conditions. It is likely that QS allows spatial and temporal control of gene expression within the biofilm by regulating gene expression in response to the production, diffusion, degradation, and perception of QS signals within the biofilm matrix.

7.6 QS in the Real World

Most investigations of QS have involved laboratory cultures of a single species, but, in reality, other species of bacteria will be present. In QS, there is a clear model of the bacteria within a population releasing the signal that is sensed, and responded to, by other bacterial cells in the population. It is not necessary that these bacteria be of the same species. In same-species signalling, we view QS as a way of regulating the expression of phenotypic traits that are favourable at high cell density. Signalling between cells of different species may allow cells to work together (e.g., *Burkholderia* and *Pseudomonas* in the cystic fibrosis lung) (McKenney et al. 1995), but also may be subject signal interference, signal interception, and signal destruction.

One striking example of the need to consider the real environment bacteria face when growing is the effect different ex vivo fluids have on the stability of acyl-HSLs. The simple parameters of pH and temperature can affect acyl-HSL stability. Lactone ring opening and the loss of signal activity occurs at alkaline pH at a rate that increases with temperature and the signal's stability is proportional to the length of the acyl side chain (Yates et al. 2002). For acyl-HSL-mediated signalling to be functional under physiological conditions in mammalian tissue fluids, signals require an *N*-acyl side chain of at least four carbons in length, and, for more alkaline conditions, e.g., as might be found in urine, significantly more signal would need to be produced to specify a quorate population.

7.6.1 The Effect of Other Species

In the real world, bacteria compete for resources in multispecies communities, from within there are produced many small molecule signals. Do bacteria only perceive the signals produced by members of the same species? Or are they able to intercept the signals from other bacteria to present a phenotype better suited to survival? Studies with two bacterial species (*Burkholderia cepacia* and *P. aeruginosa*) that are often found together in serious infections of the cystic fibrosis lung gave perhaps the first indication that bacteria of different species could communicate with acyl-HSLs and possibly work together (McKenney et al. 1995). The production of three virulence determinants of *B. cepacia* (siderophore, lipase, and protease) was increased in the presence of culture supernatants from *P. aeruginosa* in proportion to the level of QS signal they produced. Acyl-HSL signalling in the genus *Burkholderia* has since been identified and recently reviewed (Eberl 2006).

Bacteria-to-bacteria signalling can also be competitive. Study of AIP-mediated signalling in the staphylococci has identified four classes of AIP in *S. aureus* and another variant in *S. epidermidis* (reviewed by Harraghy et al. 2007), with other variants from other species likely to be identified. AIP variants are recognised by group-specific AgrC sensor kinases. The result is cross-activation of the agr-response for bacteria of the same group, and inactivation (in the main) of the agr-response between staphylococci of different groups. The molecular details of the activating and antagonistic interactions of AIPs with AgrC have been explored using synthetic AIP analogues and now form the basis of novel therapies against staphylococci that may downregulate virulence determinants and allow for a larger window for treatment of infections by this destructive pathogen (Chan et al. 2004; Harraghy et al. 2007; Mayville et al. 1999). In addition, the long acyl chain QS signal of *P. aeruginosa* 3-oxo-C12-HSL also inhibits the expression of the agr-response, downregulating the production of toxins and exoenzymes. The molecular details are not fully appreciated as yet, but seem to involve binding to a saturable receptor in the *S. aureus* membrane and inhibiting the expression of both *sarA* and *agr* (Qazi et al. 2006).

Caution has been advised, because the agr-response is double-edged, and an inhibition of toxins and exoenzymes is accompanied by an increase in immune evasion phenotypes that may have deleterious effects, promoting biofilm formation and antibiotic resistance (Harraghy et al. 2007).

It is not just other bacteria that can intercept the QS signal. The regulation of virulence factors by QS in pathogens and the subsequent detection of QS signals in infected tissue (reviewed by Pritchard 2006; Rumbaugh et al. 2007) has raised the question "Do QS signals have an effect on the host?" The answer is an emphatic yes. The QS signals, 3-oxo-C12-HSL and PQS, produced by *P. aeruginosa* both have an immune modulatory effect. Provision of 3-oxo-C12-HSL to LPS-stimulated macrophages in vitro suppresses proinflammatory cytokine secretion. T-cell proliferation and antibacterial T-helper-1 responses are suppressed. The effects of 3-oxo-C12 - HSL are thought to occur through interference with early T-cell signalling events. PQS seems to act at a later stage in T-cell signalling; being at least a 10-fold more potent inhibitor of T-cell proliferation, but not affecting the release of proinflammatory cytokine, interleukin 2 (see Pritchard et al. 2006; Rumbaugh et al. 2007). In general, immune modulation effects occur at concentrations of 3-oxo-C12-HSL below 10µM. The proposed synergistic effect of PQS and 3-oxo-C12-HSL, e.g., in infections of the cystic fibrosis lung, in which levels of QS signals approximate those found to have immune modulatory effects in vitro, may allow bacteria to dampen initial immune responses, allowing survival and proliferation to reach population sizes sufficient to establish an infection and express damaging toxins and exoenzymes (Pritchard et al. 2006). In concert with effects on the strength of the immune response, 3-oxo-C12-HSL also promotes vasodilation, suggesting that bacteria may be able to stimulate blood flow and nutrients to an area of infection (Gardiner et al. 2001).

Higher levels of 3-oxo-C12-HSL (100μM) have a further effect; they induce apoptosis via a mechanism independent of the immune modulatory effects that involves the mobilisation of intracellular calcium (Rumbaugh et al. 2007). Interception of signals by mammalian hosts of infection seems to favour the bacterium. Evidence also exists for the exploitation of bacterial acyl-HSL signals by eukaryotes. Bacterial biofilms release acyl-HSLs during their development. The release of acyl-HSLs attracts zoospores, motile reproductive stages of the green seaweed, *Ulva*, by reducing the zoospores rate of swimming and promoting settling. The effect is apparent with wild-type *V. anguillarum* bacteria producing acyl-HSLs, recombinant *E. coli* bacteria producing acyl-HSLs, but not *V. anguillarum* mutants unable to produce acyl-HSLs, unmodified *E. coli* or *V. anguillarum* expressing a recombinant *Bacillus* enzyme AiiA that is able to degrade acyl-HSLs. The final evidence came from the attraction of *Ulva* zoospores to agarose gel impregnated with synthetic acyl-HSLs and the disruption of attraction by including synthetic acyl-HSLs in the test water to mask QS signal gradients produced from bacterial biofilms (reviewed by Joint et al. 2007). The transduction of the acyl-HSL signal within the zoospore is through the influx of calcium ions (Joint et al. 2007), although the exact mechanism is unknown. The benefit for the *Ulva* to settle on a biofilm is not entirely clear, but there are numerous ecological considerations (discussed by Joint et al. 2007).

Another seaweed, the red alga, *Delisea pulchra*, produce halogenated furanones that are antagonistic to QS signals. The secretion of these substances prevents the colonisation of the plant by bacteria, by inhibiting the activation of LuxR-type proteins, and, thus, the colonisation factors they activate (see McDougald et al. 2007 for a review). Halogenated furanones are potent anti-fouling agents in the marine environment and have also been used to inhibit QS-regulon expression in pathogenic bacteria, attenuating their virulence in animal models (McDougald et al. 2007).

Finally, enzymes have been discovered in numerous bacteria and in eukaryotes that can inactivate QS signals. The first to be discovered, AiiA, is found in *Bacillus* species and is an acyl-HSL lactonase that inactivates acyl-HSLs by opening the lactone ring. Lactonases may contribute to the control of the quorum response, for example, in *Agrobacterium tumefaciens*, the AttM lactonase eliminates acyl-HSLs produced by TraI in the stationary phase of growth. Lactonases may exert a protective effect against bacteria using QS to regulate virulence factor expression, for example, the lactonases produced by human airway epithelia (reviewed by Dong and Zhang 2005). A second type of acyl-HSL-degrading enzyme is an acyl-HSL acylase, which acts by removing the acyl chain to leave HSL. The acylases have also been identified from bacterial and eukaryotic sources (Dong and Zhang 2005).

7.7 New Perspectives and Applications

The appreciation that cells within bacterial populations communicated and regulated gene expression in response to density-dependent signals was a major event in bacteriology and has led to the discovery of a whole new world of signals that affect bacterial activities. Some signals are not cell-density dependent and, therefore, do not fall into the category of QS signals. Some signals allow communication between different species of bacteria and between bacteria and species of the plant and animal kingdoms. The appreciation that bacteria also use small molecule signals to influence the behaviour of hosts for infection and symbiosis is another major event in bacteriology and may help to further elucidate the complexity of these interactions. The immune modulation activity of 3-oxo-C12-HSL is a clear example of how we may be able to exploit host effects of QS signals for clinical applications. The suppression of autoimmune inflammation in non-obese diabetic mice is one example that may translate to a pharmaceutical application in humans (Pritchard et al. 2005).

The appreciation of QS and other bacterial signalling strategies has major implications for biotechnology. There are now many publications discussing the possibility of using QS antagonists to block expression of the QS regulon and inhibit, e.g., virulence gene expression as part of an antibiotic strategy to control pathogens or as an anti-fouling strategy (Chan et al. 2004; McDougald et al. 2007; Williams 2002). The use of QS antagonists may be as soluble drugs, but also as coating on implanted medical devices or surfaces in the environment subject to biofouling (McDougald et al. 2007). Halogenated furanones (see Fig. 7.2) act via inhibition of the signal receptor/response regulator. An alternative strategy may be to inhibit signal synthesis using compounds specifically designed to inhibit signal synthases (see Sect. 7.2). Quenching of the quorum response may also be achieved by the use of enzymes that degrade the signal, a strategy that may find use in the engineering of disease-resistant transgenic plants (Dong and Zhang 2005; Zhang 2003).

Highly Recommended Readings

Ahmer BM (2004) Cell-to-cell signalling in *Escherichia coli* and *Salmonella enterica*. Mol Microbiol 52:933–945

Bainton NJ, Stead P, Chhabra SR, Bycroft BW, Salmond GPC, Stewart GSAB, Williams P (1992a) N-(3-oxohexanoyl)-L-homoserine lactone regulates carbapenem antibiotic production in *Erwinia carotovora*. Biochem J 288:997–1004

Barnard AM, Salmond GPC (2007) Quorum sensing in *Erwinia* species. Anal Bioanal Chem 387:415–423

De Kievit TR, Gillis R, Marx S, Brown C, Iglewski BH (2001) Quorum-sensing genes in *Pseudomonas aeruginosa* biofilms: their role and expression patterns. Appl Environ Microbiol 67:1865–1873

Dong YH, Zhang LH (2005) Quorum sensing and quorum-quenching enzymes. J Microbiol 43:101–109

Fuqua WC, Winans SC, Greenberg EP (1994) Quorum sensing in bacteria—the LuxR-LuxI family of cell density-responsive transcriptional regulators. J Bacteriol 176:269–275

Swift S (2003) Quorum Sensing: Approaches to identify signals and signalling genes in Gram-negative bacteria. In: Blot M (ed) Methods and Tools in Biosciences and Medicine: Prokaryotic Genomics and Genetics Birkhäuser Verlag, Basel, Switzerland, pp. 110–130

Swift S, Downie JA, Whitehead NA, Barnard AM, Salmond GPC, Williams P (2001) Quorum sensing as a population-density-dependent determinant of bacterial physiology. Adv Microb Physiol 45:199–270

Williams P (2002) Quorum sensing: an emerging target for antibacterial chemotherapy? Expert Opin Ther Targets 6:257–274

References

Abdelnour A, Arvidson S, Bremell T, Ryden C, Tarkowski A (1993) The accessory gene regulator (*agr*) controls *Staphylococcus aureus* virulence in a murine arthritis model. Infect Immun 61:3879–3885

Ahmer BM (2004) Cell-to-cell signalling in *Escherichia coli* and *Salmonella enterica*. Mol Microbiol 52:933–945

Al-Tahhan RA, Sandrin TR, Bodour AA, Maier RM (2000) Rhamnolipid-induced removal of lipopolysaccharide from *Pseudomonas aeruginosa*: effect on cell surface properties and interaction with hydrophobic substrates. Appl Environ Microbiol 66:3262–3268

Andersson RA, Eriksson AR, Heikinheimo R, Mae A, Pirhonen M, Koiv V, Hyytiainen H, Tuikkala A, Palva ET (2000) Quorum sensing in the plant pathogen *Erwinia carotovora* subsp. *carotovora*: the role of *expR*(Ecc). Mol Plant Microbe Interact 13:384–393

Ansaldi M, Marolt D, Stebe T, Mandic-Mulec I, Dubnau D (2002) Specific activation of the *Bacillus* quorum-sensing systems by isoprenylated pheromone variants. Mol Microbiol 44:1561–1573

Arvidson S, Tegmark K (2001) Regulation of virulence determinants in *Staphylococcus aureus*. Int J Med Microbiol 291:159–170

Atkinson S, Throup JP, Stewart GSAB, Williams P (1999) A hierarchical quorum-sensing system in *Yersinia pseudotuberculosis* is involved in the regulation of motility and clumping. Mol Microbiol 33:1267–1277

Bacon Schneider K, Palmer TM, Grossman AD (2002) Characterization of *comQ* and *comX*, two genes required for production of ComX pheromone in *Bacillus subtilis*. J Bacteriol 184:410–419

Bainton NJ, Stead P, Chhabra SR, Bycroft BW, Salmond GPC, Stewart GSAB, Williams P (1992a) N-(3-oxohexanoyl)-L-homoserine lactone regulates carbapenem antibiotic production in *Erwinia carotovora*. Biochem J 288:997–1004

Bainton NJ, Bycroft BW, Chhabra SR, Stead P, Gledhill L, Hill PJ, Rees CE, Winson MK, Salmond GPC, Stewart GSAB, Williams P (1992b) A general role for the *lux* autoinducer in bacterial cell signalling: control of antibiotic biosynthesis in *Erwinia*. Gene 116:87–91

Barnard AM, Salmond GPC (2007) Quorum sensing in *Erwinia* species. Anal Bioanal Chem 387:415–423

Bassler BL, Wright M, Showalter RE, Silverman MR (1993) Intercellular signalling in *Vibrio harveyi*: sequence and function of genes regulating expression of luminescence. Mol Microbiol 9:773–786

Bassler BL, Wright M, Silverman MR (1994) Multiple signalling systems controlling expression of luminescence in *Vibrio harveyi*: sequence and function of genes encoding a second sensory pathway. Mol Microbiol 13:273–286

Beenken KE, Blevins JS, Smeltzer MS (2003) Mutation of *sarA* in *Staphylococcus aureus* limits biofilm formation. Infect Immun 71:4206–4211

Beenken KE, Dunman PM, McAleese F, Macapagal D, Murphy E, Projan SJ, Blevins JS, Smeltzer MS (2004) Global gene expression in *Staphylococcus aureus* biofilms. J Bacteriol 186:4665–4684

Bell KS, Sebaihia M, Pritchard L, Holden MT, Hyman LJ, Holeva MC, Thomson NR, Bentley SD, Churcher LJ, Mungall K, Atkin R, Bason N, Brooks K, Chillingworth T, Clark K, Doggett J, Fraser A, Hance Z, Hauser H, Jagels K, Moule S, Norbertczak H, Ormond D, Price C, Quail MA, Sanders M, Walker D, Whitehead S, Salmond GPC, Birch PR, Parkhill J, Toth IK (2004) Genome sequence of the enterobacterial phytopathogen *Erwinia carotovora* subsp. *atroseptica* and characterization of virulence factors. Proc Natl Acad Sci USA 101:11105–11110

Benito Y, Kolb FA, Romby P, Lina G, Etienne J, Vandenesch F (2000) Probing the structure of RNAIII, the *Staphylococcus aureus agr* regulatory RNA, and identification of the RNA domain involved in repression of protein A expression. RNA 6:668–679

Benton BM, Zhang JP, Bond S, Pope C, Christian T, Lee L, Winterberg KM, Schmid MB, Buysse JM (2004) Large-scale identification of genes required for full virulence of *Staphylococcus aureus*. J Bacteriol 186:8478–8489

Berka RM, Hahn J, Albano M, Draskovic I, Persuh M, Cui X, Sloma A, Widner W, Dubnau D (2002) Microarray analysis of the *Bacillus subtilis* K-state: genome-wide expression changes dependent on ComK. Mol Microbiol 43:1331–1345

Bi ZX, Liu YJ, Lu CP (2007) Contribution of AhyR to virulence of *Aeromonas hydrophila* J-1. Res Vet Sci doi:10.1016/j.rvsc.2007.01.003

Blevins JS, Elasri MO, Allmendinger SD, Beenken KE, Skinner RA, Thomas JR, Smeltzer MS (2003) Role of *sarA* in the pathogenesis of *Staphylococcus aureus* musculoskeletal infection. Infect Immun 71:516–523

Bonfiglio G, Russo G, Nicoletti G (2002) Recent developments in carbapenems. Expert Opin Investig Drugs 11:529–544

Bonomo RA, Szabo D (2006) Mechanisms of multidrug resistance in *Acinetobacter* species and *Pseudomonas aeruginosa*. Clin Infect Dis 43 Suppl 2:S49–56

Booth MC, Atkuri RV, Nanda SK, Iandolo JJ, Gilmore MS (1995) Accessory gene regulator controls *Staphylococcus aureus* virulence in endophthalmitis. Invest Ophthalmol Vis Sci 36:1828–1836

Branda SS, Gonzalez-Pastor JE, Ben-Yehuda S, Losick R, Kolter R (2001) Fruiting body formation by *Bacillus subtilis*. Proc Natl Acad Sci USA 98:11621–11626

Branny P, Pearson JP, Pesci EC, Kohler T, Iglewski BH, Van Delden C (2001) Inhibition of quorum sensing by a *Pseudomonas aeruginosa dksA* homologue. J Bacteriol 183:1531–1539

Burr T, Barnard AM, Corbett MJ, Pemberton CL, Simpson NJ, Salmond GPC (2006) Identification of the central quorum sensing regulator of virulence in the enteric phytopathogen, *Erwinia carotovora*: the VirR repressor. Mol Microbiol 59:113–125

Bycroft BW, Maslen C, Box SJ, Brown A, Tyler JW (1988) The biosynthetic implications of acetate and glutamate incorporation into (3R,5R)-carbapenam-3-carboxylic acid and (5R)-carbapen-2-em-3-carboxylic acid by *Serratia* sp. J Antibiot 41:1231–1242

Caiazza NC, O'Toole GA (2003) Alpha-toxin is required for biofilm formation by *Staphylococcus aureus*. J Bacteriol 185:3214–3217

Caiazza NC, Shanks RM, O'Toole GA (2005) Rhamnolipids modulate swarming motility patterns of *Pseudomonas aeruginosa*. J Bacteriol 187:7351–7361

Cao JG, Meighen EA (1989) Purification and structural identification of an autoinducer for the luminescence system of *Vibrio harveyi*. J Biol Chem 264:21670–21676

Castang S, Reverchon S, Gouet P, Nasser W (2006) Direct evidence for the modulation of the activity of the *Erwinia chrysanthemi* quorum-sensing regulator ExpR by acylhomoserine lactone pheromone. J Biol Chem 281:29972–29987

Chai Y, Winans SC (2004) Site-directed mutagenesis of a LuxR-type quorum-sensing transcription factor: alteration of autoinducer specificity. Mol Microbiol 51:765–776

Chan WC, Coyle BJ, Williams P (2004) Virulence regulation and quorum sensing in staphylococcal infections: competitive AgrC antagonists as quorum sensing inhibitors. J Med Chem 47:4633–4641

Chan PF, Bainton NJ, Daykin MM, Winson MK, Chhabra SR, Stewart GSAB, Salmond GPC, Bycroft BW, Williams P (1995) Small molecule mediated autoinduction of antibiotic biosynthesis in the plant pathogen *Erwinia carotovora*. Biochem Soc Trans 23:127S

Chatterjee A, Cui Y, Liu Y, Dumenyo CK, Chatterjee AK (1995) Inactivation of *rsmA* leads to overproduction of extracellular pectinases, cellulases, and proteases in Erwinia carotovora subsp. carotovora in the absence of the starvation/cell density-sensing signal, N-(3-oxohexanoyl)-L-homoserine lactone. Appl Environ Microbiol 61:1959–1967

Chen X, Schauder S, Potier N, Van Dorsselaer A, Pelczer I, Bassler BL, Hughson FM (2002) Structural identification of a bacterial quorum-sensing signal containing boron. Nature 415:545–549

Cheung AL, Koomey JM, Butler CA, Projan SJ, Fischetti VA (1992) Regulation of exoprotein expression in *Staphylococcus aureus* by a locus (*sar*) distinct from *agr*. Proc Natl Acad Sci USA 89:6462–6466

Cheung AL, Eberhardt KJ, Chung E, Yeaman MR, Sullam PM, Ramos M, Bayer AS (1994) Diminished virulence of a *sar-/agr-* mutant of *Staphylococcus aureus* in the rabbit model of endocarditis. J Clin Invest 94:1815–1822

Choi SH, Greenberg EP (1991) The C-terminal region of the *Vibrio fischeri* LuxR protein contains an inducer-independent *lux* gene activating domain. Proc Natl Acad Sci USA 88:11115–11119

Chugani SA, Whiteley M, Lee KM, D'Argenio D, Manoil C, Greenberg EP (2001) QscR, a modulator of quorum-sensing signal synthesis and virulence in *Pseudomonas aeruginosa*. Proc Natl Acad Sci USA 98:2752–2757

Clarke MB, Hughes DT, Zhu C, Boedeker EC, Sperandio V (2006) The QseC sensor kinase: a bacterial adrenergic receptor. Proc Natl Acad Sci USA 103:10420–10425

Clarke SR, Foster SJ (2006) Surface adhesins of *Staphylococcus aureus*. Adv Microb Physiol 51:187–224

Comella N, Grossman AD (2005) Conservation of genes and processes controlled by the quorum response in bacteria: characterization of genes controlled by the quorum-sensing transcription factor ComA in *Bacillus subtilis*. Mol Microbiol 57:1159–1174

Core L, Perego M (2003) TPR-mediated interaction of RapC with ComA inhibits response regulator-DNA binding for competence development in *Bacillus subtilis*. Mol Microbiol 49:1509–1522

Cosby WM, Vollenbroich D, Lee OH, Zuber P (1998) Altered *srf* expression in *Bacillus subtilis* resulting from changes in culture pH is dependent on the SpoOK oligopeptide permease and the ComQX system of extracellular control. J Bacteriol 180:1438–1445

Coulthurst SJ, Williamson NR, Harris AK, Spring DR, Salmond GPC (2006) Metabolic and regulatory engineering of *Serratia marcescens*: mimicking phage-mediated horizontal acquisition of antibiotic biosynthesis and quorum-sensing capacities. Microbiology 152:1899–1911

Croxatto A, Pride J, Hardman A, Williams P, Camara M, Milton DL (2004) A distinctive dual-channel quorum-sensing system operates in *Vibrio anguillarum*. Mol Microbiol 52:1677–1689

Cui Y, Chatterjee A, Liu Y, Dumenyo CK, Chatterjee AK (1995) Identification of a global repressor gene, *rsmA*, of *Erwinia carotovora* subsp. *carotovora* that controls extracellular enzymes, N-(3-oxohexanoyl)-L-homoserine lactone, and pathogenicity in soft-rotting *Erwinia* spp. J Bacteriol 177:5108–5115

Cui Y, Chatterjee A, Hasegawa H, Chatterjee AK (2006) *Erwinia carotovora* subspecies produce duplicate variants of ExpR, LuxR homologs that activate *rsmA* transcription but differ in their interactions with N-acylhomoserine lactone signals. J Bacteriol 188:4715–4726

Daniels R, Vanderleyden J, Michiels J (2004) Quorum sensing and swarming migration in bacteria. FEMS Microbiol Rev 28:261–289

Davey ME, Caiazza NC, O'Toole GA (2003) Rhamnolipid surfactant production affects biofilm architecture in *Pseudomonas aeruginosa* PAO1. J Bacteriol 185:1027–1036

Defoirdt T, Bossier P, Sorgeloos P, Verstraete W (2005) The impact of mutations in the quorum sensing systems of *Aeromonas hydrophila*, *Vibrio anguillarum* and *Vibrio harveyi* on their virulence towards gnotobiotically cultured *Artemia franciscana*. Environ Microbiol 7:1239–1247

De Kievit TR, Gillis R, Marx S, Brown C, Iglewski BH (2001) Quorum-sensing genes in *Pseudomonas aeruginosa* biofilms: their role and expression patterns. Appl Environ Microbiol 67:1865–1873

DeLisa MP, Wu CF, Wang L, Valdes JJ, Bentley WE (2001) DNA microarray-based identification of genes controlled by autoinducer 2-stimulated quorum sensing in *Escherichia coli*. J Bacteriol 183:5239–5247

Deora R, Tseng T, Misra TK (1997) Alternative transcription factor sigmaSB of *Staphylococcus aureus*: characterization and role in transcription of the global regulatory locus *sar*. J Bacteriol 179:6355–6359

Desai JD, Banat IM (1997) Microbial production of surfactants and their commercial potential. Microbiol Mol Biol Rev 61:47–64

Diggle SP, Winzer K, Lazdunski A, Williams P, Camara M (2002) Advancing the quorum in *Pseudomonas aeruginosa*: MvaT and the regulation of *N*-acylhomoserine lactone production and virulence gene expression. J Bacteriol 184:2576–2586

Diggle SP, Cornelis P, Williams P, Camara M (2006) 4-quinolone signalling in *Pseudomonas aeruginosa*: old molecules, new perspectives. Int J Med Microbiol 296:83–91

Diggle SP, Matthijs S, Wright VJ, Fletcher MP, Chhabra SR, Lamont IL, Kong X, Hider RC, Cornelis P, Camara M, Williams P (2007) The *Pseudomonas aeruginosa* 4-quinolone signal molecules HHQ and PQS play multifunctional roles in quorum sensing and iron entrapment. Chem Biol 14:87–96

Doherty N, Holden MT, Qazi SN, Williams P, Winzer K (2006) Functional analysis of *luxS* in *Staphylococcus aureus* reveals a role in metabolism but not quorum sensing. J Bacteriol 188:2885–2897

Dong YH, Zhang LH (2005) Quorum sensing and quorum-quenching enzymes. J. Microbiol 43:101–109

D'Souza C, Nakano MM, Zuber P (1994) Identification of *comS*, a gene of the *srfA* operon that regulates the establishment of genetic competence in *Bacillus subtilis*. Proc Natl Acad Sci USA 91:9397–9401

Dubern JF, Lugtenberg BJ, Bloemberg GV (2006) The *ppuI-rsaL-ppuR* quorum-sensing system regulates biofilm formation of *Pseudomonas putida* PCL1445 by controlling biosynthesis of the cyclic lipopeptides putisolvins I and II. J Bacteriol 188:2898–2906

Dunman PM, Murphy E, Haney S, Palacios D, Tucker-Kellogg G, Wu S, Brown EL, Zagursky RJ, Shlaes D, Projan SJ (2001) Transcription profiling-based identification of *Staphylococcus aureus* genes regulated by the *agr* and/or *sarA* loci. J Bacteriol 183:7341–7353

Dunny GM, Winans SC (1999) Cell-Cell Signaling in Bacteria. ASM Press, Washington, DC.

Dunny GM, Brown BL, Clewell DB (1978) Induced cell aggregation and mating in *Streptococcus faecalis*: evidence for a bacterial sex pheromone. Proc Natl Acad Sci USA 75:3479–3483

Eberhard A, Burlingame AL, Eberhard C, Kenyon GL, Nealson KH, Oppenheimer NJ (1981) Structural identification of autoinducer of Photobacterium fischeri luciferase. Biochemistry 20:2444–2449

Eberl L (2006) Quorum sensing in the genus Burkholderia. Int J Med Microbiol 296:103–110

Eberl L, Winson MK, Sternberg C, Stewart GSAB, Christiansen G, Chhabra SR, Bycroft B, Williams P, Molin S, Givskov M (1996) Involvement of *N*-acyl-L-hormoserine lactone autoinducers in controlling the multicellular behaviour of *Serratia liquefaciens*. Mol Microbiol 20:127–136

Eberl L, Molin S, Givskov M (1999) Surface motility of *Serratia liquefaciens* MG1. J Bacteriol 181:1703–1712

Engebrecht J, Nealson K, Silverman M (1983) Bacterial bioluminescence: isolation and genetic analysis of functions from *Vibrio fischeri*. Cell 32:773–781

Eggimann P, Sax H, Pittet D (2004) Catheter-related infections. Microbes Infect 6:1033–1042

Fabret C, Quentin Y, Guiseppi A, Busuttil J, Haiech J, Denizot F (1995) Analysis of errors in finished DNA sequences: the surfactin operon of *Bacillus subtilis* as an example. Microbiology 141:345–350

Fekete A, Frommberger M, Rothballer M, Li X, Englmann M, Fekete J, Hartmann A, Eberl L, Schmitt-Kopplin P (2007) Identification of bacterial *N*-acylhomoserine lactones (AHLs) with a combination of ultra-performance liquid chromatography (UPLC), ultra-high-resolution mass spectrometry, and in-situ biosensors. Anal Bioanal Chem 387:455–467

Finney AH, Blick RJ, Murakami K, Ishihama A, Stevens AM (2002) Role of the C-terminal domain of the alpha subunit of RNA polymerase in LuxR-dependent transcriptional activation of the lux operon during quorum sensing. J Bacteriol 184:4520–4528

Flavier AB, Clough SJ, Schell MA, Denny TP (1997) Identification of 3-hydroxypalmitic acid methyl ester as a novel autoregulator controlling virulence in *Ralstonia solanacearum*. Mol Microbiol 26:251–259

Freeman JA, Bassler BL (1999a) Sequence and function of LuxU: a two-component phosphorelay protein that regulates quorum sensing in *Vibrio harveyi*. J Bacteriol 181:899–906

Freeman JA, Bassler BL (1999b) A genetic analysis of the function of LuxO, a two-component response regulator involved in quorum sensing in *Vibrio harveyi*. Mol Microbiol 31:665–677

Freeman JA, Lilley BN, Bassler BL (2000) A genetic analysis of the functions of LuxN: a two-component hybrid sensor kinase that regulates quorum sensing in *Vibrio harveyi*. Mol Microbiol 35:139–149

Freestone PP, Haigh RD, Williams PH, Lyte M (1999) Stimulation of bacterial growth by heat-stable, norepinephrine-induced autoinducers. FEMS Microbiol Lett 172:53–60

Frezza M, Castang S, Estephane J, Soulere L, Deshayes C, Chantegrel B, Nasser W, Queneau Y, Reverchon S, Doutheau A (2006) Synthesis and biological evaluation of homoserine lactone derived ureas as antagonists of bacterial quorum sensing. Bioorg Med Chem 14:4781–4791

Fuqua C, Eberhard A (1999) Signal Generation in Autoinducer Systems: Synthesis of Acylated Homoserine Lactones by LuxI-type Proteins. In: Dunny GM, Winans SC (ed) Cell-Cell Signaling in Bacteria. ASM Press, Washington, DC, pp. 211–230

Fuqua WC, Winans SC, Greenberg EP (1994) Quorum sensing in bacteria—the LuxR-LuxI family of cell density-responsive transcriptional regulators. J Bacteriol 176:269–275

Fux CA, Wilson S, Stoodley P (2004) Detachment characteristics and oxacillin resistance of *Staphyloccocus aureus* biofilm emboli in an in vitro catheter infection model. J Bacteriol 186:4486–4491

Fux CA, Shirtliff M, Stoodley P, Costerton JW (2005) Can laboratory reference strains mirror "real-world" pathogenesis? Trends Microbiol 13:58–63

Galli G, Rodriguez F, Cosmina P, Pratesi C, Nogarotto R, de Ferra F, Grandi G (1994) Characterization of the surfactin synthetase multi-enzyme complex. Biochim Biophys Acta 1205:19–28

Gao J, Stewart GC (2004) Regulatory elements of the *Staphylococcus aureus* protein A (Spa) promoter. J Bacteriol 186:3738–3748

Gardiner SM, Chhabra SR, Harty C, Williams P, Pritchard DI, Bycroft BW, Bennett T (2001) Haemodynamic effects of the bacterial quorum sensing signal molecule, *N*-(3-oxododecanoyl)-L-homoserine lactone, in conscious, normal and endotoxaemic rats. Br J Pharmacol 133:1047–1054

Giese MJ, Berliner JA, Riesner A, Wagar EA, Mondino BJ (1999) A comparison of the early inflammatory effects of an *agr-/sar-* versus a wild type strain of *Staphylococcus aureus* in a rat model of endophthalmitis. Curr Eye Res 18:177–185

Gillaspy AF, Hickmon SG, Skinner RA, Thomas JR, Nelson CL, Smeltzer MS (1995) Role of the accessory gene regulator (*agr*) in pathogenesis of staphylococcal osteomyelitis. Infect Immun 63:3373–3380

Gilson L, Kuo A, Dunlap PV (1995) AinS and a new family of autoinducer synthesis proteins. J Bacteriol 177:6946–6951

Giraudo AT, Raspanti CG, Calzolari A, Nagel R (1994) Characterization of a Tn*551*-mutant of *Staphylococcus aureus* defective in the production of several exoproteins. Can J Microbiol 40:677–681

Giraudo AT, Cheung AL, Nagel R (1997) The *sae* locus of *Staphylococcus aureus* controls exo-protein synthesis at the transcriptional level. Arch Microbiol 168:53–58

Givskov M, Ostling J, Eberl L, Lindum PW, Christensen AB, Christiansen G, Molin S, Kjelleberg S (1998) Two separate regulatory systems participate in control of swarming motility of *Serratia liquefaciens* MG1. J Bacteriol 180:742–745

Goerke C, Campana S, Bayer MG, Doring G, Botzenhart K, Wolz C (2000) Direct quantitative transcript analysis of the *agr* regulon of *Staphylococcus aureus* during human infection in comparison to the expression profile in vitro. Infect Immun 68:1304–1311

Goerke C, Fluckiger U, Steinhuber A, Zimmerli W, Wolz C (2001) Impact of the regulatory loci *agr*, *sarA* and *sae* of *Staphylococcus aureus* on the induction of alpha-toxin during device-related infection resolved by direct quantitative transcript analysis. Mol Microbiol 40:1439–1447

Gould TA, Schweizer HP, Churchill ME (2004) Structure of the *Pseudomonas aeruginosa* acyl-homoserinelactone synthase LasI. Mol Microbiol 53:1135–1146

Grimont PA, Grimont F (1978) The genus *Serratia*. Annu Rev Microbiol 32:221–248

Hanzelka BL, Stevens AM, Parsek MR, Crone TJ, Greenberg EP (1997) Mutational analysis of the *Vibrio fischeri* LuxI polypeptide: critical regions of an autoinducer synthase. J Bacteriol 179:4882–4887

Hanzelka BL, Parsek MR, Val DL, Dunlap PV, Cronan JE Jr, Greenberg EP (1999) Acylhomoserine lactone synthase activity of the *Vibrio fischeri* AinS protein. J Bacteriol 181:5766–5770

Hara O, Beppu T (1982) Mutants blocked in streptomycin production in *Streptomyces griseus*—the role of A-factor. J Antibiot 35:349–358

Harraghy N, Kerdudou S, Herrmann M (2007) Quorum-sensing systems in staphylococci as therapeutic targets. Anal Bioanal Chem 387:437–444

Hassett DJ, Ma JF, Elkins JG, McDermott TR, Ochsner UA, West SE, Huang CT, Fredericks J, Burnett S, Stewart PS, McFeters G, Passador L, Iglewski BH (1999) Quorum sensing in *Pseudomonas aeruginosa* controls expression of catalase and superoxide dismutase genes and mediates biofilm susceptibility to hydrogen peroxide. Mol Microbiol 34:1082–1093

Heinrichs JH, Bayer MG, Cheung AL (1996) Characterization of the *sar* locus and its interaction with *agr* in *Staphylococcus aureus*. J Bacteriol 178:418–423

Henke JM, Bassler BL (2004) Three parallel quorum-sensing systems regulate gene expression in *Vibrio harveyi*. J Bacteriol 186:6902–6914

Henrichsen J (1972) Bacterial surface translocation: a survey and a classification. Bacteriol Rev 36:478–503

Holden MTG, Chhabra SR, de Nys R, Stead P, Bainton NJ, Hill PJ, Manefield M, Kumar N, Labatte M, England D, Rice S, Givskov M, Salmond GPC, Stewart GSAB, Bycroft BW, Kjelleberg S, Williams P (1999) Quorum-sensing cross talk: isolation and chemical characterization of cyclic dipeptides from *Pseudomonas aeruginosa* and other Gram-negative bacteria. Mol Microbiol 33:1254–1266

Horng YT, Deng SC, Daykin M, Soo PC, Wei JR, Luh KT, Ho SW, Swift S, Lai HC, Williams P (2002) The LuxR family protein SpnR functions as a negative regulator of *N*-acylhomoserine lactone-dependent quorum sensing in *Serratia marcescens*. Mol Microbiol 45:1655–1671

Horswill AR, Stoodley P, Stewart PS, Parsek MR (2007) The effect of the chemical, biological, and physical environment on quorum sensing in structured microbial communities. Anal Bioanal Chem 387:371–380

Hwang I, Smyth AJ, Luo ZQ, Farrand SK (1999) Modulating quorum sensing by antiactivation: TraM interacts with TraR to inhibit activation of Ti plasmid conjugal transfer genes. Mol Microbiol 34:282–294

Ji G, Beavis R, Novick RP (1997) Bacterial interference caused by autoinducing peptide variants. Science 276:2027–2030

Jensen PO, Bjarnsholt T, Phipps R, Rasmussen TB, Calum H, Christoffersen L, Moser C, Williams P, Pressler T, Givskov M, Hoiby N (2007) Rapid necrotic killing of polymorphonuclear leukocytes is caused by quorum-sensing-controlled production of rhamnolipid by *Pseudomonas aeruginosa*. Microbiology 153:1329–1338

Johansson J, Cossart P (2003) RNA-mediated control of virulence gene expression in bacterial pathogens. Trends Microbiol 11:280–285

Johnson DC, Ishihama A, Stevens AM (2003) Involvement of region 4 of the sigma70 subunit of RNA polymerase in transcriptional activation of the *lux* operon during quorum sensing. FEMS Microbiol Lett 228:193–201

Joint I, Tait K, Wheeler G (2007) Cross-kingdom signalling: exploitation of bacterial quorum sensing molecules by the green seaweed *Ulva*. Philos Trans R Soc Lond B Biol Sci doi:10.1098/rstb.2007.2047

Jones S, Yu B, Bainton NJ, Birdsall M, Bycroft BW, Chhabra SR, Cox AJ, Golby P, Reeves PJ, Stephens S, Winson MK, Salmond GPC, Stewart GSAB, Williams P (1993) The *lux* autoinducer regulates the production of exoenzyme virulence determinants in *Erwinia carotovora* and *Pseudomonas aeruginosa*. EMBO J 12:2477–2482

Juhas M, Wiehlmann L, Huber B, Jordan D, Lauber J, Salunkhe P, Limpert AS, von Gotz F, Steinmetz I, Eberl L, Tummler B (2004) Global regulation of quorum sensing and virulence by VqsR in *Pseudomonas aeruginosa*. Microbiology 150:831–841

Kaplan HB, Greenberg EP (1985) Diffusion of autoinducer is involved in regulation of the *Vibrio fischeri* luminescence system. J Bacteriol 163:1210–1214

Kendall MM, Sperandio V (2007) Quorum sensing by enteric pathogens. Curr Opin Gastroenterol 23:10–15

Kiratisin P, Tucker KD, Passador L (2002) LasR, a transcriptional activator of *Pseudomonas aeruginosa* virulence genes, functions as a multimer. J Bacteriol 184:4912–4919

Kirisits MJ, Parsek MR (2006) Does *Pseudomonas aeruginosa* use intercellular signalling to build biofilm communities? Cell Microbiol 8:1841–1849

Kirke DF, Swift S, Lynch MJ, Williams P (2004) The *Aeromonas hydrophila* LuxR homologue AhyR regulates the *N*-acyl homoserine lactone synthase, AhyI positively and negatively in a growth phase-dependent manner. FEMS Microbiol Lett 241:109–117

Kleerebezem M, Quadri LE, Kuipers OP, de Vos WM (1997) Quorum sensing by peptide pheromones and two-component signal-transduction systems in Gram-positive bacteria. Mol Microbiol 24:895–904

Kong KF, Vuong C, Otto M (2006) *Staphylococcus* quorum sensing in biofilm formation and infection. Int J Med Microbiol 296:133–139

Koenig RL, Ray JL, Maleki SJ, Smeltzer MS, Hurlburt BK (2004). *Staphylococcus aureus* AgrA binding to the RNAIII-*agr* regulatory region. J Bacteriol 186:7549–7555

Kristian SA, Golda T, Ferracin F, Cramton SE, Neumeister B, Peschel A, Gotz F, Landmann R (2004) The ability of biofilm formation does not influence virulence of *Staphylococcus aureus* and host response in a mouse tissue cage infection model. Microb Pathog 36:237–245

Kuo A, Blough NV, Dunlap PV (1994) Multiple *N*-acyl-L-homoserine lactone autoinducers of luminescence in the marine symbiotic bacterium *Vibrio fischeri*. J Bacteriol 176:7558–7565

Lamb JR, Patel H, Montminy T, Wagner VE, Iglewski BH (2003) Functional domains of the RhlR transcriptional regulator of *Pseudomonas aeruginosa*. J Bacteriol 185:7129–7139

Lambalot RH, Gehring AM, Flugel RS, Zuber P, LaCelle M, Marahiel MA, Reid R, Khosla C, Walsh CT (1996) A new enzyme superfamily—the phosphopantetheinyl transferases. Chem Biol 3:923–936

Lazazzera BA, Palmer T, Quisel J, Grossman AD (1999) Cell density control of gene expression and development in *Bacillus subtilis*. In: Dunny GM, Winans SC (ed) Cell-Cell Signaling in Bacteria. ASM Press, Washington, DC, pp. 27–46

Lenz DH, Mok KC, Lilley BN, Kulkarni RV, Wingreen NS, Bassler BL (2004) The small RNA chaperone Hfq and multiple small RNAs control quorum sensing in *Vibrio harveyi* and *Vibrio cholerae*. Cell 118:69–82

Lequette Y, Greenberg EP (2005) Timing and localization of rhamnolipid synthesis gene expression in *Pseudomonas aeruginosa* biofilms. J Bacteriol 187:37–44

Lequette Y, Lee JH, Ledgham F, Lazdunski A, Greenberg EP (2006) A distinct QscR regulon in the *Pseudomonas aeruginosa* quorum-sensing circuit. J Bacteriol 188:3365–3370

Lerat E, Moran NA (2004) The evolutionary history of quorum-sensing systems in bacteria. Mol Biol Evol 21:903–913

Li H, Tanikawa T, Sato Y, Nakagawa Y, Matsuyama T (2005) *Serratia marcescens* gene required for surfactant serrawettin W1 production encodes putative aminolipid synthetase belonging to nonribosomal peptide synthetase family. Microbiol Immunol 49:303–310

Liang X, Zheng L, Landwehr C, Lunsford D, Holmes D, Ji Y (2005) Global regulation of gene expression by ArlRS, a two-component signal transduction regulatory system of *Staphylococcus aureus*. J Bacteriol 187:5486–5492

Lindum PW, Anthoni U, Christophersen C, Eberl L, Molin S, Givskov M (1998) *N*-Acyl-L-homoserine lactone autoinducers control production of an extracellular lipopeptide biosurfactant required for swarming motility of *Serratia liquefaciens* MG1. J Bacteriol 180:6384–6388

Liu Z, Stirling FR, Zhu J (2007) Temporal quorum-sensing induction regulates *Vibrio cholerae* biofilm architecture. Infect Immun 75:122–126

Lowy FD (1998) *Staphylococcus aureus* infections. N Engl J Med 339:520–532

Luo ZQ, Farrand SK (1999) Signal-dependent DNA binding and functional domains of the quorum-sensing activator TraR as identified by repressor activity. Proc Natl Acad Sci USA 96:9009–9014

Lupp C, Ruby EG (2005) *Vibrio fischeri* uses two quorum-sensing systems for the regulation of early and late colonization factors. J Bacteriol 187:3620–3629

Lupp C, Urbanowski M, Greenberg EP, Ruby EG (2003) The *Vibrio fischeri* quorum-sensing systems *ain* and *lux* sequentially induce luminescence gene expression and are important for persistence in the squid host. Mol Microbiol 50:319–331

Lyczak JB, Cannon CL, Pier GB (2000) Establishment of *Pseudomonas aeruginosa* infection: lessons from a versatile opportunist. Microbes Infect 2:1051–1060

Lynch MJ, Swift S, Kirke DF, Keevil CW, Dodd CE, Williams P (2002) The regulation of biofilm development by quorum sensing in *Aeromonas hydrophila*. Environ Microbiol 4:18–28

Lyon GJ, Novick RP (2004) Peptide signaling in *Staphylococcus aureus* and other Gram-positive bacteria. Peptides 25:1389–1403

Lyte M, Ernst S (1992) Catecholamine induced growth of gram negative bacteria. Life Sci 50:203–212

Magnuson R, Solomon J, Grossman AD (1994) Biochemical and genetic characterization of a competence pheromone from *B. subtilis*. Cell 77:207–216

Majdalani N, Vanderpool CK, Gottesman S (2005) Bacterial small RNA regulators. Crit Rev Biochem Mol Biol 40:93–113

Matsuyama T, Kaneda K, Nakagawa Y, Isa K, Hara-Hotta H, Yano I (1992) A novel extracellular cyclic lipopeptide which promotes flagellum-dependent and -independent spreading growth of *Serratia marcescens*. J Bacteriol 174:1769–1776

Mayville P, Ji G, Beavis R, Yang H, Goger M, Novick RP, Muir TW (1999) Structure-activity analysis of synthetic autoinducing thiolactone peptides from *Staphylococcus aureus* responsible for virulence. Proc Natl Acad Sci USA 96:1218–1223

McClean KH, Winson MK, Fish L, Taylor A, Chhabra SR, Camara M, Daykin M, Lamb JH, Swift S, Bycroft BW, Stewart GSAB, Williams P (1997) Quorum sensing and *Chromobacterium violaceum*: exploitation of violacein production and inhibition for the detection of *N*-acylhomoserine lactones. Microbiology 143:3703–3711

McDougald D, Rice SA, Kjelleberg S (2007) Bacterial quorum sensing and interference by naturally occurring biomimics. Anal Bioanal Chem 387:445–453

McDowell P, Affas Z, Reynolds C, Holden MT, Wood SJ, Saint S, Cockayne A, Hill PJ, Dodd CER, Bycroft BW, Chan WC, Williams P (2001) Structure, activity and evolution of the group I thiolactone peptide quorum-sensing system of *Staphylococcus aureus*. Mol Microbiol 41:503–512

McGowan S, Sebaihia M, Jones S, Yu B, Bainton N, Chan PF, Bycroft B, Stewart GSAB, Williams P, Salmond GPC (1995) Carbapenem antibiotic production in *Erwinia carotovora* is regulated by CarR, a homologue of the LuxR transcriptional activator. Microbiology 141:541–550

McGowan SJ, Sebaihia M, Porter LE, Stewart GSAB, Williams P, Bycroft BW, Salmond GPC (1996) Analysis of bacterial carbapenem antibiotic production genes reveals a novel beta-lactam biosynthesis pathway. Mol Microbiol 22:415–426

McGowan SJ, Sebaihia M, O'Leary S, Hardie KR, Williams P, Stewart GSAB, Bycroft BW, Salmond GPC (1997) Analysis of the carbapenem gene cluster of *Erwinia carotovora*: definition of the antibiotic biosynthetic genes and evidence for a novel beta-lactam resistance mechanism. Mol Microbiol 26:545–556

McKenney D, Brown KE, Allison DG (1995) Influence of *Pseudomonas aeruginosa* exoproducts on virulence factor production in *Burkholderia cepacia*: evidence of interspecies communication. J Bacteriol 177:6989–6992

Medina G, Juarez K, Soberon-Chavez G (2003a) The *Pseudomonas aeruginosa rhlAB* operon is not expressed during the logarithmic phase of growth even in the presence of its activator RhlR and the autoinducer *N*-butyryl-homoserine lactone. J Bacteriol 185:377–380

Medina G, Juarez K, Valderrama B, Soberon-Chavez (2003b) Mechanism of *Pseudomonas aeruginosa* RhlR transcriptional regulation of the *rhlAB* promoter. J Bacteriol 185:5976–5983

Meighen EA (1991) Molecular biology of bacterial bioluminescence. Microbiol Rev 55:123–142

Menkhaus M, Ullrich C, Kluge B, Vater J, Vollenbroich D, Kamp RM (1993) Structural and functional organization of the surfactin synthetase multienzyme system. J Biol Chem 268:7678–7684

Michael B, Smith JN, Swift S, Heffron F, Ahmer BM (2001) SdiA of *Salmonella enterica* is a LuxR homolog that detects mixed microbial communities. J Bacteriol 183:5733–5742

Michiels J, Dirix G, Vanderleyden J, Xi C (2001) Processing and export of peptide pheromones and bacteriocins in Gram-negative bacteria. Trends Microbiol 9:164–168

Milton DL (2006) Quorum sensing in vibrios: complexity for diversification. Int J Med Microbiol 296:61–71

Milton DL, Chalker VJ, Kirke D, Hardman A, Camara M, Williams P (2001) The LuxM homologue VanM from *Vibrio anguillarum* directs the synthesis of *N*-(3-hydroxyhexanoyl)homoserine lactone and *N*-hexanoylhomoserine lactone. J Bacteriol 183:3537–3547

Minogue TD, Wehland-von Trebra M, Bernhard F, von Bodman SB (2002) The autoregulatory role of EsaR, a quorum-sensing regulator in *Pantoea stewartii* ssp. *stewartii*: evidence for a repressor function. Mol Microbiol 44:1625–1635

Nakano MM, Corbell N, Besson J, Zuber P (1992) Isolation and characterization of *sfp*: a gene that functions in the production of the lipopeptide biosurfactant, surfactin, in *Bacillus subtilis*. Mol Gen Genet 232:313–321

Nasser W, Reverchon S (2007) New insights into the regulatory mechanisms of the LuxR family of quorum sensing regulators. Anal Bioanal Chem 387:381–390

Nasser W, Bouillant ML, Salmond G, Reverchon S (1998) Characterization of the *Erwinia chrysanthemi expI-expR* locus directing the synthesis of two N-acyl-homoserine lactone signal molecules. Mol Microbiol 29:1391–1405

Nealson KH (1977) Autoinduction of bacterial luciferase. Occurrence, mechanism and significance. Arch Microbiol 112:73–79

Nealson KH, Platt T, Hastings JW (1970) Cellular control of the synthesis and activity of the bacterial luminescent system. J Bacteriol 104, 313–322

Neidhardt FC, Curtiss III R, Ingraham JL, Lin ECC, Low KB, Magasanik B, Reznikoff WS, Riley M, Schaechter M, Umbarger HE (eds) (1996) *Escherichia coli* and *Salmonella*: cellular and molecular biology. ASM Press, Washington, DC

Neiditch MB, Federle MJ, Miller ST, Bassler BL, Hughson FM (2005) Regulation of LuxPQ receptor activity by the quorum-sensing signal autoinducer-2. Mol Cell 18:507–518

Nouwens AS, Beatson SA, Whitchurch CB, Walsh BJ, Schweizer HP, Mattick JS, Cordwell SJ (2003) Proteome analysis of extracellular proteins regulated by the *las* and *rhl* quorum sensing systems in *Pseudomonas aeruginosa* PAO1. Microbiology 149:1311–1312

Novick RP, Jiang D (2003) The staphylococcal *saeRS* system coordinates environmental signals with *agr* quorum sensing. Microbiology 149:2709–2717

Nyholm SV, McFall-Ngai MJ (2004) The winnowing: establishing the squid–vibrio symbiosis. Nature Rev Microbiol 2:632–642

Okada M, Sato I, Cho SJ, Iwata H, Nishio T, Dubnau D, Sakagami Y (2005) Structure of the *Bacillus subtilis* quorum-sensing peptide pheromone ComX. Nat Chem Biol 1:23–24

Ochsner UA, Reiser J (1995) Autoinducer-mediated regulation of rhamnolipid biosurfactant synthesis in *Pseudomonas aeruginosa*. Proc Natl Acad Sci USA 92:6424–6428

Ochsner UA, Fiechter A, Reiser J (1994) Isolation, characterization, and expression in *Escherichia coli* of the *Pseudomonas aeruginosa rhlAB* genes encoding a rhamnosyltransferase involved in rhamnolipid biosurfactant synthesis. J Biol Chem 269:19787–19795

Oggioni MR, Trappetti C, Kadioglu A, Cassone M, Iannelli F, Ricci S, Andrew PW, Pozzi G (2006) Switch from planktonic to sessile life: a major event in pneumococcal pathogenesis. Mol Microbiol 61:1196–1210

Ortori CA, Atkinson S, Chhabra SR, Camara M, Williams P, Barrett DA (2007) Comprehensive profiling of N-acylhomoserine lactones produced by *Yersinia pseudotuberculosis* using liquid chromatography coupled to hybrid quadrupole-linear ion trap mass spectrometry. Anal Bioanal Chem 387:497–511

Oscarsson J, Kanth A, Tegmark-Wisell K, Arvidson S (2006) SarA is a repressor of *hla* (alpha-hemolysin) transcription in *Staphylococcus aureus*: its apparent role as an activator of *hla* in the prototype strain NCTC 8325 depends on reduced expression of *sarS*. J Bacteriol 188:8526–8533

Pappas KM, Winans SC (2003) A LuxR-type regulator from *Agrobacterium tumefaciens* elevates Ti plasmid copy number by activating transcription of plasmid replication genes. Mol Microbiol 48:1059–1073

Parsek MR, Greenberg EP (2005) Sociomicrobiology: the connections between quorum sensing and biofilms. Trends Microbiol 13:27–33

Parsek MR, Schaefer AL, Greenberg EP (1997) Analysis of random and site-directed mutations in *rhlI*, a *Pseudomonas aeruginosa* gene encoding an acylhomoserine lactone synthase. Mol Microbiol 26:301–310

Passador L, Cook JM, Gambello MJ, Rust L, Iglewski BH (1993) Expression of *Pseudomonas aeruginosa* virulence genes requires cell-to-cell communication. Science 260:1127–1130

Pearson JP, Pesci EC, Iglewski BH (1997) Roles of *Pseudomonas aeruginosa las* and *rhl* quorum-sensing systems in control of elastase and rhamnolipid biosynthesis genes. J Bacteriol 179:5756–5767

Pearson JP, Van Delden C, Iglewski BH (1999) Active efflux and diffusion are involved in transport of *Pseudomonas aeruginosa* cell-to-cell signals. J Bacteriol 181:1203–1210

Pearson JP, Feldman M, Iglewski BH, Prince A (2000) *Pseudomonas aeruginosa* cell-to-cell signaling is required for virulence in a model of acute pulmonary infection. Infect Immun 68:4331–4334

Perego M (1997) A peptide export-import control circuit modulating bacterial development regulates protein phosphatases of the phosphorelay. Proc Natl Acad Sci USA 94:8612–8617

Perego M (1998) Kinase-phosphatase competition regulates *Bacillus subtilis* development. Trends Microbiol 6:366–370

Perego M, Higgins CF, Pearce SR, Gallagher MP, Hoch JA (1991) The oligopeptide transport system of *Bacillus subtilis* plays a role in the initiation of sporulation. Mol Microbiol 5:173–185

Pesci EC, Pearson JP, Seed PC, Iglewski BH (1997) Regulation of *las* and *rhl* quorum sensing in *Pseudomonas aeruginosa*. J Bacteriol 179:3127–3132

Pesci EC, Milbank JB, Pearson JP, McKnight S, Kende AS, Greenberg EP, Iglewski BH (1999) Quinolone signaling in the cell-to-cell communication system of *Pseudomonas aeruginosa*. Proc Natl Acad Sci USA 96:11229–11234

Pessi G, Williams F, Hindle Z, Heurlier K, Holden MT, Camara M, Haas D, Williams P. (2001) The global posttranscriptional regulator RsmA modulates production of virulence determinants and *N*-acylhomoserine lactones in *Pseudomonas aeruginosa*. J Bacteriol 183:6676–6683

Piper KR, Beck von Bodman S, Farrand SK (1993) Conjugation factor of *Agrobacterium tumefaciens* regulates Ti plasmid transfer by autoinduction. Nature 362:448–450

Pirhonen M, Flego D, Heikinheimo R, Palva ET (1993) A small diffusible signal molecule is responsible for the global control of virulence and exoenzyme production in the plant pathogen *Erwinia carotovora*. EMBO J 12:2467–2476

Pragman AA, Yarwood JM, Tripp TJ, Schlievert PM (2004) Characterization of virulence factor regulation by SrrAB, a two-component system in *Staphylococcus aureus*. J Bacteriol 186:2430–2438

Pratten J, Foster SJ, Chan PF, Wilson M, Nair SP (2001) *Staphylococcus aureus* accessory regulators: expression within biofilms and effect on adhesion. Microbes Infect 3:633–637

Pritchard DI (2006) Immune modulation by *Pseudomonas aeruginosa* quorum-sensing signal molecules. Int J Med Microbiol 296:111–116

Pritchard DI, Todd I, Brown A, Bycroft BW, Chhabra SR, Williams P, Wood P (2005) Alleviation of insulitis and moderation of diabetes in NOD mice following treatment with a synthetic

Pseudomonas aeruginosa signal molecule, *N*-(3-oxododecanoyl)-L-homoserine lactone. Acta Diabetol 42:119–122

Qazi SN, Counil E, Morrissey J, Rees CE, Cockayne A, Winzer K, Chan WC, Williams P, Hill PJ (2001) *agr* expression precedes escape of internalized *Staphylococcus aureus* from the host endosome. Infect Immun 69:7074–7082

Qazi S, Middleton B, Muharram SH, Cockayne A, Hill P, O'Shea P, Chhabra SR, Camara M, Williams P (2006) N-acylhomoserine lactones antagonize virulence gene expression and quorum sensing in *Staphylococcus aureus*. Infect Immun 74:910–919

Qin Y, Luo ZQ, Smyth AJ, Gao P, Beck von Bodman S, Farrand SK (2000) Quorum-sensing signal binding results in dimerization of TraR and its release from membranes into the cytoplasm. EMBO J 19:5212–5221

Qin Y, Su S, Farrand SK (2007) Molecular basis of transcriptional antiactivation: TraM disrupts the TraR-DNA complex through stepwise interactions. J Biol Chem doi:10.1074/jbc.M703332200

Quadri LE, Weinreb PH, Lei M, Nakano MM, Zuber P, Walsh CT (1998) Characterization of Sfp, a *Bacillus subtilis* phosphopantetheinyl transferase for peptidyl carrier protein domains in peptide synthetases. Biochemistry 37:1585–1595

Queck SY, Weitere M, Moreno AM, Rice SA, Kjelleberg S (2006) The role of quorum sensing mediated developmental traits in the resistance of *Serratia marcescens* biofilms against protozoan grazing. Environ Microbiol 8:1017–1025

Redfield RJ (2002) Is quorum sensing a side effect of diffusion sensing? Trends Microbiol 10:365–370

Recsei P, Kreiswirth B, O'Reilly M, Schlievert P, Gruss A, Novick RP (1986) Regulation of exoprotein gene expression in *Staphylococcus aureus* by agar. Mol Gen Genet 202:58–61

Rice SA, Koh KS, Queck SY, Labbate M, Lam KW, Kjelleberg S (2005) Biofilm formation and sloughing in *Serratia marcescens* are controlled by quorum sensing and nutrient cues. J Bacteriol 187:3477–3485

Rosenberg E, Ron EZ (1999) High- and low-molecular-mass microbial surfactants. Appl Microbiol Biotechnol 52:154–162

Rudner DZ, LeDeaux JR, Ireton K, Grossman AD (1991) The *spo0K* locus of *Bacillus subtilis* is homologous to the oligopeptide permease locus and is required for sporulation and competence. J Bacteriol 173:1388–1398

Rumbaugh KP (2007) Convergence of hormones and autoinducers at the host/pathogen interface. Anal Bioanal Chem 387:425–435

Rumbaugh KP, Griswold JA, Iglewski BH, Hamood AN (1999) Contribution of quorum sensing to the virulence of *Pseudomonas aeruginosa* in burn wound infections. Infect Immun 67:5854–5862

Said-Salim B, Dunman PM, McAleese FM, Macapagal D, Murphy E, McNamara PJ, Arvidson S, Foster TJ, Projan SJ, Kreiswirth BN (2003) Global regulation of *Staphylococcus aureus* genes by Rot. J Bacteriol 185:610–619

Schauder S, Shokat K, Surette MG, Bassler BL (2001) The LuxS family of bacterial autoinducers: biosynthesis of a novel quorum-sensing signal molecule. Mol Microbiol 41:463–476

Schito GC (2006) The importance of the development of antibiotic resistance in *Staphylococcus aureus*. Clin Microbiol Infect 12 Suppl 1:3–8

Schuster M, Lostroh CP, Ogi T, Greenberg EP (2003) Identification, timing, and signal specificity of *Pseudomonas aeruginosa* quorum-controlled genes: a transcriptome analysis. J Bacteriol 185:2066–2079

Schuster M, Urbanowski ML, Greenberg EP (2004) Promoter specificity in *Pseudomonas aeruginosa* quorum sensing revealed by DNA binding of purified LasR. Proc Natl Acad Sci USA 101:15833–15839

Sharp FC, Sperandio V (2007) QseA directly activates transcription of LEE1 in enterohemorrhagic *Escherichia coli*. Infect Immun 75:2432–2440

Shaw PD, Ping G, Daly SL, Cha C, Cronan JE Jr, Rinehart KL, Farrand SK (1997) Detecting and characterizing *N*-acyl-homoserine lactone signal molecules by thin-layer chromatography. Proc Natl Acad Sci USA 94:6036–6041

Simonen M, Palva I (1993) Protein secretion in *Bacillus* species. Microbiol Rev 57:109–137

Sitnikov DM, Shadel GS, Baldwin TO (1996) Autoinducer-independent mutants of the LuxR transcriptional activator exhibit differential effects on the two lux promoters of *Vibrio fischeri*. Mol Gen Genet 252:622–625

Sleeman MC, Smith P, Kellam B, Chhabra SR, Bycroft BW, Schofield CJ (2004) Biosynthesis of carbapenem antibiotics: new carbapenam substrates for carbapenem synthase (CarC). Chembiochem 5:879–882

Smith JN, Ahmer BM (2003) Detection of other microbial species by *Salmonella*: expression of the SdiA regulon. J Bacteriol 185:1357–1366

Solomon JM, Lazazzera BA, Grossman AD (1996) Purification and characterization of an extra-cellular peptide factor that affects two different developmental pathways in *Bacillus subtilis*. Genes Dev 10:2014–2024

Sperandio V, Mellies JL, Nguyen W, Shin S, Kaper JB (1999) Quorum sensing controls expression of the type III secretion gene transcription and protein secretion in enterohemorrhagic and enteropathogenic *Escherichia coli*. Proc Natl Acad Sci USA 96:15196–15201

Sperandio V, Torres AG, Giron JA, Kaper JB (2001) Quorum sensing is a global regulatory mech-anism in enterohemorrhagic *Escherichia coli* O157:H7. J Bacteriol 183:5187–5197

Sperandio V, Li CC, Kaper JB (2002a) Quorum-sensing *Escherichia coli* regulator A: a regulator of the LysR family involved in the regulation of the locus of enterocyte effacement pathogenic-ity island in enterohemorrhagic *E. coli*. Infect Immun 70:3085–3093

Sperandio V, Torres AG, Kaper JB (2002b) Quorum sensing *Escherichia coli* regulators B and C (QsebC): a novel two-component regulatory system involved in the regulation of flagella and motility by quorum sensing in *E. coli*. Mol Microbiol 43:809–821

Sperandio V, Torres AG, Jarvis B, Nataro JP, Kaper JB (2003) Bacteria-host communication: the language of hormones. Proc Natl Acad Sci USA 100:8951–8956

Stanley NR, Lazazzera BA (2004) Environmental signals and regulatory pathways that influence biofilm formation. Mol Microbiol 52:917–924

Steindler L, Venturi V (2007) Detection of quorum-sensing *N*-acyl homoserine lactone signal molecules by bacterial biosensors. FEMS Microbiol Lett 266:1–9

Stevens AM, Dolan KM, Greenberg EP (1994) Synergistic binding of the *Vibrio fischeri* LuxR transcriptional activator domain and RNA polymerase to the *lux* promoter region. Proc Natl Acad Sci USA 91:12619–12623

Stevens AM, Fujita N, Ishihama A, Greenberg EP (1999) Involvement of the RNA polymerase alpha-subunit C-terminal domain in LuxR-dependent activation of the *Vibrio fischeri* lumines-cence genes. J Bacteriol 181:4704–4707

Sullivan ER (1998) Molecular genetics of biosurfactant production. Curr Opin Biotechnol 9:263–269

Sun J, Daniel R, Wagner-Dobler I, Zeng AP (2004) Is autoinducer-2 a universal signal for inter-species communication: a comparative genomic and phylogenetic analysis of the synthesis and signal transduction pathways. BMC Evol Biol 4:36

Surette MG, Miller MB, Bassler BL (1999) Quorum sensing in *Escherichia coli*, *Salmonella typhimurium*, and *Vibrio harveyi*: a new family of genes responsible for autoinducer produc-tion. Proc Natl Acad Sci USA 96:1639–1644

Swartzman E, Silverman M, Meighen EA (1992) The *luxR* gene product of *Vibrio harveyi* is a transcriptional activator of the *lux* promoter. J Bacteriol 174:7490–7493

Swift S (2003) Quorum Sensing: Approaches to Identify Signals and Signalling Genes in Gram-Negative Bacteria. In: Blot M (ed) Methods and Tools in Biosciences and Medicine: Prokaryotic Genomics and Genetics Birkhäuser Verlag, Basel, Switzerland, pp. 110–130

Swift S, Winson MK, Chan PF, Bainton NJ, Birdsall M, Reeves PJ, Rees CE, Chhabra SR, Hill PJ, Throup JP, Bycroft BW, Salmond GPC, Williams P, Stewart GSAB (1993) A novel strategy for the isolation of *luxI* homologues: evidence for the widespread distribution of a LuxR:LuxI superfamily in enteric bacteria. Mol Microbiol 10:511–520

Swift S, Karlyshev AV, Fish L, Durant EL, Winson MK, Chhabra SR, Williams P, Macintyre S, Stewart GSAB (1997) Quorum sensing in *Aeromonas hydrophila* and *Aeromonas salmonicida*:

identification of the LuxRI homologs AhyRI and AsaRI and their cognate N-acylhomoserine lactone signal molecules. J Bacteriol 179:5271–5281

Swift S, Lynch MJ, Fish L, Kirke DF, Tomas JM, Stewart GSAB, Williams P (1999) Quorum sensing-dependent regulation and blockade of exoprotease production in *Aeromonas hydrophila*. Infect Immun 67:5192–5199

Swift S, Downie JA, Whitehead NA, Barnard AM, Salmond GPC, Williams P (2001) Quorum sensing as a population-density-dependent determinant of bacterial physiology. Adv Microb Physiol 45:199–270

Taga ME, Semmelhack JL, Bassler BL (2001) The LuxS-dependent autoinducer AI-2 controls the expression of an ABC transporter that functions in AI-2 uptake in *Salmonella typhimurium*. Mol Microbiol 42:777–793

Taga ME, Miller ST, Bassler BL (2003) Lsr-mediated transport and processing of AI-2 in *Salmonella typhimurium*. Mol Microbiol 50:1411–1427

Tang HB, DiMango E, Bryan R, Gambello M, Iglewski BH, Goldberg JB, Prince A (1996) Contribution of specific *Pseudomonas aeruginosa* virulence factors to pathogenesis of pneumonia in a neonatal mouse model of infection. Infect Immun 64:37–43

Thomson NR, Crow MA, McGowan SJ, Cox A, Salmond GPC (2000) Biosynthesis of carbapenem antibiotic and prodigiosin pigment in *Serratia* is under quorum sensing control. Mol Microbiol 36:539–556

Tortosa P, Logsdon L, Kraigher B, Itoh Y, Mandic-Mulec I, Dubnau D (2001) Specificity and genetic polymorphism of the *Bacillus* competence quorum-sensing system. J Bacteriol 183:451–460

Tu KC, Bassler BL (2007) Multiple small RNAs act additively to integrate sensory information and control quorum sensing in *Vibrio harveyi*. Genes Dev 21:221–233

Turgay K, Hahn J, Burghoorn J, Dubnau D (1998) Competence in *Bacillus subtilis* is controlled by regulated proteolysis of a transcription factor. EMBO J 17:6730–6738

Urbanowski ML, Lostroh CP, Greenberg EP (2004) Reversible acyl-homoserine lactone binding to purified *Vibrio fischeri* LuxR protein. J Bacteriol 186:631–637

Valle J, Toledo-Arana A, Berasain C, Ghigo JM, Amorena B, Penades JR, Lasa I (2003) SarA and not sigmaB is essential for biofilm development by *Staphylococcus aureus*. Mol Microbiol 48:1075–1087

Van Houdt R, Moons P, Jansen A, Vanoirbeek K, Michiels CW (2005) Genotypic and phenotypic characterization of a biofilm-forming *Serratia plymuthica* isolate from a raw vegetable processing line. FEMS Microbiol Lett 246:265–272

Van Houdt R, Givskov M, Michiels CW (2007) Quorum sensing in *Serratia*. FEMS Microbiol Rev doi:10.1111/j.1574-6976.2007.00071.x

Vannini A, Volpari C, Gargioli C, Muraglia E, Cortese R, De Francesco R, Neddermann P, Marco SD (2002) The crystal structure of the quorum sensing protein TraR bound to its autoinducer and target DNA. EMBO J 21:4393–4401

Vendeville A, Winzer K, Heurlier K, Tang C, Hardie K (2005) Making sense of metabolism: Autoinducer-2, LuxS and pathogenic bacteria. Nat Rev Microbiol 3:383–396

Visick KL, Ruby EG (2006) *Vibrio fischeri* and its host: it takes two to tango. Curr Opin Microbiol 9:632–638

Wagner VE, Bushnell D, Passador L, Brooks AI, Iglewski BH (2003) Microarray analysis of *Pseudomonas aeruginosa* quorum-sensing regulons: effects of growth phase and environment. J Bacteriol 185:2080–2095

Wagner VE, Gillis RJ, Iglewski BH (2004) Transcriptome analysis of quorum-sensing regulation and virulence factor expression in *Pseudomonas aeruginosa*. Vaccine 22 Suppl 1:S15–20

Wagner VE, Li LL, Isabella VM, Iglewski BH (2007) Analysis of the hierarchy of quorum-sensing regulation in *Pseudomonas aeruginosa*. Anal Bioanal Chem 387:469–479

Walters M, Sperandio V (2006) Quorum sensing in *Escherichia coli* and *Salmonella*. Int J Med Microbiol 296:125–131

Walters M, Sircili MP, Sperandio V (2006) AI-3 synthesis is not dependent on *luxS* in *Escherichia coli*. J Bacteriol 188:5668–5681

Wang D, Ding X, Rather PN (2001) Indole can act as an extracellular signal in *Escherichia coli*. J Bacteriol 183:4210–4216

Wang LH, He Y, Gao Y, Wu JE, Dong YH, He C, Wang SX, Weng LX, Xu JL, Tay L, Fang RX, Zhang LH (2004) A bacterial cell-cell communication signal with cross-kingdom structural analogues. Mol Microbiol 51:903–912

Wasserman HH, Keggi JJ, McKeon JE (1961) Serratamolide, a metabolic product of *Serratia*. J Am Chem Soc 83:4107–4108

Waters CM, Bassler BL (2005) Quorum sensing: cell-to-cell communication in bacteria. Annu Rev Cell Dev Biol 21:319–346

Watson WT, Minogue TD, Val DL, von Bodman SB, Churchill ME (2002) Structural basis and specificity of acyl-homoserine lactone signal production in bacterial quorum sensing. Mol Cell 9:685–694

Wei JR, Tsai YH, Horng YT, Soo PC, Hsieh SC, Hsueh PR, Horng JT, Williams P, Lai HC (2006) A mobile quorum-sensing system in *Serratia marcescens*. J Bacteriol 188:1518–1525

Welch M, Todd DE, Whitehead NA, McGowan SJ, Bycroft BW, Salmond GPC (2000) N-acyl homoserine lactone binding to the CarR receptor determines quorum-sensing specificity in *Erwinia*. EMBO J 19:631–641

White CE, Winans SC (2005) Identification of amino acid residues of the *Agrobacterium tumefaciens* quorum-sensing regulator TraR that are critical for positive control of transcription. Mol Microbiol 55:1473–1486

Whiteley M, Lee KM, Greenberg EP (1999) Identification of genes controlled by quorum sensing in *Pseudomonas aeruginosa*. Proc Natl Acad Sci USA 96:13904–13909

Whiteley M, Parsek MR, Greenberg EP (2000) Regulation of quorum sensing by RpoS in *Pseudomonas aeruginosa*. J Bacteriol 182:4356–4360

Williams P (2002) Quorum sensing: an emerging target for antibacterial chemotherapy? Expert Opin Ther Targets 6:257–274

Winson MK, Camara M, Latifi A, Foglino M, Chhabra SR, Daykin M, Bally M, Chapon V, Salmond GPC, Bycroft BW, Lazdunski A, Stewart GSAB, Williams P (1995) Multiple *N*-acyl-L-homoserine lactone signal molecules regulate production of virulence determinants and secondary metabolites in *Pseudomonas aeruginosa*. Proc Natl Acad Sci USA 92:9427–9431

Winson MK, Swift S, Fish L, Throup JP, Jorgensen F, Chhabra SR, Bycroft BW, Williams P, Stewart GSAB (1998) Construction and analysis of *luxCDABE*-based plasmid sensors for investigating *N*-acyl homoserine lactone-mediated quorum sensing. FEMS Microbiol Lett 163:185–192

Winzer K, Hardie KR, Williams P (2002a) Bacterial cell-to-cell communication: sorry, can't talk now—gone to lunch! Curr Opin Microbiol 5:216–222

Winzer K, Hardie KR, Burgess N, Doherty N, Kirke D, Holden MT, Linforth R, Cornell KA, Taylor AJ, Hill PJ, Williams P (2002b) LuxS: its role in central metabolism and the in vitro synthesis of 4-hydroxy-5-methyl-3(2H)-furanone. Microbiology. 148:909–922

Xavier KB, Bassler BL (2005) Regulation of uptake and processing of the quorum-sensing autoinducer AI-2 in *Escherichia coli*. J Bacteriol 187:238–248

Xavier KB, Miller ST, Lu W, Kim JH, Rabinowitz J, Pelczer I, Semmelhack MF, Bassler BL (2007) Phosphorylation and processing of the quorum-sensing molecule autoinducer-2 in enteric bacteria. ACS Chem Biol 2:128–136

Yao Y, Martinez-Yamout MA, Dickerson TJ, Brogan AP, Wright PE, Dyson HJ (2006) Structure of the *Escherichia coli* quorum sensing protein SdiA: activation of the folding switch by acyl homoserine lactones. J Mol Biol 355:262–273

Yarwood JM, McCormick JK, Schlievert PM (2001) Identification of a novel two-component regulatory system that acts in global regulation of virulence factors of *Staphylococcus aureus*. J Bacteriol 183:1113–1123

Yarwood JM, McCormick JK, Paustian ML, Kapur V, Schlievert PM (2002) Repression of the *Staphylococcus aureus* accessory gene regulator in serum and in vivo. J Bacteriol 184:1095–1101

Yarwood JM, Bartels DJ, Volper EM, Greenberg EP (2004) Quorum sensing in *Staphylococcus aureus* biofilms. J Bacteriol 186:1838–1850

Yates EA, Philipp B, Buckley C, Atkinson S, Chhabra SR, Sockett RE, Goldner M, Dessaux Y, Camara M, Smith H, Williams P (2002) N-acylhomoserine lactones undergo lactonolysis in a pH-, temperature-, and acyl chain length-dependent manner during growth of *Yersinia pseudotuberculosis* and *Pseudomonas aeruginosa*. Infect Immun 70:5635–5646

Zhang LH (2003) Quorum quenching and proactive host defense. Trends Plant Sci 8:238–244

Zhang RG, Pappas T, Brace JL, Miller PC, Oulmassov T, Molyneaux JM, Anderson JC, Bashkin JK, Winans SC, Joachimiak A (2002) Structure of a bacterial quorum-sensing transcription factor complexed with pheromone and DNA. Nature 417:971–974

Zhu J, Winans SC (2001) The quorum-sensing transcriptional regulator TraR requires its cognate signaling ligand for protein folding, protease resistance, and dimerization. Proc Natl Acad Sci USA 98:1507–1512

8
Environmental Sensing and the Role of Extracytoplasmic Function Sigma Factors

Bronwyn G. Butcher, Thorsten Mascher, and John D. Helmann(✉)

Abstract Bacteria continually modulate gene expression in response to changing environmental conditions. Although many transcription regulatory pathways respond to internal signal molecules or intermediary metabolites, other systems have evolved to sense signals from the environment. For example, extracytoplasmic function (ECF) sigma (σ) factors are typically regulated by a transmembrane anti-σ factor that is, in turn, regulated by changes in the environment. Here, we focus on the best-studied ECF σ factors from *Escherichia coli* and *Bacillus subtilis*. These are regulated by signals such as the presence of misfolded secretory proteins, iron chelate complexes, or antibiotics active on the membrane or cell wall. In the presence

John D. Helmann
Department of Microbiology, Cornell University, Ithaca, NY 14853-8101, USA
jdh9@cornell.edu

W. El-Sharoud (ed.) *Bacterial Physiology: A Molecular Approach.*
© Springer-Verlag Berlin Heidelberg 2008

of an inducing stimulus, the σ factor is released and directs RNA polymerase to transcribe appropriate target genes. Recent results for both *E. coli* σ^E and *B. subtilis* σ^W indicate that the cognate anti-σ factors are themselves inactivated by sequential proteolysis involving an initial cleavage event external to the membrane (site-1 cleavage), regulated intramembrane proteolysis (RIP) (site-2 cleavage), and, finally, degradation of the residual anti-σ domain within the cytosol. Bacteria often contain multiple ECF σ paralogs that respond to distinct sets of signals. Comparative genomic approaches and consideration of genome context indicate that many ECF σ factors belong to conserved functional groups, although many of these groups have yet to be investigated experimentally. Unraveling the functions of the multiple ECF σ factors present in many genomes provides an enormous challenge for future research.

8.1   Introduction

The regulation of gene expression in response to environmental signals is a universal feature of living systems and underlies the process of adaptation. In bacteria, the signaling pathways linking environmental signals to alterations in gene expression are often fairly simple, particularly when viewed in comparison to the complex mammalian signal transduction networks that have challenged cell biologists. For example, many bacterial genes are regulated by what have recently been termed "one-component" regulators, in which a single protein acts both to sense a change in the environment through a sensor domain and to effect a genetic response, often as a DNA-binding transcription factor (Ulrich et al. 2005). At the next level of complexity are the ubiquitous two-component regulatory systems (TCS) that contain separate but interacting sensing and regulatory proteins (Mascher et al. 2006). Bacterial TCS sense signals with a sensor kinase protein, often located in the cell membrane, that phosphorylates the response regulator (RR). The phosphorylated RR may bind specific target sites in the DNA to either activate or repress gene expression. TCS allow the sensor protein to directly monitor the external environment and then relay this information across the membrane to regulate the activity of a soluble, DNA binding protein.

Although one-component regulatory systems and TCS suffice for much of the signal processing in bacteria, in several cases, the regulatory cascades are more elaborate than, and rival in complexity, those found in eukaryotes. One notable example is the regulation of the general stress response of *Bacillus subtilis* (Price 2000). In this system, an alternative σ subunit for RNA polymerase, σ^B, is regulated by an anti-σ, which, in turn, is regulated by an anti-anti-σ. The activity of this anti-anti-σ is controlled by reversible phosphorylation. Activation of this stress response occurs when either of two phosphatases removes phosphate bound to the anti-anti-σ. These phosphatases are controlled by an entire series of other proteins that sense, by mechanisms that are still unclear, either energy stress or a variety of environmental stresses.

The stress responses controlled by the extracytoplasmic function (ECF) subfamily of σ factors are of intermediate complexity. In those systems in which the regulatory pathways are best understood, the activity of the ECF σ factors is controlled by an anti-σ factor (Helmann 2002). Anti-σ factors may themselves directly sense stress or they may be on the receiving end of a signaling cascade. Similar to TCS, regulation by ECF σ factors often serves to tie changes in gene expression to changes in the external environment. Alterations in the cell envelope (which includes the cytoplasmic membrane, cell wall, and Gram-negative outer membrane) may lead to the functional inactivation of a membrane-localized anti-σ factor and the consequent activation of the ECF σ factor-dependent genes.

Historically, the realization that σ factor substitution could function as a regulatory mechanism resulted from studies of *Bacillus* phages in the mid-1970s by Jan Pero and colleagues (Fox et al. 1976). The presence of alternative σ factors in non-phage systems also first came to light in *B. subtilis* from biochemical analyses of RNA polymerase (Losick and Pero 1981). It is now appreciated that many of the alternative σ factors in *B. subtilis* participate in the complex developmental program associated with endospore formation and others control genes for flagellar-based motility, the general stress response, and other stress responses (Helmann and Moran 2002).

The isolation of genes encoding various σ factors revealed the presence of two apparently unrelated protein families often referred to as the σ^{70} and σ^{54} families (Paget and Helmann 2003). In most bacteria, there is a single essential σ factor of the σ^{70} family and multiple alternative σ factors, most of which are also related to σ^{70}. The σ^{70} family has been further divided into groups based on patterns of sequence conservation. In general, σ^{70} family proteins contain two highly conserved regions (designated regions 2 and 4) that are involved directly in promoter recognition, and larger members of this family also contain conserved regions 1 and 3. The primary σ factor (so-called because it is essential and is responsible for most transcription in the cell under most growth conditions) defines group 1 and closely related alternative σ factors define group 2. In *Escherichia coli*, RpoS (σ^S) (see Chap. 11) is considered a group 2 σ factor. More distantly related σ factors that, nevertheless, share sequence identity with the group 1 and 2 σ factors (particularly in conserved regions 2, 3, and 4) define group 3. This group includes the *E. coli* flagellar σ factor (σ^F) and the *B. subtilis* sporulation and motility σ factors. The ECF σ factors define a still more divergent group (group 4) in which there is conservation of regions 2 and 4, but no discernible region 3. As a result, ECF σ factors are typically smaller in size than those of groups 1 to 3. There are hints that there are still other, even more divergent σ factors, and a group 5 has been proposed to accommodate the divergent regulators associated with toxin gene expression in Clostridia (Helmann 2002; Dupuy and Matamouros 2006).

The original description of the ECF σ subfamily resulted from two convergent lines of research. The *E. coli* σ^E protein was described biochemically as a factor enabling transcription of the heat shock σ factor (the group 3 σ^{32} protein) at temperatures greater than 42°C (Erickson and Gross 1989). Isolation and sequencing of the corresponding gene, *rpoE*, led to the realization that this protein was a relatively divergent member of the σ^{70} family (Raina et al. 1995; Rouviere et al. 1995). At approximately the same time, a related alternative σ factor, also named σ^E, was isolated from *Streptomyces coelicolor*. Sequence comparisons led Lonetto and coworkers to predict that these two proteins were founding members of a new and divergent group (now group 4) of alternative σ factors (Lonetto et al. 1994). Importantly, several other regulatory proteins, not previously recognized to function as σ factors, were found to be clearly related in sequence. In nearly all cases in which a function could be ascribed, the evidence suggested that these diverse proteins controlled genes induced in response to stresses occurring in the cell envelope. Thus, Lonetto et al. (1994) proposed that these σ factors be named ECF (for extracytoplasmic function) σ factors.

In this chapter, we review in some detail the best-understood model systems for the ECF σ factors, including σ^E and the FecI protein of *E. coli*, and the multiple ECF σ factors of *B. subtilis*. Studies of other ECF σ factors are also briefly summarized. Comparisons of these initial model systems have revealed a number of features that seem to be common to many members of the ECF subfamily. These conserved features include (1) highly similar −35 recognition elements, including a nearly invariant "AAC" motif, (2) positive autoregulation, (3) regulation by a transmembrane anti-σ factor (often encoded by the gene immediately downstream of the ECF σ factor gene), and (4) inactivation of the anti-σ factor by a proteolytic cascade. We conclude with a genomic perspective on both the distribution and classification of ECF σ factors in the bacteria. This analysis makes clear that the systems studied to date represent but a small and nonrepresentative fraction of the ECF subfamily; there are large clusters of related ECF σ factors that have yet to be investigated experimentally.

8.2   *E. coli* RpoE and its Regulators

E. coli σ^E has been extensively studied, and a detailed model for regulation of this ECF σ factor has been elucidated (Alba and Gross 2004) (Fig. 8.1). This σ factor was purified from RNA polymerase and shown to be required for the expression at high temperatures of *rpoH* (encoding a group 3 alternative σ factor, σ^{32}) and *degP* (*htrA*), which encodes a periplasmic protease essential for growth above 42°C (Erickson and Gross 1989). These early experiments led researchers to propose that σ^E functioned as a supplemental heat shock σ-factor that was involved in the maintenance of σ^{32} levels under conditions of severe stress. Expression of the σ^E gene, *rpoE*, is autoregulated, and strains lacking this gene have decreased viability at elevated temperatures (Raina et al. 1995; Rouviere et al. 1995). It is now appreciated that *rpoE* itself is essential for viability (De Las Penas et al. 1997a), and that these strains apparently carried suppressor mutations.

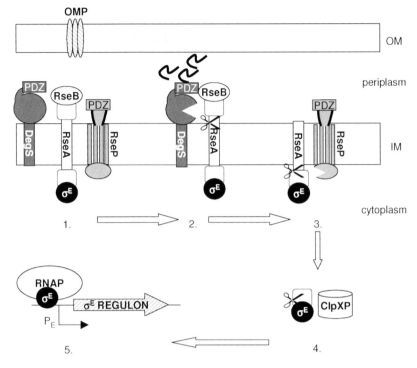

Fig. 8.1 This detailed model of σ^E regulation by proteolysis serves as the paradigm for activation of many ECF σ factors. (*1*) A membrane-spanning anti-σ factor (RseA) inhibits the activity of σ^E factor by preventing interaction with RNA polymerase (*RNAP*). (*2*) In the presence of misfolded OMPs, the site-1 protease DegS is activated and cleaves the periplasmic region of RseA. (*3*) This, in turn, activates cleavage of RseA by the site-2 protease, RseP, resulting in release of the cytoplasmic region of the anti-σ bound to σ^E. This is an example of RIP. (*4*) The remaining portion of RseA is degraded by the cytoplasmic protease, ClpXP, releasing σ^E. (*5*) σ^E interacts with RNAP, driving expression of the σ^E regulon. OM, outer membrane; IM, inner membrane

8.2.1   Inducing Signals for σ^E Activity

A significant insight into the biological role of σ^E emerged from the observation that σ^E activity increases when outer membrane proteins (OMPs) are overproduced (Mecsas et al. 1993). Induction of σ^E activity by OMPs is increased in cells expressing an OmpC variant (OmpCtd) that is secreted through the cytoplasmic membrane but not efficiently incorporated into the outer membrane. Thus, the inducing signal was proposed to be an accumulation of unfolded or misfolded OMPs in the periplasm (Mecsas et al. 1993). This hypothesis was strengthened by the observation that σ^E activity was induced in strains with defects in the folding of periplasmic proteins (Raina et al. 1995; Missiakas et al. 1996; Rouviere and Gross 1996). Other conditions known to induce the σ^E regulon include hyperosmotic shock (Bianchi and Baneyx 1999) and alterations to the lipopolysaccharide (Tam and Missiakas 2005).

8.2.2   Regulators of σ^E Activity

The *rpoE* gene encoding σ^E is part of a four-gene operon that also includes the *rseA*, *rseB*, and *rseC* (<u>r</u>egulator of <u>s</u>igma <u>E</u>) genes (De Las Penas et al. 1997b; Missiakas et al. 1997). RseA acts as an anti-σ factor because (1) disruption of *rseA* results in high levels of σ^E expression independent of known inducing conditions; (2) overexpression of RseA in vivo and addition of RseA in vitro inhibits σ^E activity; and (3) RseA and σ^E form a tight complex (De Las Penas et al. 1997b; Missiakas et al. 1997). RseA is a cytoplasmic membrane protein with a single membrane-spanning segment linking a cytoplasmic N-terminal domain, which interacts with and inhibits σ^E activity, and a periplasmic C-terminal domain (Campbell et al. 2003). A crystal structure demonstrates that the N-terminal region of RseA interacts with both regions 2 and 4 of σ^E (Campbell et al. 2003).

RseB is localized to the periplasm and interacts with the periplasmic domain of RseA, thereby, increasing its stability (De Las Penas et al. 1997b; Cezairliyan and Sauer 2007). The function of RseC is less clear. Deletions of this putative membrane-bound protein result in a small reduction in σ^E activity, although RseC interacts with RseA and RseB in a yeast two-hybrid assay (Missiakas et al. 1997). However, RseC also seems to have other cellular functions; it is involved in reducing the oxidized SoxR transcriptional regulator, which mediates the response to oxidative stress (Koo et al. 2003), and, in *Salmonella typhimurium*, *rseC* plays an unspecified a role in thiamine biosynthesis (Beck et al. 1997).

8.2.3   Role of Proteolysis in σ^E Activity

Ades and coworkers (1999) found that overexpression of OmpC leads to a decrease in RseA levels (because of a decrease in half-life), whereas σ^E levels gradually increase. This led to a model in which the RseA anti-σ is degraded in response to stress, thereby releasing σ^E into the cytoplasm (Ades et al. 1999). The current model for σ^E activation involves three sequential cleavage events (Fig. 8.1): a site-1 cleavage in the periplasmic domain of RseA, a site-2 cleavage in the transmembrane portion, and degradation of the remaining cytosolic fragment by ClpXP.

A survey of the known extracytoplasmic proteases revealed that DegS, a member of the HtrA/DegP serine protease family, initiates the proteolysis cascade required for σ^E activation. Indeed, DegS is essential in *E. coli* because of its role in providing the cell with active σ^E, which is required for viability (Alba et al. 2001). DegS contains an N-terminal transmembrane domain, which anchors the protein in the cytoplasmic membrane and is required for activity, with the catalytic domain localized on the periplasmic side of the membrane (Alba et al. 2001).

Cleavage of RseA by DegS is necessary, but not sufficient, for activation of the σ^E stress response because, even after DegS cleavage, the remaining portion of RseA is sufficient to inhibit σ^E (Alba et al. 2002). The truncated RseA protein is next cleaved within the membrane by RseP (formerly YaeL) (Alba et al. 2002). RseP is a cytoplasmic membrane protease homologous to the site-2-protease (S2P) family of proteins. S2P proteins play regulatory roles in a variety of systems that involve regulated intramembrane proteolysis (RIP), a process in which regulated cleavages are introduced within the transmembrane segments of membrane-localized proteins (Makinoshima and Glickman 2006). Site-2 cleavage of RseA results in release from the membrane of a complex containing the cytoplasmic region of the RseA anti-σ bound to σ^E. For the σ^E to be free to interact with RNA polymerase, the remaining portion of RseA needs to be degraded by ClpXP (Flynn et al. 2004; Chaba et al. 2007). ClpX is a hexameric-ring ATPase that binds to, denatures, and translocates substrate proteins into the degradation chamber of the ClpP peptidase. Kinetic analyses indicate that the initial DegS step is rate limiting for σ^E activation. The cytosolic proteases are also critical for this pathway because the rate of disso-ciation of the released σ^E:RseA complex is extremely slow in the absence of further degradation (Chaba et al. 2007).

Clues to the process by which extracytoplasmic stress activates this proteolytic cascade emerged from the observation that both DegS and RseP contain PDZ domains. These domains generally recognize peptide sequences and are involved in protein–protein interactions. A deletion of the PDZ domain of DegS results in increased basal expression of σ^E, suggesting that the PDZ domain negatively regulates the signal transduction pathway (Walsh et al. 2003). A purified frag-ment of DegS consisting of the PDZ domain and its flanking sequences was found to bind preferentially to peptides ending with YYF-COOH. Significantly, many OMPs end with YXF sequences, suggesting that OMP C termini may inter-act with DegS. To test this idea, peptide sequences from the C terminus of OmpC, which ends with YQF, were fused to the soluble periplasmic protein cytochrome-b$_{562}$ and this hybrid protein increased expression from a σ^E-dependent promoter. This induction was not observed in a strain expressing a DegS protein lacking the PDZ domain. Thus, the PDZ domain in DegS inhibits protease activity and this inhibition is relieved when this domain binds to the C termini of unfolded OMPs (Walsh et al. 2003). Structural analyses suggest that free DegS is inactive because its catalytic triad is inappropriately positioned, and after peptide binding to the PDZ domain, there is a conformational change that activates the protease (Wilken et al. 2004).

The central periplasmic domain of RseP also contains a PDZ domain, and deletion of this domain results in unregulated cleavage of full-length RseA. This indicates that this domain is involved in preventing site-2 cleavage of RseA before site-1 cleavage by DegS has occurred (Kanehara et al. 2003; Bohn et al. 2004). A glutamine-rich region of RseA, also located within the periplasm, is required for avoidance of RseP cleavage (Kanehara et al. 2003). The role of the PDZ domain of RseP is unknown; it might interact with the glutamine-rich region of RseA or with an unidentified third protein. Although the protease cascade regulated σ^E activity is the best understood, it is likely that similar regulatory cascades control the activity of many different ECF σ factors.

8.2.4   Defining the σ^E Regulon

To understand the function of ECF σ factors, it is critical to be able to define the complete set of genes (regulon) under their control. As noted above (Sect. 8.1), the first two σ^E activated genes to be identified were *htrA* (*degP*) and *rpoH*, encoding a periplasmic protease for the removal of misfolded proteins and the alternative σ factor σ^{32} that controls the cytoplasmic heat shock response. σ^E also directs the expression of its own operon. However, these genes represent only a small fraction of the genes controlled by σ^E.

The σ^E regulon has been investigated using a variety of molecular approaches. A combination of proteomic (two-dimensional gel electrophoresis) and genetic approaches led to the prediction of approximately 43 σ^E-controlled proteins, including factors that act directly on misfolded proteins in the periplasm or are involved in the synthesis of lipopolysaccharides (Dartigalongue et al. 2001). Analysis of a plasmid library for promoters activated on induction of σ^E identified 27 promoters, including several not previously observed in earlier studies (Rezuchova et al. 2003). However, in several cases, it was unclear whether these target genes were direct targets for activation by σ^E or whether they might represent indirect effects.

The most comprehensive analysis of the σ^E regulon to date is that of Rhodius and colleagues (2006). First, σ^E-dependent genes were identified by transcriptome analysis of messenger RNA (mRNA) levels in a strain overexpressing σ^E. A total of 96 genes organized in 50 transcriptional units were identified, of which, 42 were induced and 8 were repressed. Many (28 of the 42) of the induced operons contained σ^E-dependent promoters that could be confirmed by start-site mapping. Promoter sequence alignments were used to build a position weight matrix (PWM) to search the *E. coli* genome for additional σ^E promoters. This approach identified 15 additional candidate promoters, 13 of which were confirmed by in vitro transcription or in vivo promoter assays. Altogether, the σ^E regulon of *E. coli* is now thought to consist of at least 49 promoters. The optimized PWM provides a useful tool and successfully identifies 37 of the 49 verified target promoters in *E. coli*.

In an extension of their genomic analysis, Rhodius et al. (2006) applied the same PWM algorithm search to predict σ^E promoters from eight organisms closely related to *E. coli*. The resulting predictions define an "extended σ^E regulon" consisting of 89 unique transcriptional units from across all nine genomes. Nineteen of these are highly conserved (i.e., found in six or more genomes) and are involved in the synthesis, assembly, or homeostasis of lipopolysaccharides and OMPs. The remaining genes are often only present in a subset of the genomes and may be involved in adaptation of the bacterium to the eukaryotic host or other aspects of pathogenesis (Rhodius et al. 2006).

Although these studies have revealed the core elements of the σ^E regulon in enteric bacteria, it is likely that there are still additional target genes to be discovered. Recent results suggest that σ^E also activates the synthesis of small, noncoding RNAs (sRNA) that act as antisense RNA for multiple OMP transcripts (Figueroa-Bossi et al. 2006; Papenfort et al. 2006; Udekwu and Wagner 2007). At least two distinct sRNA species are involved in this response, *micA* and *rybB*, and the resulting translational repression requires the RNA chaperone, Hfq. This sRNA-mediated response leads to the degradation of numerous mRNAs, including those for *ompX, ompF, ompA, lpp,* and *ompC* (Guisbert et al. 2007). This additional layer of complexity allows the cell to rapidly shut off the translation of new OMPs under conditions that impede protein folding in the periplasm.

8.3 The FecI-Type ECF σ Factors

The second *E. coli* ECF σ factor, FecI, is the prototype for a subset of ECF σ factors that regulate the synthesis or uptake of iron-chelating compounds known as siderophores. Under conditions of iron limitation, *E. coli* will synthesize siderophores to chelate and, thereby, solubilize ferric iron salts. In addition, *E. coli* can use ferric citrate as an iron source. The ferric citrate transport operon (*fecABCDE*) is repressed by the iron-sensing repressor, Fur, and is, therefore, only expressed under iron-starvation conditions. In addition to iron regulation, the ferric citrate uptake genes are subject to substrate induction; they are only transcribed to a high level in the presence of ferric citrate (Van Hove et al. 1990). The first clues to the unusual mode of substrate induction emerged from the findings that ferric citrate did not have to enter the cell to induce expression. Induction can be mediated by periplasmic ferric citrate, and this requires two genes (*fecI* and *fecR*) located immediately upstream of the ferric citrate transport operon. With the identification of the ECF σ family of proteins, it was rapidly appreciated that FecI functions as a σ factor designated σ^{FecI} (Harle et al. 1995). FecR is a transmembrane protein and is required for transcriptional activation of the ferric citrate transport operon (Ochs et al. 1995). The *fecIR* operon is also regulated by the Fur repressor, and, therefore, only expressed under iron-limiting conditions (Angerer and Braun 1998).

Extensive genetic and biochemical analyses of the FecIR system, primarily by the Braun laboratory, have led to a novel model for transmembrane signaling (Fig. 8.2). The process of activating *fecA* operon transcription begins with the FecA protein itself. FecA is an outer membrane, beta-barrel protein that binds diferric citrate and transports this compound into the periplasm in response to a signal from TonB, an inner membrane protein that senses the energized state of the inner membrane. FecA also communicates the presence of bound diferric citrate to FecR (Braun and Mahren 2005; Braun et al. 2006).

In the absence of FecR, σ^{FecI} is only poorly able to activate *fecA* operon transcription. This suggests that FecR somehow functions to "activate" σ^{FecI}. This is clearly distinct from the anti-σ role usually associated with the regulation of ECF σ factors. Activation of σ^{FecI} can also be achieved by expression of just the N-terminal cytoplasmic domain of FecR (FecR$_{1-85}$), which is apparently in an active form (Braun et al. 2006). The requirement for FecR to activate σ^{FecI} is not well understood in molecular terms. FecR may transiently associate with σ^{FecI} and facilitate the binding of this σ factor to RNA polymerase core enzyme. Alternatively, it is

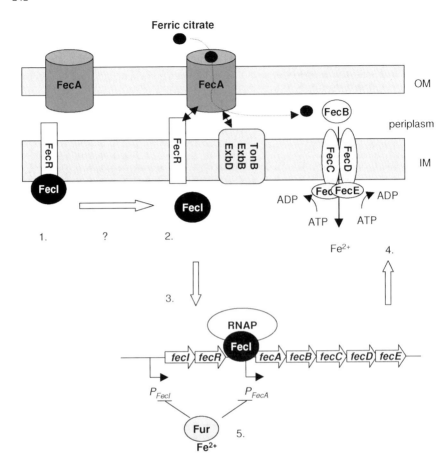

Fig. 8.2 Induction of the ferric-citrate uptake system in *E. coli* is regulated by σ^{Fec}. Under iron-limiting conditions, Fur repression of *fecIR* and *fecABCDE* is lifted and low levels of these proteins are synthesized. (*1*) In the absence of ferric citrate, FecR and FecI are colocalized to the membrane and FecI is inactive. (*2*) FecA transports ferric citrate across the outer membrane (*OM*) and, in doing so, interacts with FecR, resulting in activation and release of σ^{Fec}. (*3*) σ^{Fec} binds to RNAP and drives expression of the *fecABCDE* operon. (*4*) The *fecBCDE* genes encode an ABC transporter that transports ferric-citrate across the inner membrane (*IM*). (*5*) As the intracellular iron needs are met, the Fur regulator will again repress expression of these genes.

possible that FecR stably alters the conformation of σ^{Fec} or modifies the protein in some way. However, no covalent changes in σ^{Fec} have been identified. It is also possible that σ^{Fec} is simply unstable in cells lacking FecR and that binding with FecR protects σ^{Fec} against proteolytic degradation. According to this model, FecR might function like a classic anti-σ factor to inhibit σ^{Fec} activity and then release active σ on stimulation by signal (Braun et al. 2006).

The role of FecR-like proteins in regulating σ^{Fec}-like factors may differ between organisms. For example, the regulation of heme uptake in *Bordetella* spp. is analogous to that observed for ferric citrate induction in *E. coli*; binding of heme to the outer membrane transporters signals to the cytoplasmic membrane FecR homolog, which, in turn, is required for activation of the cognate σ factor. The result is substrate induction of the heme uptake system (Kirby et al. 2004). Conversely, in *Serratia marcescens*, the FecR ortholog HasS functions as an anti-σ factor. In this organism, iron starvation leads to the production of HasA, a heme-binding protein. The heme-loaded HasA hemophore interacts with the outer membrane transport protein HasR. HasR apparently distinguishes between heme-loaded and free HasA and sends a signal to the cytoplasmic membrane protein, HasS, to trigger the release and activity of the σ^{HasI} protein (Biville et al. 2004). Many other species also harbor one or more ECF σ factors that belong to the FecI family (see Sect. 8.6.5).

8.4 *B. subtilis* ECF σ Factors and Their Regulons

The soil-dwelling spore-forming bacterium *B. subtilis* possesses at least 17 σ factors, including 7 members of the ECF σ family: σ^X, σ^W, σ^M, σ^Y, σ^V, σ^Z, and σ^{YlaC} (Helmann and Moran 2002). Several studies have addressed the roles and regulation of σ^X, σ^W, and σ^M, but much less is known regarding the remaining ECF σ factors (Helmann 2002). This section summarizes the current understanding of ECF σ factors found in this model Gram-positive bacterium.

8.4.1 σ^X and Cell Envelope Modifications

The seven ECF σ factors encoded by the model Gram-positive bacterium *B. subtilis* were unknown before the sequencing of the genome; none had been identified genetically or biochemically. The first ECF σ to be studied in *B. subtilis* was σ^X. The *sigX* gene is co-transcribed with *rsiX*, which encodes its cognate anti-σ factor. The *sigX* operon is transcribed from both σ^A- and σ^X-dependent promoters during logarithmic and early stationary phase growth (Huang et al. 1997). To identify other candidate σ^X-dependent promoters, saturation mutagenesis of the σ^X-dependent promoter (P_X) was used to define bases important for activity, and the resulting consensus sequence was used to scan the *B. subtilis* genome. Sites were identified preceding genes implicated in cell division (*divIC*), cell surface metabolism (*csbB*, *lytR*), and regulation (*abh*, *rapD*). A subsequent microarray- and consensus search-based study identified an additional four σ^X-dependent operons: *dltABCDE*, *pssAybfMpsd*, *pbpX*, and *ywnJ* (Cao and Helmann 2004). Several of these genes encode proteins related to cell envelope composition or metabolism, including LytR (a negative regulator of autolysin activity), CsbB (a membrane-bound glucosyl transferase), and PbpX (a penicillin-binding protein).

The identification of σ^X-regulated genes led to the specific hypothesis that σ^X regulates modification of the cell envelope. The Gram-positive cell wall is characterized by a thick cell wall consisting of peptidoglycan, wall teichoic acids (WTA), and lipoteichoic acids (LTA) and has an overall negative charge. The products of the *dltABCDE* operon are responsible for the D-alanylation of the WTA and LTA, and this introduces positively charged amino groups, thereby reducing the net negative charge of the wall. The teichoic acids of *B. subtilis* strain 168 consist of an alternating glycerol phosphate polymer, yet, in the closely related W23 strain, the glycerol is replaced with ribitol. Interestingly, the genes responsible for this ribitol-based teichoic acid biosynthesis have been shown to be under the control of σ^X and σ^M (Minnig et al. 2003). The cytoplasmic membrane also displays a net negative charge because of the abundance of anionic phospholipids. However, in *B. subtilis*, the neutral lipid phosphatidylethanolamine (PE) makes up as much as 30% of the membrane. PssA and Psd catalyze the synthesis of PE and, therefore, changes in the expression of these enzymes alter the composition and charge of the membrane. Changes in cell surface charge have been shown to affect both autolysis and resistance to cationic antimicrobial peptides (CAMPs). *B. subtilis sigX* mutants have increased rates of autolysis and are more sensitive to the CAMP, nisin (Cao and Helmann 2004), as expected based on the altered charge of the cell envelope.

Many of the promoters regulated by σ^X in vivo are also recognized by σ^W and/or σ^M (Qiu and Helmann 2001). The significant overlap in the promoter specificity of ECF σ factors in *B. subtilis* complicates the determination of their regulons. The picture that has emerged is that each ECF σ factor may control distinct but overlapping sets of genes that are activated in response to different signals sensed via the unique regulation mechanisms controlling each σ factor (Helmann 2002).

8.4.2 σ^W and Antibiotic Resistance

The best-characterized *B. subtilis* ECF σ factor is σ^W (Helmann 2006). Similar to *sigX*, *sigW* is transcribed together with its anti-σ, *rsiW*, from an autoregulated promoter (P_W). To gain insights into the likely function of σ^W, the regulon was defined using a combination of approaches. For example, by searching for sequences similar to P_W, a large number of candidate promoters were identified (Huang et al. 1999). These were then tested for σ^W-dependent transcriptional activity both in vivo and in vitro. To complement this *in silico* approach, DNA microarray technology was used to provide a direct comparison of genes expressed in wild-type, *sigW*, and *rsiW* mutant strains (Cao et al. 2002a). Finally, a new technique was introduced in which total chromosomal DNA was transcribed with purified RNA polymerase in the presence and absence of σ^W. Activation of σ^W-dependent sites led to the synthesis of labeled transcripts that were then used to identify the corresponding target genes by hybridization to a DNA microarray (Cao et al. 2002a). This approach, designated run-off transcription/microarray analysis (ROMA), has proven useful in defining the regulons of other σ factors and DNA activator proteins (Cao et al. 2003).

Comparison of the data from these various approaches led to a final compilation of approximately 30 promoter sites (controlling ~60 genes) activated by σ^W (Cao et al. 2002a). Many of these 60 genes encode membrane proteins, proteins involved in detoxification, transporters, or small peptides. This led to the hypothesis that σ^W may function as part of an antibiosis stress response (the production of and protection against antibiotics). This hypothesis was strengthened by the finding that σ^W and its regulon is induced by antibiotics that inhibit cell wall biosynthesis such as vancomycin and cephalosporin C (Cao et al. 2002b) as well as the mammalian cationic peptides LL37 and PG-1 (Pietiainen et al. 2005). However, the disruption of *sigW* did not result in increased sensitivity to these antibiotics. The σ^W regulon is also induced by alkaline shock, but a *sigW* mutant is not impaired in growth or survival at alkaline pH (Wiegert et al. 2001). This lack of correlation between the identified inducing conditions and the phenotype of the *sigW* mutant frustrated initial attempts to confidently assign a role for σ^W.

The first confirmation of a role for σ^W in antibiotic resistance came from the characterization of the σ^W-dependent *fosB* gene, which provides resistance to fosfomycin, an antibiotic produced by *Streptomyces* species (Cao et al. 2001). To explore the hypothesis that σ^W provides resistance to antibiotics made by other soil bacteria, wild-type and *sigW* mutant strains were tested for growth inhibition by compounds produced by closely related *Bacillus* species. Remarkably, in all cases in which an inhibitory compound was produced by the competitor, the *sigW* mutant was more sensitive (Butcher and Helmann 2006). For example, growth of the *sigW* mutant is inhibited by coculture in the presence of *B. amyloliquefaciens* FZB42, and the σ^W-controlled genes *ybdST* and *fosB* provide resistance to unknown compounds produced by this organism. Similarly, the σ^W-dependent *yqeZyqfAB* operon provides resistance to sublancin (a lantibiotic encoded by the *sunA* gene located on the *B. subtilis* SPβ prophage).

σ^W also provides resistance to an antimicrobial protein (SdpC) produced by sporulating *B. subtilis* cells. SdpC is expressed when nutrient limitation is first sensed and has been shown to kill sibling cells that have not yet sensed nutrient limitation (Claverys and Havarstein 2007). This peptide presumably functions to delay the onset of sporulation by providing a nutrient source for the surviving cells (Gonzalez-Pastor et al. 2003; Ellermeier et al. 2006). Immunity to SdpC is provided by SdpI, which is encoded by the convergently transcribed *sdpRI* operon. Indeed, disruption of this operon by transposon mutagenesis results in upregulation of σ^W expression (Turner and Helmann 2000). Deletion of *sigW* in strains lacking *sdpI* results in a greatly increased sensitivity to SdpC, indicating that the SpdI- and σ^W-dependent processes are partially redundant (Butcher and Helmann 2006). In fact, one of the σ^W-dependent proteins that provides resistance to SdpC is YfhL, a homolog of SdpI. In addition, the σ^W-dependent *yknWXYZ* operon (encoding an ATP-binding cassette [ABC] transporter) also contributes to SdpC resistance. Taken together, these results demonstrate that the σ^W regulon provides resistance to antibiotics and, thereby, enables *B. subtilis* to compete for and maintain its niche in the soil environment.

The pathway responsible for the regulation of σ^W activity bears similarity to the σ^E signal transduction cascade. The σ^W anti-σ, RsiW, contains a single membrane-spanning domain with an extracytoplasmic C terminus and cytoplasmic N terminus that interacts directly with σ^W (Schobel et al. 2004). RsiW is degraded on alkaline shock, suggesting that RsiW interacts with and inhibits the activity of σ^W and that this inhibition is relieved by proteolysis. A screen of known proteases in *B. subtilis* revealed that RasP (a homolog of *E. coli* RseP) is responsible for site-2 cleavage (RIP) of RsiW (Schobel et al. 2004). Similar to σ^E, release of σ^W from the soluble N-terminal fragment of RsiW present after site-2 cleavage is dependent on the ClpXP cytoplasmic protease complex (Zellmeier et al. 2006).

Although the downstream proteases involved in RIP activation of σ^W are similar to those involved in the σ^E cascade, the protease performing site-1 proteolysis is unique. The protease responsible for site-1 cleavage of RsiW is PrsW (formerly YpdC), a recently identified protein unrelated to DegS (Ellermeier and Losick 2006; Heinrich and Wiegert 2006). PrsW contains five transmembrane helices with two large extracytoplasmic loops (Heinrich and Wiegert 2006). Mutations of conserved catalytic residues inactivated PrsW, thereby preventing activation of σ^W (Ellermeier and Losick 2006). PrsW was discovered independently by the Losick and Weigert laboratories. Ellermeier and Losick characterized suppressor mutations that occur when the SdpC immunity protein SdpI is absent. These mutations were found to map to *rsiW*, *ysdB*, and *prsW* and, in each case, enhanced expression of σ^W, which controls expression of SdpC resistance determinants (see Sect. 8.4.2). The isolated *prsW* suppressor mutation was a gain-of-function (dominant) allele that resulted in constitutive activation of σ^W (Ellermeier and Losick 2006). PrsW was also discovered in a screen to identify genes that result in accumulation of RsiW; in a *prsW* mutant, RsiW is not degraded even under inducing conditions (Heinrich and Wiegert 2006).

The nature of the signal sensed during σ^W activation, and whether this is sensed via PrsW or another protein are still unknown. By analogy with the σ^E system, the likely candidate for the sensor of stress is the site-1 protease, PrsW. Interestingly, all constitutively active mutants of PrsW led to substitution of negatively charged residues in the extracytoplasmic loops with neutral or positively charged residues (Ellermeier and Losick 2006). Thus, these residues were hypothesized to act as a receptor patch for antimicrobial peptides.

8.4.3   Other B. subtilis ECF σ Factors

In addition to σ^X and σ^W, *B. subtilis* encodes five additional ECF σ factors, and the total number of ECF σ factors varies even among closely related *Bacillus* spp. The remaining five ECF σ factors of *B. subtilis* are not yet well understood. The *sigM* gene is co-transcribed with *yhdL* and *yhdK*, which encode negative regulators of σ^M activity (Horsburgh and Moir 1999). This operon is maximally expressed at early to mid-logarithmic growth from both a σ^A- and a σ^M-dependent promoter. The *sigM* gene is essential for growth in high salt, and expression from the σ^M promoter is increased by salt stress. σ^M is also induced by low pH, heat, paraquat, ethanol, bacitracin, vancomycin, and fosfomycin (Thackray and Moir 2003). Deletion of the *yhdLK* genes was not possible because overexpression of σ^M is detrimental to the cell. Overexpression of YhdL results in lower levels of *sigM* expression, consistent with its role as a negative regulator (Horsburgh and Moir 1999). YhdL is predicted to have one transmembrane spanning domain and an ATP/GTP binding motif, whereas YhdK is a small hydrophobic protein with at least one transmembrane domain. These proteins presumably function as an anti-σ complex to regulate σ^M activity.

Although a comprehensive study of the σ^M regulon has not been published, analysis of the published members indicates that their functions are diverse. σ^M contributes to the expression of a number of genes with likely roles in detoxification (Thackray and Moir 2003). The *bcrC* bacitracin resistance gene is inducible by antibiotics, and this induction is dependent on σ^M (Cao and Helmann 2002). The *sigM* mutant is sensitive to oxidative stress caused by some superoxide-generating compounds such as paraquat. This σ^M-dependent paraquat resistance is primarily dependent on *yqjL*, which encodes a member of the α/β hydrolase family of proteins (Cao et al. 2005).

The *sigY* gene is co-transcribed with five other genes (*yxlCDEFG*), and this operon is positively autoregulated. Deletions of *yxlC*, *yxlD*, or *yxlE* all resulted in increased σ^Y activity, indicating that all three genes play a role in negative regulation of σ^Y (Cao et al. 2003). Additionally, the N terminus of YxlC was shown by yeast two-hybrid assays to interact with σ^Y, suggesting that this protein may function as an anti-σ factor or as part of an anti-σ complex (Yoshimura et al. 2004). The *yxlCDE* genes encode small proteins with predicted transmembrane domains, whereas *yxlFG* encodes an ABC transporter. Only one additional target for σ^Y was identified by ROMA—*ybgB*. Interestingly, YbgB has homology to the SdpI immunity protein (Butcher and Helmann 2006). The signals resulting in σ^Y activity have not been identified, yet expression from the *sigY* promoter was found to be induced late in stationary phase during growth in nitrogen starvation medium (Tojo et al. 2003). Although the function of the σ^Y regulon is unknown, the fact that it controls expression of an ABC transporter and a possible bacteriocin immunity protein suggests that σ^Y plays a role in protection against an as yet unidentified antimicrobial peptide (Butcher and Helmann 2006).

Similar to most ECF σ factors, *sigV* and the gene encoding the cognate anti-σ (*rsiV*, formerly *yrhM*) are co-transcribed and autoregulated (Zellmeier et al. 2005). No expression is observed under standard growth conditions, and removal of RsiV only resulted in low levels of σ^V (below detection levels of immunoblots) after many generations. Microarray analysis has been performed under three different conditions: (1) RNA was collected 40 generations after removal of RsiV (Zellmeier et al., 2005), (2) *sigV* was overexpressed under the control of a xylose-inducible promoter (Zellmeier et al. 2005), and (3) *sigV* was overexpressed under the control of the IPTG-inducible P_{spac} promoter (Asai et al. 2003). The putative σ^V regulons identified by these three methods were only partly overlapping. A list of 13 σ^V-controlled genes that were found in at least two of the three methods was compiled. However, many of these genes are known to also be controlled by σ^W, σ^M, and/or σ^X (Zellmeier et al. 2005). Because the promoter sequences recognized by the

B. subtilis ECF σ factors are similar, artificial overexpression of these σ factors might relax specificity, resulting in detection of genes that would not normally be expressed by the specific σ factor being studied. Therefore, the in vivo relevance of the σ^V-controlled genes identified above is still uncertain.

The final two *B. subtilis* ECF σ factors are unusual. Little is known about σ^Z, which is expressed from a monocistronic operon. No cognate anti-σ has been identified. σ^{YlaC} is encoded by the third gene in a four-gene operon (*ylaABCD*) and *ylaD* encodes the anti-σ. Expression of *ylaABCD* is induced via a σ^{YlaC}-dependent promoter, whereas a second internal Spx-dependent promoter results in *ylaCD* expression under oxidative stress (Matsumoto et al. 2005). A recent study suggested that σ^{YlaC} contributes to oxidative stress resistance (Ryu et al. 2006).

8.5   ECF σ Factors in Other Bacteria

Although, in this chapter, we have focused in detail on *E. coli* and *B. subtilis* as model systems, there is a wealth of information emerging regarding the roles of ECF σ factors in other bacteria. Among Gram-negative bacteria, ECF σ factors are well known for their role in iron uptake physiology in *Pseudomonas spp.*, which often contain multiple ECF σ factors related to σ^{FecI} (Visca et al. 2002). However, the plant pathogen *P. syringae* also contains several ECF σ factors of a distinct class (Oguiza et al. 2005). In the opportunistic human pathogen *P. aeruginosa*, expression of polysaccharide capsule (an important virulence factor) is controlled by the AlgT(AlgU) σ factor (Wood et al. 2006). Recently, the activity of σ^{AlgT} was found to be induced by antibiotics in a process requiring proteolytic inactivation of its cognate anti-σ in a cascade reminiscent of that documented for *E. coli* σ^E (Wood et al. 2006). ECF σ factors are notably abundant in the human gut symbiont, *Bacteroides thetaiotamicron*. This Gram-negative, obligate anaerobe contains at least 50 ECF σ factors hypothesized to help coordinate the uptake and metabolism of various dietary polysaccharides (Xu et al. 2004).

Among the Gram-positive bacteria, ECF σ factors have figured prominently in the postgenomic analyses of *Mycobacterium* spp. The roles of the 10 ECF σ factors encoded by *M. tuberculosis* have been recently reviewed (Manganelli et al. 2004; Rodrigue et al. 2007). It is notable that several are apparently involved in pathogenesis because mutant strains are attenuated in virulence. Recently, chromatin immunoprecipitation was used to help define several ECF σ factor regulons (Rodrigue et al. 2007). The results confirm that many ECF σ factors are positively autoregulated, but also suggest that some ECF σ factors control expression of other factors. Interestingly, differences in the expression of two immunodominant antigens under σ^K control likely accounts for some of the phenotypic differences between *M. bovis* (a cattle pathogen) and *M. tuberculosis* (a human pathogen) (Husson 2006; Said-Salim et al. 2006).

The record for the most ECF σ factors encoded in a single bacterial genome is currently held by *S. coelicolor*, a Gram-positive soil organism with a large and complex genome. The roles of the greater than 60 ECF σ factors in this organism are largely unknown, although significant progress has been made in defining the roles of several. The σ^R regulon coordinates a disulfide stress response and is regulated by an unusual, cytosolic anti-σ factor, RsrA (Paget et al. 1998; Zdanowski et al. 2006). RsrA is the prototype for a subfamily of anti-σ factors containing a tightly associated zinc ion (the ZAS: *z*inc-binding *a*nti-*s*igma family). The σ^E regulon functions in cell envelope homeostasis and is activated by cell envelope-active antibiotics (Paget et al. 1999). Unusually for ECF σ factors, transcription of the *sigE* operon is activated by a co-transcribed TCS: the CseBC system (*cse*, *c*ontrol of *s*igma-*E*). Activity of the CseBC TCS is negatively regulated by CseA, a lipoprotein localized to the external face of the cell membrane (Hutchings et al. 2006). Research into the roles of the many other ECF σ factors in this organism is still in the early stages.

These examples provide merely a few highlights representing the roles of ECF σ factors in diverse systems. The availability of an ever-increasing number of bacterial genome sequences, together with powerful methods for global analysis, promises to rapidly advance the scope of our understanding of ECF σ factors and their roles in bacterial physiology. Some hints regarding what might await can be gleaned from a genomic survey of the distributions of ECF σ factors in sequenced bacterial genomes.

8.6   Comparative Genomics of ECF σ Factors

Considering their abundance and overall importance, only a surprisingly small number of ECF σ factors have been addressed experimentally. Until very recently, comparative genomics analyses of ECF σ factors were limited to known types, i.e., the FecI- or RpoE-like, or restricted to a limited group of closely related bacteria, such as the pseudomonades, or the *E. coli/Salmonella* Typhimurium group (Visca et al. 2002; Oguiza et al. 2005; Rhodius et al. 2006). In several cases, genomes have been analyzed for their repertoire of σ factors (Manganelli et al. 2004; Xu et al. 2004; Waagmeester et al. 2005), and a census of σ factors from 240 finished bacterial genomes was reported (Kill et al. 2005). However, these analyses have not provided insights into the likely role of the ECF σ family members beyond the assumption that they likely respond to some type of extracytoplasmic stress. To try to develop hypotheses regarding the possible roles of ECF σ factors, comparative genomic analyses have recently been applied to group ECF σ factors based both on sequence and genomic context.

8.6.1   Total Numbers and Overall Distribution

Approximately 4,000 ECF σ factors can be found in the databases (as of December 2006). The total numbers of ECF σ factors per genome varies greatly and is influenced by both genome complexity and lifestyle. The first approximation is intuitive; larger genome size and habitat complexity result in a higher number of ECF σ factors. Bacteria especially rich in ECF σ factors, as well as some model organisms addressed in this chapter, are listed in Table 8.1. Overall, the numbers of ECF σ factors per genome is typically lower than that for TCS, ranging from 0 to approximately 50. No ECF σ factors are found in the genera *Borrelia*, *Campylobacter*, *Chlamydia/Chlamydophila* (the complete phylum *Chlamydiae*), *Desulfovibrio*, and some *Firmicutes* genera, i.e. *Lactobacillus*, *Staphylococcus*, and *Streptococcus* (with the exception of *S. agalactiae* and its close relatives, which harbor one ECF σ factor). Most of these bacteria are characterized by a reduced genome size and a stable environment. *E. coli* encodes 23 TCS, but only 2 ECF σ factors (Table 8.1). An even larger discrepancy can be found in the cyanobacterium *Anabaena variabilis*, which harbors approximately 100 TCS, but only 1 ECF σ factor. The phyla *Bacteroidetes* and *Planctomycetes*, on the other end, are particularly ECF rich (40–50 per genome), and, here, their numbers equal those of the TCS (Table 8.1).

8.6.2   Sequence Similarity and Genomic Context Conservation

ECF σ factors can be grouped based on sequence similarity (multiple sequence alignments and cluster analysis), genomic context conservation, and sequence conservation/domain architecture of the corresponding anti-σ factors. Using this approach, it is possible to identify previously unrecognized members of known groups, and also define new subgroups of ECF σ factors (Staroń and Mascher, unpublished observations). Examples of the types of groupings that emerge from these analyses are illustrated in Fig. 8.3, and their distribution in selected organisms is listed in Table 8.1. Ultimately, it is anticipated that these various subgroups may correlate with function, yet this hypothesis remains to be verified. Significantly, many predicted ECF σ factors, including several in well-characterized organisms, do not belong to any of these groups, and understanding their functions will require direct experimental approaches.

8.6.3   σ^E (rpoE)-Like ECF σ Factors

These proteins are exemplified by the paradigm of *E. coli* σ^E (Fig. 8.3a). The σ factors in this family studied to date have a diversity of functions, but may be united by a common regulatory mechanism, involving proteolytic destruction of the cognate, membrane-bound anti-σ (Fig. 8.1). The crystal structure has been determined for

Table 8.1 Phylogenomic distribution and conservation of ECF σ factors in selected bacteria[a]

Phylum/organism	Primary habitat	Genome size (Mb)	TCS[b]			RpoE	FecI	ECF σ factors[c]		
			HK	RR	Total			TCS	CMD	Other
Acidobacteria										
Acidobacteria	Soil	5.6	56	107	24	7	—	—	1	—
Actinobacteria										
Frankia sp.	Root nodules	7.5	54	49	29	2	—	—	5	10
Mycobacterium avium	Soil/OP	4.8	15	17	15	2	—	—	3	3
M. tuberculosis	Throat/lung	4.4	11	12	10	2	—	—	1	5
Nocardia farcinica	Soil/OP	6.0	20	30	24	2	—	—	6	5
Streptomyces avermitidis	Soil	9.0	59	69	45	2	—	—	5	17
S. coelicolor	Soil	8.7	74	81	44	2	—	—	11	17
Symbiobacterium thermophilum	Compost	3.6	24	34	18	11	—	—	—	2
Bacteroidetes										
Bacteroides fragilis	Gut	5.2	43	42	42	1	26	—	—	15
B. thetaiotaomicron	Gut	6.3	65	67	48	—	23	—	—	15
Cyanobacteria										
Anabaena nostoc	Aquatic	6.4	97	122	1	1	—	—	—	—
Firmicutes										
Bacillus anthracis	Soil	5.2	39	46	16	2	—	—	1	12
B. subtilis	Soil	4.2	23	35	7	2	—	—	—	3
Desulfitobacterium hafniense	Soil									
Planctomycetes										
Rhodopirellula baltica	Aquatic	7.1	35	63	50	—	—	—	—	—
Proteobacteria										
Agrobacterium tumefaciens (α)	Plants	2.8+2.1	35	51	12	—	—	2	—	8
Bradyrhizobium japonicum (α)	Root nodules	9.1	79	115	17	—	—	1	1	11
Caulobacter crescentus (α)	Soil	4.0	54	63	15	—	1	2	1	7
Nitrobacter winogradskyi (α)	Soil	3.4	22	33	15	—	5	1	1	2
Rhodopseudomonas palustris (α)	Aquatic/soil	5.4	53	71	20	—	2	1	1	7

(continued)

Table 8.1 (continued)

Phylum/organism	Primary habitat	Genome size (Mb)	TCS[b]			ECF σ factors[c]					
			HK	RR	Total	RpoE	FecI	TCS	CMD	Other
Nitrosomonas europaea (β)	Soil	2.8	10	20	23	1	14	—	—	—
Escherichia coli (γ)	Gut	4.6–5.6	23	38	2	1	1	—	—	—
Pseudomonas aeruginosa (γ)	Soil	6.3	53	83	19	1	9	—	—	6
P. fluorescens (γ)	Plant	7.1	68	113	28	1	16	—	1	8
P. putida (γ)	Soil	6.2	58	89	19	1	10	—	—	5
Saccharophagus degradans (γ)	Aquatic	5.1	37	77	14	2	—	—	—	1
Myxococcus xanthus (δ)	Soil	9.1	137	160	39	5	—	—	1	1
Spirochaetes										
Leptospira interrogans	Aquatic/OP	4.3+0.4	36	39	11	—	—	—	—	—

[a]All organisms with more than 10 ECF σ factors are given (with the exception of some closely related species, where only one representative is given, i.e., *B. anthracis* for *B. cereus* and *B. thuringiensis*). Additionally, model bacteria crucial for the investigation of ECF σ factors, as covered in this chapter, are also listed irrespective of the total numbers of ECF σ factors (e.g., *B. subtilis* or *E. coli*). HK, histidine kinase; RR, response regulator; RpoE/FecI-like ECF σ factors as defined in the text and indicated in Fig. 8.1a and 8.1b, respectively; TCS, conserved ECF σ factors linked to RRs and histidine kinase by genomic context clustering and co-evolution (Fig. 8.1c); CMD, conserved ECF σ factors linked to carboxymuconolactone decarboxylases (CMD) or oxidoreductases (Fig. 8.1d); other, number of additional conserved subgroups, the discrepancy between the total number and the sum of the four conserved groups listed indicates the presence of ECF σ factors not related to any of the conserved subgroups; OP, opportunistic pathogen; +, presence of two chromosomes; −, range in case of significant divergence between different genome sequences from one species.

[b]Numbers of TCS are based on the information in the MiST (at http://genomics.ornl.gov/mist/) or GenomeAtlas (at http://www.cbs.dtu.dk/services/GenomeAtlas/index.php) databases.

[c]The total numbers of ECF σ factors are derived from published data or GenomeAtlas (at http://www.cbs.dtu.dk/services/GenomeAtlas/index.php). Data on the ECF classification are derived from comprehensive comparative genomics analyses, as described in the text (Staroń and Mascher, unpublished observations). See Fig. 8.3 and text for details.

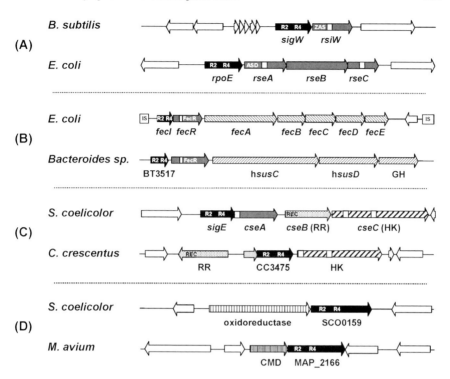

Fig. 8.3 Genomic context and characteristic features of conserved groups of ECF σ factors. Genes are represented by *arrows*, with shadings to highlight similar functions. Gene names or the function of the corresponding gene products are given below. *Black arrows*, ECF σ factors; the localization of the conserved regions 2 and 4 is indicated within the arrows. *Dark grey arrows*, (putative) anti-σ factors; conserved domains are labeled, putative transmembrane regions are represented by *white boxes*. *White arrows*, unrelated flanking genes. The genomic regions are drawn to scale, and the *lines* represent 4.6 kb, with the exception of the regions in (**b**), which correspond to 9.6 kb each. **a** Genomic context of σ^E (RpoE)-like ECF σ factors. **b** Genomic context of σ^{FecI}-like ECF σ factors. **c** Genomic context of ECF σ factors from α-proteobacteria that are functionally linked to TCS. Note that the SigE-CseABC system from *S. coelicolor* is not a member of this conserved group, in contrast to the ECF σ factors represented by CC3475 in the second line. **d** Genomic context of ECF σ factors linked to carboxymuconolactone decarboxylase or oxidoreductases, respectively. *ASD*, anti-σ domain; *CMD*, carboxymuconolactone decarboxylase; *FecR*, domain found in FecR-like anti-σ factors; *GH*, glycosyl hydrolase; *HK*, histidine kinase; *IS*, insertion sequence; *R2/4*, Sigma70_region2 or Sigma70_region4, respectively; *REC*, receiver domain; *RR*, response regulator; *ZAS*, zinc-binding anti-σ domain

the complex of the cytoplasmic N terminus of RseA in complex with *E. coli* σ^E (Campbell et al. 2003). Weak but significant sequence similarity shows that the RseA σ factor-binding region is a conserved domain found at the N terminus of a surprisingly large number of anti-σ factors (Heidi Sofia, personal communication). This so-called anti-σ domain (ASD) sometimes harbors a zinc-binding motif, and the corresponding anti-σ factors, such as *B. subtilis* RsiW (Fig. 8.3a) or RsrA from

S. coelicolor, are referred to as ZAS proteins (Zdanowski et al. 2006). Often, more than one putative (membrane-anchored) regulator of ECF σ factor activity is encoded within the operon, as is the case for *E. coli rpoE* (De Las Penas et al. 1997b; Missiakas et al. 1997). The crystal structure of σ^E region 4 bound to a promoter region fragment has also been determined, and provides insights into the high conservation of the "AAC" motif characteristic of many ECF σ factor-regulated promoters (Lane and Darst 2006).

8.6.4   σ^{FecI}-Like ECF σ Factors

Genes for these related σ factors are also organized in an operon with, and normally located upstream of, a gene encoding the cognate anti-σ factor or regulatory protein. The latter harbors an extracytoplasmic FecR-domain, named after the eponymous protein from *E. coli* (Van Hove et al. 1990). This domain is important for stimulus perception in the periplasm and interacts with an OMP (Enz et al. 2003). The operon encoding the σ:anti-σ pair is not autoregulated, but is located next to (usually upstream of) the corresponding target genes (Fig. 8.3b). Two major groups of FecI-like ECF σ factors can be differentiated based on sequence similarity and genomic context. "Classic" FecI-like proteins, present in the genomes of proteobacteria (Table 8.1), are involved in the regulation of iron uptake and/or siderophore biosynthesis. Their downstream genes often encode transport systems for ferric citrate of ferri-siderophore complexes (Fig. 8.3b). A second subgroup of FecI-like σ factors is found in large numbers in the phylum *Bacteroidetes*. Typically, these σ factors are present in genetic modules that consist of two operons; the σ:anti-σ pair is encoded upstream of genes for homologs of the polysaccharide-binding OMPs SusC and SusD (Reeves et al. 1997), one or more glycosylhydrolases, and other enzymes involved in carbohydrate metabolism (Fig. 8.3b). This genetic system is thought to enable the inducible synthesis of enzymes involved in using complex polysaccharides present in the gut habitat (Xu et al. 2004).

8.6.5   Functional Connections between TCS and ECF σ Factors

Not all ECF σ factors are linked to membrane-anchored anti-σ factors. One well-investigated example is σ^E from *S. coelicolor* (Fig. 8.3c; Sect. 8.5), in which the activity of the σ factor is regulated by a TCS, CseBC, at the level of the transcription of its structural gene (Paget et al. 1999). Comparative genomics revealed the presence of a large and novel group of ECF σ factors, which is conserved in α-proteobacteria and characterized by the presence of an unusual type of RR replacing the anti-σ factor (Table 8.1; Fig. 8.3c). The receiver domain characteristic of RR proteins is located in the C-terminal half of the proteins, in contrast to almost all other RRs (Fig. 8.3c). These lack a typical output, i.e., DNA binding domain, thereby excluding a transcriptional control mechanism of σ factor activity as described above (Sect. 8.6.3). Instead, the N-terminal half of the RR protein shows remarkable sequence similarity to its corresponding ECF σ factor (Staroń and Mascher, unpublished observations). Another distinguishing feature of these ECF σ factors is the presence of one to three genes encoding predicted sensor histidine kinases in vicinity to the σ factors and RR (Fig. 8.3c). These correlations suggest a model in which the ECF σ factor is kept inactive via direct interaction with a unique type of RR. In the presence of a stimulus, a sensor histidine kinase activates its cognate RR by phosphotransfer, which subsequently releases the ECF σ factor, resulting in expression of its target genes. If this model is correct, it represents a new mode for regulation of ECF σ activity. However, the role of σ factors of this class, and the corresponding inducing signals, are still unknown.

8.6.6   ECF σ Factors Are Perhaps Related to Oxidative Stress

The genomic context of another novel group of ECF σ factors, conserved in five different phyla (Acidobacteria, Actinobacteria, Cyanobacteria, Firmicutes, and Proteobacteria) is depicted in Fig. 8.3d. ECF σ factors classified to this group share a conserved C-terminal extension of approximately 100 amino acids and often occur more than once in a genome (up to 11 in *S. coelicolor*). The genomic context of these ECF σ factors is conserved; a gene encoding a putative carboxy-muconolactone decarboxylase (CMD protein) or an oxidoreductase is located adjacent to the σ factor (Fig. 8.3d). Carboxymuconolactone decarboxylases are normally involved in protocatechuate catabolism, but their roles in these ECF-regulated systems are currently unknown. The only ECF σ factor in this group that has been experimentally investigated is SigJ of *M. tuberculosis*, which is involved in H_2O_2 resistance (Hu et al. 2004). Thus, this group of ECF σ factors may be involved in oxidative stress resistance, although the validity of this extrapolation awaits experimental confirmation.

8.7   Concluding Remarks

The ECF σ factors are a large and important family of regulatory proteins found in diverse bacterial species. Their roles encompass many aspects of bacterial physiology, with the only unifying feature being their general involvement in processes occurring in the cell envelope (i.e., ECF). Even here, however, there are exceptions; the *S. coelicolor* $σ^R$ regulon is primarily involved in cytoplasmic disulfide homeostasis.

Ongoing work, in several laboratories, is currently focused on defining the roles and regulation of various ECF σ factors. Even for the best-characterized model systems, such as *E. coli* σ^E and *B. subtilis* σ^W, there are still many questions regarding the precise nature of the inducing signals and how these are sensed by the first (rate-limiting) protease that ultimately leads to the release the ECF σ from its inactive complex. Another area of focus is the potential overlap between the regulons controlled by multiple ECF σ factors present in a single organism. This has already created challenges in *B. subtilis* (with only seven ECF σ factors) and is likely to be an even greater obstacle in organisms such as *S. coelicolor* (>60) and *B. thetaiotamicron* (>50). Finally, comparative genomic analyses suggest that some groups of ECF σ factors may be controlled by as yet uncharacterized mechanisms involving novel types of regulatory interactions. Work in all of these areas will ultimately be required to better understand these important regulators. Interest in ECF σ factors is further stimulated by the observation that these proteins are often important in human pathogens (Bashyam and Hasnain, 2004), and genes under their control can contribute to antibiotic resistance (Helmann 2006).

Acknowledgments We thank Anna Staroń, Heidi J. Sofia, and the members of the Helmann laboratory for critical reading of the manuscript. Preparation of this manuscript was supported by grants from the National Institutes of Health (GM47446 to JDH) and from the Deutsche Forschungsgemeinschaft (DFG) and the Fonds der Chemischen Industrie (FCI) (to TM).

Highly Recommended Readings

Alba BM, Gross CA (2004) Regulation of the *Escherichia coli* σ^E-dependent envelope stress response. Mol Microbiol 52:613–619

Braun V, Mahren S (2005) Transmembrane transcriptional control (surface signalling) of the *Escherichia coli* Fec type. FEMS Microbiol Rev 29:673–684

Helmann JD (2002) The extracytoplasmic function (ECF) σ factors. Adv Microb Physiol 46:47–110

Helmann JD (2006) Deciphering a complex genetic regulatory network: the *Bacillus subtilis* σ^W protein and intrinsic resistance to antimicrobial compounds. Sci Prog 89:243–266

Lonetto MA, Brown KL, Rudd KE, Buttner MJ (1994) Analysis of the *Streptomyces coelicolor sigE* gene reveals the existence of a subfamily of eubacterial RNA polymerase σ factors involved in the regulation of extracytoplasmic functions. Proc Natl Acad Sci USA 91:7573–7577

Rhodius VA, Suh WC, Nonaka G, West J, Gross CA (2006) Conserved and variable functions of the σ^E stress response in related genomes. PLoS Biol 4:e2

References

Ades SE, Connolly LE, Alba BM, Gross CA (1999) The *Escherichia coli* σ^E-dependent extracytoplasmic stress response is controlled by the regulated proteolysis of an anti-σ factor. Genes Dev 13:2449–2461

Alba BM, Gross CA (2004) Regulation of the *Escherichia coli* σ^E-dependent envelope stress response. Mol Microbiol 52:613–619

Alba BM, Leeds JA, Onufryk C, Lu CZ, Gross CA (2002) DegS and YaeL participate sequentially in the cleavage of RseA to activate the sigma(E)-dependent extracytoplasmic stress response. Genes Dev 16:2156–2168

Alba BM, Zhong HJ, Pelayo JC, Gross CA (2001) *degS* (*hhoB*) is an essential *Escherichia coli* gene whose indispensable function is to provide σ^E activity. Mol Microbiol 40:1323–1333

Angerer A, Braun V (1998) Iron regulates transcription of the *Escherichia coli* ferric citrate transport genes directly and through the transcription initiation proteins. Arch Microbiol 169:483–490

Asai K, Yamaguchi H, Kang CM, Yoshida K, Fujita Y, Sadaie Y (2003) DNA microarray analysis of *Bacillus subtilis* σ factors of extracytoplasmic function family. FEMS Microbiol Lett 220:155–160

Bashyam MD, Hasnain SE (2004) The extracytoplasmic function sigma factors: role in bacterial pathogenesis. Infect Genet Evol 4:301–308

Beck BJ, Connolly LE, De Las Penas A, Downs DM (1997) Evidence that *rseC*, a gene in the *rpoE* cluster, has a role in thiamine synthesis in *Salmonella typhimurium*. J Bacteriol 179:6504–6508

Bianchi AA, Baneyx F (1999) Hyperosmotic shock induces the σ^{32} and σ^E stress regulons of *Escherichia coli*. Mol Microbiol 34:1029–1038

Biville F et al. (2004) Haemophore-mediated signalling in *Serratia marcescens*: a new mode of regulation for an extra cytoplasmic function (ECF) σ factor involved in haem acquisition. Mol Microbiol 53:1267–1277

Bohn C, Collier J, Bouloc P (2004) Dispensable PDZ domain of *Escherichia coli* YaeL essential protease. Mol Microbiol 52:427–435

Braun V, Mahren S (2005) Transmembrane transcriptional control (surface signalling) of the *Escherichia coli* Fec type. FEMS Microbiol Rev 29:673–684

Braun V, Mahren S, Sauter A (2006) Gene regulation by transmembrane signaling. Biometals 19:103–113

Butcher BG, Helmann JD (2006) Identification of *Bacillus subtilis* σ^W-dependent genes that provide intrinsic resistance to antimicrobial compounds produced by Bacilli. Mol Microbiol 60:765–782

Campbell EA et al. (2003) Crystal structure of *Escherichia coli* σ^E with the cytoplasmic domain of its anti-σ RseA. Mol Cell 11:1067–1078

Cao M, Bernat BA, Wang Z, Armstrong RN, Helmann JD (2001) FosB, a cysteine-dependent fosfomycin resistance protein under the control of σ^W, an extracytoplasmic-function σ factor in *Bacillus subtilis*. J Bacteriol 183:2380–2383

Cao M, Helmann JD (2002) Regulation of the *Bacillus subtilis bcrC* bacitracin resistance gene by two extracytoplasmic function σ factors. J Bacteriol 184:6123–6129

Cao M, Helmann JD (2004) The *Bacillus subtilis* extracytoplasmic-function σ^X factor regulates modification of the cell envelope and resistance to cationic antimicrobial peptides. J Bacteriol 186:1136–1146

Cao M, Kobel PA, Morshedi MM, Wu MF, Paddon C, Helmann JD (2002a) Defining the *Bacillus subtilis* σ^W regulon: a comparative analysis of promoter consensus search, run-off transcription/ macroarray analysis (ROMA), and transcriptional profiling approaches. J Mol Biol 316:443–457

Cao M, Moore CM, Helmann JD (2005) *Bacillus subtilis* paraquat resistance is directed by σ^M, an extracytoplasmic function σ factor, and is conferred by YqjL and BcrC. J Bacteriol 187:2948–2956

Cao M et al. (2003) Regulation of the *Bacillus subtilis* extracytoplasmic function protein σ^Y and its target promoters. J Bacteriol 185:4883–4890

Cao M, Wang T, Ye R, Helmann JD (2002b) Antibiotics that inhibit cell wall biosynthesis induce expression of the *Bacillus subtilis* σ^W and σ^M regulons. Mol Microbiol 45:1267–1276

Cezairliyan BO, Sauer RT (2007) Inhibition of regulated proteolysis by RseB. Proc Natl Acad Sci USA 104:3771–3776

Chaba R, Grigorova IL, Flynn JM, Baker TA, Gross CA (2007) Design principles of the proteolytic cascade governing the σ^E-mediated envelope stress response in *Escherichia coli*: keys to
graded, buffered, and rapid signal transduction. Genes Dev 21:124–136

Claverys JP, Havarstein LS (2007) Cannibalism and fratricide: mechanisms and raisons d'etre.
Nat Rev Microbiol 5:219–229

Dartigalongue C, Missiakas D, Raina S (2001) Characterization of the *Escherichia coli* σ^E regulon.
J Biol Chem 276:20866–20875

De Las Penas A, Connolly L, Gross CA (1997a) σ^E is an essential σ factor in *Escherichia coli*. J
Bacteriol 179:6862–6864

De Las Penas A, Connolly L, Gross CA (1997b) The σ^E-mediated response to extracytoplasmic
stress in *Escherichia coli* is transduced by RseA and RseB, two negative regulators of σ^E. Mol
Microbiol 24:373–385

Dupuy B, Matamouros S (2006) Regulation of toxin and bacteriocin synthesis in Clostridium
species by a new subgroup of RNA polymerase σ factors. Res Microbiol 157:201–205

Ellermeier CD, Hobbs EC, Gonzalez-Pastor JE, Losick R (2006) A three-protein signaling pathway governing immunity to a bacterial cannibalism toxin. Cell 124:549–559

Ellermeier CD, Losick R (2006) Evidence for a novel protease governing regulated intramembrane proteolysis and resistance to antimicrobial peptides in *Bacillus subtilis*. Genes Dev
20:1911–1922

Enz S, Brand H, Orellana C, Mahren S, Braun V (2003) Sites of interaction between the FecA and
FecR signal transduction proteins of ferric citrate transport in *Escherichia coli* K-12. J
Bacteriol 185:3745–3752

Erickson JW, Gross CA (1989) Identification of the σ^E subunit of *Escherichia coli* RNA polymerase: a second alternate σ factor involved in high-temperature gene expression. Genes Dev
3:1462–1471

Figueroa-Bossi N, Lemire S, Maloriol D, Balbontin R, Casadesus J, Bossi L (2006) Loss of Hfq
activates the σ^E-dependent envelope stress response in *Salmonella enterica*. Mol Microbiol
62:838–852

Flynn JM, Levchenko I, Sauer RT, Baker TA (2004) Modulating substrate choice: the SspB adap-
tor delivers a regulator of the extracytoplasmic-stress response to the AAA+ protease ClpXP
for degradation. Genes Dev 18:2292–2301

Fox TD, Losick R, Pero J (1976) Regulatory gene 28 of bacteriophage SPO1 codes for a phage-
induced subunit of RNA polymerase. J Mol Biol 101:427–433

Gonzalez-Pastor JE, Hobbs EC, Losick R (2003) Cannibalism by sporulating bacteria. Science
301:510–513

Guisbert E, Rhodius VA, Ahuja N, Witkin E, Gross CA (2007) Hfq modulates the σ^e-mediated
envelope stress response and the σ^{32}-mediated cytoplasmic stress response in *Escherichia coli*.
J Bacteriol 189:1963–1973

Harle C, Kim I, Angerer A, Braun V (1995) Signal transfer through three compartments: transcription initiation of the *Escherichia coli* ferric citrate transport system from the cell surface.
EMBO J 14:1430–1438

Heinrich J, Wiegert T (2006) YpdC determines site-1 degradation in regulated intramembrane
proteolysis of the RsiW anti-σ factor of *Bacillus subtilis*. Mol Microbiol 62:566–579

Helmann JD (2002) The extracytoplasmic function (ECF) σ factors. Adv Microb Physiol
46:47–110

Helmann JD (2006) Deciphering a complex genetic regulatory network: the *Bacillus subtilis* σ^W
protein and intrinsic resistance to antimicrobial compounds. Sci Prog 89:243–266

Helmann JD, Moran CP, Jr. (2002) RNA Polymerase and σ Factors. In: Sonenshein AL, Losick
R (eds) *Bacillus subtilis* and Its Relatives: From Genes to Cells. ASM Press, Washington
D.C., pp 289–312

Horsburgh MJ, Moir A (1999) σ^M, an ECF RNA polymerase σ factor of *Bacillus subtilis* 168, is
essential for growth and survival in high concentrations of salt. Mol Microbiol 32:41–50

Hu Y, Kendall S, Stoker NG, Coates AR (2004) The Mycobacterium tuberculosis *sigJ* gene controls sensitivity of the bacterium to hydrogen peroxide. FEMS Microbiol Lett 237:415–423

Huang X, Decatur A, Sorokin A, Helmann JD (1997) The *Bacillus subtilis* σ^X protein is an extracytoplasmic function σ factor contributing to survival at high temperature. J Bacteriol 179:2915–2921

Huang X, Gaballa A, Cao M, Helmann JD (1999) Identification of target promoters for the *Bacillus subtilis* extracytoplasmic function σ factor, σ^W. Mol Microbiol 31:361–371

Husson RN (2006) Leaving on the lights: host-specific derepression of *Mycobacterium tuberculosis* gene expression by anti-σ factor gene mutations. Mol Microbiol 62:1217–1219

Hutchings MI, Hong HJ, Leibovitz E, Sutcliffe IC, Buttner MJ (2006) The σ^E cell envelope stress response of *Streptomyces coelicolor* is influenced by a novel lipoprotein, CseA. J Bacteriol 188:7222–7229

Kanehara K, Ito K, Akiyama Y (2003) YaeL proteolysis of RseA is controlled by the PDZ domain of YaeL and a Gln-rich region of RseA. EMBO J 22:6389–6398

Kill K et al. (2005) Genome update: σ factors in 240 bacterial genomes. Microbiology 151:3147–3150

Kirby AE, King ND, Connell TD (2004) RhuR, an extracytoplasmic function sigma factor activator, is essential for heme-dependent expression of the outer membrane heme and hemoprotein receptor of Bordetella avium. Infect Immun 72:896–907

Koo MS et al. (2003) A reducing system of the superoxide sensor SoxR in *Escherichia coli*. EMBO J 22:2614–2622

Lane WJ, Darst SA (2006) The structural basis for promoter -35 element recognition by the group IV σ factors. PLoS Biol 4:e269

Lonetto MA, Brown KL, Rudd KE, Buttner MJ (1994) Analysis of the *Streptomyces coelicolor sigE* gene reveals the existence of a subfamily of eubacterial RNA polymerase σ factors involved in the regulation of extracytoplasmic functions. Proc Natl Acad Sci USA 91:7573–7577

Losick R, Pero J (1981) Cascades of Sigma factors. Cell 25:582–584

Makinoshima H, Glickman MS (2006) Site-2 proteases in prokaryotes: regulated intramembrane proteolysis expands to microbial pathogenesis. Microbes Infect 8:1882–1888

Manganelli R, Provvedi R, Rodrigue S, Beaucher J, Gaudreau L, Smith I (2004) Sigma factors and global gene regulation in *Mycobacterium tuberculosis*. J Bacteriol 186:895–902

Mascher T, Helmann JD, Unden G (2006) Stimulus perception in bacterial signal-transducing histidine kinases. Microbiol Mol Biol Rev 70:910–938

Matsumoto T, Nakanishi K, Asai K, Sadaie Y (2005) Transcriptional analysis of the *ylaABCD* operon of *Bacillus subtilis* encoding a σ factor of extracytoplasmic function family. Genes Genet Syst 80:385–393

Mecsas J, Rouviere PE, Erickson JW, Donohue TJ, Gross CA (1993) The activity of σ^E, an *Escherichia coli* heat-inducible σ factor, is modulated by expression of outer membrane proteins. Genes Dev 7:2618–2628

Minnig K, Barblan JL, Kehl S, Moller SB, Mauel C (2003) In *Bacillus subtilis* W23, the duet $\sigma^X\sigma^M$, two σ factors of the extracytoplasmic function subfamily, are required for septum and wall synthesis under batch culture conditions. Mol Microbiol 49:1435–1447

Missiakas D, Betton JM, Raina S (1996) New components of protein folding in extracytoplasmic compartments of *Escherichia coli* SurA, FkpA and Skp/OmpH. Mol Microbiol 21:871–884

Missiakas D, Mayer MP, Lemaire M, Georgopoulos C, Raina S (1997) Modulation of the *Escherichia coli* σ^E (RpoE) heat-shock transcription-factor activity by the RseA, RseB and RseC proteins. Mol Microbiol 24:355–371

Ochs M, Veitinger S, Kim I, Welz D, Angerer A, Braun V (1995) Regulation of citrate-dependent iron transport of *Escherichia coli*: FecR is required for transcription activation by FecI. Mol Microbiol 15:119–132

Oguiza JA, Kiil K, Ussery DW (2005) Extracytoplasmic function σ factors in *Pseudomonas syringae*. Trends Microbiol 13:565–568

Paget MS, Helmann JD (2003) The σ^{70} family of σ factors. Genome Biol 4:203

Paget MS, Kang JG, Roe JH, Buttner MJ (1998) σ^R, an RNA polymerase σ factor that modulates expression of the thioredoxin system in response to oxidative stress in *Streptomyces coelicolor* A3(2). EMBO J 17:5776–5782

Paget MS, Leibovitz E, Buttner MJ (1999) A putative two-component signal transduction system regulates σ^E, a sigma factor required for normal cell wall integrity in *Streptomyces coelicolor* A3(2). Mol Microbiol 33:97–107

Papenfort K, Pfeiffer V, Mika F, Lucchini S, Hinton JC, Vogel J (2006) σ^E-dependent small RNAs of *Salmonella* respond to membrane stress by accelerating global *omp* mRNA decay. Mol Microbiol 62:1674–1688

Pietiainen M, Gardemeister M, Mecklin M, Leskela S, Sarvas M, Kontinen VP (2005) Cationic antimicrobial peptides elicit a complex stress response in *Bacillus subtilis* that involves ECF-type σ factors and two-component signal transduction systems. Microbiology 151:1577–1592

Price CW (2000) Protective Function and Regulation of the General Stress Response in *Bacillus subtilis* and Related Gram-Positive Bacteria. In: Storz G, Hengge-Aronis R (eds) Bacterial Stress Responses. ASM Press, Washington, D.C., pp 179–197

Qiu J, Helmann JD (2001) The -10 region is a key promoter specificity determinant for the *Bacillus subtilis* extracytoplasmic-function σ factors σ^X and σ^W. J Bacteriol 183:1921–1927

Raina S, Missiakas D, Georgopoulos C (1995) The *rpoE* gene encoding the σ^E (σ^{24}) heat shock σ factor of *Escherichia coli*. EMBO J 14:1043–1055

Reeves AR, Wang GR, Salyers AA (1997) Characterization of four outer membrane proteins that play a role in utilization of starch by *Bacteroides thetaiotaomicron*. J Bacteriol 179:643–649

Rezuchova B, Miticka H, Homerova D, Roberts M, Kormanec J (2003) New members of the *Escherichia coli* σ^E regulon identified by a two-plasmid system. FEMS Microbiol Lett 225:1–7

Rhodius VA, Suh WC, Nonaka G, West J, Gross CA (2006) Conserved and variable functions of the σ^E stress response in related genomes. PLoS Biol 4:e2

Rodrigue S, Brodeur J, Jacques PE, Gervais AL, Brzezinski R, Gaudreau L (2007) Identification of mycobacterial σ factor binding sites by chromatin immunoprecipitation assays. J Bacteriol 189:1505–1513

Rouviere PE, De Las Penas A, Mecsas J, Lu CZ, Rudd KE, Gross CA (1995) *rpoE*, the gene encoding the second heat-shock σ factor, σ^E, in *Escherichia coli*. EMBO J 14:1032–1042

Rouviere PE, Gross CA (1996) SurA, a periplasmic protein with peptidyl-prolyl isomerase activity, participates in the assembly of outer membrane porins. Genes Dev 10:3170–3182

Ryu HB, Shin I, Yim HS, Kang SO (2006) YlaC is an extracytoplasmic function (ECF) σ factor contributing to hydrogen peroxide resistance in *Bacillus subtilis*. J Microbiol 44:206–216

Said-Salim B, Mostowy S, Kristof AS, Behr MA (2006) Mutations in *Mycobacterium tuberculosis* Rv0444c, the gene encoding anti-SigK, explain high level expression of MPB70 and MPB83 in *Mycobacterium bovis*. Mol Microbiol 62:1251–1263

Schobel S, Zellmeier S, Schumann W, Wiegert T (2004) The *Bacillus subtilis* σ^W anti-sigma factor RsiW is degraded by intramembrane proteolysis through YluC. Mol Microbiol 52:1091–1105

Tam C, Missiakas D (2005) Changes in lipopolysaccharide structure induce the σ^E-dependent response of *Escherichia coli*. Mol Microbiol 55:1403–1412

Thackray PD, Moir A (2003) σ^M, an extracytoplasmic function σ factor of *Bacillus subtilis*, is activated in response to cell wall antibiotics, ethanol, heat, acid, and superoxide stress. J Bacteriol 185:3491–3498

Tojo S et al. (2003) Organization and expression of the *Bacillus subtilis sigY* operon. J Biochem (Tokyo) 134:935–946

Turner MS, Helmann JD (2000) Mutations in multidrug efflux homologs, sugar isomerases, and antimicrobial biosynthesis genes differentially elevate activity of the σ^X and σ^W factors in *Bacillus subtilis*. J Bacteriol 182:5202–5210

Udekwu KI, Wagner EG (2007) σ^E controls biogenesis of the antisense RNA MicA. Nucleic Acids Res 35:1279–1288

Ulrich LE, Koonin EV, Zhulin IB (2005) One-component systems dominate signal transduction in prokaryotes. Trends Microbiol 13:52–56

Van Hove B, Staudenmaier H, Braun V (1990) Novel two-component transmembrane transcription control: regulation of iron dicitrate transport in *Escherichia coli* K-12. J Bacteriol 172:6749–6758

Visca P, Leoni L, Wilson MJ, Lamont IL (2002) Iron transport and regulation, cell signalling and genomics: lessons from *Escherichia coli* and Pseudomonas. Mol Microbiol 45:1177–1190

Waagmeester A, Thompson J, Reyrat JM (2005) Identifying σ factors in *Mycobacterium smegmatis* by comparative genomic analysis. Trends Microbiol 13:505–509

Walsh NP, Alba BM, Bose B, Gross CA, Sauer RT (2003) OMP peptide signals initiate the envelope-stress response by activating DegS protease via relief of inhibition mediated by its PDZ domain. Cell 113:61–71

Wiegert T, Homuth G, Versteeg S, Schumann W (2001) Alkaline shock induces the *Bacillus subtilis* s^W regulon. Mol Microbiol 41:59–71

Wilken C, Kitzing K, Kurzbauer R, Ehrmann M, Clausen T (2004) Crystal structure of the DegS stress sensor: How a PDZ domain recognizes misfolded protein and activates a protease. Cell 117:483–494

Wood LF, Leech AJ, Ohman DE (2006) Cell wall-inhibitory antibiotics activate the alginate biosynthesis operon in *Pseudomonas aeruginosa*: Roles of σ(AlgT) and the AlgW and Prc proteases. Mol Microbiol 62:412–426

Xu J, Chiang HC, Bjursell MK, Gordon JI (2004) Message from a human gut symbiont: sensitivity is a prerequisite for sharing. Trends Microbiol 12:21–28

Yoshimura M, Asai K, Sadaie Y, Yoshikawa H (2004) Interaction of *Bacillus subtilis* extracytoplasmic function (ECF) σ factors with the N-terminal regions of their potential anti-σ factors. Microbiology 150:591–599

Zdanowski K et al. (2006) Assignment of the zinc ligands in RsrA, a redox-sensing ZAS protein from *Streptomyces coelicolor*. Biochemistry 45:8294–8300

Zellmeier S, Hofmann C, Thomas S, Wiegert T, Schumann W (2005) Identification of σ^V-dependent genes of *Bacillus subtilis*. FEMS Microbiol Lett 253:221–229

Zellmeier S, Schumann W, Wiegert T (2006) Involvement of Clp protease activity in modulating the *Bacillus subtilis* σ^W stress response. Mol Microbiol 61:1569–1582

9
Extracellular Sensors and Extracellular Induction Components in Stress Tolerance Induction

Robin J. Rowbury

Robin J. Rowbury
Biology Department, University College London, Gower Street London WCIE 6BT, UK
and 3, Dartmeet Court, Poundbury, Dorchester, Dorset DTI 3SH, UK
rrowbury@tiscali.co.uk

W. El-Sharoud (ed.) *Bacterial Physiology: A Molecular Approach.*   263
© Springer-Verlag Berlin Heidelberg 2008

Abstract This chapter reviews how physical stresses and extracellular chemical stresses are sensed and how sensing triggers stress responses. Intracellular sensors switch on many inducible systems, but, for many stress responses, detection of extracellular chemical stresses is by extracellular sensors (extracellular sensing components [ESCs]), and these ESCs are activated by stress, in the absence of organisms, to extracellular induction components (EICs). As well as inducing stress responses in the producing organisms, EICs can (because they are small molecules), act as alarmones, giving advanced warning to unstressed organisms of impending stress challenges and preparing them to resist such challenges. EICs bring about these effects by diffusing to regions not yet subjected to stress, and warning organisms there of impending stress. Thus, they act pheromonally, and this intercellular communication (cross-talk) gives early warning of stress. The ESC–EIC pair for a specific response are of very similar sizes and properties, but ESCs cannot induce stress tolerance, unless activated. All ESCs examined to date occur in more than one structural form, the nature of this depends on conditions prevailing during synthesis; each form shows a distinct activation profile, leading to a second type of early warning against stress. Distinct ESCs sense physical stresses, e.g., thermal stress and UV stress. Such physical stresses activate ESCs to EICs, and the latter can give early warning to unstressed organisms of impending physical stresses and prepare these organisms to resist such stresses by inducing them to stress tolerance. It should be noted that at least five ESCs act as biological thermometers, detecting increasing temperature, with gradual increases leading to gradual ESC activation. Similarly, at least eight ESCs act as pH sensors, sensing pH changes and being gradually activated by increasing acidification, increasing alkalinization, or both.

9.1   Introduction: Potentially Lethal Stresses that Affect Enterobacteria

9.1.1   Enterobacteria Face Numerous Chemical and Biological Stressors and Many Physical Stressing Conditions

A wide range of potentially lethal chemical, physical, and biological inhibitory agents and conditions affect bacteria in foods and food preparation procedures; in domestic, commercial, and hospital environments; in the natural environment, especially natural waters; and in the human and animal body (Russell 1984; Elliott and Colwell 1985). Many such agents and conditions play a major role in determining whether organisms survive in these environments. The major stresses (Table 9.1) to which organisms can be exposed include pH extremes (Raja et al. 1991; Rowbury et al. 1989, 1996), potentially lethal levels of oxidizing agents (Ananthaswamy and Eisenstark 1977; Kullick et al. 1995; Mongkolsuk et al. 1997) and osmotic pressure (Csonka 1989; Wood 1999), extremes of temperature (Mackey and Derrick 1982; Mackey 1984; van Bogelen and Neidhardt 1990; Wolffe 1995; Rowbury and Goodson 2001), high levels of irradiation, particularly UV (Walker 1984; Rowbury 2004b), and potentially lethal levels of certain metal ions (Khazaeli and Mitra 1981); biological agents such as bacteriophages and colicins can also cause challenges. This chapter mainly describes and discusses stress from lethal chemical agents and lethal physical conditions.

In natural waters, for organisms exposed to many of the above stressors, stressing agents build up gradually to potentially lethal levels and then fall in concentration as dilution occurs with unpolluted waters (Rowbury et al. 1989). Such gradual building up of stress agents from low levels to potentially noxious levels occurs in some other natural locations and in foods and in the animal and human body.

Survival of organisms after exposure to stressors depends on tolerance levels to the stressing agents or conditions. Organisms have inherent stress tolerances allowing them to withstand low levels of noxious agents (Humphrey et al. 1995); high

Table 9.1 Responses to stresses that affect enterobacteria in the environment; in foods or food preparation; in domestic, commercial, or hospital situations; or in the animal or human body

Stress	Tolerance or sensitivity responses induced by mild stress
Chemical stresses	
Acidity	Inducible acid tolerance, alkali sensitivity, salt tolerance, UV tolerance, H_2O_2 tolerance, and thermotolerance
Alkalinity	Inducible alkali tolerance, UV tolerance, alkylhydroperoxide tolerance, thermotolerance, and acid sensitivity
Oxidizing agents	Inducible tolerance to, e.g., H_2O_2 and O_2^-
Toxic metal ions	Inducible tolerance to metal ions, to acid, and to heat
Electrophiles	Inducible tolerance to electrophiles
Alkylating agents	Inducible tolerance to alkylating agents
Mutagens	Inducible tolerance to mutagens
Nutritional stress	
Starvation for carbon	Induced thermotolerance, osmotolerance, and tolerance to acid, alkali, salt, and H_2O_2
Physical stresses	
Osmotic stress	Inducible tolerance to high osmolality, to oxidizing agents, and to heat
Elevated temperature	Inducible tolerance to heat, to acid, to alkali, and to UV irradiation
Low temperature	Resistance to low temperatures, reduced thermotolerance
UV irradiation	Inducible tolerance to UV irradiation, to heat, and to acid, alkali, and Cu^{2+}
Biological stressing agents	
Antibiotics	Induced tolerance to, e.g., tetracycline
Bacteriophages	Induced tolerance to, e.g., phage Me1

levels of inherent tolerance seem linked to increased virulence (Humphrey et al. 1996). As concentrations of stressors increase, however, inherent tolerance may be overwhelmed and organisms killed unless they are able to trigger inducible stress tolerances (Samson and Cairns 1977; Demple and Halbrook 1983; Mackey and Derrick 1986; Rowbury et al. 1989), as shown in Table 9.1. Inhibitor build up is often rapid, therefore, inducible responses must be triggered immediately when inhibitors appear and, in cases in which the inhibitor is particularly lethal and its build up rapid, organisms may need to anticipate the appearance of lethal levels of the agents. To ensure that induction is triggered immediately, it is probable that early stages of induction will have evolved to put in place rapidly acting processes, and processes that allow stressor build up to be anticipated.

9.2 Classic Views on How Stimuli Are Sensed, and Novel Ideas Regarding Sensing Chemical and Biological Stressors

Inducible processes in enterobacteria are thought to be switched on in the following general way (Neidhardt et al. 1990). The stimulus crosses the outer membrane (OM), enters the cell or cell compartment, and, on interacting with an intracellular sensor, activation occurs, which produces a signal generally leading to enhanced transcription (Fig. 9.1a). Most commonly in enterobacteria, such as *Escherichia coli*, the sensor is in the cytoplasmic membrane (CM) and is transmembrane, with

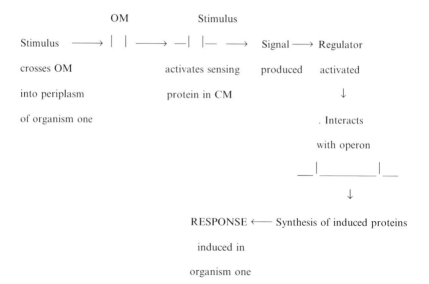

Fig. 9.1 a The mechanism, proposed for most inducible responses, involving intracellular sensor and exclusively intracellular reactions and components. **b** Novel inducible stress response induction involving extracellular components (ESCs and EICs)

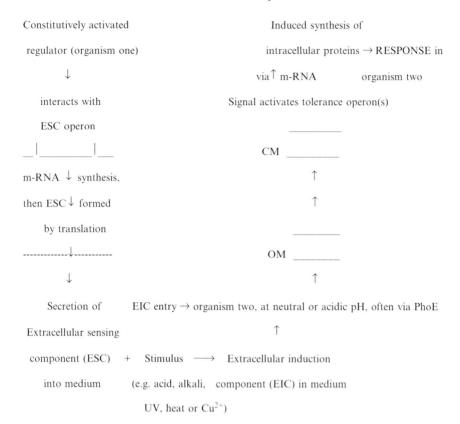

Fig. 9.1 (continued)

a sensing group that extends into the periplasm (Igo and Silhavy 1988; Aiba et al. 1990; Neidhardt et al. 1990). When the stimulus (e.g., response inducer) associates with this sensing group, another group (on the sensor protein) on the cytoplasmic side of the membrane is modified, and this leads to induction. Such a mechanism occurs, e.g., for many inducible catabolic processes (Chen and Amster-Choder 1999), for nutrient uptake (Wanner 1987), and for the regulation of some amino acid biosynthetic pathways (Artz and Holzschu 1982; Somerville 1982). Some osmotic stress-induced processes involve intracellular sensors (Igo and Silhavy 1988; Aiba et al. 1990; Neidhardt et al. 1990; Wood 1999) and, therefore, some have claimed that stress-induced responses involve intracellular sensors. This is probably true for some stressors that build up intracellularly, but the present chapter reports on numerous responses (Table 9.2) induced by extracellular stresses that are switched on by activation of extracellular sensors (extracellular sensing components [ESCs]), which are secreted to the medium; these responses also use extracellular induction components (EICs) for response induction (Fig. 9.1b). Similar ESCs sense physical stresses such as thermal stress and UV irradiation.

Table 9.2 Stress responses that need ESCs and EICs, and the size, nature, and properties of these components

Inducible stress response	Size, properties, and chemical nature of ESCs and EICs
1. Acid tolerance induced at pH 5.0	ESC and EIC are fairly small[a], heat-stable proteins
2. Alkali sensitisation induced at pH 5.5	ESC and EIC are very small[b], very heat-stable proteins
3. Thermotolerance induction	ESC and EIC are both fairly small[a] proteins, the EIC is rather heat-sensitive, the ESC more resistant to heat
4. Induced tolerance to UV irradiation	The ESC is a small protein[c], the EIC somewhat larger, both are rather heat labile; the EIC is more protease-sensitive than the ESC
5. Alkali tolerance induction	Two ESC–EIC pairs, one pair proteins, one pair nonprotein components
6. Alkylhydroperoxide tolerance induction	ESC and EIC are small[c], rather heat labile, nonprotein components
7. Acid sensitivity induced at pH 9.0	Two EICs involved; probably one protein, one nonprotein; ESC not tested
8. Acid tolerance induced at pH 7.0 by sugars and salts	ESCs and EICs needed; most EICs are proteins, no information on the nature of ESCs
9. Acid tolerance induced at pH 7.0 by amino acids	ESCs and EICs needed; EICs often proteins, but that involved in L-proline induction of acid tolerance is not a protein
10. Acid sensitivity induced by salt	EIC is involved in sensitisation
11. Induced sensitisation to acid by L-leucine	EIC is involved in sensitisation
12. Induced tolerance to phage Me1	ESC and EIC involved
13. Induced tolerance to Cu^{2+}	ESC and EIC needed, EIC is a heat-stable protein

[a]Less than 30,000 molecular weight but greater than 10,000 molecular weight.
[b]Less than 5,000 molecular weight.
[c]Less than 10,000 molecular weight.

9.2.1 Intracellular Sensors and Detection of Chemical Stressing Agents

Many stress responses involve the sensing of extracellular stress agents and physical stressing conditions by ESCs, which are activated by the stress (Rowbury 2001a, b). In cases in which a chemical stressor arises intracellularly, however, detection may also be intracellular. Thus, the intracellular OxyR and SoxR components sense intracellularly produced peroxides and superoxide (Kullick et al. 1995), and some studies have suggested that activation of the intracellular Fur by H$^+$ (Foster and Moreno 1999) leads to induction of some acid-tolerance processes (see Fig. 9.2). Similar sensing mechanisms detect the levels of intracellular alkalinity and intracellular Na$^+$, with the intracellular NhaA and NhaR involved in sensing in this case (Padan et al. 1999). Others have claimed that heat sensing occurs intracellularly, with ribosomes or DnaK

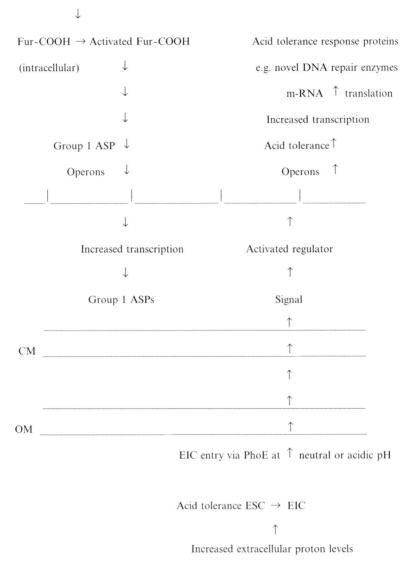

Fig. 9.2 Diagram showing how responses to acidity may be induced in enterobacteria. It is proposed that intracellular acidity is first sensed by the intracellular Fur (the proton being sensed near the carboxyl terminus, and the sensor is, therefore, termed here Fur-COOH). Activated Fur is opined to interact with group 1 ASP operons, to induce group one ASPs. Extracellular acidity is sensed by the extracellular ESC and converts this ESC to EIC, which attaches to a receptor and then passes into the cell, via PhoE, leading to a signal that causes induction of, e.g., novel DNA repair enzymes. ASP, acid-shock protein

being the sensor (van Bogelen and Neidhardt 1990; McCarty and Walker 1991); there is no clear evidence for this, however, and increasing temperature is undoubtedly sensed by an ESC (Rowbury and Goodson 2001), which acts as an extracellular thermometer (Rowbury, 2003a, b; 2005a), with resulting stress responses being triggered by the specific EIC formed by thermal activation of the ESC.

9.3   Outside Is Better Than Inside: Extracellular Components Are Involved in the Switching On of Stress Responses, as Exemplified by the Acid-Tolerance Response

In 1997, the author of this chapter opined that E. coli might use external components (ECs) in response induction, particularly in cases in which stimuli (e.g., stressors) were external. Initial tests examined triggering of acid tolerance by a low external pH stimulus. Enterobacteria shifted from neutral pH to mildly acidic pH values (4.5–6.0) become acid tolerant within 15 to 30 min at 37°C (Rowbury et al. 1989; Foster and Hall 1990; Raja et al. 1991; Rowbury 1997). The initial premise of the 1997 studies was that if an EC was formed at acidic pH and was responsible for tolerance induction, neutralised cell-free filtrates should contain this component and be able to induce acid tolerance in unstressed cells at neutral pH. In agreement with this, it was found that, associated with acid tolerance induction at pH 5.0, an extracellular protein (Table 9.2) accumulated in media (Rowbury and Goodson 1998; Rowbury et al. 1998; Rowbury 1999a; Rowbury and Goodson 2001). Two main strands of evidence established that acid tolerance induction is absolutely dependent on this extracellular component. First, during induction at pH 5.0, treatments that removed extracellular components (e.g., continuous filtration) (Rowbury 1999a) or destroyed such components (e.g., protease treatment) stopped induction, suggesting that an extracellular component was involved (Rowbury 1999a); this component was termed an extracellular induction component (EIC). Second, neutralised filtrates from pH 5.0-grown cultures, when added to pH 7.0-grown cultures, induced them to acid tolerance at pH 7.0, and treatments (e.g., continuous filtration or protease addition) that removed extracellular molecules or destroyed extracellular proteins inactivated such filtrates (Rowbury and Goodson 1998; Rowbury et al. 1998; Rowbury 1999a). Thus, EICs (in cell-free filtrates from pH 5.0 cultures or from pH 7.0 cultures, activated at pH 5.0) induce acid tolerance even in unstressed cells (Table 9.3). Inactivation of filtrates by protease suggested that the EIC was a protein, and further studies showed it to be of relatively low molecular weight (Table 9.2). The finding that, if protease was added during pH 5.0 induction of tolerance, this prevented the response, strongly suggested that EIC was essential for the original tolerance response and that it remained fully extracellular throughout induction.

Originally, it was assumed that the EIC was synthesised de novo at acidic pH, but, in 1999, the author of this chapter wondered whether there might be an EC needed for acid tolerance induction present in media at neutral pH, i.e., synthesised in nonstressing conditions. This proved to be so; cell-free medium filtrates from cultures grown at pH 7.0 were found to contain an EC that was activated at pH 5.0

Table 9.3 Acid tolerance induction; involvement of ESCs and EICs[a]

Filtrate component	Pretreatment of filtrate component before activation	Activation of filtrate component	Organisms incubated with or without filtrate component	Acid tolerance, % survival ± SEM after challenge
None	NA	NA	pH 7.0 grown	0.7 ± 0.1%
None	NA	NA	pH 5.0 grown	46.0 ± 2.0%
ESC	None	None	pH 7.0 grown	0.8 ± 0.1%
ESC	None	At pH 5.0	pH 7.0 grown	19.3 ± 1.8%
ESC	With protease	At pH 5.0	pH 7.0 grown	0.8 ± 0.1%
ESC	At 75°C	At pH 5.0	pH 7.0 grown	13.7 ± 1.7%
ESC	At 100°C	At pH 5.0	pH 7.0 grown	1.1 ± 0.21%
ESC	None	With UV, 90 s	pH 7.0 grown	37.2 ± 6.4%
ESC	None	At 50°C	pH 7.0 grown	34.2 ± 1.5%
ESC	None	At 55°C	pH 7.0 grown	43.3 ± 2.35%
EIC	None	NA	pH 7.0 grown	21.0 ± 2.5%
EIC	With protease	NA	pH 7.0 grown	2.0 ± 0.9%
EIC	At 75°C	NA	pH 7.0 grown	10.0 ± 0.6%
EIC	At 100°C	NA	pH 7.0 grown	2.2 ± 0.71%

[a]Strain 1829 ColV, I-K94 was grown to mid-log phase at the stated pH, and cell-free filtrates were prepared by passage through Gelman syringe filters with 0.2-μm-pore membranes (filtrate from pH 7.0-grown culture contains ESC; pH 5.0-grown filtrate contains EIC). Filtrates (neutralised, if required) were treated as described and activated if required, followed, if necessary, by neutralisation. UV treatment was with a Philips 6 W TUV tube placed 15 cm from the filtrate. Filtrates were incubated (1:1 filtrate to culture) with pH 7.0-grown mid-log phase cultures of the same strain at pH 7.0 for 60 min, and washed organisms were challenged at pH 3.0 for 7 min. NA, not applicable; SEM, standard error of the mean.

(in the absence of organisms) to an EIC that would induce acid tolerance (Figs. 9.1 and 9.2) in organisms at pH 7.0 (Rowbury and Goodson 1999a, b). This extracellular component in filtrates from pH 7.0 cultures was behaving as an extracellular sensor and was named an extracellular sensing component (ESC). Similar to the EIC, this ESC proved to be a small protein. Activation of the ESC to EIC occurs at pH 5.0 (Rowbury and Goodson 1999a), but mild heat treatment and UV irradiation (see Fig. 9.3 and Table 9.3) also activated the sensor, once again in the absence of organisms (Rowbury and Goodson 1999a, b). The ESC is a stress sensor in that (1) it is present in media from stressed and unstressed cells, i.e., stress is not needed for synthesis; (2) it detects stress and is activated by stress (several stresses can bring this about); (3) activation produces a component (EIC) that, on its own, can induce the acid-tolerance response (Figs 9.1 and 9.2).

9.3.1 Synthesis, Nature, and Properties of the Acid-Tolerance ESC and EIC

The acid-tolerance ESC has several properties that make it an amazing pH sensor. First, its synthesis, which occurs in unstressed organisms, is by a unique process, not needing normal protein synthesis, in which it resembles cold-shock protein

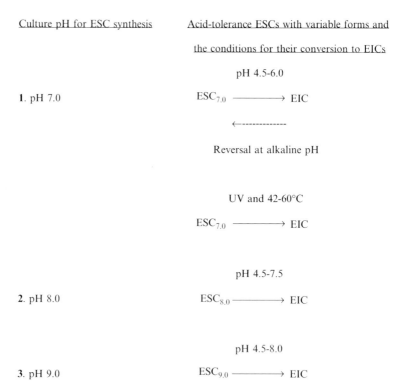

Culture pH for ESC synthesis   Acid-tolerance ESCs with variable forms and
                                 the conditions for their conversion to EICs

                                              pH 4.5-6.0
1. pH 7.0 $ESC_{7.0}$ ───────→ EIC

                                      ←-------------

                                    Reversal at alkaline pH

                                      UV and 42-60°C
                                  $ESC_{7.0}$ ───────→ EIC

                                              pH 4.5-7.5
2. pH 8.0 $ESC_{8.0}$ ───────→ EIC

                                              pH 4.5-8.0
3. pH 9.0 $ESC_{9.0}$ ───────→ EIC

Fig. 9.3 Growth-modified forms of the acid-tolerance ESC and their activation to EIC. A series of distinct acid-tolerance ESCs (shown here as $ESC_{7.0}$, $ESC_{8.0}$, and $ESC_{9.0}$) are formed at different pH values, as indicated here, and the conditions that activate them to EICs are also given. Only the ESC formed at pH 7.0 (shown here as $ESC_{7.0}$) has been tested for activation by heat and UV irradiation. It is also indicated here that the EIC formed from the $ESC_{7.0}$ can be converted back to the ESC by, e.g., exposure to alkali

synthesis (Rowbury and Goodson 1999a; Etchegaray and Inouye 1999). Thus, this important stress sensor forms under full protein synthesis inhibition. Second, this ESC is extracellular, and is immediately activated when higher levels of protons come into the environment, i.e., there is no delay (as occurs for intracellular sensors) or membrane impediment while the stimulus penetrates to its sensor. Third, ESC responses to potentially lethal agents are such that it is exquisitely sensitive to activation by protons but very resistant to irreversible inactivation by other noxious agents (Rowbury 2001a). The former property means that the ESC is very readily activated in the medium, whereas its robustness in the presence of potentially lethal environmental agents means that it survives for long periods in the medium; the acid-tolerance EIC resembles its ESC in this latter respect. Additionally, this ESC (and its cognate EIC) is a small molecule, which can, accordingly, readily diffuse from the region of synthesis to other regions, including those that have not yet faced acidity. If this diffusion is to organisms that, for genetical or physiological reasons, fail to produce ESCs, then there can be cross-feeding, provided that the ESC is later activated to an EIC. These ESCs, when cross-feeding nonproducers, in this way, are acting pheromonally. Also, as described below (Sect. 9.7), ESCs are variable, depending on the synthesis conditions, and the most suitable form (i.e., that best able to protect from acid killing) is produced under any particular set of conditions. Clearly, the acid-tolerance ESC is a pH sensor, but it can also act as a thermometer, detecting increasing temperature, and being activated by increasing temperature to an EIC (Rowbury 2003a, b; Rowbury 2005a, b).

It was initially demonstrated that the protease P4531 completely inactivated the acid-tolerance EIC, and also destroyed the ESC (Rowbury and Goodson 1998; Rowbury 1999a; Rowbury and Goodson 1999a). Subsequent studies have shown that other proteases also destroy these ECs. The effects of proteases were not caused by impurities in these enzymes because protease inhibitors reduced the extent of inactivation of the extracellular components. Additionally, RNase had no effect on the two ECs, whereas the small effect of DNase was probably caused by protease impurities in this enzyme.

The process of ESC activation seems to involve simply an interaction between the ESC and protons. Thus, no protein synthesis is needed for the ESC to EIC conversion, and the easy reformation of ESC from EIC in purified preparations (Fig. 9.3) suggests that no parts of the ESC molecule are removed during activation.

9.3.2 Effects of Genetic Lesions and Metabolites on Tolerance Induction Confirm the Obligate Involvement of ESCs and EICs in this Process

There is more evidence that acid tolerance cannot be triggered unless ESCs are produced and are activated to EICs. First, *cysB* mutants are defective in acid-tolerance induction (Rowbury 1997), but studies of cross-feeding show that ESCs (provided that they are activated at pH 5.0) and EICs, diffusing from *cysB*+ strains into *cysB* cultures (see below Sect. 9.3.4), can restore tolerance induction to the latter strains (Rowbury 1999a). Second, metabolites and nutrients that inhibit tolerance induction act via their inhibitory effects on ESC synthesis (this also blocks EIC formation, because EICs only form by ESC activation). First, phosphates inhibit acid habituation; we now know that phosphate stops formation of both the ESC and EIC (Rowbury 1999a; Rowbury and Goodson 1999a; Rowbury 2001a); this may not, however, be the only reason for its effects on habituation, because phosphate will also compete with the EIC for passage through the PhoE porin. Phosphate stops acid killing and, therefore, in the presence of phosphate, acid tolerance induction is not needed; presumably, the organism has evolved accordingly. Secondly, 3´, 5´-cyclic AMP (cAMP) was earlier shown to stop tolerance induction (Rowbury and Goodson 1997); it acts by inhibiting ESC synthesis and, therefore, stops EIC formation. Levels of cAMP are high when amounts of fermentable sugars are low.

Less acid will be produced by fermentation in such situations, and this may have led to the organisms evolving to switch off tolerance induction by cAMP.

Regulatory effects of cAMP generally involve direct interaction with DNA. Both cAMP and bicarbonate (which also blocks ESC synthesis), however, show an indirect effect on ESC formation (Rowbury 1999a, 2001a), their presence leading to production of an extracellular inhibitory agent that blocks ESC synthesis. The inhibitor produced from cAMP is proteinaceous.

Accordingly, it is now certain that the above extracellular sensor (ESC) and its EIC are absolutely essential for acid habituation, functioning as outlined diagrammatically in Figs. 9.1 and 9.2.

9.3.3 Regarding the Role of the PhoE Porin in Acid Tolerance Induction by the Specific EIC

E. coli strain AB1157 is unusual in that acid tolerance induction (habituation) at 37°C and pH 5.0 is unusually slow (75 min for full habituation compared with 12–20 min in most strains). This abnormality seems to be dependent on the phoE lesion of strain AB1157, because its phoE⁺ derivative shows normal habituation; apparently, therefore, a lesion in phoE slows down induction of stress tolerance. Studies of synthesis of the acid-tolerance ESC and of its conversion to the EIC show that these processes occur at the same rate in the phoE mutant as in the phoE⁺ parent, i.e., synthesis and activation of ESC are normal in strain AB1157. In contrast, tolerance induction by added EIC is greatly reduced in this strain. PhoE is an OM pore and can take up protonated molecules; it, therefore, seems likely that the EIC (which is a protonated form of ESC) uses the PhoE pore for entry, which is followed by tolerance induction.

9.3.4 EICs Involved in Acid Tolerance Induction Are Diffusible and Can Cross-Feed

The acid-tolerance EICs can diffuse away from the site of synthesis, and EICs needed for other responses also do this. First, if a cysB⁺ culture is placed in one vessel, surrounding a cysB culture (unable to make EIC) in a second vessel, with the cultures separated from each other by a 0.2-μm-pore membrane, the cysB culture is cross-fed by diffusion of the EIC into it from the cysB⁺ culture, and becomes acid-tolerant (Rowbury 1999a). Second, EIC in a producing culture will diffuse away into a large volume of pure broth if separated from such broth by a 0.2-μm-pore membrane. The dilution of the EIC that occurs because of the large volume of pure broth inhibits the response in the producing culture (Rowbury 1999a). Third, with small EICs (molecular weight <10 kDa), diffusion away of the

EIC through dialysis membranes can inhibit EIC-induced responses (this applies to the EIC for alkali sensitisation and to the EIC that induces alkylhydroperoxide tolerance). In addition, if a culture that has produced EIC at pH 5.0 is neutralised and separated by a 0.2-μm-pore membrane from the same volume of unstressed culture, the EIC diffuses from the first culture and induces acid tolerance in the unstressed culture. Similarly, the thermally produced thermotolerance EIC diffuses through membranes into unstressed cultures, leading to thermotolerance. With smaller EICs, e.g., the alkali sensitisation EIC, diffusion from a neutralised EIC-containing culture can occur into an unstressed culture in dialysis tubing suspended in the first culture, leading to EIC induction of the appropriate stress response.

Accordingly, many EICs are diffusible and if, in the environment, they diffuse to nearby unstressed regions that face impending stress, they give early warning to organisms there, i.e., they act as extracellular signalling molecules and early warning alarmones as well as inducing stress tolerance (Rowbury 2001b). Again, the cross-talk between producing and responding organisms means that EICs in the former culture are acting pheromonally.

9.4 Extracellular Components; Further Regarding Properties and Significant Effects

9.4.1 The Major Significance of ESCs and EICs: They Enable Producing Organisms to Give Early Warning of Stress

Strikingly, and contrary to all previous ideas, a large class of extracellular sensor-induced responses has recently been found (Table 9.2). Those responses studied are all stress responses, but it is possible that other inducible responses will be found that are switched on by extracellular sensors. The stress responses studied to date behave like the acid-tolerance response in that they have pairs of extracellular components (generally small proteins) essential for triggering the response. As with the acid-tolerance response, the first component for each response is an extracellular sensor, an ESC, that detects the stress in the medium and is activated by it to the EIC, which associates with (and enters) the cell, leading to response induction (Figs. 9.1 and 9.2). The ESC, accordingly, is both an extracellular stress sensor and an EIC precursor (Rowbury and Goodson 1999a, b; Lazim and Rowbury 2000; Rowbury 2001a, b; Rowbury and Goodson 2001; Rowbury 2004b). Because some groups refer to sensors that are attached to the cell surface, to the OMs, or even to the outer surface of the CMs as extracellular, we emphasise here that the ESCs and EICs are components that are completely free in the growth medium, i.e., not attached in any way to the organisms.

Certain intracellularly produced stress-induced components have been referred to as alarmones (Bochner et al. 1984), although it is not clear how they can cross-feed other cells. The extracellular sensors described here can, however, readily diffuse, cross-feed, and lead to early warning of extracellular stress and very rapid response inductions (compared with response inductions switched on by intracellular sensors) for several reasons:

1. The stressor (stimulus) does not need to enter the cell to activate the response, it can interact with the ESC in the medium; the ESC, accordingly, detects the stressor immediately as it appears, and EIC production begins immediately.
2. Activated (protonated) sensor (EIC) interacts with organisms either during stress or in the absence of stress. The latter situation is of particular significance, with respect to early warning, because activated sensor can induce the response in cells before they are exposed to the stress. This results from the small EIC molecules diffusing from the site of activation, i.e., diffusing from where the stimulus is, and affecting cells in the region that the EIC diffuses to that are not yet facing acid challenge. There can, therefore, be "cross-talk" between organisms that have been exposed to stress and those that have not been exposed to stress (but for which stress is impending); the EICs acting as alarmones and intercellular communicators, i.e., pheromonally, leading to response induction in a location where there are unstressed organisms; by the time the stress has built-up to potentially lethal level in this latter location, the response (e.g., acid tolerance) will be in place (Rowbury 2001b).
3. For at least some ESCs, structure can be modified by cultural conditions (see below Sect. 9.7). For example, an acidity sensor formed at neutral pH is activated, triggering the acid-tolerance response, at low external pH. If the sensor is formed at pH 8.0 or 9.0, however, it is synthesised in a modified form (Fig. 9.3) that can be activated at slightly above neutrality (Rowbury and Goodson 1999a). Thus, for organisms growing at pH 9.0, if the medium rapidly becomes acidified, the response can be switched on at near neutrality, i.e., the response is very rapid rather than being delayed until pH 6.0 is reached.

9.4.2 Some ESCs Inactivate Chemical or Biological Stressors

It is now well established that because ESCs are formed in the absence of stress and interact directly with stressors, some of them confer a second form of stress tolerance by directly inactivating inhibitory stressors (i.e., interaction of ESC with stressor not only activates ESC to EIC but also converts the potentially lethal stressor to a harmless compound or agent). For example, ESC involved in the inducible bacteriophage Me1 tolerance response (Tables 9.1 and 9.2), after interaction with the phage, is not only converted to an EIC, but also inactivates the phage. The same may be true for other biological agents on interaction with their stress sensors, and ability of sensors to counteract chemical stress agents, which is the basis for the activity of the "protectants" of Nikolaev (1996, 1997), needs to be tested further.

9.5 Responses, Other Than Acid Tolerance Induction, Switched On by Extracellular Sensors and EICs

Triggering of three classes of stress response have, at the present time, been shown to depend on extracellular components, namely: (1) responses to several chemical stresses, especially pH stresses, (2) responses to a few biological stresses (Table 9.2), and (3) several responses to physical stresses; each involves ESCs and EICs.

9.5.1 Acid Sensitisation at Alkaline pH and the Role of an EIC

Organisms shifted from pH 7.0 to pH 9.0 rapidly become acid sensitive. Sensitisation seems to involve two stages; one is dependent on protein synthesis, the other independent of protein synthesis. Sensitisation is only induced if EICs are formed at pH 9.0 (Rowbury 1999a). Strikingly, the presence of protease at pH 9.0 reduces sensitisation and partially inactivates EIC-containing filtrates (Rowbury 1999a). Some sensitisation occurs under these conditions, however, and some EIC remains after protease treatment, suggesting that there may be two EICs formed at pH 9.0, one a protein and the other a nonprotein component. It seems likely that the two stages of acid sensitisation depend on the respective functioning of these two EICs (Table 9.2).

9.5.2 Responses Affecting Tolerance to Potentially Lethal Alkalinisation

9.5.2.1 Alkali Sensitisation at Acidic pH

Cultures shifted from pH 7.0 to mild acidity (pH 5.5–6.0) rapidly become alkali sensitive. The induction process is very distinct from that leading to acid tolerance induction. Sensitisation at pH 5.5 needs an EIC because removal or destruction of ECs at pH 5.5 abolishes induction. The EIC in neutralised filtrates also induces alkali sensitivity in organisms at pH 7.0 (Rowbury and Hussain 1998). Although this EIC is a protein, it is very heat stable (survives heating in a boiling water bath for 15 min), and fractionation shows a molecular weight of less than 5,000 Da (in accord with this EIC being dialysable). Synthesis of EIC shows dependence on the Hns, Him, and Fur gene products (Rowbury and Hussain 1998), which accords with the finding that these three gene products are needed for alkali sensitisation at pH 5.5.

The alkali sensitisation process is also dependent on a specific EIC precursor, which is present in media from pH 7.0-grown cultures; this precursor is an ESC on the basis that it has the properties of an acidity sensor, being converted by protons (in the absence of organisms) to EIC. This ESC resembles its cognate EIC in most respects (Table 9.4), but cannot induce sensitisation on addition to organisms at pH 7.0. This ESC (and its corresponding EIC) is distinct in properties from the acid-tolerance ESC, although both function as acidity sensors.

Table 9.4 Nature and properties of the alkali sensitisation ESC[a]

Filtrate component	Filtrate treatment before activation	Conditions for ESC activation	Organisms incubated with or without ESC	Percent survival ± SEM after alkali challenge
None	NA	NA	pH 7.0 grown	23.2 ± 1.3%
None	NA	NA	pH 5.5 grown	5.2 ± 0.6%
ESC	None	At pH 7.0	pH 7.0 grown	26.4 ± 1.5%
ESC	None	At pH 5.5	pH 7.0 grown	4.3 ± 1.3%
ESC	With protease	At pH 5.5	pH 7.0 grown	24.9 ± 1.2%
ESC	With RNase	At pH 5.5	pH 7.0 grown	4.8 ± 0.46%
ESC	With DNase	At pH 5.5	pH 7.0 grown	9.3 ± 0.27%
ESC	At 75°C	At pH 5.5	pH 7.0 grown	7.2 ± 0.53%
ESC	At 100°C	At pH 5.5	pH 7.0 grown	5.2 ± 1.7%
ESC	By dialysis	At pH 5.5	pH 7.0 grown	20.2 ± 0.6%

[a]Organisms (strain 1829) were grown to mid-log phase at the stated pH and neutralised, if required. Cell-free filtrates were prepared from cultures grown at pH 7.0, as for Table 9.3. These filtrates were treated as indicated and activated as stated, followed by neutralisation. Mid-log phase organisms grown at the stated pH were incubated with filtrate (1:1 culture to filtrate) for 60 min at pH 7.0. After washing, alkali challenge was at pH 10.5 for 8 min. NA, not applicable.

9.5.2.2   Role of Extracellular Components in Alkali Tolerance Induced at pH 9.0

Organisms transferred to pH 9.0 from pH 7.0 become alkali tolerant within a short period; tolerant organisms surviving exposure to pH 10.5 to 11.0 (Rowbury et al. 1989, 1996). Induction at pH 9.0 is associated with the presence of extracellular components, these being essential for response induction at this pH. These ECs can also induce alkali tolerance in organisms at pH 7.0. Accordingly, these components are alkali-tolerance EICs (Table 9.5), and because they induce alkali tolerance in unstressed cells, it is likely that, in the environment, they could diffuse to nearby regions not exposed to alkalinity and there act as alarmones, warning unstressed cells of the impending noxious stress and preparing them to resist it.

The above-mentioned EICs (Rowbury 1999a) arise from alkali-sensing components, present in filtrates from organisms grown at pH 7.0; these are converted to EIC at pH 8.0 to 9.0, i.e., they behave as alkali-sensing ESCs (Table 9.5). Substantial evidence suggests that there are two pairs of ESCs and two pairs of EICs involved in alkali tolerance induction (Tables 9.2 and 9.5).

9.5.3   Role of ESCs and EICs in Thermotolerance Induction

Cultures of *E. coli* grown at 45°C are more thermotolerant than those grown at 30°C (Table 9.6). If ESCs and EICs were involved in thermal induction of thermotolerance, one would expect cell-free filtrates from 45°C-grown cultures to contain specific thermotolerance EICs and to be able to induce thermotolerance in 30°C-grown cultures. This proved to be the case (Table 9.6); 30°C-grown cultures,

Table 9.5 Extracellular components induce alkali tolerance[a]

Filtrate used for alkali tolerance induction	Filtrate treatment before activation	Activation conditions for filtrate	Cultures used for tolerance induction	Percent survival after challenge for culture preincubated with or without filtrate
None	NA	NA	pH 7.0 grown	0.9 ± 0.12%
None	NA	NA	pH 9.0 grown	20.4 ± 0.7%
pH 9.0 filtrate	None	None	pH 7.0 grown	17.6 ± 0.5%
Filtered pH 9.0 broth[b]	None	None	pH 7.0 grown	0.4 ± 0.2%
pH 7.0 filtrate	None	None	pH 7.0 grown	0.6 ± 0.2%
pH 7.0 filtrate	None	At pH 9.0	pH 7.0 grown	15.5 ± 1.8%
pH 7.0 filtrate	With protease	At pH 9.0	pH 7.0 grown	12.0 ± 3.0%
pH 7.0 filtrate	At 75°C	At pH 9.0	pH 7.0 grown	12.1 ± 4.8%
pH 7.0 filtrate	At 100°C	At pH 9.0	pH 7.0 grown	7.3 ± 0.3%
pH 7.0 filtrate	By dialysis	At pH 9.0	pH 7.0 grown	12.8 ± 0.2%

[a]Strain 1829 was grown to mid-log phase at the stated pH, and filtrates prepared as for Table 9.3; filtrates were treated and activated as stated, with neutralisation if necessary. All filtrates were incubated (1:1 filtrate plus pH 7.0-grown mid-phase culture) at pH 7.0 for 60 min at 37°C, washed and challenged with pH 11.0 broth for 5 min. NA, not applicable.

[b]Filtrate from pure broth was also used.

Table 9.6 Induction of thermotolerance needs an EIC[a]

Thermotolerance-inducing filtrate	Filtrate treatment before incubation with culture	Culture incubated withor without filtrate	Survival % ± SEM after heat treatment (49°C, 5 min)
None	NA	30°C grown	0.3 ± 0.06%
None	NA	45°C grown	22.5 ± 1.5%
From 30°C-grown culture	None	30°C grown	0.27 ± 0.03%
From 45°C-grown culture	None	30°C grown	25.1 ± 2.0%
From 45°C-grown culture	With protease	30°C grown	0.7 ± 0.08%
From 45°C-grown culture	At 75°C	30°C grown	3.5 ± 0.3%
From 45°C-grown culture	By dialysis	30°C grown	17.2 ± 2.0%

[a]Cell-free filtrates were prepared from strain 1829ColV, K-94 grown in pH 7.0 broth as for Table 9.3. After stated treatments, filtrates were incubated (1:1) for 60 min with mid-log phase 30°C-grown cultures of the same strain. Washed organisms from mixtures and control cultures were challenged at 49°C for 5 min (this low temperature was used because mid-log phase ColV⁺ strains are particularly heat sensitive), with plating on nutrient agar to assess survival. NA, not applicable.

preincubated with filtrate from a 45°C-grown culture, showed greatly increased survival after thermal challenge compared with the original 30°C-grown culture control. Trivial reasons for tolerance induction by the filtrates were excluded. First, filtrates were free of viable organisms, excluding tolerance in the filtrate plus culture mixture resulting from growth (during incubation) of thermotolerant organisms from the filtrate. Second, filtrates failed to alter growth rates of the cultures or their final pH values. Reduced growth rates or altered final pH values sometimes lead to increased thermotolerance.

Pretreatment of active filtrate with protease greatly reduced its ability to induce tolerance (Table 9.6), implicating a proteinaceous EIC as the tolerance inducer. Heat treatment at 75°C almost fully inactivated this EIC; this contrasts with results for the acid-tolerance EIC, which is rather resistant at this temperature. The thermotolerance EIC, however, resembled the acid-tolerance EIC in being relatively nondialysable (Table 9.6); filtrates lost approximately one third of their thermotolerance-inducing ability if dialysed before testing, indicating that the thermotolerance EIC is likely to be of approximately 10,000 Da molecular weight.

Although filtrates from 30°C cultures failed to induce thermotolerance, they contain a thermotolerance ESC, because, if exposed to 45°C, in the absence of organisms, they gain the ability to induce thermotolerance (Table 9.7). The ESC resembles EIC in being relatively nondialysable, but it is more heat resistant and more resistant to protease than the EIC.

It should be noted that as temperature gradually increases, there is a gradual increase in conversion of ESC to EIC, with the ESC acting as an extracellular thermometer (Rowbury 2003a, b; 2005a). Strikingly, however, the thermotolerance

Table 9.7 An ESC, which is a protein, senses increased temperature[a]

Thermotolerance-inducing filtrate	Filtrate treatment before activation	Conditions for filtrate activation	Cultures incubated with or without filtrate	Percent survival ± SEM after thermal stress
None	NA	NA	30°C grown	0.35 ± 0.06%
None	NA	NA	45°C grown	21.0 ± 0.2%
From 30°C-grown culture	None	None	30°C grown	0.25 ± 0.3%
From 30°C-grown culture	None	At 45°C	30°C grown	13.4 ± 1.5%
From 30°C-grown culture	With protease	At 45°C	30°C grown	4.2 ± 1.4%
From 30°C-grown culture	At 75°C	At 45°C	30°C grown	8.8 ± 0.3%
From 30°C-grown culture	By dialysis	At 45°C	30°C grown	8.6 ± 0.26%

[a]Strain 1829 ColV, I-K94 was grown in pH 7.0 broth and filtrates prepared (see Table 9.3). After the stated treatments and activation (if necessary), filtrates were incubated with mid-log phase cultures (1:1 filtrate to culture) for 60 min at 30°C, before challenging washed suspensions as for Table 9.6 NA, applicable.

ESC can also act as a pH sensor, detecting gradual changes in pH and being gradually activated by them (Rowbury 2002). Activation can occur at both acidic and alkaline pH values, and also occurs after UV irradiation and exposure to Cu^{2+}.

9.5.4   UV Tolerance Induction Processes Involving Extracellular Components

9.5.4.1   RecA and LexA in UV Tolerance Induction

There has been controversy for many years regarding how UV irradiation is sensed and how this leads to switching on of the SOS response. An early stage of tolerance induction is undoubtedly RecA activation. RecA, however, is not the UV sensor, rather it has been thought that RecA detects DNA damage and that such damage activates RecA and sets in train processes leading to response induction. Strikingly, activation of RecA involves its proteolytic cleavage, and such cleavage confers protease activity on the resulting RecA fragment. Activated RecA then cleaves a few proteins, most importantly, LexA—this protein is a repressor of many operons relating to the SOS response, and, accordingly, LexA cleavage by activated RecA switches on this response.

It is now well established that, for organisms to stand the best chance of resisting lethal stresses, they must have evolved processes to give early warning of such stresses, i.e., stimulus detection must not only lead to response induction, but to production of components that diffuse away to other regions to switch on tolerance in unstressed organisms. There is no evidence that components produced by DNA breakdown, after DNA damage, can act in this way, but we now know that UV-induced EICs can do so.

9.5.4.2   Early Warning of UV Stress Involving EIC Functioning

It is now known that UV tolerance-related EICs occur in UV-irradiated cultures (Rowbury 2004b). Thus, cell-free filtrates from cultures exposed to low UV doses (such cultures rapidly become UV tolerant on incubation) can induce UV tolerance in unstressed cultures and this results from functioning of a proteinaceous EIC, because protease inactivates the filtrates (Table 9.8). The EIC in the original UV-irradiated culture is essential for its UV tolerance, because protease added to this culture during incubation prevents it from becoming UV tolerant. Filtrates from unirradiated cultures do not contain a UV tolerance-related EIC, but they do contain the cognate ESC, because, if such filtrates are briefly irradiated and incubated, they become able to confer tolerance, i.e., an EIC appears in them (Table 9.8). Obviously, if organisms in the environment have been exposed to UV irradiation, then

Table 9.8 UV tolerance induction needs a specific ESC and a specific EIC[a]

Culture filtrate from culture grown at 37°C and pH 7.0	Treatment of filtrate before activation	Treatment of filtrate after activation	Percent survival ± SEM after UV-treatment for 37°C-grown organisms pre-incubated ± filtrate
Not activated	None	None	0.06 ± 0.003%
30-s UV activated	None	None	0.2 ± 0.06%
60-s UV activated	None	None	7.9 ± 1.1%
90-s UV activated	None	None	8.3 ± 1.2%
90-s UV activated	With protease	None	5.2 ± 0.54%
90-s UV activated	None	With protease	0.13 ± 0.1%
90-s UV activated	At 75°C	None	0.6 ± 0.06%
90-s UV activated	None	At 75°C	0.5 ± 0.06%
90-s UV activated	By dialysis	None	1.5 ± 0.23%
90-s UV activated	None	By dialysis	6.9 ± 0.13%

[a]Filtrates, containing UV-tolerance ESCs, were from strain 1829 ColV, I-K94 grown to mid-log phase at 37°C and pH 7.0, and were activated with UV. Activated filtrates (treated as stated before or after activation) were incubated with the same culture (1:1 filtrate to culture) for 60 min at 37°C before UV challenge for 120 seconds. Survivor numbers were assessed on nutrient agar as for Table 9.3.

organisms in nearby regions that are, so far, unstressed, but face impending UV exposure, can be subjected to EICs diffusing from regions already facing UV irradiation; these EICs will give early warning and induce tolerance to the UV.

9.5.5 Inducible Tolerance to Alkylhydroperoxides

Cultures shifted from neutral pH to pH 9.0 become tolerant to alkylhydroperoxides (Rowbury 1997; Lazim and Rowbury 2000). A specific EIC forms at alkaline pH and is essential for induction, because continuous filtration or dialysis during incubation at alkaline pH abolishes tolerance (Lazim and Rowbury 2000). The EICs present in the neutralised filtrates from pH 9.0 cultures allow them to induce alkylhydroperoxide tolerance in pH 7.0 cultures. An EIC precursor (i.e., a specific alkali-sensing ESC) is present in pH 7.0 cultures. Thus the EIC is not synthesised *de novo* at pH 9.0, but arises by activation of the specific ESC (Table 9.2). Their susceptibility to removal by dialysis establishes that both the EIC and its cognate ESC are small molecules. Both are more heat labile (being readily destroyed at 75°C) than the corresponding extracellular components involved in acid tolerance induction. Unusually, these alkylhydroperoxide tolerance-related ECs do not seem to be proteins because they are insensitive to proteases.

The alkali-sensing ESC seems to be larger than the EIC (Lazim and Rowbury 2000), because the EIC readily passes through dialysis membranes, whereas the ESC is only partially removed by dialysis. Accordingly, the ESC may be an oligomer of the EIC, depolymerising at alkaline pH, or activation of ESC at pH 9.0 might involve cleavage to EIC and a second component.

9.5.6 Tolerance to Copper in Plasmid-Free Cultures Is Dependent on Extracellular Components

Usually, it has been assumed that copper tolerance is only encoded by plasmids in enterobacteria. We now know, however, that *E. coli* strains have a chromosomally encoded Cu^{2+} tolerance process that involves ECs. To establish this, *E. coli* ED1829, a plasmid-free strain, was grown with low levels of Cu^{2+} (88 µg/ml) or in copper-free medium. Many organisms from the culture containing Cu^{2+} (~20% of these organisms) were able to grow on medium containing 275 µg/ml Cu^{2+}, whereas none of those grown in copper-free medium formed colonies on this medium. Filtrate from the culture grown without copper did not confer copper tolerance (no organisms incubated with this filtrate gained the ability to form colonies on 275 µg/ml Cu^{2+}-containing media), but the filtrate from the copper-containing culture (dialysed to remove Cu^{2+}) conferred such tolerance (~16% of organisms preincubated with this filtrate gained the ability to grow on media containing 275 µg/ml Cu^{2+}). Accordingly, such a filtrate contained a copper tolerance-inducing EIC (Rowbury 2001a). This EIC is apparently a heat-stable protein, because protease inactivated the filtrate, whereas heating or treating the filtrate with RNase or DNase had no effect. Filtrate from the culture grown without copper did not contain a copper-tolerance EIC, but it did contain a copper-tolerance ESC, because incubation of this filtrate with Cu^{2+} led to production of a copper-tolerance EIC (Fig. 9.1). The ESC for copper tolerance is similar to other ESCs in being synthesised in the absence of stressor and in being converted to EIC by the stress (Cu^{2+}) in the absence of organisms. The EIC for copper tolerance resembles others in being a heat-resistant protein. It is also nondialysable.

9.6 Stress Cross-Tolerance Responses

9.6.1 Cross-Tolerance Responses Exemplified by the Effects of UV Irradiation

The best-known stress responses are those that lead to increased tolerance to a specific stress after exposure to the same stress. Stress cross-tolerance responses involve induced tolerance to a range of stresses after exposure to a different stress. Let us consider why cross-tolerance responses after UV irradiation may have evolved. Irradiation with UV causes thymine dimers to appear in DNA. On exposure to low doses of UV, inherent levels of DNA repair enzymes are sufficient to repair the dimers and virtually no killing occurs. At these low levels of UV, the irradiation activates the UV tolerance ESC (Table 9.8), and the resulting specific EIC induces further DNA repair mechanisms. Irradiation also activates ESCs for thermal stress tolerance induction and for acid and alkali tolerance induction (Table 9.9).

Table 9.9 UV irradiation activates a range of stress response ESCs[a]

Nature of ESC under test	Conditions for activation	Percent survival ± SEM after challenge by			
		Acid	Alkali	Heat	UV
Acid-tolerance ESC	None	0.7 ± 0.06%	NA	NA	NA
Acid-tolerance ESC	UV, 15 s	7.5 ± 0.3%	NA	NA	NA
Acid-tolerance ESC	UV, 30 s	10.3 ± 0.1	NA	NA	NA
Acid-tolerance ESC	UV, 60 s	34.1 ± 0.3%	NA	NA	NA
Acid-tolerance ESC	UV, 90 s	54.0 ± 1.9%	NA	NA	NA
Thermotolerance ESC	None	NA	NA	0.67 ± 0.07%	NA
Thermotolerance ESC	UV, 90 s	NA	NA	36.5 ± 1.1%	NA
Alkali tolerance ESC	None	NA	0.33 ± 0.09%	NA	NA
Alkali tolerance ESC	UV, 300 s	NA	20.9 ± 0.7%	NA	NA
UV tolerance ESC	None	NA	NA	NA	0.07 ± 0.01%
UV tolerance ESC	UV, 90 s	NA	NA	NA	10.1 ± 0.9%

[a]ESC-containing filtrates were prepared from strain 1829 ColV, I-K94 (acid-, heat-, or UV-tolerance studies) or strain 1829 (alkali-tolerance study) grown to mid-log phase in pH 7.0 broth at 37°C (30°C for thermotolerance tests). After activation (if appropriate), filtrates were mixed (1:1 filtrate to culture) with unstressed cultures of the same strains, followed by 60 min incubation at 37°C (or 30°C for thermotolerance tests) and mixtures were stress challenged (for thermotolerance, 5 min, 49°C; acid tolerance, 7 min, pH 3.0; alkali tolerance, 5 min, pH 11.0; UV tolerance, 120-second exposure). Survival after challenge was assessed as for Table 9.3, with incubation at 37°C (or 30°C for thermotolerance tests). All UV treatments used the same apparatus and conditions as for Table 9.3. NA, not applicable

As the dose of irradiation increases, the number of dimers introduced also increases, and inherent repair systems would soon be overwhelmed, but the induced UV tolerance enzymes will, for a while, cope with the dimer increase. Eventually, as greatly increased numbers of dimers are introduced, even the induced level of UV tolerance repair enzymes becomes inadequate. This is where the cross-tolerance responses come into their own. Because the main lethal effects of heat, acid, and alkali are on DNA, the UV-induced cross-tolerances to these stresses produce extra DNA repair enzymes. Accordingly, each cross-tolerance process that is induced by UV helps to protect the organisms from high UV doses. Cross-tolerance responses to other stressors have evolved for similar reasons.

9.7 Altered Structures of ESCs After Changes in Growth Conditions Leads to a Novel Early Warning System

Studies with filtrates from pH 7.0-grown cultures show that the acid tolerance-related ESC is converted to EIC at pH values from 4.5 to 6.0, with no activation at above pH 6.0. When, however, this ESC is formed at higher pH values, e.g., pH 9.0, it can be activated to EIC at pH values as high as 7.5 to 8.0 (Rowbury and Goodson 1999a; Rowbury 2001a). Three other responses dependent on ESCs and EICs show similar behaviours. Thus, ESC for the alkali sensitisation response shows altered responsiveness depending on synthesis pH, with ESC formed during growth at pH 7.0 being converted to EIC at pH 5.5 to 6.0 (but not at more alkaline pH values), whereas, for ESC synthesised at pH 8.5, activation to EIC can occur at pH values as high as 7.5 to 8.0.

The ESCs involved in alkali tolerance induction also show modified pH responsiveness depending on culture pH. If synthesised at neutral pH, these ESCs show activation to their cognate EICs only at high pH values. In contrast, ESCs formed at acidic pH can be activated at neutrality (Table 9.10). Similar behaviour occurs for the alkylhydroperoxide tolerance response, i.e., ESC formed at acidic pH can be activated at lower pH than ESC formed at neutral pH (Lazim and Rowbury 2000).

ESC structure has obviously evolved so that the structural form produced depends on cultural conditions, with such structures ensuring that responsiveness for an ESC is such that early warning of the stress can be given. Thus, organisms growing at alkaline pH but facing rapid acidification would be overwhelmed by protons if acid tolerance were not induced until the pH reached 6.0. Sensor modification at pH 9.0 allows ESC to EIC to occur at pH 7.5 to 8.0, with induction of tolerance beginning at this higher pH (Fig. 9.3).

Table 9.10 Modification of the pH responsiveness of the alkali tolerance sensor at pH 5.5[a]

pH for sensor synthesis	pH for sensor activation	Percent survival ± SEM after challenge with alkali for pH 7.0 grown cells preincubated with activated sensor
7.0	7.0	0.8 ± 0.16%
7.0	7.5	3.4 ± 0.35%
7.0	8.0	11.3 ± 0.71%
7.0	9.0	16.9 ± 0.86%
5.5	6.5	3.8 ± 1.4%
5.5	7.0	8.1 ± 1.0%
5.5	7.5	15.8 ± 1.7%
5.5	8.0	21.9 ± 3.1%

[a]*E. coli* 1829 was grown to mid-log phase at the stated pH and cell-free filtrates prepared as for Table 9.3. Filtrates were, activated, in the absence of organisms, at the indicated pH. Filtrates were, after neutralisation, if required, incubated with mid-log phase pH 7.0-grown organisms (1:1 culture to filtrate) at pH 7.0 for 60 min. After washing, mixtures were challenged at pH 11.0 for 5 min.

9.8 Death Is not the End: Protection of Enterobacteria from Chemical and Physical Stress by Dead Cultures

Many cases of disease are caused by ingestion of pathogenic bacteria that have survived and multiplied in food. It is worrying that many such organisms can tolerate chemical and physical stresses.

Although ESCs are exquisitely sensitive to activation by low doses of stress, both ESCs and EICs are resistant to irreversible inactivation by high doses of noxious agents or conditions (Rowbury and Goodson 1999b; Rowbury 2001a). Accordingly, the present author (Rowbury, 2000) considered it likely that killed cultures containing ESCs or EICs before death might protect living cultures from stress (i.e., ECs in these killed cultures, which had resisted the killing processes, might induce living cultures to stress tolerance). Tests established this to be the case. First, pH 7.0-grown cultures of *E. coli*, killed by exposure to high pH, protected living acid-sensitive organisms from lethal acidity provided that they were incubated at pH 5.0 (followed by neutralisation) after alkali killing. Similarly, pH 5.0-grown cultures, if neutralised and alkali killed, conferred acid tolerance on the same living cultures. Cultures killed by lethal acidity, UV-irradiation, or by lethal chemicals were (after removal of the lethal agent, where necessary) able to confer an acid-tolerance legacy on living acid-sensitive organisms (Rowbury 2000).

Acid tolerance might have occurred in the living organisms because the killed cultures caused an acidic pH in the mixture or a reduced growth rate for the mixture; pH changes during incubation of living cultures were, however, the same in the presence or absence of the killed culture, and killed cultures did not reduce growth rates. Additionally, use of genetically marked strains showed that the acid-tolerant organisms, present after activation by killed cultures, derive from the living culture rather than being survivors of the killing process (Rowbury 2000).

Killed cultures also conferred both alkali tolerance and alkali sensitivity on living organisms (Rowbury 2000). First, $2.7 \pm 0.6\%$ of log-phase pH 7.0-grown strain 1829 survived 5 min at pH 11.0, whereas, after preincubation of this culture for 60 min at pH 7.0 with a UV irradiation-killed pH 7.0-grown culture, activated at pH 9.0 (i.e., containing alkali-tolerance EICs), the percent survival on alkali challenge rose to $39.4 \pm 1.2\%$. Second, $19.3 \pm 0.82\%$ of log-phase pH 7.0-grown strain 1829 survived 8 min at pH 10.5, whereas the culture became more alkali sensitive after incubation with a heat-killed pH 7.0-grown culture of the same strain, activated at pH 5.5 (i.e., containing alkali-sensitisation EICs); incubation of living plus killed culture for 60 min at pH 7.0 produced a culture with only $5.2 \pm 0.61\%$ of organisms surviving pH 10.5 for 8 min.

9.9 Quorum-Sensed Processes: Are They Similar to ESC–EIC-Controlled Processes?

Another major group of responses that involve functioning of extracellular components occur in Gram-negative bacteria. These are the quorum-sensing controlled group of responses, for which the extracellular components are *N*-acyl-L-homoserine lactones

(AHLs) or related compounds. A well-known system is that of *Vibrio fischeri*, in which, at high cell densities, luminescence is switched on by quorum sensing of AHLs accumulated in the organisms and the medium. At certain intracellular levels, AHLs trigger the response (Gray et al. 1994; Swift et al. 1996). The AHL-controlled systems and those switched on by ESC–EIC pairs show several major differences and no appreciable similarities. First, quorum-sensed responses involve internal sensors, e.g., in the *Vibrio*, when enough AHL has accumulated in the periplasm, the AHL activates the CM protein, LuxR, and activation switches on the response. This contrasts with the stress responses, which are switched on when extracellular stress activates an extracellular sensor, i.e., the ESC. Another major difference between the quorum-sensed systems and ESC–EIC-controlled ones is that the signalling molecule, the AHL, for quorum-sensed systems, is synthesised, *de novo*, in response to growth conditions that eventually trigger induction. In contrast, in the stress responses, the ESC is synthesised in the absence of the stimulus (the stress) and is converted to EIC by the stress in the absence of organisms. A final difference between the two types of system is that the quorum-sensed responses are triggered by an accumulating level of AHL, and large amounts, enough to trigger the response, are only synthesised at high culture density. In contrast, ESC–EIC-induced responses occur at very low cell densities.

9.10   Conclusions

9.10.1   ESC: EIC-Dependent Systems Are Distinct from Other Systems that Involve Accumulation of Extracellular Components

This chapter described a novel group of stimulus sensing and response induction processes, distinct from most other response inductions in needing ECs. Other inducible systems involve ECs, but those described here are distinct in that the stimulus sensor is extracellular (Rowbury and Goodson 1999a, b; Rowbury 1999b; Lazim and Rowbury 2000; Rowbury 2004a), a feature not found in quorum-sensing systems and related processes.

The systems described here are induced by extracellular stress and involve pairs of ECs; the ESC, which is produced by stressed and unstressed cells, detects the stress and is converted by it to the EIC; the EIC interacts with receptors on the cells, entering the cells and inducing the response; this EC can induce the response in unstressed or stressed organisms.

9.10.2   Features of the ESC–EIC-Dependent Systems that Trigger Pheromone-Regulated Stress Responses

1. A stressor sensing component (ESC) occurs that is extracellular, therefore, there is an immediate response to external stressor.

2. EICs are present in media during induction of responses and are absolutely required for induction.

3. EICs can induce their response in unstressed organisms.

4. EICs that are in cultures exposed to stress arise from ESCs that are present in unstressed cultures, and interact with the stress to produce the EIC. The EICs act as primary signalling molecules initiating response induction.

5. ESCs are exquisitely sensitive to activation by stressors, but both ESCs and EICs are very resistant to most environmental inhibitors and inhibitory conditions.

6. The ESC occurs in several forms (Rowbury and Goodson 1999a; Lazim and Rowbury 2000; Rowbury 2001a, b), the form synthesised being that which can most rapidly respond to stress and lead to tolerance (Fig. 9.3). In some cases, this property allows a response to a chemical agent (e.g., protons) before that agent becomes stressing (e.g., before pH becomes acidic), this characteristic provides an ideal early warning system against stress.

7. At least five ESCs function as biological thermometers, detecting increasing temperature, and being gradually activated to EICs as temperature gradually increases (Rowbury 2003b).

8. Many ESCs (at least eight) act as pH sensors, detecting changes in proton concentration, with some being gradually activated (to EIC) by gradually increasing acidification, others by gradual alkalinization, and a few by both.

9. After attaching to specific receptors, all tested EICs cross into organisms to induce the response. For several EICs, the PhoE porin forms the route of entry.

10. Some responses seem to need two ESCs and two EICs for full induction.

11. Because ESCs and EICs are highly resistant to irreversible inactivation by most lethal chemical and physical agents and conditions, even killed cultures can confer stress tolerance onto living organisms that subsequently enter the environment, i.e., EICs in cultures killed by several means induce stress responses in living unstressed cultures and both ESCs and EICs in killed cultures induce stress responses in cultures unable to produce the ESC–EIC pair.

12. The EICs are small molecules (Rowbury et al. 1998; Hussain et al. 1998; Rowbury 1999a; Lazim and Rowbury 2000; Rowbury 2004a) and can, therefore, diffuse away, even passing to areas not thus far exposed to stress.

13. The EIC induces tolerance in unstressed cells (Rowbury and Goodson 1998). This property, together with the diffusibility of the EIC, allows EICs to act as extracellular alarmones, i.e., they are intercellular communicating molecules, which allow cross-talk between stressed and unstressed populations, with the latter being induced to stress tolerance before being exposed to the stress.

14. EICs can cross-feed both unstressed EC producers and cultures unable to produce ESCs and EICs whereas ESCs can cross-feed the latter.

15. Taken together, the above characteristics provide early warning systems against external stress, giving populations the best chance of detecting stresses and rapidly putting in place tolerance mechanisms that allow survival in the presence of lethal agents.

Highly Recommended Readings

Fratamico PM, Bhunia AK, Smith JL (eds) (2005) Foodborne pathogens: microbiology and molecular biology. Caister Academic Press, Wymondham, Norfolk

Neidhardt FC, Ingraham JL, Schaechter M (1990) Physiology of the bacterial cell. Sinauer Associates Inc, Sunderland, Mass

Poole RK (ed) (1999) Bacterial responses to pH. Novartis Foundation. Wiley, New York

Rowbury RJ (2001) Extracellular sensing components and extracellular induction component alarmones give early warning against stress in *Escherichia coli*. Adv Microbial Physiol 44: 215–257

Rowbury RJ (2005) Intracellular and extracellular components as bacterial thermometers and early warning against thermal stress. Sci Prog 88: 71–99

Rowbury RJ, Goodson M (2001) Extracellular sensing and signalling pheromones switch-on thermotolerance and other stress responses in *Escherichia coli*. Sci Prog 84: 205–233

Yousef AE, Juneja VK (eds) (2003) Microbial stress adaptation and food safety. CRC Press, Boca Raton

References

Aiba H, Mizuno T, Mizushima S (1990) Transfer of phosphoryl group between two regulatory proteins involved in osmoregulatory expression of the *ompF* and *ompC* genes in *Escherichia coli*. J Biol Chem 264: 8563–8567

Ananthaswamy HN, Eisenstark A (1977) Repair of hydrogen peroxide induced single strand breaks in *Escherichia coli*. J Bacteriol 130: 187–191

Artz SW, Holzschu D (1982) Histidine biosynthesis and its regulation. In: Hermann KM, Somerville RL (Eds) Amino Acids and Genetic Regulation. Adison Wesley, Reading, MA, pp 379–404

Bochner BR, Lee PC, Wilson SW, Cutler CW, Ames BN (1984) ApppppA and related adenylated nucleotides are synthesized as a consequence of oxidation stress. Cell 37: 227–232

van Bogelen RA, Neidhardt FC (1990) Ribosomes as sensors of heat and cold shock in *Escherichia coli*. Proc Natl Acad Sci USA 87: 5589–5593

Chen Q, Amster-Choder O (1999) BglF, the *Escherichia coli* β-glucosidase permease and sensor of the *bgl* system: domain requirements of the different catalytic activities. J Bacteriol 181: 462–468

Csonka LN (1989) Physiological and genetic responses of bacteria to osmotic stress. Microbiol Revs 53: 121–147

Demple B, Halbrook J (1983) Inducible repair of oxidative damage in *Escherichia coli*. Nature 304: 466–468

Elliott EL, Colwell RR (1985) Indicator organisms for estuarine and marine waters. FEMS Microbiol Revs 32: 61–79

Etchegaray J-P, Inouye M (1999) CspA, CspB and CspG, major cold-shock proteins of *Escherichia coli* are induced at low temperatures under conditions that completely block protein synthesis. J Bacteriol 181: 1827–1830

Foster JW, Hall HK (1990) Adaptive acidification tolerance response in *Salmonella typhimurium*. J Bacteriol 172: 771–778

Foster JW, Moreno M (1999) Inducible acid tolerance mechanisms in enterobacteria. In: Bacterial Responses to pH. Novartis Found Symp 221: 55–69

Gray KM, Passador L, Inglewski BH, Greenberg EP (1994) Interchangeability and specificity of components from the quorum-sensing regulatory systems of *Vibrio fischeri* and *Pseudomonas aeruginosa*. J Bacteriol 176: 3076–3080

Humphrey TJ, Slater E, McAlpine K, Rowbury RJ, Gilbert RJ (1995) *Salmonella enteritidis* PT4 isolates more tolerant of heat, acid or hydrogen peroxide also survive longer on surfaces. Appl Environ Microbiol 61: 3161–3164

Humphrey TJ, Williams A, McAlpine K, Lever MS, Guard-Petter J, Cox, JM (1996) Isolates of *Salmonella enterica* Enteritidis PT4 with enhanced heat and acid tolerance are more virulent in mice and more invasive in chickens. Epidemiol Infect 117: 79–88

Hussain NH, Goodson M, Rowbury RJ (1998) Intercellular communication and quorum sensing in micro-organisms. Sci Prog 81: 69–80

Igo MM, Silhavy TJ (1988) EnvZ, a transmembrane sensor of *Escherichia coli* is phosphorylated in vitro. J Bacteriol 170: 5971–5973

Khazaeli MB, Mitra RS (1981) Cadmium-binding component in *Escherichia coli* during accomodation to low levels of this ion. Appl Environ Microbiol 41: 46–50

Kullik I, Toledano MB, Tartaglia LA, Storz G (1995) Mutation analysis of the redox-sensitive transcriptional regulator OxyR: regions important for oxidation and transcriptional activation. J Bacteriol 177: 1275–1284

Lazim Z, Rowbury RJ (2000) An extracellular sensor and an extracellular induction component are required for alkali induction of alkyl hydroperoxide tolerance in *Escherichia coli*. J Appl Microbiol 89: 651–656

Mackey BM (1984) Lethal and sub-lethal effects of refrigeration, freezing and freeze-drying on microorganisms. Symp Soc Appl Bacteriol 12: 45–75

Mackey BM, Derrick CM (1982) The effect of sub-lethal injury by heating, freezing, drying and gamma radiation on the duration of the lag-phase of *Salmonella typhimurium*. J Appl Bacteriol 53: 243–251

Mackey BM, Derrick CM (1986) Changes in the heat resistance of *Salmonella typhimurium* during heating at rising temperatures. Letts Appl Microbiol 4: 13–16

McCarty JS, Walker GC (1991) DnaK as a thermometer: threonine-199 is site of autophosphorylation and is critical for ATPase activity. Proc Natl Acad Sci USA 88: 9513–9517

Mongkolsuk S, Loprasert S, Whangsuk W, Fuangthong M, Atichartpongkun S (1997) Characterisation of transcription organisation and analysis of unique expression patterns of an alkyl hydroperoxide reductase C gene (*ahpC*) and the peroxide regulator operon *ahpF-oxyR-orfX* from *Xanthomonas campestris* pv. phaseoli. J Bacteriol 179: 3950–3955

Neidhardt FC, Ingraham JL, Schaechter M (1990) Physiology of the Bacterial Cell. Sinauer Associates Inc: Sunderland, MA

Nikolaev YA (1996) General protective effect of exometabolite(s) produced by tetracycline-treated *Escherichia coli*. Microbiol (Moscow) 65: 749–752

Nikolaev YA (1997) Involvement of exometabolites in stress adaptation of *Escherichia coli*. Microbiol (Moscow) 66: 38–41

Padan E, Gerchman Y, Rimon A, Rothman A, Dover N, Carmel-Harel O (1999) The molecular mechanism of regulation of the NhaA Na$^+$/H$^+$ antiporter of *Escherichia coli*, a key transporter in the adaptation to Na$^+$ and H$^+$. In: Bacterial Responses to pH. Novartis Found Symp 221: 183–199

Raja N, Goodson M, Smith DG, Rowbury RJ (1991) Decreased DNA damage and increased repair of acid-damaged DNA in acid-habituated *Escherichia coli*. J Appl Bacteriol 70: 507–511

Rowbury RJ (1997) Regulatory components, including integration host factor, CysB and H-NS, that influence pH responses in *Escherichia coli*. Letts Appl Microbiol 24: 319–328

Rowbury RJ (1999a) Acid tolerance induced by metabolites and secreted proteins and how tolerance can be counteracted. In: Bacterial Responses to pH. Novartis Found Symp 221: 93–111

Rowbury RJ (1999b) Extracellular sensors and inducible protective mechanisms. Trends in Microbiol 7: 345–346

Rowbury RJ (2000) Killed cultures of *Escherichia coli* can protect living organisms from acid stress. Microbiol 146:1759–1760

Rowbury RJ (2001a) Cross-talk involving extracellular sensors and extracellular alarmones gives early warning to unstressed *Escherichia coli* of impending lethal chemical stress and leads to induction of tolerance responses. J Appl Microbiol 90: 677–695

Rowbury RJ (2001b) Extracellular sensing components and extracellular induction component alarmones give early warning against stress in *Escherichia coli*. Adv Microbial Physiol 44: 215–257

Rowbury RJ (2002) Microbial disease: recent studies show that novel extracellular components can enhance microbial resistance to lethal host chemicals and increase virulence. Sci Prog 85: 1–11

Rowbury RJ (2003a) Physiology and molecular basis of stress adaptation, with particular reference to the subversion of stress adaptation and to the involvement of extracellular components in adaptation. In: Yousef AE, Juneja VK (eds) Microbial Stress Adaptation and Food Safety. CRC Press LLC, Boca Raton, FL, pp 247–302

Rowbury RJ (2003b) Extracellular proteins as enterobacterial thermometers. Sci Prog 86: 139–156

Rowbury RJ (2004a) Enterobacterial responses to external protons, including responses that involve early warning against stress and the functioning of extracellular pheromones, alarmones and varisensors. Sci Prog 87: 193–225

Rowbury RJ (2004b) UV radiation-induced enterobacterial responses, other processes that influence UV tolerance and likely environmental significance. Sci Prog 87: 313–332

Rowbury RJ (2005a) Intracellular and extracellular components as bacterial thermometers, and early warning against thermal stress. Sci Prog 88: 71–99

Rowbury RJ (2005b) Stress responses of foodborne pathogens, with specific reference to the switching on of such responses. In: Fratamico PN, Bhunia AK, Smith JL (eds) Foodborne Pathogens Microbiology and Molecular Biology. Caister Academic Press. Wymondham, Norfolk, UK, pp 77–97

Rowbury RJ, Goodson M (1997) Metabolites and other agents which abolish the CysB-regulated acid tolerance induced by low pH in *Escherichia coli*. Rec Res Devel Microbiol 1: 1–12

Rowbury RJ, Goodson M (1998) Induction of acid tolerance at neutral pH in log-phase *Escherichia coli* by medium filtrates from organisms grown at acidic pH. Letts Appl Microbiol 26: 447–451

Rowbury RJ, Goodson M (1999a) An extracellular acid stress-sensing protein needed for acid tolerance induction in *Escherichia coli*. FEMS Microbiol Letts 174: 49–55

Rowbury RJ, Goodson M (1999b) An extracellular stress-sensing protein is activated by heat and u.v. irradiation as well as by mild acidity, the activation producing an acid tolerance-inducing protein. Letts Appl Microbiol 29: 10–14

Rowbury RJ, Goodson M (2001) Extracellular sensing and signalling pheromones switch-on thermotolerance and other stress responses in *Escherichia coli*. Sci Prog 84: 205–233

Rowbury RJ, Hussain NH (1998) The role of regulatory gene products in alkali sensitisation by extracellular medium components in *Escherichia coli*. *Letts Appl Microbiol* 27: 193–197

Rowbury RJ, Goodson M, Whiting GC (1989) Habituation of *Escherichia coli* to acid and alkaline pH and its relevance for bacterial survival in chemically polluted natural waters. Chem Ind 1989: 685–686

Rowbury RJ, Lazim Z, Goodson M (1996) Regulatory aspects of alkali tolerance induction. Letts Appl Microbiol 22: 429–432

Rowbury RJ, Hussain NH, Goodson M (1998) Extracellular proteins and other components as obligate intermediates in the induction of a range of acid tolerance and sensitisation responses in *Escherichia coli*. FEMS Microbiol Letts 166: 283–288

Russell AD (1984) Potential sites of damage in micro-organisms exposed to chemical or physical agents. In: Andrew MHE, Russell AD (eds) The Revival of Injured Microbes. Academic Press, San Diego, CA pp 1–18

Samson L, Cairns J (1977) A new pathway for DNA repair in *Escherichia coli*. Nature 267: 281–283

Somerville RL (1982) Tryptophan: biosynthesis, regulation and large scale production. In: Hermann KM, Somerville RL (eds) Amino Acids and Genetic Regulation. Adison Wesley, Reading, MA, pp 351–378

Swift S, Throup JP, Salmond GPC, Stewart GSAB (1996) Quorum sensing: a population-density component in the determination of bacterial phenotype. TIBS 21: 214–219

Walker GC (1984) Mutagenesis and inducible responses to deoxyribonucleic acid damage in *Escherichia coli*. Microbiol Revs 48: 60–93

Wanner BL (1987) Phosphate regulation of gene expression in *Escherichia coli*. In: Neidhardt FC, Ingraham JL, Low KB, Magasanik B, Schaechter M, Umbarger HE (eds) *Escherichia coli* and *Salmonella typhimurium*: Cellular and Molecular Biology, Vol.2. American Society for Microbiology, Washington D.C., pp 1326–1333
Wolffe AP (1995) The cold-shock response in bacteria. Sci Prog 78: 301–310
Wood JM (1999) Signals and membrane-based sensors. Microbiol Molec Biol Revs 63: 230–262

10
Ribosome Modulation Factor

Gordon W. Niven(✉) and Walid M. El-Sharoud

Abstract Ribosome modulation factor (RMF) is a ribosome-associated protein in *Escherichia coli* that is synthesised under stringent control during stationary phase and during periods of slow growth. The binding of RMF seems to make ribosomes, and in particular ribosomal RNA (rRNA), more resistant to degradation. Comparison of RMF-deficient mutant strains and the parent strain suggest that RMF contributes to the survival of *E. coli* under environmental extremes, such as conditions of heat, cold, acid, and osmotic stress. RMF may bind to inactive ribosomes to produce a stable resting state. Because binding blocks the peptidyl transferase site, ribosomes with associated RMF are not capable of protein synthesis. It has, therefore, been further suggested that RMF binding serves to inactivate excess ribosomes to make protein synthesis more efficient under conditions of slow growth. RMF binding is associated with the formation of 100 S ribosome dimers that are observed after sucrose density centrifugation of extracts of stationary-phase cultures. Dimer formation is generally seen as an inherent feature of RMF function, but evidence that this may not be the case is discussed in this review. The exact function and mechanism of action of RMF, therefore, remain to be fully elucidated.

Gordon W. Niven
Dstl, Porton Down, Salisbury, Wiltshire SP4 0JQ, UK
gwniven@dstl.gov.uk

W. El-Sharoud (ed.) *Bacterial Physiology: A Molecular Approach.*
© Springer-Verlag Berlin Heidelberg 2008

10.1   Introduction

From the earliest days of their studies, students of microbiology are made familiar with the growth phases of microbial batch cultures: lag, exponential, stationary, and death. However, the smooth and sequential transition between ordered growth phases observed under laboratory conditions is largely atypical of bacteria growing in the wider environment. In nature, many of the environments encountered by bacteria are nutrient limited and/or physically harsh, with potentially damaging extremes of pH, temperature, and osmotic stress. The convention of attributing bacterial cultures to discrete phases serves to highlight that bacteria can spend much of their time either adapting their metabolism to enable growth under changed conditions, or attempting to survive under conditions that do not permit growth at all. The physiological mechanisms by which cells are protected under potentially damaging conditions, and can react to changes in their environment, are, thus, central to survival and proliferation.

The fate and behaviour of the ribosome is a key element when considering the adaptation of bacterial cells to new or extreme environments. First, efficient ribosome function is essential to produce the new proteins that are often required to ensure growth and survival. This may be, for example, the synthesis of enzymes required to metabolise an available nutrient or the production of stress proteins in response to heat or cold shock. Second, the ribosome is particularly vulnerable to disruption and damage under harsh conditions. Ribosomes are large and complex multicomponent structures and their function is highly dependent on the conformation of the proteins and nucleic acids of which they are composed. Any factors that can alter macromolecular conformation can result in disruption of the interactions that maintain ribosome integrity, or are essential for the protein synthesis process. Such conformational changes can be a direct result of environmental conditions, such as heating, or can be a secondary effect of transient loss of membrane integrity (see, for example, Davis et al. 1986; VanBogelen and Neidhardt 1990; Tolker-Nielsen and Molin 1996; Niven et al. 1999; Bayles et al. 2000; El-Sharoud 2004b).

This review focuses on a single protein, ribosome modulation factor (RMF), which has been implicated in the survival of *Escherichia coli* under nongrowing conditions and environmental stress. It will demonstrate that RMF plays a central role in the maintenance of ribosome structure and function in *E. coli* under potentially lethal conditions. However, the physiological activities of RMF are poorly understood, and the exact nature and purpose of its interaction with the ribosome remains to be elucidated. Undoubtedly, it represents only one small aspect of the varied and complex mechanisms by which ribosome activity is regulated in bacterial cells (see, for example, Agafonov et al. 1999; Agafonov et al. 2001; Maki et al. 2000; El-Sharoud 2004b), but a greater understanding of its function will give further insight into the nature of the cell's interaction with its environment.

10.2   Discovery of RMF and Ribosome Dimerisation

RMF was first discovered in *E. coli* by Wada et al. (1990) as a small protein (55 amino acid residues) associated with stationary-phase ribosome dimers. They initially observed ribosome dimers, designated 100 S particles, after sucrose density gradient centrifugation of cell extracts that had been prepared using cells that had not been washed in buffer before lysis. Ribosome dimerisation was demonstrated to be associated with stationary-phase cultures, and they were converted back to 70 S monomers after transfer of cells to fresh medium (Fig. 10.1). It was, therefore, proposed that 100 S ribosome dimers were a storage form of ribosomes and suggested that such ribosomes may be more resistant to degradation during stationary phase.

100 S particles had previously been observed in extracts of *E. coli* 30 years earlier, but had received relatively little attention since that time. Tissiéres and Watson (1958) and Tissiéres et al. (1959) observed that the sedimentation coefficients of "ribonucleoprotein particles" were dependent on the concentration of Mg^{2+} in the buffer, with 100 S particles observed at 5 mM Mg^{2+}. Progressively lower concentrations of Mg^{2+} resulted in dissociation of 100 S particles to 70 S, and 70 S to 50 S and 30 S. This enabled the structural relationship between these bodies to be understood. Confirmation of the ribosome as the site of protein synthesis by McQuillen et al. (1959) enabled McCarthy (1960) to make the association between 100 S particles, stationary phase, and protein synthetic activity. He demonstrated conversion of 100 S ribosomes to 70 S particles and resumption of protein synthesis on addition of glucose

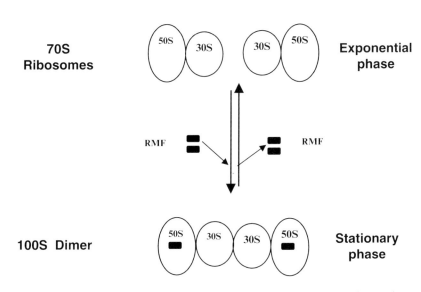

Fig. 10.1   Dimerisation of *E. coli* 70 S ribosomes by RMF binding during the stationary phase and dissociation of 100 S dimers on reculturing cells and resumption of exponential growth

to a stationary-phase culture. He also observed conversion of 70 S ribosomes to 100 S on entry to stationary phase, and described the dimers as a ribosome "resting state."

In addition to "rediscovering" ribosome dimerisation in stationary phase, Wada et al. (1990) attempted to elucidate the mechanism by which this conversion was carried out. They, thus, identified RMF (then referred to as "protein E") as a basic protein exclusively associated with 100 S particles, an observation that was dependent on the use of radical-free highly reducing two-dimensional polyacrylamide gel electrophoresis. Further, they identified and sequenced the gene encoding RMF (now designated *rmf*) and confirmed the molecular weight and isoelectric point of the protein as 6,475 and 11.3, respectively.

10.3   RMF Enhances Cell Survival Under Stressful Conditions

The suggestion that ribosomes should have a specific conformation that represents a storage form raises the question of why such a form should exist in stationary-phase cells. The assumption must be that it confers some survival advantage, or perhaps that it is essential for survival. The first evidence that RMF played a role in stationary-phase survival was provided by Yamagishi et al. (1993). They constructed an RMF-deficient mutant strain of *E. coli* W3110 (designated strain HMY15) in which the *rmf* gene was disrupted by insertion of a gene encoding chloramphenicol resistance. The viability of this mutant strain was shown to decrease more rapidly during stationary phase, being reduced by approximately 4 log units after 5 days compared with less than 1 log unit for the parent strain. Similar observations were made by Apirakaramwong et al. (1998) and Raj et al. (2002) using RMF-deficient mutant strains derived from *E. coli* strain C600. Wada et al. (2000) also demonstrated reduced stationary-phase viability in *E. coli* strain Q13, which was identified as deficient in dimer formation. Because *rmf* messenger RNA (mRNA) was detected in this strain, it was speculated that the deficiency may be caused by impaired RMF binding to ribosomes.

Evidence of the role of RMF in protecting cells against potentially damaging environmental stress was provided by Garay-Arroyo et al. (2000). They were interested in "late embryogenesis abundant" (LEA) proteins in higher plants, which are highly hydrophilic proteins produced in response to water deficit. A database search aimed at identifying potential LEA analogues with similar physical properties in the genomes of other organisms revealed several candidates, including RMF from *E. coli*. RMF is not likely to be an LEA analogue because these are usually composed of unstructured random coils, whereas RMF is predicted to consist of a coiled structure over 50% of its length (Garay-Arroyo et al. 2000). However, this study included the demonstration of increased synthesis of RMF mRNA in response to osmotic shock induced during exponential-phase growth by addition of 0.4 M NaCl. In addition, using strain HMY15, the *rmf*::Cmʳ mutant strain of Yamagishi et al. (1993), they showed reduced growth and increased death compared with the parent strain in response to osmotic shock.

The same RMF-deficient mutant strain was used by Niven (2004) to investigate the influence of RMF on exposure to heat stress. Stationary-phase cultures were subjected to heating at 50°C, which proved to be lethal to the mutant strain and resulted in a 5-log decrease in viability within 40 min. In comparison, the parent strain suffered no reduction in viability under the same conditions. Bacteria are generally more vulnerable to stress during exponential phase than stationary phase, and this was shown to be the case in this study. On heating exponential-phase cultures of the parent strain, viability was reduced by 2-log units within 40 min, and 4-log units within 100 min. The RMF-deficient mutant strain was not more vulnerable than the parent strain when subjected to heat stress during exponential phase, and showed similar death kinetics under these conditions. These observations highlighted the association of RMF with stationary phase, and contrasted with those of Garay-Arroyo et al. (2000), who demonstrated a role of RMF in osmotic stress resistance in exponential phase.

El-Sharoud and Niven (2007), again using strain HMY15 (Yamagishi et al. 1993), demonstrated that RMF was also implicated in survival under acid stress during stationary phase. When placed in medium adjusted to pH 3.0 using HCl, the viability of the mutant strain was reduced by 3 log units after 5 h, compared with less than 1 log unit for the parent strain. The difference in survival between the mutant and parent strains was more pronounced in the presence of organic acids when 30 mM lactic acid or acetic acid were added at pH 3.0 (El-Sharoud 2004a). In exponential-phase cultures, the growth rates of both strains were influenced to a similar extent by acidification of the medium with HCl across a range of pH values (El-Sharoud and Niven, 2005). However, greater *rmf* expression was observed during growth under acidic conditions. This was estimated using a strain carrying an *rmf-lacZ* fusion (Yamagishi et al. 1993) and measuring β-galactosidase activity. Because the growth pH influenced *rmf* expression, further experiments were performed to determine whether this then had an effect on cell resistance to lethal acid shock at pH 2.5. The results were similar for the RMF-deficient strain and the parent strain. Both displayed increased acid resistance after growth at reduced pH in accordance with the acid habituation response (Rowbury et al. 1989; Rowbury 1997). It, thus, seems that RMF was not involved in this acid stress-specific protective response.

In the case of cold stress, strain HMY15 proved to be more vulnerable than the parent strain when cultures grown at 37°C were transferred to 4°C during exponential-phase growth (Niven, unpublished data). Interestingly, no such difference was seen for stationary-phase cultures. It is possible to speculate that cellular mechanisms not associated with RMF made the cells more resistant to cold shock during stationary phase. The influence of RMF was, therefore, only observed in the absence of such stationary phase-specific mechanisms during exponential phase.

The studies described demonstrate that a lack of functional RMF increases the vulnerability of *E. coli* to the damaging influence of physical and chemical stresses, and to increased cell death during stationary phase. In view of the variety of stress conditions examined, the influence of RMF seems to be general rather than specific for particular stress conditions.

10.4   RMF Increases Ribosome Stability

It is clear that the synthesis of RMF on entry into stationary phase or during environmental stress plays some role in maintaining cell viability. The hypothesis that the function of RMF is to protect ribosomes was put forward by Fukuchi et al. (1995), who suggested that RMF "may function as an anti-degradation factor." They observed that accumulation of spermidine in a mutant strain of *E. coli* deficient in spermidine acetyltransferase (SAT) resulted in reduced synthesis of RMF during stationary phase and more rapid loss of ribosomes. Wada (1998) also speculated that RMF may "protect ribosomes from degradation by proteases and nucleases induced in the stationary phase."

Niven (2004) used differential scanning calorimetry (DSC) to demonstrate that RMF may contribute to making ribosomes physically more robust. In this technique, samples are heated according to a predefined temperature gradient. To maintain the programmed temperature gradient, the energy input must be varied to compensate for any endothermic or exothermic reactions that occur in the sample. Such reactions are, thus, visualised as peaks and troughs in the resulting thermogram plot of energy input against sample temperature. In the case of whole live bacterial cells, the major thermogram peaks have been attributed to ribosome degradation, the relative temperatures of these peaks being indicative of the thermal stability of the ribosomes in vivo (Mackey et al. 1991; Teixeira et al. 1997). When this method was applied to exponential-phase cells of the RMF-deficient mutant strain HMY15 and the parent strain, the main endothermic maxima were located at 68.4°C and 68.1°C, respectively. However, although the peak maximum was similar for stationary-phase cells of the parent strain (68.0°C), the temperature was significantly lower for strain HMY15 (66.7°C). These data suggested that ribosomes become more vulnerable to denaturation by heat on entry into stationary phase, and that this vulnerability is countered by RMF.

Application of DSC analysis to chilled cells also offered evidence that the vulnerability of the mutant strain during exponential phase could be attributed to ribosome instability (Niven, unpublished data). When stationary-phase cultures of HMY15 and the parent strain, and exponential-phase cultures of the parent strain were incubated for 24 h at 4°C, the thermograms were similar to controls examined before cold stress (Fig. 10.2). In the case of exponential-phase cultures of the RMF-deficient mutant strain, the thermogram peak maximum temperature was substantially reduced. This indicated that the ribosomes of the exponential-phase cells were in a significantly less stable conformation after exposure to cold stress in the absence of RMF.

In the study of the influence of RMF on heat stressed cells, increased degradation of ribosomal RNA (rRNA) was observed in heated cultures of the RMF-deficient mutant strain compared with the parent strain (Niven 2004). This suggested that RMF may function to protect the RNA, although it could not be excluded that the reduction in rRNA was a result of viability loss rather than a cause of it. More substantial evidence that the role of RMF may specifically relate to the protection of rRNA was provided by El-Sharoud and Niven (2007). On exposure of stationary-phase cultures

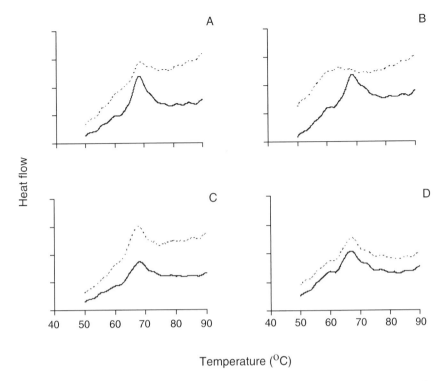

Heat flow

Temperature (^{O}C)

Fig. 10.2 DSC thermograms of whole cell cultures of *E. coli* strains W3110 and HMY15 before chilling (*solid line*) and after (*broken line*) incubation for 24 h at 4°C. Shown are W3110 in exponential phase (**a**), HMY15 in exponential phase (**b**), W3110 in stationary phase (**c**), and HMY15 in stationary phase (**d**).

to acid stress at pH 3.0, the viability of the RMF-deficient strain decreased more rapidly than that of the parent strain, despite a similarly substantial loss of intact ribosome particles in both strains. Analysis of rRNA demonstrated that, although the RNA of the parent strain remained intact, that of the mutant strain was highly degraded. The hypothesis was, therefore, presented that recovery of the cells from this stress was dependent on the regeneration of functional ribosomes from their macromolecular components, and that RMF functioned to reduce rRNA degradation rather than preserving intact ribosome particles.

10.5 RMF Binding Results in Ribosome Conformational Changes

Since its discovery, RMF has been associated with the dimerisation of 70 S ribosomes to form 100 S particles, and this phenomenon is generally regarded as intimately linked to RMF function. Dimers have, therefore, come to embody the ribosome storage form that contributes to the preservation of cell viability under harsh conditions. Indeed, Wada (1998) and Wada et al. (2000) described four distinct periods of stationary phase in *E. coli* that were defined by the dimerisation state of the ribosomes. There can be little doubt that RMF causes the formation of the ribosome dimers that are observed on sucrose density gradient centrifugation of bacterial cell lysates. RMF was originally isolated specifically from 100 S particles (Wada et al. 1990), and it has not been detected associated with other ribosome forms or free in the cytoplasm (for example, Wada et al. 1990; Wada et al. 2000; Izutsu et al. 2001; Yoshida et al. 2004). Both natural and synthetic RMF have been shown to catalyse the formation of dimers from 70 S particles in vitro (Wada et al. 1995), and no dimers are formed by RMF-deficient mutant strains (Yamagishi et al. 1993; Apirakaramwong et al. 1998; Raj et al. 2002).

The hypothesis that RMF serves to protect rRNA from degradation rather than preserve intact ribosome particles per se seems to contradict the idea that dimerisation is central to the mechanism of RMF function. Niven (2004) suggested that increased ribosome stability in the presence of RMF was not dependent on the dimeric structure. This was based on the observation that parent strain cultures were more resistant to heat stress than an RMF-deficient mutant strain despite the rapid dissociation of the 100 S dimers on heating. Indeed, the rate of dissociation of 100 S dimers seemed to be greater than that of 70 S monomers. Similar observations were made in relation to acid stress by El-Sharoud and Niven (2007). Further, ribosome dimers have not been observed in exponential-phase cells, even under circumstances in which RMF activity has been indicated, such as during acid stress (El-Sharoud and Niven, 2005) and cold stress (Niven, unpublished data). Stronger evidence for this hypothesis was produced by Ueta et al. (2005), who constructed a YhbH-deletion mutant that synthesised RMF and showed a similar viability to the parent strain during stationary phase and yet did not produce 100 S dimers. A key question, therefore, in elucidating the function and mechanism of action of RMF, is to reconcile the very strong evidence that RMF causes dimerisation with those observations that suggest that dimerisation may not be essential to its function.

Ultimately, this question cannot be fully resolved until it is possible to view ribosome dimers in vivo. To date, the only method available to observe dimerisation is to separate ribosome particles in cell-free extracts, which ,by their nature, do not reproduce the conditions inside the living cell. It has long been known that 70 S ribosomes from exponential-phase cells can spontaneously dimerise in the absence of RMF at high Mg^{2+} concentrations (Tissiéres et al. 1959). Because ribosome conformation is highly dependent on the chemical environment, it is difficult to be certain that the RMF-induced dimerisation observed in vitro reflects the ribosome state in the cell. However, it is clear that RMF binds to the ribosome and, thus, induces conformational changes that influence ribosome structure and function. Whether dimerisation is the purpose of these conformational changes or an artefact is a moot point.

Initial calculations suggested a single molecule of bound RMF per 100 S dimer (Wada et al. 1990; Wada et al. 1995), but a value of two molecules of RMF per dimer was later given by Wada (1998). This was further refined by Wada et al. (2000) to vary in accordance with different putative periods of stationary phase with ratios of RMF molecules per 70 S particle in dimerised ribosomes ranging from 0.8 to 0.9 in stage I, to 1.2 to 1.4 in stage III. Given the inherent difficulty in accurately quantifying both RMF protein and ribosome concentrations, it is reasonable to assume that there is generally a single molecule of RMF associated with each 70 S unit (Fig. 10.1). On dissociation of ribosome dimers by decreasing the Mg^{2+} concentration, Wada et al. (1990) recovered RMF associated with the 50 S fraction, indicating that the protein may be bound to that subunit.

Yoshida et al. (2002) characterised the RMF binding site in more detail by preparing an RMF-deficient mutant strain and complementing the mutation using a plasmid expressing His-tagged RMF. Ribosomes were subjected to protein–protein cross-linking, followed by specific isolation of conjugates carrying the His-tag label. This resulted in identification of ribosomal proteins S13, L2, and L13 as being associated with the site of RMF binding. It was suggested that this placed the binding site close to the peptidyl transferase active site. In addition, because L2 and L13 are 50 S proteins, and S13 is a 30 S subunit protein, RMF is presumably located at a site between the two ribosomal subunits. Further studies by Yoshida et al. (2004) aimed to characterise the interaction between RMF and 23 S rRNA by dimethyl sulphate (DMS) footprinting. Ribosome dimers were subjected to reaction with DMS, which modifies adenosine and cytidine moieties of RNA. Identification of areas protected from such modification in the presence of bound RMF indicated areas of the 23 S rRNA that interact with RMF. These data led to the conclusion that bound RMF covers the peptidyl transferase centre and the entrance of the peptide exit tunnel.

From the earliest studies of RMF, it has been known that ribosome dimers were inactive and that RMF, thus, inhibited protein synthesis. This has been a central factor in the hypothesis that the function of RMF is to induce dimer formation as a ribosome storage form. Inhibition of an in vitro protein synthesis system by added natural or synthetic RMF was reported by Wada et al. (1995). This was accompanied by concomitant inhibition of amino acyl-transfer RNA (tRNA) binding. These observations are now understandable in the context of RMF physically blocking critical catalytic sites in the ribosome structure, as described in the previous paragraph.

El-Sharoud and Niven (2005) compared the peptide chain elongation rates of an RMF-deficient strain and the parent strain during growth under a range of different pH conditions. This was performed by determining the time from the addition of an inducer of *lacZ* expression to the first appearance of β-galactosidase enzyme activity. Although the elongation rates were similar for the parent strain at each of the examined growth pH levels, those of the mutant strain were significantly lower when cells were grown at low pH. In view of the increased *rmf* expression at low pH, this may seem at odds with the body of evidence stating that RMF inhibits protein synthesis. However, the shorter enzyme induction time was indicative of a faster protein synthesis process rather than higher levels of protein synthetic activity. It was, therefore, speculated that RMF may increase the efficiency of protein synthesis under conditions of slow growth by inactivating excess ribosomes. A surfeit of active ribosomes may result in competition for available protein synthesis factors. Resulting ribosome stalling could, thus, cause hydrolysis of mRNA and the disruption of translation, with a resulting waste of energy and resources. The inactivation of excess ribosomes may, therefore, overcome this by maintaining the balance between the various protein synthesis components, thus enabling the cell to synthesise those proteins most needed during a period of physiological stress.

Electron micrographs of 100S dimers in vitro show pairs of 70S ribosomes attached via the 30S subunits (Wada 1998; Yoshida et al. 2002). Because RMF binds at a site that encompasses both 30S and 50S subunits, it is unlikely that RMF binding facilitates ribosome dimerisation simply by acting as a "bridge" at the 30S point of attachment. Rather, it would seem that RMF binding causes conformational changes in the ribosome that result in the formation of the complementary areas of the 30S subunits that are required for dimerisation. Interestingly, the DMS-footprinting studies of Yoshida et al. (2004) showed that the DMS modification of some residues of 23S rRNA were enhanced on RMF binding rather than protected. This is likely to be indicative of conformational changes in the rRNA molecule that result in increased exposure of some areas to modification.

10.6 The Regulation of RMF Synthesis and Activity

RMF is often referred to in the literature as a protein expressed in stationary phase, or as a stationary phase-specific factor. The above discussion demonstrates that expression of the *rmf* gene has also been observed in exponential phase under certain conditions. Initial studies by Wada et al. (1990) isolated RMF protein specifically from 100S dimers purified from stationary-phase cultures and demonstrated the rapid conversion of dimers back to 70S ribosomes on culturing in fresh medium. This, however, was the case when cultures were grown under laboratory conditions that were favourable for growth. Yamagishi et al. (1993) subsequently showed that expression of the *rmf* gene was inversely proportional to growth rate when reduced growth rates were obtained by nutrient limitation. In addition, Garay-Arroyo et al. (2000) detected increased synthesis of RMF mRNA during exponential phase in response to osmotic shock, and El-Sharoud and Niven (2005) showed that *rmf* expression was inversely proportional to growth rate under acidic conditions. It is, therefore, clear that RMF is associated with low growth rates rather than stationary phase per se.

This is further highlighted by the observation that RMF synthesis is not regulated by the stationary phase-specific sigma factor, σ^S (Yamagishi et al. 1993; Izutsu et al. 2001). Rather, it is a component of the so-called "stringent response"— a global regulatory system that causes wide-ranging changes in metabolism in response to nutrient starvation and environmental stress (Neidhardt et al. 1990; Hou et al. 1999; Hogg et al. 2004). It is mediated by the cellular levels of the nucleotides ppGpp and pppGpp, synthesised by phosphorylation of GDP and GTP, respectively. It is thought that the process is initiated by high levels of free (non-amino acylated) tRNA, which links the process intimately with the status of protein synthesis in the cell. Izutsu et al. (2001) investigated the influence of global regulators on *rmf* expression using stains of *E. coli* expressing a *lacZ* gene controlled by the *rmf* promoter. Activity of the *rmf* promoter was determined by measuring β-galactosidase activity in strains that were defective in the regulators sigma factor S, cAMP, FIS, H-NS, IHF, OmpR, and ppGpp. Expression of the gene fusion was reduced during stationary phase to 10% in the *relA/spoT* mutant strain (ppGpp). Further, *rmf* transcription correlated with cellular ppGpp levels during amino acid starvation. In the *relA/spoT*-defective mutant strain, reduction in growth rate did not cause an increase in *rmf* transcription, indicating that ppGpp directly influenced expression, rather than some other growth rate-related factors.

However, not all evidence published on the regulation of RMF is in agreement with its positive regulation by cellular ppGpp concentration. Fukuchi et al. (1995) produced a mutant strain of *E. coli* that was deficient in SAT. Elevated cellular concentrations of spermidine in stationary phase caused by growth in the presence of exogenous spermidine resulted in reduced viability, decreased synthesis of RMF, lower ribosome dimerisation, reduced protein synthesis, and lower levels of ribosomes. However, the increased intracellular levels of spermidine were also associated with increased concentrations of ppGpp in stationary-phase cultures (Apirakaramwong et al. 1999). The possible role of ppGpp in RMF regulation was investigated by introducing to the SAT-deficient mutant strain a plasmid carrying a gene encoding ppGpp biosynthesis (*relA*) under the control of an inducible promoter. When ppGpp synthesis was induced during stationary phase, RMF levels declined, suggesting that RMF was *negatively* regulated by the cellular ppGpp concentration. This would seem to confound the hypothesis of Izutsu et al. (2001) described above, that RMF synthesis is mediated within the "stringent response."

A possible resolution of this apparent conflict may be that the influence of exogenous spermidine on RMF and the protein synthesis system was the result of a decrease in the cellular Mg^{2+} concentration. Fukuchi et al. (1995) reported that both RMF and outer membrane protein C (OmpC, a component of the cation-specific porin) were reduced when the SAT-deficient strain was grown in the presence of spermidine. The reduced viability of an *rmf*-deletion mutant strain during stationary phase, along with the decreased levels of protein synthesis and concentrations of ribosomes and RNA, were exacerbated in an *OmpC-rmf* double mutant (Apriakaramwong et al. 1998). These effects were overcome or reduced by addition of Mg^{2+} to the culture. It is, therefore, possible that less RMF was observed in the SAT-deficient strain in the presence of spermidine, despite elevated levels of ppGpp, because Mg^{2+} deficiency caused by reduced OmpC resulted in ribosome destabilisation accompanied by rapid degradation of free RMF in the cytoplasm (see Yoshida et al. 2004, described in the next paragraph). It has long been understood that Mg^{2+} concentration has a strong influence on ribosome conformation and enhances stability (for example, see Gesteland 1966; Sabo and Spirin 1971; Noll and Noll 1976). Interestingly, a mutant strain deficient in OmpC alone showed similar characteristics to the parent strain with regard to stationary-phase viability and protein synthesis (Apirakaramwong et al. 1998). The available data, therefore, suggest that the action of RMF can overcome ribosome instability caused by Mg^{2+} deficiency in vivo. This may be an important factor in explaining the ability of

RMF to protect the cell under conditions that damage membrane integrity, such as heating and chilling.

It has recently been demonstrated that RMF synthesis is also regulated on a translational level via the stability of the mRNA (Aiso et al. 2005). An extremely long half-life of 24 min was estimated for *rmf* mRNA in early stationary phase, increasing to 120 min after further culture incubation. After inoculation into fresh medium, the half-life decreased to 5 min. The postinoculation degradation of the mRNA was suppressed by rifampin, indicating that de novo RNA synthesis was required to mediate the degradation process. Highly conserved terminal hairpin loops in the predicted mRNA secondary structure were suggested to be responsible for the stability of the molecule. It was also proposed that RNase E was involved in the degradation process on entry into exponential phase because the increased degradation was suppressed in an *rne-131* mutant strain. Further, the observations of Aiso et al. (2005) confirmed previous suggestions by some authors that free RMF in the cytoplasm is rapidly degraded (for example, Fukuchi et al. 1995; Wada et al. 2000). Although RMF protein and mRNA were detectable in stationary-phase cells, within 1 min of inoculation into fresh medium, no RMF protein was detectable, despite approximately 90% of *rmf* mRNA remaining. Yoshida et al. (2004) predicted a "simple elongated structure" for RMF that would contribute to its sensitivity to proteolysis. This may offer an explanation why RMF protein has not been detected in exponential-phase cells, and a mechanism for the rapid reversal of its effects when stationary-phase cells are inoculated into fresh medium.

Being one element of a globally regulated system, RMF is not acting in isolation in influencing cell physiology in general or the structure and function of the ribosome specifically. During periods of profound change in cell activity, a wide range of coordinated and independent processes are occurring. Maki et al. (2000) identified two ribosome-associated proteins in stationary-phase cultures using a similar technique to that which resulted in the discovery of RMF. Whereas YfiA was detected in both 70S ribosomes and 100S dimers, YhbH was detected exclusively in 100S particles. It was subsequently shown that disruption of the *yfiA* gene resulted in increased levels of ribosome dimerisation in stationary phase, whereas no dimers were detected in *yhbH*-deleted mutants (Ueta et al. 2005). It was, therefore, postulated that these proteins have opposite functions in regulating ribosome dimerisation. Because they have high degrees of sequence homology (40%), it is possible that they bind to the same site on the 30S ribosome. Their relative amounts may, thus, have an influence in determining the degree of dimerisation.

10.7   The Wider Significance of RMF

Almost all work that has been reported on RMF has been conducted in *E. coli*, which raises the question of its wider significance in microbiology. Do other bacteria have equivalent systems for regulating protein synthesis and/or protecting ribosomes during periods of slow growth and, if not, how do they survive in its absence? Wada et al. (2000) observed RMF in 19 different *E. coli* strains, whereas the existence of RMF analogues has been indicated in other bacterial genera, including *Salmonella, Serratia, Proteus,* and *Pseudomonas* (Wada 1998; Yoshida et al. 2004). A search of the National Center for Biotechnology Information (NCBI) nucleotide database was conducted for this review using the BLAST search engine. This revealed genes with substantially similar sequences to the *rmf* protein-coding region reported by Yamagishi et al. (1993) in various strains and species of *E. coli, Shigella, Salmonella,* and *Erwinia*. Homology with partial areas of genes from species of *Yersinia, Vibrio,* and *Sodalis* were also noted. In view of the wide range of entire bacterial genomes that have now been sequenced and were included in the search, this would seem to indicate a relatively narrow distribution of *rmf* analogues. A similar search for analogous proteins using the primary amino acid sequence of RMF resulted in a greater number of hits. The full result of this search is shown in Fig. 10.3. It can be seen that the Enterobacteriaceae are well represented and various other families are included. However, many groups of bacteria that have been widely studied are absent, again suggesting a limited distribution of proteins analogous to RMF. It is interesting to note that the RMF protein sequences are highly conserved within the limited number, but wide range, of bacteria revealed by the search.

There are a variety of physiological mechanisms among bacteria that can contribute to survival during stationary phase or under harsh environmental conditions. One clear example is sporulation; the formation of so-called "viable nonculturable" states is perhaps more controversial (Kell et al. 1998; Mukamolova et al. 2003). In *E. coli* and other bacteria, changes to membrane structure occur in stationary phase that also contribute to a more robust physiology (Marr and Ingraham 1962; Chang and Cronan 1999). However, in view of the relatively limited distribution of RMF, it would be interesting to ascertain whether those bacteria that lack RMF have alternative mechanisms for ribosome protection that are absent from *E. coli*, or if they are generally more vulnerable under nongrowing or stressful conditions.

Because RMF is important for cell viability, a more practical consideration from a human perspective is whether or not it offers a potential target for antibiotic action. Disruption of RMF activity, although not directly lethal, would clearly make cells more vulnerable to the stressful conditions encountered during the infection process. However, any antibiotic with this mechanism of action would have limited application because of the narrow range of organisms that could be targeted in this way. Alternatively, the structure of RMF may offer some scope for the design of peptide-based antibiotics that could bind to the ribosome and inhibit protein synthesis. Wada et al. (1995) reported that they failed to generate RMF-overproducing mutant strains of *E. coli* and speculated that such a genetic manipulation may be lethal. This suggests that a stable RMF analogue that could be taken up into the bacterial cytoplasm may be bacteriostatic or bacteriocidal.

Some studies of RMF-deficient mutant strains have indicated that such strains may have some relevance to the biotechnological production of metabolites. Chao et al. (1996) examined the influence of *rmf* disruption on gene expression from a pH-inducible expression system and a *lac* promoter system. They reported a two-fold increase in recombinant protein production for the former, but not the latter.

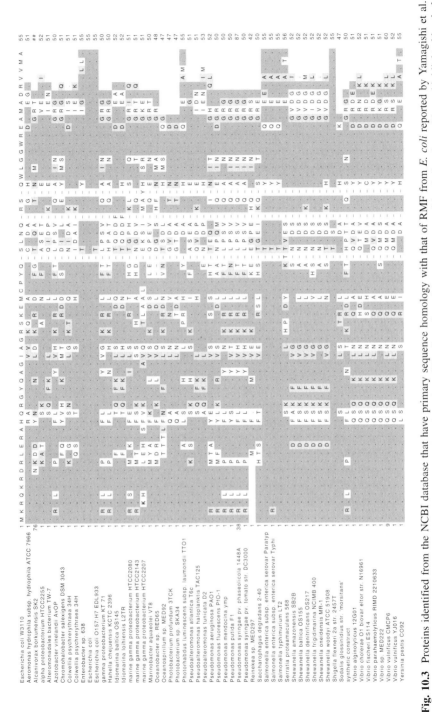

Fig. 10.3 Proteins identified from the NCBI database that have primary sequence homology with that of RMF from *E. coli* reported by Yamagishi et al. (1993). Amino acids are identified by their single letter codes where they differ from that of the test sequence. Identical amino acids are highlighted by *dark shading*, with amino acids that differ from the test sequence but are of a similar chemical type highlighted in *lighter shading*

Imaizumi et al. (2005) similarly demonstrated that *rmf* disruption resulted in a higher rate of lysine production, and higher specific production, in a lysine-over-producing strain of *E. coli*. These studies indicate that RMF-deficient mutant strains may be of particular interest in fed-batch fermentations or encapsulated cell systems in which growth rates are low. Imaizumi et al. (2006) further provided evidence that RMF deficiency had a wider influence on gene expression in *E. coli* than might be supposed. They analysed gene expression in an *rmf*-disrupted strain under excess and limited phosphate availability. The data suggested that RMF deficiency resulted in a pattern of gene expression that was similar to that caused by phosphate starvation in the parent strain. Of the 30 most downregulated genes in the mutant strain, 19 encoded ribosomal proteins. *RpoA*, which encodes the α-subunit of the RNA polymerase core enzyme, was also downregulated. The upregulated genes included Pho regulon genes that were previously thought to be induced only in response to phosphate starvation. Production of a recombinant acid phosphatase that also catalyses nucleoside phosphorylation was shown to be enhanced by *rmf* disruption, which may have industrial significance for the production of flavour enhancers. Chao et al. (2001) demonstrated that the characteristics of the *rmf* promoter may also have biotechnological applications. Expression of β-galactosidase from the *rmf* promoter was shown to result in growth phase- and growth rate-dependent synthesis of the enzyme. High levels of β-galactosidase synthesis were achieved by a combination of nutrient limitation and temperature down-shift.

Because RMF influences cell survival under extreme environmental conditions, it is reasonable to suppose that it may also influence the action of bacteria as pathogens. It would be interesting to know whether RMF directly influences pathogenicity and acts as a virulence factor, but no data on this are currently available as far as the authors are aware. It is, however, clear that RMF could be relevant to human infections by facilitating the survival of pathogens in the environment. For example, it may increase the resistance of food-borne bacteria to preservation techniques such as pasteurisation and fermentation. We conducted a study to examine the influence of RMF on the survival of *E. coli* in lactic fermentations. When *E. coli* was co-inoculated with *Lactococcus lactis* subsp. *lactis* (a commonly used cheese starter bacterium) in reconstituted skim milk and incubated at 30°C, both the *E. coli* parent stain and the RMF deletion mutant strain HMY15 grew at slower rates compared with their corresponding single cultures. The presence of *E. coli* did not influence the growth of *L. lactis* or the fermentative acid production, with the culture pH falling to a value of 4.3 in each case (El-Sharoud and Niven, unpublished data). After incubation for 24 h under these conditions, no reduction in viability of either *E. coli* strain was observed. However, on further incubation of the stationary phase culture at 4°C (to represent cold-storage of a fermented product), the viable count of the *E. coli* parent strain decreased from approximately 10^6 to 10^5 CFU/ml over 7 days, whereas that of the mutant strain decreased to a viable count of 10 CFU/ml. A similar experiment was performed to study the effect of RMF on the survival of *E. coli* during the preparation and cold storage of yoghurt. Milk was co-inoculated with *E. coli* W3110 or HMY15, and a dairy mixed starter consisting of *Streptococcus thermophilus* and *Lactobacillus delbrueckii* subsp. *bulgaricus*. In this case, both *E. coli* strains were capable of growth until the milk coagulated, and viable counts increased by 1.5 log units. Thereafter, the viability of the RMF-deficient mutant strain decreased abruptly by 1 log unit. On further incubation at 4°C, the viability of the mutant strain culture decreased rapidly compared with no substantial change in the viability of the parent strain (El-Sharoud and Niven, unpublished data). These observations demonstrated that RMF can be relevant to the survival of *E. coli* on contamination of fermented foods. They also indicate that the protection conferred by the presence of RMF relates particularly to survival under cold-storage conditions.

10.8 Conclusions

In view of the influence that inactivation of the *rmf* gene has on the survival of *E. coli*, it seems to serve an important function in the cell. However, the exact nature of this function has yet to be clarified. Indeed, the number of questions that remain to be answered are perhaps one of the features that make RMF an intriguing subject for study. Is the synthesis of RMF a response to inhibition of protein synthesis or a cause of it? Is it a regulator of protein synthesis or a protective factor of ribosomes? How does RMF protect ribosomes; can reducing rRNA degradation aid the regeneration of intact ribosomes? What is the significance of ribosome dimerisation; is it an inherent part of RMF function or an artefact of ribosome analysis in vitro? Why is RMF confined to a relative narrow group of bacteria?

One of the important aspects of modern bacterial physiology is the realisation that cell activities and components cannot be viewed purely in isolation, but that a full understanding of structure and function can only be appreciated from a view that places individual components in the context of a coordinated and complex whole. With RMF, that perspective remains elusive, but it is perhaps only just out of reach.

Highly Recommended Readings

El-Sharoud WM (2004) Ribosome inactivation for preservation: concepts and reservations. Science Progress 87: 137–152

El-Sharoud WM, Niven GW (2005) The activity of ribosome modulation factor during growth of *Escherichia coli* under acidic conditions. Arch Microbiol 184: 18–24

El-Sharoud WM, Niven GW (2007) The influence of ribosome modulation factor on the survival of stationary-phase *Escherichia coli* during acid stress. Microbiol 153: 247–253

Izutsu K, Wada A, Wada C (2001) Expression of ribosome modulation factor (RMF) in *Escherichia coli* requires ppGpp. Genes to Cells 6: 665–676

Niven GW (2004) Ribosome modulation factor protects *Escherichia coli* during heat stress, but this may not be dependent on ribosome dimerisation. Arch Microbiol 182: 60–66

Wada A (1998) Growth phase coupled modulation of *Escherichia coli* ribosomes. Genes to Cells 3: 203–208

Wada A, Yamazaki Y, Fujita N, Ishihama A (1990) Structure and probably genetic location of a "ribosome modulation factor" associated with 100 S ribosomes in stationary-phase *Escherichia coli* cells. Proc Natl Acad Sci USA 87: 2657–2661

Wada A, Igarashi K, Yoshimura S, Aimoto S, Ishihama A (1995) Ribosome modulation factor: stationary growth phase-specific inhibitor of ribosome functions from *Escherichia coli*. Biochem Biophys Res Comm 214: 410–417

Wada A, Mikkola R, Kurland CG, Ishihama A (2000) Growth phase-coupled changes of the ribosome profile in natural isolates and laboratory strains of *Escherichia coli*. J Bacteriol 182: 2893–2899

Yamagishi M, Matsushima H, Wada A, Sakagami M, Fujita N, Ishihama A (1993) Regulation of the *Escherichia coli rmf* gene encoding the ribosome modulation factor: growth phase- and growth-rate dependent control. EMBO J 12: 625–630

References

Aiso T, Yoshida H, Wada A, Ohki R (2005) Modulation of mRNA stability participates in stationary-phase-specific expression of ribosome modulation factor. J Bacteriol 187: 1951–1958

Agafonov DE, Kolb VA, Nazimove IV Spirin As (1999) A protein residing at the subunit interface of the bacteria ribosome. Proc Natl Acad Sci USA 96: 12345–12349

Agafonov DE, Kolb VA, Spirin AS (2001) Ribosome-associated protein that inhibits translation at the aminoacyl-tRNA binding stage. Eur Mol Biol Org Rep 2: 399–402

Apirakaramwong A, Fukuchi J, Kashiwagi K, Kakinuma Y, Ito E, Ishihama A, Igarashi K (1998) Enhancement of cell death due to decrease in Mg^{2+} uptake by OmpC (cation-selective porin) deficiency in ribosome modulation factor-deficient mutant. Biochem Biophys Res Comm 251: 482–487

Apirakaramwong A, Kashiwagi K, Raj VS, Sakata K, Kakinuma Y, Ishihama A, Igarashi K (1999) Involvement of ppGpp, ribosome modulation factor, and stationary phase-specific sigma factor σ^S in the decrease in cell viability caused by spermidine. Biochem Biophys Res Comm 264: 643–647

Bayles DO, Tunick MH, Foglia TA, Miller AJ (2000) Cold shock and its effect on ribosomes and thermal tolerance in *Listeria monocytogenes*. Appl Env Microbiol 66: 4351–4355

Chang Y-Y, Cronan JE (1999) Membrane cyclopropane fatty acid content is a major factor in acid resistance of *Escherichia coli*. Mol Microbiol 33: 249–259

Chao YP, Bennett GN, San KY (1996) Genetic manipulation of stationary-phase genes to enhance recombinant protein production in *Escherichia coli*. Biotech Bioeng 50: 636–642

Chao YP, Wen CS, Chiang CJ, Wang JJ (2001) Construction of the expression vector based on the growth phase- and growth rate-dependent *rmf* promoter: use of cell growth rate to control the expression of cloned genes in *Escherichia coli*. Biotech Lett 23: 5–11

Davis BD, Luger SM, Tai PC (1986) Role of ribosome degradation in the death of starved *Escherichia coli* cells. J Bacteriol 166: 439–445

El-Sharoud WM (2004a) Effect of ribosome modulation factor on the behaviour of *Escherichia coli* under acid stress and milk fermentation conditions. PhD thesis, The University of Reading, UK

El-Sharoud WM (2004b) Ribosome inactivation for preservation: concepts and reservations. Science Progress 87: 137–152

El-Sharoud WM, Niven GW (2005) The activity of ribosome modulation factor during growth of *Escherichia coli* under acidic conditions. Arch Microbiol 184: 18–24

El-Sharoud WM, Niven GW (2007) The influence of ribosome modulation factor on the survival of stationary-phase *Escherichia coli* during acid stress. Microbiol 153: 247–253

Fukuchi J, Kashiwagi K, Yamagishi M, Ishihama A, Igarashi K (1995) Decrease in cell viability due to accumulation of spermidine in spermidine acetyltransferase-deficient mutant of *Escherichia coli*. J Biol Chem 270: 18831–18835

Garay-Arroyo A, Colmenero-Flores JM, Garciarrubio A, Covarrubias AA (2000) Highly hydrophilic proteins in prokaryotes and eukaryotes are common during conditions of water deficit. J Biol Chem 275: 5668–5674

Gesteland RF (1966) Unfolding of *Escherichia coli* ribosomes by removal of magnesium. J Mol Biol 18: 356–371

Hogg T, Mechold U, Malke H, Cashel M, Hilgenfeld R (2004) Conformational antagonism between opposing active sites in a bifunctional RelA/SpoT homolog modulates (p)ppGpp metabolism during the stringent response. Cell 117: 57–68

Hou Z, Cashel M, Froman HJ, Honzatko RB (1999) Effectors of the stringent response target the active site of *Escherichia coli* adenylosuccinate synthetase. J Biol Chem 274: 17505–17510

Imaizumi A, Takikawa R, Koseki C, Usuda Y, Yasueda H, Kojima H, Matsui K, Sugimoto S (2005) Improved production of L-lysine by disruption of stationary phase-specific *rmf* gene in *Escherichia coli*. J Biotechnol 117: 111–118

Izutsu K, Wada A, Wada C (2001) Expression of ribosome modulation factor (RMF) in *Escherichia coli* requires ppGpp. Genes Cells 6: 665–676

Kell DB, Kaprelyants AS, Weichart DH, Harwood CR, Barer MR (1998) Viability and activity in readily culturable bacteria: a review and discussion of the practical issues. Antonie Van Leeuwenhoek 73: 169–187

Maki Y, Yoshida H, Wada A (2000) Two proteins, YfiA and YhbH, associated with resting ribosomes in stationary phase *Escherichia coli*. Genes Cells 5: 965–974

Mackey BM, Miles CA, Parsons SE, Seymore DA (1991) Thermal denaturation of whole cells and cell components of *Escherichia coli* examined by differential scanning calorimetry. J Gen Microbiol 137: 2361–2374

Marr AG, Ingraham JL (1962) Effect of temperature on the composition of fatty acids in *Escherichia coli*. J Bacteriol 84: 1260–1267

McCarthy BJ (1960) Variations in bacterial ribosomes. Biochim Biophys Acta 39: 563–564

McQuillen K, Roberts RB, Britten RJ (1959) Synthesis of nascent protein by ribosomes in *Escherichia coli* Proc Natl Acad Sci USA 45: 1437–1447

Mukamolova GV, Kaprelyants AS, Kell DB, Young M (2003) Adoption of the transiently non-culturable state—a bacterial survival strategy? Adv Microb Physiol 47: 65–129

Neidhardt FC, Ingraham JL, Schaechter M (1990) Physiology of the Bacterial Cell. Sunderland, MA: Sinauer Associates Inc

Niven GW (2004) Ribosome modulation factor protects *Escherichia coli* during heat stress, but this may not be dependent on ribosome dimerisation. Arch Microbiol 182: 60–66

Niven GW, Miles CA, Mackey BM (1999) The effects of hydrostatic pressure on ribosome conformation in *Escherichia coli*: an in vivo study using differential scanning calorimetry. Microbiol 145: 419–425

Noll M, Noll H (1976) Structural dynamics of bacterial ribosomes. V. Magnesium-dependent dissociation of tight couples into subunits: measurements of dissociation constants and exchange rates. J Mol Biol 105: 111–130

Raj VS, Füll C, Yoshida M, Sakata K, Kashiwagi K, Ishihama A, Igarashi K (2002) Decrease in cell viability in an RMF, σ^{38}, and OmpC triple mutant of *Escherichia coli*. Biochem Biophys Res Comm 299: 252–257

Rowbury RJ (1997) Regulatory components, including integration host factor, CysB and H-NS, that influence pH response in *Escherichia coli*. Lett Appl Microbiol 24:319–328

Rowbury RJ, Goodson M, Whiting GC (1989) Habituation of *Escherichia coli* to acid and alkaline pH and its relevance for bacterial survival in chemically polluted waters. Chem Ind 109: 685–686

Sabo B, Spirin AS (1971) Dissociation of 70S monoribosomes of *Escherichia coli* in relation to ionic strength, pH, and temperature. Mol Biol 4: 509–511

Teixeira P, Castro H, Mohacsi-Farkas C, Kirby R (1997) Identification of sites of injury in *Lactobacillus bulgaricus* during heat stress. J Appl Microbiol 83: 219–226

Tissières A, Watson JD (1958) Ribonucleoprotein particles from *Escherichia coli*. Nature 182: 778–780

Tissières A, Watson JD, Schessinger D, Hollingworth (1959) Ribonucleoprotein particles for *Escherichia coli*. J Mol Biol 1: 221–233

Tolker-Nielsen T, Molin S (1996) Role of ribosome degradation in the death of heat-stressed *Salmonella typhimurium*. FEMS Microbiol Lett 142: 155–160

Ueta M, Yoshida H, Wada C, Baba T, Mori H, Wada A (2005) Ribosome binding proteins YhbH and YfiA have opposite functions during 100S formation in the stationary phse of *Escherichia coli*. Genes Cells 10: 1103–1112

VanBogelen RA, Neidhardt FC (1990) Ribosomes as sensor of heat and cold shock in *Escherichia coli*. Proc Natl Acad Sci USA 87: 5589–5593

Wada A (1998) Growth phase coupled modulation of *Escherichia coli* ribosomes. Genes Cells 3: 203–208

Wada A, Yamazaki Y, Fujita N, Ishihama A (1990) Structure and probably genetic location of a "ribosome modulation factor" associated with 100S ribosomes in stationary-phase *Escherichia coli* cells. Proc Natl Acad Sci USA 87: 2657–2661

Wada A, Igarashi K, Yoshimura S, Aimoto S, Ishihama A (1995) Ribosome modulation factor: stationary growth phase-specific inhibitor of ribosome functions from *Escherichia coli*. Biochem Biophys Res Comm 214: 410–417

Wada A, Mikkola R, Kurland CG, Ishihama A (2000) Growth phase-coupled changes of the ribosome profile in natural isolates and laboratory strains of *Escherichia coli*. J Bacteriol 182: 2893–2899

Yamagishi M, Matsushima H, Wada A, Sakagami M, Fujita N, Ishihama A (1993) Regulation of the *Escherichia coli rmf* gene encoding the ribosome modulation factor: growth phase- and growth-rate dependent control. EMBO J 12: 625–630

Yoshida H, Maki Y, Kato H, Fujisawa H, Izutsu K, Wada C, Wada A (2002) The ribosome modulation factor (RMF) binding site on the 100S ribosome of *Escherichia coli*. J Biochem 132: 983–989

Yoshida H, Yamamoto H, Uchiumi T, Wada A (2004) RMF inactivates ribosomes by covering the peptidyl transferase centre and entrance of the peptide exit tunnel. Genes Cells 9: 271–278

11
The Role of RpoS in Bacteria

Tao Dong, Charlie Joyce, and Herb E. Schellh

Abstract   Bacterial adaptation to changing conditions and to the host environment requires coordinated changes in gene expression that permit more efficient use of metabolites and increased survival. An important form of gene control is through the use of alternative sigma factors that direct RNA polymerase to recognize a distinct group of genes. One such sigma factor is RpoS, which is widely present in many Gram-negative bacteria. RpoS is important for adaptation under nutrient-limited conditions, but, interestingly, *rpoS* mutants may have a selective growth advantage under such conditions. In this chapter, we review the factors that control RpoS induction,

Herb E. Schellhorn
Department of Biology, McMaster University, Canada
Schell@mcmaster.ca

W. El-Sharoud (ed.) *Bacterial Physiology: A Molecular Approach.*
© Springer-Verlag Berlin Heidelberg 2008

ttenuation, including *hns, dsrA*, and other regulators. The nature
gulon, including the factors it controls and their potential metabolic
ptation, is discussed in terms of the physiological challenges faced by
during stress. We review the recent contribution of global gene expression
performed with microarrays and genetic screening protocols to exhaustively
acterize this important regulon. Finally, we consider the specific roles of RpoS in
athogenicity in the mammalian host and in natural bacterial isolates.

11.1 Introduction

Bacteria are the most abundant life form on Earth and can be found in nearly every environmental niche in the world, from hot springs to cold deep-sea beds. This wide distribution is largely because of an important feature of bacteria, their extreme adaptability, which allows bacteria to survive adverse environmental conditions in the natural habitat. Two distinct strategies are used: cellular differentiation into specialized structures, e.g., sporulation in Gram-positive bacteria (reviewed in Piggot and Hilbert 2004), and specific induction of stress-responsive genes that improves the metabolic fitness of the vegetative cell, as found in most Gram-negative bacteria. Spores are extremely stable and are able to endure extreme stresses such as UV light and heat exposure. When they meet favorable growth conditions, spores become active and start proliferation. In nonsporulating bacteria, however, stress response and adaptation depend on the stimulated expression of stress response-governing regulators and their regulated genes.

Stress response regulators can be divided into two types: (1) specific regulators that are induced only under particular stress conditions and only control genes required for dealing with this specific stress (e.g., OxyR in oxidative stress [Farr and Kogoma 1991]), and (2) general stress regulators that are induced in response to multiple environmental signals and activate the expression of large regulons that include genes required not only for the proximal stress condition but also genes for other potential stress conditions. Therefore, general stress regulators likely provide a preventative mechanism that prepares the cell for concurrently experienced stresses compared with specific regulators that mainly activate genes only for protection against the current stress and repair of damage. In this chapter, we review one of the best-studied general stress regulators, RpoS, an alternative sigma factor, and its role in cell adaptation in the primary model organism of Gram-negative bacteria, *Escherichia coli*.

In *E. coli*, RNA polymerase holoenzyme is composed of five subunits ($\alpha\alpha\beta\beta'\sigma$) (for reviews see Ishihama 2000; Gross et al. 1998; Helmann and Chamberlin 1988). The $\alpha\alpha\beta\beta'$ subunits are assembled as a core enzyme that, though self-sufficient for transcription elongation, requires sigma factors to specifically bind to promoter regions and initiate the transcription process. There are seven known sigma factors—RpoD, RpoN, RpoS, RpoH, RpoF, RpoE, and FecI—with each regulating the transcription of particular genes in response to corresponding environmental conditions, except for RpoD and RpoS (reviewed in Ishihama 2000). RpoD is the

Table 11.1 Sigma factors and their corresponding functions

Sigma subunit	Molecular weight	Primary function
RpoD	70 kDa	Growth-related genes
RpoS	38 kDa	Stationary phase and stress response genes
RpoN	54 kDa	Nitrogen regulation and related stress response genes
RpoH	32 kDa	Heat shock and related stress response genes
RpoF	28 kDa	Flagella-chemotaxis genes
RpoE	24 kDa	Extreme heat shock and extracytoplasmic genes
FecI	19 kDa	Ferric citrate transport and extracytoplasmic genes

This table lists the seven known sigma factors in *E. coli*. For a review on the function of sigma factors, see Ishihama (2000).

vegetative sigma factor responsible for the transcription of most genes in fast-growing cells, whereas the alternative RpoS is highly induced during the transition from exponential phase to stationary phase or under general stress conditions. RpoS functions as a general stress response regulator, whereas other alternative sigma factors are specific for certain stress conditions (Table 11.1).

RpoS was originally identified in several independent contexts as a regulator of the expression of phosphatase (Touati et al. 1986) and of catalase (Loewen and Triggs 1984) and in protection from near-UV light (Sammartano et al. 1986). It soon became clear that RpoS likely controlled the expression of many other genes as well, and these genes, as a group, were expressed primarily in postexponential phase. Historically, gene regulation studies have been performed using exponential-phase cultures under "balanced" growth conditions, and this probably led to a delayed appreciation of the importance of genes expressed when the cell is in a slow growth state.

11.2 Distribution of RpoS

The RpoS-controlled stress response mechanism is a conserved function among the gamma- and beta- proteobacteria. Homologous *rpoS* sequences from 29 genera can be identified by searching *rpoS* against all the annotated genomes in the TIGR database (Fig. 11.1). The phylogeny of RpoS is similar to that of the 16S ribosomal RNA (rRNA) tree, indicating the conserved nature of RpoS in these species (unpublished data). Although it is still not clear how RpoS arose during evolution, it is likely that RpoS evolved by duplication from RpoD, with which RpoS shares 59% identity in gene sequence. Furthermore, many RpoS-regulated genes can be transcribed by RpoD-associated holoenzyme in vitro, suggesting a strong functional similarity. In contrast, many RpoS-regulated genes, including *aldB*, *katE*, *gabP*, and *osmY*, have likely been laterally transferred between species (unpublished data). This is consistent with the idea that horizontal transfer is more likely to have occurred for operational genes, such as those involved in

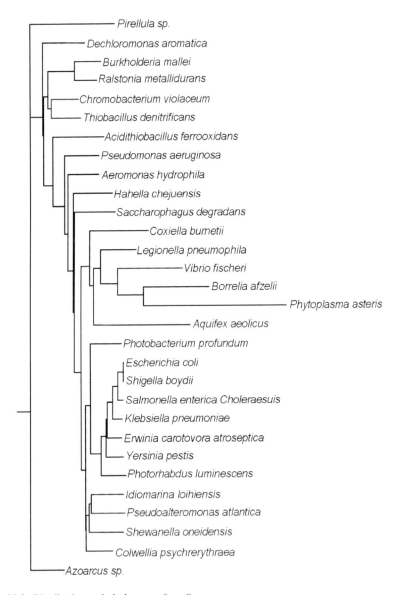

Fig. 11.1 Distribution and phylogeny of *rpoS*

amino acid synthesis and other housekeeping functions, rather than informational genes that are involved in transcription, translation, and regulation processes (Simonson et al. 2005). How genes are recruited into the RpoS regulon remains unclear. One possible scenario is that if an external gene with a promoter region that is poorly recognized by existing sigma factors is introduced into a bacterium's genome, and if the expression of this new gene can confer a growth advantage under selective pressure, either promoter mutations or sigma factor mutations may be selected that enhance the expression of this gene. Although promoter mutations are more frequently observed under laboratory conditions than structural changes of a sigma factor, and, thus, are more likely to occur, one major disadvantage of such a mutation is that constant expression of this newly recruited gene may impose an excessive biosynthetic energy cost. Therefore, sigma factor mutations resulting in specialization are preferred so that alternative sigma factors, such as RpoS, are evolved. However, this hypothesis requires further experimental evidence and testing.

11.3   Regulation of RpoS Expression and Function

RpoS is induced during the transition from exponential phase to stationary phase or in response to various stress conditions, followed by the activation of RpoS-regulated general stress response machinery, resulting in a series of physiological and morphological changes (Hengge-Aronis 2002a; Lange and Hengge-Aronis 1991a). Therefore, the expression of RpoS must be strictly regulated because inappropriate expression of the large regulon would likely have deleterious consequences for the cell. Indeed, many regulatory factors have been identified to regulate RpoS expression at the transcriptional, translational, and posttranslational levels (Fig 11.2) (reviewed in Hengge-Aronis 2002a; Ishihama 2000).

11.3.1   Transcriptional Regulation

The *rpoS* gene shares two promoters with its upstream gene *nlpD*, generating polycistronic *nlpD-rpoS* messenger RNAs (mRNAs) that are independent of environmental stimuli and may provide a low, but constant, level of *rpoS* transcript throughout growth (Lange and Hengge-Aronis 1994b). However, the major *rpoS* promoter lies inside the *nlpD* gene, and this promoter is primarily responsible for *rpoS* induction. Growth rate reduction (Ihssen and Egli 2004), guanosine 3′,5′-bispyrophosphate (ppGpp) (Gentry et al. 1993), polyphosphate (Shiba et al. 1997), and acetate (Schellhorn and Stones 1992) positively regulate *rpoS* transcription, whereas Fis (Hirsch and Elliott 2005) is a negative regulator that can bind to the *rpoS* promoter region to block transcription. The cAMP–CRP molecule regulates *rpoS* transcription in a growth-phase dependent manner—a negative regulator in exponential phase but a positive regulator in stationary phase (Lange and Hengge-Aronis 1994a). Although expression of *rpoS* is greatly reduced in a ppGpp-deficient strain (Gentry et al. 1993), how ppGpp enhances *rpoS* expression is still not clear. It was shown that ppGpp is important for *rpoS* transcriptional elongation (Lange

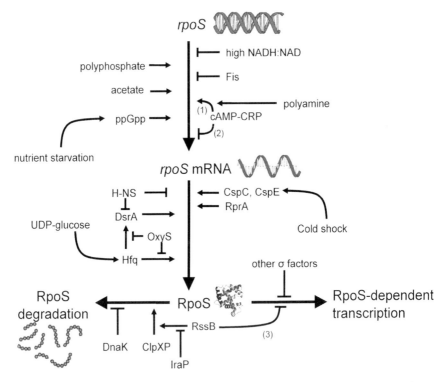

Fig. 11.2 Factors involved in regulation of RpoS at various levels. (1) Active during stationary phase; (2) active during exponential phase; (3) active under protease-limited conditions

et al. 1995), whereas another study showed that overproduction of ppGpp has little effect on the abundance of *rpoS* transcripts (Brown et al. 2002). The stimulation of *rpoS* expression by ppGpp is growth-phase independent (Hirsch and Elliott 2002). Fis, a global transcriptional factor (see Chap. 6), inhibits *rpoS* transcription by directly binding to the *rpoS* promoter region. Fis levels are growth-phase dependent. At the onset of stationary phase, Fis disappears and the transcription of *rpoS* is induced (Hirsch and Elliott 2005). Polyphosphate is an important inorganic molecule that is produced by many bacterial species and functions as a phosphate reservoir, a cation chelator, and a regulatory factor (for a review, see Kornberg et al. 1999). Polyphosphate-free mutants are stress sensitive and are impaired in survival in stationary phase (Rao and Kornberg 1996). Consistently, it has been found that polyphosphate stimulates *rpoS* transcription and this stimulation is likely to be independent of ppGpp (Shiba et al. 1997). High cellular NADH to NAD ratio also attenuates *rpoS* transcription, although the mechanism is not known (Sevcik et al. 2001).

11.3.2 Translational Regulation

Translational control of *rpoS* is also important for stationary phase RpoS induction (Lange and Hengge-Aronis 1994a). At onset of stationary phase, the *rpoS* mRNA level increases 10-fold, whereas the RpoS protein level increases 30-fold (Lange and Hengge-Aronis 1994a). Translation of *rpoS* is stimulated by cold shock (Sledjeski et al. 1996), polyamines (Yoshida et al. 2002), high osmolarity (Muffler et al. 1996), CspC and CspE (Phadtare and Inouye 2001), DsrA (Sledjeski et al. 1996; Majdalani et al. 1998), RprA (Majdalani et al. 2001; Majdalani et al. 2002), OxyS (Zhang et al. 1998), Hfq (Zhang et al. 1998), HU (Balandina et al. 2001), H-NS (Barth et al. 1995), and UDP-glucose (Bohringer et al. 1995).

The major *rpoS* transcript starts in the middle of the upstream gene *nlpD*, generating an untranslated 567-nt segment at the 5′ end of the mRNA product (Lange et al. 1995; Takayanagi et al. 1994). This leader region folds into a hairpin structure that stabilizes the *rpoS* transcript and blocks ribosome binding to prevent translation initiation (Cunning et al. 1998; Brown and Elliott 1997). Therefore, the translation of *rpoS* is blocked until positive regulatory factors such as Hfq, DsrA, RprA, and HU bind to the 5′-end hairpin to expose the translation initiation region (Cunning et al. 1998; Zhang et al. 1998; Majdalani et al. 1998; Majdalani et al. 2001). Hfq is an RNA binding protein that plays an important role in stabilizing small RNA regulators, as discussed below. Binding of Hfq to *rpoS* mRNA may also allow other regulators to be recruited to initiate the translation. Depletion of UDP-glucose stimulates RpoS expression in exponential phase in an Hfq-dependent manner (Bohringer et al. 1995). The cold-shock proteins CspC and CspE can also stimulate RpoS expression by stabilizing the *rpoS* mRNA (Phadtare and Inouye 2001). Overexpression of CspC or CspE increases the abundance of *rpoS* mRNA transcripts by approximately fourfold, whereas deletion of *cspC* and *cspE* leads to a fourfold decrease in *rpoS* transcript level (Phadtare et al. 2006). Similarly, HU, a nucleoid protein, stimulates *rpoS* translation, possibly by altering the secondary structure of *rpoS* mRNA (Balandina et al. 2001).

H-NS and OxyS are negative regulators for *rpoS* translation. H-NS is a universal repressor in gene regulation that forms nucleo–protein complexes with target genes (Barth et al. 1995; Yamashino et al. 1995) (see Chap. 6 for more details). Because H-NS can bind directly to *rpoS* mRNA, it may function as an RNA chaperone to alter the secondary structure of the *rpoS* transcript. H-NS may also inhibit the binding between DsrA and *rpoS* mRNA. By contrast, OxyS competes with DsrA and *rpoS* mRNA for binding to Hfq (Zhang et al. 1998).

Small regulatory RNAs can modulate gene expression by base pairing with target RNAs. More than 60 small RNAs have now been identified in *E. coli* (reviewed in Majdalani et al. 2005). Among them, DsrA and RprA are positive regulators of *rpoS,* whereas OxyS is an inhibitor of *rpoS* transcription. The 87-nt-long *dsrA* forms a secondary structure with three stem loops, part of which can complementarily bind to the antisense region at the 5′ end of *rpoS* mRNA to release the transcription initiation region for ribosome binding (Majdalani et al. 1998; Majdalani et al. 2002). The *dsrA* gene is expressed at temperatures below 30°C (Repoila and Gottesman 2001; Repoila et al. 2003) and is protected from nuclease degradation by direct interaction with Hfq (Moll et al. 2003; Brescia et al. 2003). It is possible that Hfq alters the structures of both *dsrA* and *rpoS* mRNA to facilitate the binding between DsrA and *rpoS*. There is another binding domain in DsrA for *hns* mRNA that inhibits *hns* mRNA translation (Lease et al. 1998). The competition for DsrA binding between *rpoS* mRNA and *hns* mRNA may partly explain the negative impact on *rpoS* translation by H-NS. Although the factor(s) inducing RprA are not known, this RNA likely acts like DsrA because of their structural similarities. OxyS is induced under oxidative stress conditions (Altuvia et al. 1997; Zhang et al. 1998) and downregulates *rpoS* translation by binding to Hfq and blocking Hfq interaction to both *rpoS* and DsrA (Zhang et al. 1998). When RpoS is induced during oxidative stress, negative regulation of *rpoS* by OxyS may provide a fine-tuning system to prevent overexpression of RpoS-regulated genes induced by oxidative stress, such as *katE*, a highly RpoS-dependent gene encoding hydrogen peroxidase II (Altuvia et al. 1997; Zhang et al. 1998).

11.3.3   Posttranslational Regulation

In addition to the plethora of identified transcriptional and translational modulators, regulated proteolysis of RpoS is a key (and perhaps the most important) mechanism for maintaining low levels of this sigma factor in optimal growth conditions and for rapidly increasing its levels during adaptation to starvation. In exponential phase, the RpoS protein is unstable, with a half-life of 1.4 min, whereas, during stationary phase, the half-life of RpoS significantly increases to 20 min (Lange and Hengge-Aronis 1994a; Lange and Hengge-Aronis 1994b). This increased stability alone can result in a significant increase in RpoS levels, underscoring the importance of posttranslational regulation in RpoS control.

In exponential phase cells, RpoS is specifically targeted for proteolysis by RssB (Muffler et. al., 1996; Schweder et. al., 1996) (also known as SprE [Pratt and Silhavy 1996] and MviA [Bearson et al. 1996]) which directs the protein to the ATP-dependent ClpXP protease (Zhou et al. 2001). RssB is a response regulator whose affinity to RpoS is modulated by phosphorylation at D58 of the conserved N-terminal receiver domain. This affinity, although likely important for targeted degradation, is probably not a sufficient condition, because increases in RpoS during starvation occur in strains carrying a mutation eliminating the D58 phosphorylation site (Peterson et al. 2004). Levels of RssB increase as cells enter stationary phase, whereas levels of the ClpXP are fairly constant. Other factors can also play a key role. In contrast to RssB, DnaK inhibits RpoS proteolysis because RpoS stability significantly decreases in *dnaK* mutants (Rockabrand et al. 1998).

A small, recently identified protein (Bougdour et al. 2006), IraP, inhibits the activity of RssB under low-phosphate conditions and can, thus, stabilize RpoS under some but not all conditions (because it is not effective during glucose starvation). The complex nature of proteolytic control of RpoS, thus, allows cells during starvation to more quickly increase levels of this regulator in response to specific starvation signals, independent of increases in transcription or stabilization of transcript (see section 11.3.2). The metabolic imperative for evolving this type of control is likely driven by the necessity of increasing the efficiency of de novo expression of the large RpoS regulon during starvation. While in the starvation state, the limited availability of precursors and energy sources may reduce the effectiveness of earlier controls at the transcription/translation level.

In view of the large number of factors that participate in the RssB–ClpXP pathway, what is the proximal signal that causes RpoS stabilization? Recently, the accumulation of oxidized, misfolded proteins has been shown to be an important physiological signal that is indirectly sensed by the RssB–ClpXP pathway (Fredriksson et al. 2007). Accordingly, ribosomal stalling, caused by depletion of the pool of charged transfer RNAs (tRNAs) during starvation, results in a large increase in misfolded protein, which, under aerobic conditions, readily oxidize and become substrates for ClpXP. The competition by alternative proteolysis substrates reduces the availability of ClpXP to participate in the RssB-targeted degradation of RpoS. Consistent with this model, experimental perturbation of degradation through inactivation of ClpXP, mutation of the conserved RssB aspartate phosphorylation site (Peterson et al. 2004), decreasing the fidelity of translation (Fredriksson et al. 2007), or increasing the availability of competing misfolded protein substrates for ClpXP (Fredriksson et al. 2007) all result in increased levels of RpoS through stabilization. An appealing aspect of this model is that a single metabolic state, the stalled ribosome, signals both an increase in expression of the RpoS regulon and, through the stringent response (Gourse et al. 1996), attenuation of the major RpoD-dependent operons, such as the ribosomal operons.

11.3.4   Sigma Factor Competition

Transcription of RpoS-regulated genes requires a relatively high number of RNA core polymerases associated with RpoS to initiate the process. However, even in stationary phase, in which RpoS is induced, the protein level of RpoS is only approximately 30% of that of RpoD, and the binding affinity to the RNA polymerase core enzyme of RpoD is 16-fold higher than that of RpoS (Maeda et al. 2000). Furthermore, the concentration of RpoD itself already exceeds the amount of freely available RNA polymerase core enzyme (not actively participating in transcription), as shown in both exponential phase (Grigorova et al. 2006; Ishihama 2000) and stationary phase cells (Ishihama 2000). Therefore, there is an excess of sigma factors competing for the limited amount of core enzyme. Overexpression of RpoS reduces the expression of RpoD-regulated promoters, whereas mutations in *rpoS* have the opposite effect (Farewell et al. 1998a).

Because of the poor affinity and low cellular concentration of RpoS, even in stationary phase, it is intriguing that RpoS-regulated genes are effectively expressed in this phase. A combination of other factors, including anti-sigma factors, promoter preference, and physiological changes, are involved in this phenomenon (Typas et al. 2007). Anti-sigma factors like Rsd (regulator of sigma D) and 6S RNA seem to play an important role in the competition between RpoD and RpoS. Rsd, induced at the onset of stationary phase, can directly bind to RpoD, causing inhibition of RpoD and leading to enhanced expression of RpoS-regulated genes (Jishage and Ishihama 1998). 6S RNA can also directly bind to RpoD, reducing its activity (Wassarman and Storz 2000; Wassarman and Saecker 2006). Physiological changes during the transition from exponential phase to stationary phase include the accumulation of small metabolites, such as glutamate, acetate, and trehalose, which may contribute, either directly or indirectly, to an enhanced RpoS competition for core polymerase. The alarmone ppGpp, induced in stationary phase, also plays an important role in the activation of not only RpoS but also RpoE, RpoH, and RpoN (Laurie et al. 2003; Bernardo et al. 2006; Costanzo and Ades 2006; Jishage et al. 2002). However, ppGpp stimulates both the expression and activity of RpoS (Kvint et al. 2000), while increasing the activity only of RpoE, RpoH, and RpoN (Sze and Shingler 1999; VanBogelen and Neidhardt 1990; Costanzo and Ades 2006; see reviews, Gourse et al. 2006; Nystrom 2004).

Transcription of RpoS-regulated genes may also require additional *trans*-acting regulators such as Crl, a low temperature-induced regulatory protein that binds directly to RpoS-holoenzyme to stimulate RpoS activity (Pratt and Silhavy 1998; Bougdour et al. 2004; Farewell et al. 1998a). There are 63 proteins that are co-regulated by RpoS and Crl (Lelong et al. 2007).

11.3.5 Summary of Regulation

All of these regulatory factors clearly support a central role for RpoS as a multisignal receiver in the stress response circuit. Cell adaptation to growth conditions is a complicated process involving signal receiver factors to sense the environmental changes, signal transducers and processors, and response effectors. As a general adaptation regulator, RpoS may be considered as both a multisignal receiver that can sense the environmental changes indirectly through these specific regulators and a processor that can transduce these signals into functional effectors to ensure survival or growth fitness. For example, low temperature (below 30°C) triggers *dsrA* induction, which, in turn, stimulates *rpoS* translation, resulting in the expression of many genes, including the curli biosynthesis genes, *csgD* and *csgAB,* which are controlled by RpoS (Romling et al. 1998). In this temperature-related regulation route, DsrA plays a thermosensor role to input the cold-shock signal into the RpoS regulatory network. RpoS then both receives the cold-shock signal and initiates downstream changes for cell adaptation or protection against this harmful input, thereby protecting the cell from the cold—enhancing the cell's prospect of survival.

11.4 RpoS, a Master Regulator in Stress Response and Adaptation

Global expression studies, including microarray analysis and mutational screens, have greatly improved our understanding of the RpoS regulon in *E. coli* (Lacour and Landini 2004; Patten et al. 2004; Vijayakumar et al. 2004; Weber et al. 2005). More than 10% of the *E. coli* genome is controlled by RpoS, most of which is involved in stress response, such as nutrient limitation (Notley and Ferenci 1996), resistance to DNA damage (Khil and Camerini-Otero 2002), osmotic shock (Cheung et al. 2003), high hydrostatic pressure (Robey et al. 2001), oxidative stress (Schellhorn and Hassan 1988), ethanol resistance (Farewell et al. 1998b), adaptive mutagenesis (Lombardo et al. 2004), acid stress (Lin et al. 1996), and biofilm formation (Schembri et al. 2003), underlining the importance of RpoS in cell adaptation.

11.4.1 Acid Response

RpoS mutants are very sensitive to low pH challenge (Small et al. 1994), implicating RpoS in acid resistance (AR). *E. coli* has three known AR systems, including RpoS-dependent oxidative, arginine-, and glutamate-dependent decarboxylase AR systems (Castanie-Cornet et al. 1999; Price et al. 2000; Audia et al. 2001). RpoS is essential for the oxidative system, but only partially required for the other two AR systems (Lin et al. 1996). The arginine AR system, including the arginine decarboxylase (*adiA*) and the regulator CysB, requires the addition of arginine during the acid challenge. The *adiA* gene is positively controlled by RpoS (Vijayakumar et al. 2004). The glutamate AR system includes two glutamate decarboxylase genes, *gadA* and *gadB*, and is RpoS-dependent in stationary phase induction but RpoS-independent in acid induction (Castanie-Cornet et al. 1999).

11.4.2 Oxidative Stress

E. coli may face oxidative stress caused by cytotoxic partially reactive oxygen species (ROS) including the superoxide anion radical (O_2^-), hydrogen peroxide (H_2O_2), and the hydroxyl radical (HO^-). ROS may be generated as endogenous respiratory products or externally secreted by other competing bacteria or host immune systems. Two catalases, HPI and HPII, encoded by *katG* and *katE*, respectively, exist in *E. coli*. Both *katE* and *katG* are regulated by RpoS in stationary phase and under some stress conditions (reviewed in Schellhorn 1995). KatG is also regulated by OxyR, an important factor in the oxidative stress response (Ivanova et al. 1994; Christman et al. 1989). KatE is the major catalase in stationary phase and is highly RpoS dependent (Schellhorn and Hassan 1988). Another antioxidative cell defense system involves glutaredoxin or thioredoxin that can reduce cytosolic disulfide in *E. coli* (Holmgren 2000; Ritz and Beckwith 2001). There are three glutaredoxins, encoded by *grxA*, *grxB*, and *grxC*; two thioredoxins, encoded by *trxA* and *trxC*; and an NrdH redoxin that shows a glutaredoxin-like sequence but a thioredoxin-like activity (Holmgren 1989; Jordan et al. 1997). It has been shown that the *grxB* gene is positively regulated by RpoS and ppGpp in stationary phase, whereas *trxA* is ppGpp dependent but RpoS independent (Potamitou et al. 2002). Expression of *nrdH* is high in early exponential phase and drops significantly during the transition from mid-exponential phase to stationary phase in rich media (Monje-Casas et al. 2001) and is RpoS independent throughout growth (Monje-Casas et al. 2001).

11.4.3   Anaerobic Growth

In continuous culture, anaerobic growth of *E. coli* leads to twofold lower levels of RpoS than under aerobic conditions (King and Ferenci 2005). Consequently, the expression of RpoS-regulated genes decreases, and cells become more stress sensitive (King and Ferenci 2005). However, wild-type strains can out-compete *rpoS* mutants under anaerobic conditions, indicating that RpoS expression, although low, confers a growth advantage under anaerobiosis (King and Ferenci 2005). AR of anaerobically grown wild-type and *rpoS* mutant strains is similar for cells adapted to acidic conditions (Small et al. 1994). However, wild-type strains adapted to alkaline pH are more acid resistant than *rpoS* mutants (Small et al. 1994).

11.4.4   Biofilm Formation

Many bacteria in the natural environment are found as free-living individual cells (planktonic) or in large sessile communities (biofilms). Biofilm growth requires a significant change in gene expression, resulting in the morphological adaptation from planktonic individuals to sessile communities. The highly organized structure of a bacterial biofilm can confer cell resistance against antimicrobial agents and host immune defenses (Nickel et al. 1985; Stewart and Costerton 2001). RpoS is important in biofilm formation, because mutations in RpoS cause lower biofilm cell density and an alteration in biofilm structure when cells are grown to stationary phase in minimal media (Adams and McLean 1999). It has been reported that RpoS mutants failed to establish a biofilm within 42 hours, and at least 30 RpoS-regulated genes are involved in biofilm formation (Schembri et al. 2003). However, several other studies show that RpoS negatively regulates biofilm formation in exponential growth in rich complex media (Corona-Izquierdo and Membrillo-Hernandez 2002; Domka et al. 2006). Therefore, it is likely that RpoS regulation of biofilm formation is growth-phase dependent and involves other factors. Extracellular structures, such as cellulose and curli fimbriae that are important for biofilm formation (Romling 2005), are positively regulated by RpoS, possibly through regulating the production of a small signaling molecule c-di-GMP (Weber et al. 2006). However, this RpoS-dependent regulation is operant only at low temperatures (Romling 2005; Weber et al. 2006).

11.4.5   Adaptive Mutagenesis

RpoS is important for adaptive point mutations and amplifications in stationary phase (Bjedov et al. 2003; Lombardo et al. 2004). Stress-induced mutation plays an important role in bacterial evolution by increasing the adaptive mutation rate, resulting in a growth advantage in stationary phase or other stress conditions (Bjedov et al. 2003). In stationary phase, RpoS protects DNA from damage caused by oxidative stress and UV irradiation through increased expression of Dps, a nonspecific DNA binding protein that can condense the chromosome into a compacted nucleoprotein complex (Nair and Finkel 2004). However, once the damage occurs in the chromosome, RpoS activates a mechanism to bypass the error without correction through upregulating the DinB error-prone DNA polymerase (Layton and Foster 2003) or downregulating the methyl-directed mismatch repair (MMR) system mediated by MutS, MutL, and MutH (Feng et al. 1996).

DinB, encoding DNA polymerase IV, repairs DNA lesions with reduced fidelity, leading to an increased mutation rate (Kim et al. 1997). DinB is upregulated by RpoS in stationary phase (Layton and Foster 2003). Inactivation of MMR can lead to high rates of mutation and recombination (Li et al. 2003), and, in stationary phase, the level of MutS protein drops by at least 10-fold compared with that in exponential phase (Feng et al. 1996). Many other RpoS-regulated genes, including *xthA* (encoding exonuclease III), *aidB* (methylation damage repair), and *ftsQZ* (cell division), are possibly involved in the point mutation and amplification mechanism.

11.4.6   Pathogenesis

RpoS controls expression of virulence factors in several organisms, including *Salmonella* ser. Typhimurium (Fang et al. 1992), *Yersinia enterolitica* (Iriarte et al. 1995), *Pseudomonas aeruginosa* (Suh et al. 1999), and *Vibrio cholerae* (Yildiz and Schoolnik 1998). RpoS enhances not only the bacterial defense system, by regulating acid and oxidative resistance, but also the expression of virulence genes. In *Salmonella*, RpoS controls the expression of *spv* virulence genes that are located on a pathogenicity plasmid (Fang et al. 1992). RpoS is an important virulence factor in many, although not all, pathogenic *E. coli* strains. For example, RpoS stimulates the invasion of brain microvascular endothelial cells by some *E. coli* K1 strains (Wang and Kim 1999). RpoS also controls the expression of several genes on the locus of enterocyte effacement (LEE) pathogenicity island, which is responsible for forming the characteristic attaching and effacing (AE) lesions (Elliott et al. 1998). LEE carries 41 genes in five polycistronic operons (LEE1–LEE5) (Nataro and Kaper 1998), encoding a type III secretion system (TTSS), an intimin Eae (Nataro and Kaper 1998) and intimin-receptor Tir (Kenny et al. 1997), and secreted effector proteins (Elliott et al. 1998). RpoS positively controls the expression of both the LEE3 operon, encoding part of the structural and regulatory components of the TTSS (Nataro and Kaper 1998), and Tir, which is important for *E. coli* adherence (Sperandio et al. 1999). Some other genes on LEE, however, are downregulated by RpoS (Iyoda and Watanabe 2005; Tomoyasu et al. 2005; Laaberki et al. 2006). RpoS is also required for full expression of *csgA* and *csgB* encoding proteins for curli formation, an important cell surface structure implicated in pathogenesis (Uhlich et al. 2002), and *rfaH*, a primary virulence regulator in *Salmonella* and *E. coli* that modulates the biosynthesis of cell surface structures (Bittner et al. 2004; Creeger et al. 1984).

11.4.7 Negative Regulation by RpoS

As a sigma factor, RpoS is expected to have a positive regulatory role in transcription. Surprisingly, however, the expression of many genes is higher in *rpoS* mutants, and this negative regulation may help explain why, under some circumstances, there is a strong selection for loss of *rpoS* function (Zambrano et al. 1993). RpoS negatively regulates genes involved in flagellum biosynthesis, the entire tricarboxylic acid (TCA) cycle, and a cluster of genes in the Rac prophage region (Patten et al. 2004). The flagellum genes, including those encoding structural components, such as FliC, and others specifying regulatory factors, such as FliA, are required for cell motility. A second group of RpoS downregulated genes, those in the TCA cycle, are important for active metabolism and energy production (Fig. 11.3). Because RpoS is induced in stationary phase, a nutrient-limiting condition, preservation of energy sources by repressing cell motility and energy consumption pathways may be important for long-term survival. Consistent with this, the viability of *rpoS* mutants is significantly lower than that of wild-type strains after long-term incubation (Lange and Hengge-Aronis 1991b).

As a transcription sigma factor, the negative regulatory role of RpoS may result from two possibilities: (1) sigma factor competition for core RNA polymerase; (2) a negative intermediate regulator in the RpoS regulon. The second explanation is supported by the fact that some repressors are found within the RpoS regulon, such as FNR (Patten et al. 2004). The sigma factor competition model is also supported by a great deal of evidence (see Sect. 11.3.4).

Nonspecific binding of sigma factors and RNA core polymerase to DNA sequences, including not only promoter regions but also within open reading frames, has been described by many studies using both in vitro and in vivo models (von Hippel et al. 1974; deHaseth et al. 1978; Grigorova et al. 2006). Although these studies mainly focus on mechanisms releasing the core RNA polymerase from nonspecific binding sites, this nonspecific binding feature may also confer on sigma factors, in this case, RpoS, the ability to block active gene transcription as a repressor. However, this hypothesis requires experimental validation.

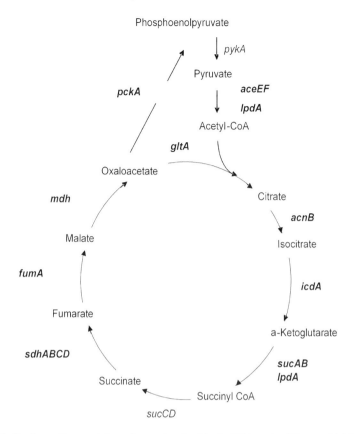

Fig. 11.3 RpoS negative regulation of genes involved in the TCA cycle. All genes highlighted in *bold* are negatively regulated by RpoS. This figure is modified from Patten et al. 2004

11.5 Consensus RpoS-Controlled Promoter Sequence

It is interesting to note that there are only minor differences in consensus promoter sequences recognized by RpoS and RpoD. Promoters preferentially activated by RpoD exhibit a consensus −10 region (TATAAT), whereas RpoS controls a more degenerate promoter sequence featuring a −10 region (TAYACT), a cytosine at −13, a TG motif at positions −14 to −18, and an A/T-rich region downstream of the −10 region (Typas et al. 2007). The cytosine at −13 is considered a specific marker for RpoS-regulated promoters and can be found in most RpoS-regulated genes (Weber et al. 2005). The −35 region in RpoS-dependent promoters is also more flexible than that in RpoD-controlled promoters, although the consensus sequence for each −35 region is the same (TTGACA) (Hengge-Aronis 2002b; Lacour and Landini 2004). In addition, a random promoter test shows that the strength of RpoS-regulated promoters can be altered by modulating the sequence from −37 to −14 upstream of the −10 region (Miksch et al. 2005).

However, not all known RpoS-dependent promoters have these conserved features. The lack of a single conserved and specific RpoS-dependent promoter sequence may allow RpoS to selectively transcribe genes under specific conditions through the action of other additional regulators. As a result, these other regulators could then modulate the stress response of the cell as needed for the particular stress stimulus.

11.6   Role of Polymorphisms of *rpoS* in Cell Adaptation

Despite the clear importance of RpoS for the survival of cells under stress conditions, *rpoS* polymorphisms, as well as variations in the nearby *mutS-rpoS* region, are common in both laboratory strains and natural isolates of *E. coli* (Herbelin et al. 2000; King et al. 2004). For example, there is significant variation at codon 33 even among K-12 strains; an amber mutation TAG is present in some K-12 strains, whereas CAG encoding glutamate or GAG encoding glutamine are found in other strains (Atlung et al. 2002). Many clinical isolates of pathogenic *E. coli* are sensitive to acid challenge because they possess *rpoS*-null mutations (Bhagwat et al. 2005). One possible explanation for *rpoS* polymorphisms is that mutations in *rpoS* may confer a growth advantage under carbon source starvation (Farrell and Finkel 2003). The *rpoS* mutants can out-compete wild-type strains and take over the population of cells after prolonged incubation (Zambrano et al. 1993). In addition, growth on succinate and other TCA cycle intermediates can specifically select for *rpoS* mutations in *E. coli* K-12 strains (Chen et al. 2004). The selective pressure may result from the negative regulation of RpoS on the TCA cycle genes, as previously mentioned. The consequence of such selection for *rpoS* mutations is to support better growth on these nonpreferred carbon sources, although at the expense of the cell's capability to cope with stress. This mutational switch to turn off a large set of genes may be important for *E. coli* cells to be competitive in a complex environment, such as the mammalian intestine, which is host to at least 1,000 different bacterial species (Sears 2005).

11.7   A New Role for RpoS in Exponential Phase

Although there are approximately 400 genes in the RpoS regulon, a key role of RpoS may have been greatly overlooked until recently, because it has become clear that RpoS is also important in exponential phase cells (Rahman et al. 2006). A potential exponential phase role for RpoS was revealed during a screening for promoters that depended on RpoS. Several genes, including *osmY* (encoding for an osmolarity response protein) and *aidB* (encoding for Isovaleryl-CoA dehydrogenase), are upregulated by RpoS in exponential phase as well as in stationary phase (Schellhorn et al. 1998; Vijayakumar et al. 2004). A recent microarray study examining *E. coli* metabolism changes during growth, corroborated with enzymatic assays, has further identified 72 genes whose expression are affected by *rpoS* mutations in exponential phase cells (Rahman et al. 2006). RpoS seems to control a distinct set of genes in exponential phase compared with the set in stationary phase, because only a small number of overlapping genes are regulated by RpoS under both conditions (Schellhorn et al. 1998; Vijayakumar et al. 2004; Rahman et al. 2006). This indicates that many RpoS-regulated genes in stationary phase may also be controlled by other coacting factors or environmental conditions, including growth rate, nutrient availability, and pH. For example, in exponential phase, *rpoS* mutants can grow as fast as the wild-type strain in Luria-Bertani (LB)-rich media under aerobic conditions. Therefore, many growth rate-dependent genes will hardly be detected. Overall, RpoS is likely to play a fine-tuning regulatory role in early exponential phase to adjust gene expression in the preparation for any potential stress, distinct from its role of active protection for cell survival in stationary phase cells.

11.8 Applications, Conclusions, and Future Prospects

Why study stationary phase? It has become clear that many genes important for host adaptation are expressed under poor growth conditions. Thus, identifying key processes required for this may help identify new potential targets for novel antimicrobials. Industrially, many bacterial fermentation products are produced primarily in the stationary phase of growth and, therefore, a better understanding of important factors during this period may help improve production yields.

Studies on RpoS and stress response may have importance in food safety and public health. Pathogenic *E. coli* strains can survive the high acidic conditions that they may encounter during food preparation and in human stomach fluids during digestion. This survival ability largely depends on RpoS (for example, see Dineen et al. 1998) and its controlled stress response systems aiding survival under acidic, osmotic, or high hydrostatic pressure conditions. The heterogeneity of *rpoS* is likely responsible for the difference in virulence of pathogenic *E. coli* strains (Bhagwat et al. 2005; Bhagwat et al. 2006). In exponential phase, when RpoS levels are low, cells are more sensitive to disinfectants, such as ClO_2 (Lisle et al. 1998), UVC (Morton and Haynes 1969), mild heat (Elliker and Frazier 1938), and sunlight (Gourmelon et al. 1997), compared with stationary-phase cells. RpoS is also important for cellular resistance to UVA light, sunlight, and thermal disinfection (Berney et al. 2006)

RpoS is clearly one of the most important regulators in the bacterial stress response and in cell adaptation by its controlling the expression of a large regulon involved in various functions. However, there are still several aspects to be investigated. For example, the RpoS regulon, as identified through microarray studies, contains a large number of genes encoding hypothetical proteins or proteins with unknown functions. Further characterization of these unknown functions will provide additional important information regarding RpoS promoter organization, potentially expand on known RpoS-regulated functions (e.g., biofilm formation) and may reveal new areas in which RpoS plays a role.

In summary, adaptation renders bacteria the flexibility to survive adverse environmental changes through adjusting gene expression patterns, and the master stress response regulator, RpoS, plays an essential role in this process.

Acknowledgments We gratefully acknowledge members of the Schellhorn laboratory, Prof. Regine Hengge, and Dr. Eberhard Klauck for suggestions during the preparation of the manuscript. Research in the authors' laboratory is supported by the Canadian Institutes for Health Research (CIHR) and the Natural Sciences and Engineering Research Council (NSERC) of Canada. C.J. was supported, in part, by an NSERC Undergraduate Summer Research Award.

Highly Recommended Readings

Gourse RL, Ross W, Rutherford ST (2006) General pathway for turning on promoters transcribed by RNA polymerases containing alternative sigma factors. J Bacteriol 188:4589–4591

Helmann JD, Chamberlin MJ (1988) Structure and function of bacterial sigma factors. Annu Rev Biochem 57:839–872

Hengge-Aronis R (2002) Signal transduction and regulatory mechanisms involved in control of the sigma(S) (RpoS) subunit of RNA polymerase. Microbiol Mol Biol Rev 66:373–395

Herbelin CJ, Chirillo SC, Melnick KA, Whittam TS (2000) Gene conservation and loss in the *mutS-rpoS* genomic region of pathogenic *Escherichia coli*. J Bacteriol 182:5381–5390

Ishihama A (2000) Functional modulation of *Escherichia coli* RNA polymerase. Annu Rev Microbiol 54:499–518

King T, Ishihama A, Kori A, Ferenci T (2004) A regulatory trade-off as a source of strain variation in the species *Escherichia coli*. J Bacteriol 186:5614–5620

Lombardo MJ, Aponyi I, Rosenberg SM (2004) General stress response regulator RpoS in adaptive mutation and amplification in *Escherichia coli*. Genetics 166:669–680

Maeda H, Fujita N, Ishihama A (2000) Competition among seven *Escherichia coli* sigma subunits: relative binding affinities to the core RNA polymerase. Nucleic Acids Res 28:3497–3503

Nystrom T (2004) Growth versus maintenance: a trade-off dictated by RNA polymerase availability and sigma factor competition? Mol Microbiol 54:855–862

Typas A, Becker G, Hengge R (2007) The molecular basis of selective promoter activation by the sigma subunit of RNA polymerase. Mol Microbiol 63:1296–1306

References

Adams JL, McLean RJ (1999) Impact of *rpoS* deletion on *Escherichia coli* biofilms. Appl Environ Microbiol 65:4285–4287

Altuvia S, Weinstein-Fischer D, Zhang A, Postow L, Storz G (1997) A small, stable RNA induced by oxidative stress: role as a pleiotropic regulator and antimutator. Cell 90:43–53

Atlung T, Nielsen HV, Hansen FG (2002) Characterisation of the allelic variation in the *rpoS* gene in thirteen K12 and six other non-pathogenic *Escherichia coli* strains. Mol Genet Genomics 266:873–881

Audia JP, Webb CC, Foster JW (2001) Breaking through the acid barrier: an orchestrated response to proton stress by enteric bacteria. Int J Med Microbiol 291:97–106

Balandina A, Claret L, Hengge-Aronis R, Rouviere-Yaniv J (2001) The *Escherichia coli* histone-like protein HU regulates *rpoS* translation. Mol Microbiol 39:1069–1079

Barth M, Marschall C, Muffler A, Fischer D, Hengge-Aronis R (1995) Role for the histone-like protein H-NS in growth phase-dependent and osmotic regulation of sigma S and many sigma S-dependent genes in *Escherichia coli*. J Bacteriol 177:3455–3464

Bearson SM, Benjamin WH Jr, Swords WE, Foster JW (1996) Acid shock induction of RpoS is mediated by the mouse virulence gene mviA of Salmonella typhimurium. J Bacteriol 178:2572–2579

Bernardo LM, Johansson LU, Solera D, Skarfstad E, Shingler V (2006) The guanosine tetraphosphate (ppGpp) alarmone, DksA and promoter affinity for RNA polymerase in regulation of sigma-dependent transcription. Mol Microbiol 60:749–764

Berney M, Weilenmann HU, Ihssen J, Bassin C, Egli T (2006) Specific growth rate determines the sensitivity of *Escherichia coli* to thermal, UVA, and solar disinfection. Appl Environ Microbiol 72:2586–2593

Bhagwat AA, Chan L, Han R, Tan J, Kothary M, Jean-Gilles J, Tall BD (2005) Characterization of enterohemorrhagic *Escherichia coli* strains based on acid resistance phenotypes. Infect Immun 73:4993–5003

Bhagwat AA, Tan J, Sharma M, Kothary M, Low S, Tall BD, Bhagwat M (2006) Functional heterogeneity of RpoS in stress tolerance of enterohemorrhagic *Escherichia coli* strains. Appl Environ Microbiol 72:4978–4986

Bittner M, Saldias S, Altamirano F, Valvano MA, Contreras I (2004) RpoS and RpoN are involved in the growth-dependent regulation of *rfaH* transcription and O antigen expression in *Salmonella enterica serovar typhi*. Microb Pathog 36:19–24

Bjedov I, Tenaillon O, Gerard B, Souza V, Denamur E, Radman M, Taddei F, Matic I (2003) Stress-induced mutagenesis in bacteria. Science 300:1404–1409

Bohringer J, Fischer D, Mosler G, Hengge-Aronis R (1995) UDP-glucose is a potential intracellular signal molecule in the control of expression of sigma S and sigma S-dependent genes in *Escherichia coli*. J Bacteriol 177:413–422

Bougdour A, Lelong C, Geiselmann J (2004) Crl, a low temperature-induced protein in *Escherichia coli* that binds directly to the stationary phase sigma subunit of RNA polymerase. J Biol Chem 279:19540–19550

Bougdour A, Wickner S, Gottesman S (2006) Modulating RssB activity: IraP, a novel regulator of sigma(S) stability in *Escherichia coli*. Genes Dev 20:884–897

Brescia CC, Mikulecky PJ, Feig AL, Sledjeski DD (2003) Identification of the Hfq-binding site on DsrA RNA: Hfq binds without altering DsrA secondary structure. RNA 9:33–43

Brown L, Elliott T (1997) Mutations that increase expression of the *rpoS* gene and decrease its dependence on *hfq* function in *Salmonella typhimurium*. J Bacteriol 179:656–662

Brown L, Gentry D, Elliott T, Cashel M (2002) DksA affects ppGpp induction of RpoS at a translational level. J Bacteriol 184:4455–4465

Castanie-Cornet MP, Penfound TA, Smith D, Elliott JF, Foster JW (1999) Control of acid resistance in *Escherichia coli*. J Bacteriol 181:3525–3535

Chen G, Patten CL, Schellhorn HE (2004) Positive selection for loss of RpoS function in *Escherichia coli*. Mutat Res 554:193–203

Cheung KJ, Badarinarayana V, Selinger DW, Janse D, Church GM (2003) A microarray-based antibiotic screen identifies a regulatory role for supercoiling in the osmotic stress response of *Escherichia coli*. Genome Res 13:206–215

Christman MF, Storz G, Ames BN (1989) OxyR, a positive regulator of hydrogen peroxide-inducible genes in *Escherichia coli* and *Salmonella typhimurium*, is homologous to a family of bacterial regulatory proteins. Proc Natl Acad Sci USA 86:3484–3488

Corona-Izquierdo FP, Membrillo-Hernandez J (2002) A mutation in *rpoS* enhances biofilm formation in *Escherichia coli* during exponential phase of growth. FEMS Microbiol Lett 211:105–110

Costanzo A, Ades SE (2006) Growth phase-dependent regulation of the extracytoplasmic stress factor, sigmaE, by guanosine 3′,5′-bispyrophosphate (ppGpp). J Bacteriol 188:4627–4634

Creeger ES, Schulte T, Rothfield LI (1984) Regulation of membrane glycosyltransferases by the *sfrB* and *rfaH* genes of *Escherichia coli* and *Salmonella typhimurium*. J Biol Chem 259:3064–3069

Cunning C, Brown L, Elliott T (1998) Promoter substitution and deletion analysis of upstream region required for *rpoS* translational regulation. J Bacteriol 180:4564–4570

deHaseth PL, Lohman TM, Burgess RR, Record MT Jr (1978) Nonspecific interactions of *Escherichia coli* RNA polymerase with native and denatured DNA: differences in the binding behavior of core and holoenzyme. Biochemistry 17:1612–1622

Dineen SS, Takeuchi K, Soudah JE, Boor KJ (1998) Persistence of *Escherichia coli* O157:H7 in dairy fermentation systems. J Food Prot 61:1602–1608

Domka J, Lee J, Wood TK (2006) YliH (BssR) and YceP (BssS) regulate *Escherichia coli* K-12 biofilm formation by influencing cell signaling. Appl Environ Microbiol 72:2449–2459

Dufraigne C, Fertil B, Lespinats S, Giron A, Deschavanne P (2005) Detection and characterization of horizontal transfers in prokaryotes using genomic signature. Nucleic Acids Res 33:e6

Elliker PR, Frazier WC (1938) Influence of time and temperature of incubation on heat resistance of *Escherichia coli*. J Bacteriol 36:83–98

Elliott SJ, Wainwright LA, McDaniel TK, Jarvis KG, Deng YK, Lai LC, McNamara BP, Donnenberg MS, Kaper JB (1998) The complete sequence of the locus of enterocyte effacement (LEE) from enteropathogenic *Escherichia coli* E2348/69. Mol Microbiol 28:1–4

Fang FC, Libby SJ, Buchmeier NA, Loewen PC, Switala J, Harwood J, Guiney DG (1992) The alternative sigma factor *katF* (*rpoS*) regulates *Salmonella* virulence. Proc Natl Acad Sci USA 89:11978–11982

Farewell A, Kvint K, Nystrom T (1998a) Negative regulation by RpoS: a case of sigma factor competition. Mol Microbiol 29:1039–1051

Farewell A, Kvint K, Nystrom T (1998b) *uspB*, a new sigmaS-regulated gene in *Escherichia coli* which is required for stationary-phase resistance to ethanol. J Bacteriol 180:6140–6147

Farr SB, Kogoma T (1991) Oxidative stress responses in Escherichia coli and Salmonella typhimurium. Microbiol Rev 55:561–585

Farrell MJ, Finkel SE (2003) The growth advantage in stationary-phase phenotype conferred by *rpoS* mutations is dependent on the pH and nutrient environment. J Bacteriol 185:7044–7052

Feng G, Tsui HC, Winkler ME (1996) Depletion of the cellular amounts of the MutS and MutH methyl-directed mismatch repair proteins in stationary-phase *Escherichia coli* K-12 cells. J Bacteriol 178:2388–2396

Fredriksson A, Ballesteros M, Peterson CN, Persson O, Silhavy TJ, Nystrom T (2007) Decline in ribosomal fidelity contributes to the accumulation and stabilization of the master stress response regulator sigmaS upon carbon starvation. Genes Dev 21:862–874

Gentry DR, Hernandez VJ, Nguyen LH, Jensen DB, Cashel M (1993) Synthesis of the stationary-phase sigma factor sigmaS is positively regulated by ppGpp. J Bacteriol 175:7982–7989

Gourmelon M, Touati D, Pommepuy M, Cormier M (1997) Survival of *Escherichia coli* exposed to visible light in seawater: analysis of *rpoS*-dependent effects. Can J Microbiol 43:1036–1043

Gourse RL, Gaal T, Bartlett MS, Appleman JA, Ross W (1996) rRNA transcription and growth rate-dependent regulation of ribosome synthesis in Escherichia coli. Annu Rev Microbiol 50:645–677

Gourse RL, Ross W, Rutherford ST (2006) General pathway for turning on promoters transcribed by RNA polymerases containing alternative sigma factors. J Bacteriol 188:4589–4591

Grigorova IL, Phleger NJ, Mutalik VK, Gross CA (2006) Insights into transcriptional regulation and sigma competition from an equilibrium model of RNA polymerase binding to DNA. Proc Natl Acad Sci USA 103:5332–5337

Gross CA, Chan C, Dombroski A, Gruber T, Sharp M, Tupy J, Young B (1998) The functional and regulatory roles of sigma factors in transcription. Cold Spring Harb Symp Quant Biol 63:141–155

Helmann JD, Chamberlin MJ (1988) Structure and function of bacterial sigma factors. Annu Rev Biochem 57:839–872

Hengge-Aronis R (2002a) Signal transduction and regulatory mechanisms involved in control of the sigma(S) (RpoS) subunit of RNA polymerase. Microbiol Mol Biol Rev 66:373–395

Hengge-Aronis R (2002b) Stationary phase gene regulation: what makes an *Escherichia coli* promoter sigma(S)-selective? Curr Opin Microbiol 5:591–595

Herbelin CJ, Chirillo SC, Melnick KA, Whittam TS (2000) Gene conservation and loss in the *mutS-rpoS* genomic region of pathogenic *Escherichia coli*. J Bacteriol 182:5381–5390

Hirsch M, Elliott T (2002) Role of ppGpp in *rpoS* stationary-phase regulation in *Escherichia coli*. J Bacteriol 184:5077–5087

Hirsch M, Elliott T (2005) Stationary-phase regulation of RpoS translation in *Escherichia coli*. J Bacteriol 187:7204–7213

Holmgren A (1989) Thioredoxin and glutaredoxin systems. J Biol Chem 264:13963–13966

Holmgren A (2000) Antioxidant function of thioredoxin and glutaredoxin systems. Antioxid Redox Signal 2:811–820

Ihssen J, Egli T (2004) Specific growth rate and not cell density controls the general stress response in *Escherichia coli*. Microbiology 150:1637–1648

Iriarte M, Stainier I, Cornelis GR (1995) The *rpoS* gene from *Yersinia enterocolitica* and its influence on expression of virulence factors. Infect Immun 63:1840–1847

Ishihama A (2000) Functional modulation of *Escherichia coli* RNA polymerase. Annu Rev Microbiol 54:499–518

Ivanova A, Miller C, Glinsky G, Eisenstark A (1994) Role of *rpoS* (*katF*) in *oxyR*-independent regulation of hydroperoxidase I in *Escherichia coli*. Mol Microbiol 12:571–578

Iyoda S, Watanabe H (2005) ClpXP protease controls expression of the type III protein secretion system through regulation of RpoS and GrlR levels in enterohemorrhagic *Escherichia coli*. J Bacteriol 187:4086–4094

Jishage M, Ishihama A (1998) A stationary phase protein in *Escherichia coli* with binding activity to the major sigma subunit of RNA polymerase. Proc Natl Acad Sci USA 95:4953–4958

Jishage M, Kvint K, Shingler V, Nystrom T (2002) Regulation of sigma factor competition by the alarmone ppGpp. Genes Dev 16:1260–1270

Jordan A, Aslund F, Pontis E, Reichard P, Holmgren A (1997) Characterization of *Escherichia coli* NrdH. A glutaredoxin-like protein with a thioredoxin-like activity profile. J Biol Chem 272:18044–18050

Kenny B, DeVinney R, Stein M, Reinscheid DJ, Frey EA, Finlay BB (1997) Enteropathogenic *E. coli* (EPEC) transfers its receptor for intimate adherence into mammalian cells. Cell 91:511–520

Khil PP, Camerini-Otero RD (2002) Over 1000 genes are involved in the DNA damage response of *Escherichia coli*. Mol Microbiol 44:89–105

Kim SR, Maenhaut-Michel G, Yamada M, Yamamoto Y, Matsui K, Sofuni T, Nohmi T, Ohmori H (1997) Multiple pathways for SOS-induced mutagenesis in *Escherichia coli*: an overexpression of *dinB/dinP* results in strongly enhancing mutagenesis in the absence of any exogenous treatment to damage DNA. Proc Natl Acad Sci USA 94:13792–13797

King T, Ferenci T (2005) Divergent roles of RpoS in *Escherichia coli* under aerobic and anaerobic conditions. FEMS Microbiol Lett 244:323–327

King T, Ishihama A, Kori A, Ferenci T (2004) A regulatory trade-off as a source of strain variation in the species *Escherichia coli*. J Bacteriol 186:5614–5620

Kornberg A, Rao NN, Ault-Riche D (1999) Inorganic polyphosphate: a molecule of many functions. Annu Rev Biochem 68:89–125

Kvint K, Farewell A, Nystrom T (2000) RpoS-dependent promoters require guanosine tetraphosphate for induction even in the presence of high levels of sigma(s). J Biol Chem 275:14795–14798

Laaberki MH, Janabi N, Oswald E, Repoila F (2006) Concert of regulators to switch on LEE expression in enterohemorrhagic *Escherichia coli* O157:H7: Interplay between Ler, GrlA, HNS and RpoS. Int J Med Microbiol 296:197–210

Lacour S, Landini P (2004) SigmaS-dependent gene expression at the onset of stationary phase in *Escherichia coli*: function of sigmaS-dependent genes and identification of their promoter sequences. J Bacteriol 186:7186–7195

Lange R, Fischer D, Hengge-Aronis R (1995) Identification of transcriptional start sites and the role of ppGpp in the expression of *rpoS*, the structural gene for the sigma S subunit of RNA polymerase in *Escherichia coli*. J Bacteriol 177:4676–4680

Lange R, Hengge-Aronis R (1991a) Growth phase-regulated expression of bolA and morphology of stationary-phase Escherichia coli cells are controlled by the novel sigma factor sigma S. J Bacteriol 173:4474–4481

Lange R, Hengge-Aronis R (1991b) Identification of a central regulator of stationary-phase gene expression in *Escherichia coli*. Mol Microbiol 5:49–59

Lange R, Hengge-Aronis R (1994a) The cellular concentration of the sigma S subunit of RNA polymerase in *Escherichia coli* is controlled at the levels of transcription, translation, and protein stability. Genes Dev 8:1600–1612

Lange R, Hengge-Aronis R (1994b) The *nlpD* gene is located in an operon with *rpoS* on the *Escherichia coli* chromosome and encodes a novel lipoprotein with a potential function in cell wall formation. Mol Microbiol 13:733–743

Laurie AD, Bernardo LM, Sze CC, Skarfstad E, Szalewska-Palasz A, Nystrom T, Shingler V (2003) The role of the alarmone (p)ppGpp in sigma N competition for core RNA polymerase. J Biol Chem 278:1494–1503

Layton JC, Foster PL (2003) Error-prone DNA polymerase IV is controlled by the stress-response sigma factor, RpoS, in *Escherichia coli*. Mol Microbiol 50:549–561

Lease RA, Cusick ME, Belfort M (1998) Riboregulation in *Escherichia coli*: DsrA RNA acts by RNA:RNA interactions at multiple loci. Proc Natl Acad Sci USA 95:12456–12461

Lelong C, Aguiluz K, Luche S, Kuhn L, Garin J, Rabilloud T, Geiselmann J (2007) The Crl-RpoS regulon of *Escherichia coli*. Mol Cell Proteomics 6:648–659

Li B, Tsui HC, LeClerc JE, Dey M, Winkler ME, Cebula TA (2003) Molecular analysis of *mutS* expression and mutation in natural isolates of pathogenic *Escherichia coli*. Microbiology 149:1323–1331

Lin J, Smith MP, Chapin KC, Baik HS, Bennett GN, Foster JW (1996) Mechanisms of acid resistance in enterohemorrhagic Escherichia coli. Appl Environ Microbiol 62:3094–3100

Lisle JT, Broadaway SC, Prescott AM, Pyle BH, Fricker C, McFeters GA (1998) Effects of starvation on physiological activity and chlorine disinfection resistance in *Escherichia coli* O157: H7. Appl Environ Microbiol 64:4658–4662

Loewen PC, Triggs BL (1984) Genetic mapping of *katF*, a locus that with *katE* affects the synthesis of a second catalase species in *Escherichia coli*. J Bacteriol 160:668–675

Lombardo MJ, Aponyi I, Rosenberg SM (2004) General stress response regulator RpoS in adaptive mutation and amplification in *Escherichia coli*. Genetics 166:669–680

Maeda H, Fujita N, Ishihama A (2000) Competition among seven *Escherichia coli* sigma subunits: relative binding affinities to the core RNA polymerase. Nucleic Acids Res 28:3497–3503

Majdalani N, Chen S, Murrow J, St John K, Gottesman S (2001) Regulation of RpoS by a novel small RNA: the characterization of RprA. Mol Microbiol 39:1382–1394

Majdalani N, Cunning C, Sledjeski D, Elliott T, Gottesman S (1998) DsrA RNA regulates translation of RpoS message by an anti-antisense mechanism, independent of its action as an antisilencer of transcription. Proc Natl Acad Sci USA 95:12462–12467

Majdalani N, Hernandez D, Gottesman S (2002) Regulation and mode of action of the second small RNA activator of RpoS translation, RprA. Mol Microbiol 46:813–826

Majdalani N, Vanderpool CK, Gottesman S (2005) Bacterial small RNA regulators. Crit Rev Biochem Mol Biol 40:93–113

Miksch G, Bettenworth F, Friehs K, Flaschel E (2005) The sequence upstream of the −10 consensus sequence modulates the strength and induction time of stationary-phase promoters in *Escherichia coli*. Appl Microbiol Biotechnol 69:312–320

Moll I, Afonyushkin T, Vytvytska O, Kaberdin VR, Blasi U (2003) Coincident Hfq binding and RNase E cleavage sites on mRNA and small regulatory RNAs. RNA 9:1308–1314

Monje-Casas F, Jurado J, Prieto-Alamo MJ, Holmgren A, Pueyo C (2001) Expression analysis of the *nrdHIEF* operon from *Escherichia coli*. Conditions that trigger the transcript level in vivo. J Biol Chem 276:18031–18037

Morton RA, Haynes RH (1969) Changes in the ultraviolet sensitivity of *Escherichia coli* during growth in batch cultures. J Bacteriol 97:1379–1385

Muffler A, Traulsen DD, Lange R, Hengge-Aronis R (1996) Posttranscriptional osmotic regulation of the sigma(S) subunit of RNA polymerase in *Escherichia coli*. J Bacteriol 178:1607–1613

Nair S, Finkel SE (2004) Dps protects cells against multiple stresses during stationary phase. J Bacteriol 186:4192–4198

Nataro JP, Kaper JB (1998) Diarrheagenic *Escherichia coli*. Clin Microbiol Rev 11:142–201

Nickel JC, Ruseska I, Wright JB, Costerton JW (1985) Tobramycin resistance of Pseudomonas aeruginosa cells growing as a biofilm on urinary catheter material. Antimicrob Agents Chemother 27:619–624

Notley L, Ferenci T (1996) Induction of RpoS-dependent functions in glucose-limited continuous culture: what level of nutrient limitation induces the stationary phase of *Escherichia coli*? J Bacteriol 178:1465–1468

Nystrom T (2004) Growth versus maintenance: a trade-off dictated by RNA polymerase availability and sigma factor competition? Mol Microbiol 54:855–862

Patten CL, Kirchhof MG, Schertzberg MR, Morton RA, Schellhorn HE (2004) Microarray analysis of RpoS-mediated gene expression in *Escherichia coli* K-12. Mol Genet Genomics 272:580–591

Peterson CN, Ruiz N, Silhavy TJ (2004) RpoS proteolysis is regulated by a mechanism that does not require the SprE (RssB) response regulator phosphorylation site. J Bacteriol 186:7403–7410

Phadtare S, Inouye M (2001) Role of CspC and CspE in regulation of expression of RpoS and UspA, the stress response proteins in *Escherichia coli*. J Bacteriol 183:1205–1214

Phadtare S, Tadigotla V, Shin WH, Sengupta A, Severinov K (2006) Analysis of *Escherichia coli* global gene expression profiles in response to overexpression and deletion of CspC and CspE. J Bacteriol 188:2521–2527

Piggot PJ, Hilbert DW (2004) Sporulation of *Bacillus subtilis*. Curr Opin Microbiol 7:579–586

Potamitou A, Neubauer P, Holmgren A, Vlamis-Gardikas A (2002) Expression of *Escherichia coli* glutaredoxin 2 is mainly regulated by ppGpp and sigmaS. J Biol Chem 277:17775–17780

Pratt LA, Silhavy TJ (1996) The response regulator SprE controls the stability of RpoS. Proc Natl Acad Sci USA 93:2488–2492

Pratt LA, Silhavy TJ (1998) Crl stimulates RpoS activity during stationary phase. Mol Microbiol 29:1225–1236

Price SB, Cheng CM, Kaspar CW, Wright JC, DeGraves FJ, Penfound TA, Castanie-Cornet MP, Foster JW (2000) Role of *rpoS* in acid resistance and fecal shedding of *Escherichia coli* O157: H7. Appl Environ Microbiol 66:632–637

Rahman M, Hasan MR, Oba T, Shimizu K (2006) Effect of *rpoS* gene knockout on the metabolism of *Escherichia coli* during exponential growth phase and early stationary phase based on gene expressions, enzyme activities and intracellular metabolite concentrations. Biotechnol Bioeng 94:585–595

Rao NN, Kornberg A (1996) Inorganic polyphosphate supports resistance and survival of stationary-phase *Escherichia coli*. J Bacteriol 178:1394–1400

Repoila F, Gottesman S (2001) Signal transduction cascade for regulation of RpoS: temperature regulation of DsrA. J Bacteriol 183:4012–4023

Repoila F, Majdalani N, Gottesman S (2003) Small non-coding RNAs, coordinators of adaptation processes in *Escherichia coli*: the RpoS paradigm. Mol Microbiol 48:855–861

Ritz D, Beckwith J (2001) Roles of thiol-redox pathways in bacteria. Annu Rev Microbiol 55:21–48

Robey M, Benito A, Hutson RH, Pascual C, Park SF, Mackey BM (2001) Variation in resistance to high hydrostatic pressure and *rpoS* heterogeneity in natural isolates of *Escherichia coli* O157:H7. Appl Environ Microbiol 67:4901–4907

Rockabrand D, Livers K, Austin T, Kaiser R, Jensen D, Burgess R, Blum P (1998) Roles of DnaK and RpoS in starvation-induced thermotolerance of *Escherichia coli*. J Bacteriol 180:846–854

Romling U (2005) Characterization of the rdar morphotype, a multicellular behaviour in Enterobacteriaceae. Cell Mol Life Sci 62:1234–1246

Romling U, Bian Z, Hammar M, Sierralta WD, Normark S (1998) Curli fibers are highly conserved between *Salmonella typhimurium* and *Escherichia coli* with respect to operon structure and regulation. J Bacteriol 180:722–731

Sammartano LJ, Tuveson RW, Davenport R (1986) Control of sensitivity to inactivation by H2O2 and broad-spectrum near-UV radiation by the *Escherichia coli katF* locus. J Bacteriol 168:13–21

Schellhorn HE (1995) Regulation of hydroperoxidase (catalase) expression in *Escherichia coli*. FEMS Microbiol Lett 131:113–119

Schellhorn HE, Audia JP, Wei LI, Chang L (1998) Identification of conserved, RpoS-dependent stationary-phase genes of *Escherichia coli*. J Bacteriol 180:6283–6291

Schellhorn HE, Hassan HM (1988) Transcriptional regulation of *katE* in *Escherichia coli* K-12. J Bacteriol 170:4286–4292

Schellhorn HE, Stones VL (1992) Regulation of *katF* and *katE* in *Escherichia coli* K-12 by weak acids. J Bacteriol 174:4769–4776

Schembri MA, Kjaergaard K, Klemm P (2003) Global gene expression in *Escherichia coli* biofilms. Mol Microbiol 48:253–267

Sears CL (2005) A dynamic partnership: Celebrating our gut flora. Anaerobe 11:247–251

Sevcik M, Sebkova A, Volf J, Rychlik I (2001) Transcription of *arcA* and *rpoS* during growth of *Salmonella typhimurium* under aerobic and microaerobic conditions. Microbiology 147:701–708

Shiba T, Tsutsumi K, Yano H, Ihara Y, Kameda A, Tanaka K, Takahashi H, Munekata M, Rao NN, Kornberg A (1997) Inorganic polyphosphate and the induction of *rpoS* expression. Proc Natl Acad Sci USA 94:11210–11215

Simonson AB, Servin JA, Skophammer RG, Herbold CW, Rivera MC, Lake JA (2005) Decoding the genomic tree of life. Proc Natl Acad Sci USA 102 Suppl 1:6608–6613

Sledjeski DD, Gupta A, Gottesman S (1996) The small RNA, DsrA, is essential for the low temperature expression of RpoS during exponential growth in *Escherichia coli*. EMBO J 15:3993–4000

Small P, Blankenhorn D, Welty D, Zinser E, Slonczewski JL (1994) Acid and base resistance in *Escherichia coli* and *Shigella flexneri*: role of *rpoS* and growth pH. J Bacteriol 176:1729–1737

Sperandio V, Mellies JL, Nguyen W, Shin S, Kaper JB (1999) Quorum sensing controls expression of the type III secretion gene transcription and protein secretion in enterohemorrhagic and enteropathogenic *Escherichia coli*. Proc Natl Acad Sci USA 96:15196–15201

Stewart PS, Costerton JW (2001) Antibiotic resistance of bacteria in biofilms. Lancet 358:135–138

Suh SJ, Silo-Suh L, Woods DE, Hassett DJ, West SE, Ohman DE (1999) Effect of *rpoS* mutation on the stress response and expression of virulence factors in *Pseudomonas aeruginosa*. J Bacteriol 181:3890–3897

Sze CC, Shingler V (1999) The alarmone (p)ppGpp mediates physiological-responsive control at the sigma 54-dependent Po promoter. Mol Microbiol 31:1217–1228

Takayanagi Y, Tanaka K, Takahashi H (1994) Structure of the 5′ upstream region and the regulation of the *rpoS* gene of *Escherichia coli*. Mol Gen Genet 243:525–531

Tomoyasu T, Takaya A, Handa Y, Karata K, Yamamoto T (2005) ClpXP controls the expression of LEE genes in enterohaemorrhagic *Escherichia coli*. FEMS Microbiol Lett 253:59–66

Touati E, Dassa E, Boquet PL (1986) Pleiotropic mutations in *appR* reduce pH 2.5 acid phosphatase expression and restore succinate utilisation in CRP-deficient strains of *Escherichia coli*. Mol Gen Genet 202:257–264

Typas A, Becker G, Hengge R (2007) The molecular basis of selective promoter activation by the sigma subunit of RNA polymerase. Mol Microbiol 63:1296–1306

Uhlich GA, Keen JE, Elder RO (2002) Variations in the *csgD* promoter of *Escherichia coli* O157: H7 associated with increased virulence in mice and increased invasion of HEp-2 cells. Infect Immun 70:395–399

VanBogelen RA, Neidhardt FC (1990) Ribosomes as sensors of heat and cold shock in *Escherichia coli*. Proc Natl Acad Sci USA 87:5589–5593

Vijayakumar SR, Kirchhof MG, Patten CL, Schellhorn HE (2004) RpoS-regulated genes of *Escherichia coli* identified by random *lacZ* fusion mutagenesis. J Bacteriol 186:8499–8507

von Hippel PH, Revzin A, Gross CA, Wang AC (1974) Non-specific DNA binding of genome regulating proteins as a biological control mechanism: I. The lac operon: equilibrium aspects. Proc Natl Acad Sci USA 71:4808–4812

Wassarman KM, Saecker RM (2006) Synthesis-mediated release of a small RNA inhibitor of RNA polymerase. Science 314:1601–1603

Wassarman KM, Storz G (2000) 6S RNA regulates *E. coli* RNA polymerase activity. Cell 101:613–623

Weber H, Pesavento C, Possling A, Tischendorf G, Hengge R (2006) Cyclic-di-GMP-mediated signalling within the sigma network of *Escherichia coli*. Mol Microbiol 62:1014–1034

Weber H, Polen T, Heuveling J, Wendisch VF, Hengge R (2005) Genome-wide analysis of the general stress response network in *Escherichia coli*: sigmaS-dependent genes, promoters, and sigma factor selectivity. J Bacteriol 187:1591–1603

Yamashino T, Ueguchi C, Mizuno T (1995) Quantitative control of the stationary phase-specific sigma factor, sigma S, in *Escherichia coli*: involvement of the nucleoid protein H-NS. EMBO J 14:594–602

Yildiz FH, Schoolnik GK (1998) Role of *rpoS* in stress survival and virulence of *Vibrio cholerae*. J Bacteriol 180:773–784

Yoshida M, Kashiwagi K, Kawai G, Ishihara A, Igarashi K (2002) Polyamines enhance synthesis of the RNA polymerase sigma 38 subunit by suppression of an amber termination codon in the open reading frame. J Biol Chem 277:37139–37146

Zambrano MM, Siegele DA, Almiron M, Tormo A, Kolter R (1993) Microbial competition: *Escherichia coli* mutants that take over stationary phase cultures. Science 259:1757–1760

Zhang A, Altuvia S, Tiwari A, Argaman L, Hengge-Aronis R, Storz G (1998) The OxyS regulatory RNA represses *rpoS* translation and binds the Hfq (HF-I) protein. EMBO J 17:6061–6068

Zhou Y, Gottesman S, Hoskins JR, Maurizi MR, Wickner S (2001) The RssB response regulator directly targets sigma(S) for degradation by ClpXP. Genes Dev 15:627–637

12
Phenotypic Variation and Bistable Switching in Bacteria

Wiep Klaas Smits(✉), Jan-Willem Veening, and Oscar P. Kuipers

Abstract   Microbial research generally focuses on clonal populations. However, bacterial cells with identical genotypes frequently display different phenotypes under identical conditions. This microbial cell individuality is receiving increasing attention in the literature because of its impact on cellular differentiation, survival under selective conditions, and the interaction of pathogens with their hosts. It is becoming clear that stochasticity in gene expression in conjunction with the architecture of the gene network that underlies the cellular processes can generate phenotypic variation. An important regulatory mechanism is the so-called positive feedback, in which a system reinforces its own response, for instance by stimulating the production of an activator. Bistability is an interesting and relevant phenomenon, in which two distinct subpopulations of cells showing discrete levels of gene expression coexist in a single culture. In this chapter, we address techniques and approaches used to establish phenotypic variation, and relate three well-characterized examples of bistability to the molecular mechanisms that govern these processes, with a focus on positive feedback.

Wiep Klaas Smits

Department of Genetics, University of Groningen, Groningen Biomolecular Sciences and Biotechnology Institute, Kerklaan 30, 9751NN, Haren, The Netherlands.
Present address: Department of Biology, Building 68-530, Massachusetts Institute of Technology, Cambridge, MA 02139, USA.
smitswk@gmail.com

W. El-Sharoud (ed.) *Bacterial Physiology: A Molecular Approach.*
© Springer-Verlag Berlin Heidelberg 2008

12.1 Phenotypic Variation

The potential of bacteria to thrive in a variety of ecological niches strongly depends on their genetic content. However, within these niches, environmental conditions such as pH, salinity, temperature, and the availability of nutrients can strongly fluctuate. In addition, bacteria may be faced with interspecies or intraspecies competition in the form of, for instance, antimicrobial peptides.

To survive these kinds of stresses, bacteria adapt their genetic programs, resulting in change of phenotypes to meet the fluctuating environmental conditions. Environmental signals are commonly interpreted by cells via a system of sensor and regulator proteins that result in a change in the transcription profiles (transcriptome) and/or protein expression patterns (proteome) of the cells (other sensing systems also exist in bacterial cells; see Chaps. 7, 8, and 9 for details on some of these systems). These adaptive changes generally last for the duration of the stress, but in some cases may persist for many generations.

Microbial research has traditionally considered bacterial cultures as homogeneous, because they are a clonal population derived from a common ancestor. Before single-cell analyses of bacterial populations, many heterogeneous processes may have gone undetected or may have been discarded as an artifact of the methods used. Single-cell analyses (see Sect. 12.2) show that cells in a culture demonstrate a remarkable level of variation, which roughly can be divided into two categories.

First of all, bacterial populations exhibit a unimodal distribution in gene expression for most genes, because of random fluctuations in transcription and translation (in other words, there is a normal distribution of expression levels around the mean; Fig. 12.1a, curve 2). These fluctuations are referred to as noise (Ozbudak et al. 2002). This general statistical argument already implies that there can be significant differences in levels of gene expression between individual cells.

In addition, several cases have been identified in which only a subpopulation of cells of a culture demonstrates an adaptive phenotype (Smits et al. 2006). This heterogeneity could result in an increased spectrum of niches that the bacteria can use, and, therefore, has the potential to increase the overall fitness of the species. When two clear subpopulations can be identified, each with a normal distribution in gene expression levels, the population is referred to as bistable (Fig. 12.1a and b, curve 3). Bistability has also been described on the phenotypic level as a situation in which two clearly distinguishable cell types are present within the clonal population (Dubnau and Losick 2006).

Consider a situation in which a certain gene is twofold upregulated under a typical stress condition, as measured by a culture-wide reporter such as beta-galactosidase activity. It is possible that the gene demonstrates twofold upregulation in all cells, but if the gene were expressed in a subpopulation of only 10% of the culture, the fold increase in expression in these specific cells would be closer to 11. Clearly, these findings demonstrate that bulk measurements have to be interpreted with care.

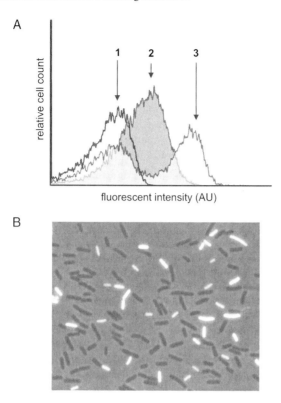

Fig. 12.1 Phenotypic variation. **a** Flow cytometric analysis of bacterial populations carrying a GFP-reporter fusion. Nonfluorescent cells give rise to curve 1. Phenotypic variation is apparent for the populations represented by curves 2 and 3. In the case of curve 2, the underlying distribution in fluorescence is monomodal. Note that some cells may seem nonfluorescent because they overlap with curve 1, depending on the detection limit of the system. Curve 3 represents a bistable expression pattern. Note that the one of the stable states corresponds to nonfluorescent cells. AU, arbitrary units with permission from the publisher. **b** Fluorescence microscopic image of a *B. subtilis* strain harboring a competence-specific GFP reporter. Note two subpopulations of cells, white cells express the reporter, whereas the black cells do not. This figure was adapted from Smits et al. (2006).

12.2 Reporters and Techniques for Single-Cell Analyses

Although culture-wide reporter assays can be informative, the example above demonstrates the necessity to determine whether a response measured in bulk experiments is representative for the events in single cells. To get statistically significant data, large numbers of cells need to be quantified. To this end, several techniques have been developed.

The simplest way to examine single cells is by fixing cells on a (coated) glass slide, followed by visualization using light microscopy. Phase-contrast microscopy suffices in some cases to reveal phenotypic variability. Sporulating cells of *Bacillus subtilis* can readily be distinguished from nonsporulating cells within an isogenic culture by normal phase-contrast microscopy because sporulating cells contain a readily visible endospore during the later stages of this process, whereas nonsporu-lating cells do not. However, many subpopulations are not characterized by pheno-types that are distinguishable by phase-contrast microscopy.

As a result, most single-cell reporters rely on fluorescence or luminescence (see Davey and Kell 1996; Kasten 1993). Fluorescence is a physicochemical property of certain molecules and compounds, and can be used to visualize individual bio-chemical, genetic, or physiological properties (Barer and Harwood 1999; Joux and Lebaron 2000). With the use of specific filters coupled to a light source, such as a mercury or xenon lamp, excitation of fluorophores and detection of fluorescent signals at certain wavelengths can be performed. Microscopic images may be cap-tured using CCD cameras. One of the most powerful ways to examine a large number of cells on their single-cell properties is by flow cytometry. In a flow cytometer, cells pass an intense light source (a laser) one by one, and the data on fluorescence and scatter (which depends on, for instance, particle size) is collected. Depending on the flow cytometer and the settings used, signals from more than 1,000 individual cells per second can easily be determined, making it an extremely valuable technique for time series analyses. Because of its sensitivity and resolution, this is the preferred method to establish whether a heterogeneous culture demon-strates a monomodal or bimodal (bistable) distribution in gene expression levels (see Sect. 12.5.2 and Chung et al. 1995; Smits et al. 2005; Veening et al. 2005).

A battery of fluorescent dyes is currently available to visualize biochemical or physiological properties of cells that may differ from cell to cell (see Brehm-Stecher and Johnson 2004). These dyes include substances that nonspecifically stain DNA (e.g., DAPI, ethidium bromide, and Hoechst), membranes/phospholip-ids (e.g., FM4-64 and Nile-red) or the cell wall (e.g., Gram staining or fluorescently labeled vancomycin) (Daniel and Errington 2003). In addition, a number of probes can be used to measure individual intracellular pH or redox state of a cell (Nebe-von-Caron et al. 2000; Vives-Rego et al. 2000; Haugland 2002). Using Nile-Red and SYTO-13 stains, phenotypic variability with respect to polyhydroxyalkanoates has been reported for *Pseudomonas aeruginosa* (Vidal-Mas et al. 2001).

Fluorogenic substrates for the classic beta-galactosidase reporter gene, for instance, have successfully been used to demonstrate heterogeneity in the expres-sion of a developmental regulator in *Myxococcus xanthus* (Russo-Marie et al. 1993), as well as to show the characteristics of sporulation bistability in *B. subtilis* (Chung et al. 1995; Chung and Stephanopoulos 1995). In addition, they have been used to study the dynamics of gene expression in a single cell using a microfluidic device (Cai et al. 2006).

When conjugated to a fluorescent dye, macromolecules such as antibodies or nucleic acids can be used to visualize protein localization (immunofluorescence) or the presence of specific RNA or DNA types (fluorescence in situ hybridization [FISH]) (for reviews, see Amann et al. 1995; Davey and Kell 1996; Moter and Gobel 2000; Amann and Ludwig 2000; Brehm-Stecher and Johnson 2004). In short, labeled probes are introduced to the cell, where they bind to target sequences. The high sen-sitivity of the method makes it possible to detect single molecules, and this method has been used to demonstrate transcriptional bursting in *E. coli* (Golding and Cox 2004; Golding et al. 2005). Alternatively, in situ reverse transcriptase (RT) polymerase chain reaction (PCR) can be used to amplify the signals in a quantitative manner. Using this technique, nongenetic heterogeneity was demonstrated for *Salmonella* (Tolker-Nielsen et al. 1998) and *Methanosarcina* (Lange et al. 2000), for instance. It has to be noted that RT-PCR can also be used to demonstrate genetic differences between single cells, as, for instance, in the case of phase-variable phenotypes.

All of the fluorescence-based techniques described above rely on the introduction of a probe into the cells, which might pose problems. A major breakthrough in molecular biology has been the use of fluorescent proteins in vivo. The green fluorescent protein (GFP), which was originally isolated from the jellyfish *Aequorea victoria*, offers a noninvasive reporter for gene activity and is now commonly used by microbiologists (Tsien 1998; Southward and Surette 2002).

Modified variants of the protein with distinct excitation and emission spectra have been developed for use in bacterial research (Shaner et al. 2005; Shaner et al. 2004), allowing the visualization of more than one protein or promoter activity simultaneously.

Using GFP and its spectral derivatives to study transcriptional regulation, phenotypic variation was demonstrated, for instance, for the expression of the structural gen e for colicin K in *E. coli*, and genomic islands in *Salmonella* (Hautefort et al. 2003) and *Pseudomonas* (Sentchilo et al. 2003a; Sentchilo et al. 2003b). In addition, it was shown that an exponentially growing culture of *B. subtilis* consists of two subpopulations that differ in the expression of swarming/motility genes (Kearns and Losick 2005).

It has to be noted that when the *gfp* gene is translationally coupled to a gene of interest, it can be used to ascertain the subcellular localization of the encoded protein. Such studies have led to many groundbreaking discoveries in bacteria, including the presence of a so-called "replisome" (Lemon and Grossman 1998) and the presence of a dynamic bacterial cytoskeleton (Carballido-Lopez 2006). Additionally, they may reveal another layer of complexity with respect to phenotypic heterogeneity, because the subcellular localization of certain proteins may differ between subpopulations. It was shown, for instance, that RecA localizes to the polar competence machinery of *B. subtilis* cells, whereas it is associated in nucleoids in noncompetent cells (Kidane and Graumann 2005). Because competence occurs in a subpopulation of cells (see Sect. 12.5.2), the localization pattern of RecA is similarly heterogeneous.

Bioluminescence reporters based on the *lux* system, although in principle suitable for single-cell analyses, are less frequently used because the resolution is inferior to fluorescence. Their primary applications are in biosensors, host–pathogen interaction, and the study of circadian rhythms (Greer III and Szalay 2002).

Some cellular characteristics are more easily assessed by alternative techniques. The study of surface properties of single cells is greatly facilitated by microcapillary electrophoresis (reviewed in Brehm-Stecher and Johnson 2004). Surface properties may depend on cell age, cell cycle status, or cell type, and are, therefore, of interest to a broad field of researchers (Glynn Jr et al. 1998). Using capillary electrophoresis, it was shown that cultures of *Enterococcus faecalis* demonstrate heterogeneity in surface charge, which affects adhesion and biofilm formation (van Merode et al. 2006a; van Merode et al. 2006b).

Finally, cell density centrifugation deserves to be mentioned. Although technically not a single-cell technique, subpopulations with discernible buoyant density can be separated and analyzed after recovery. This type of analysis has revealed subpopulations during growth of *E. coli* (Makinoshima et al. 2002) and *Vibrio parahaemolyticus* (Nishino et al. 2003), and genetic competence of *B. subtilis* (Hadden and Nester 1968; Cahn and Fox 1968; Dean and Douthit 1974; Haijema et al. 2001).

12.3   Sources of Phenotypic Variation

The origins of different cell types in a culture can be as diverse as the phenotypes the cells display, and can be both genetic and nongenetic in nature.

In the stationary-growth phase, bacteria can display *adaptive mutagenesis*. Several genetic determinants for this process have been identified, such as DNA duplications and the expression of genes involved in DNA metabolism and genome integrity. When the mutations affect certain bacterial properties, they may lead to a growth advantage of certain cells over others (Finkel and Kolter 1999). As a result, one phenotype may overgrow the other, potentially leading to the dynamic coexistence of multiple phenotypes.

Another well-characterized mechanism to generate phenotypic variation is phase variation (for review, see Henderson et al. 1999). Phase-variation phenotypes show a binary pattern, in which genes are expressed (ON) or not expressed (OFF). Transitions between the two states are random, and occur at a relatively high frequency ($>10^{-5}$ per generation). Phase variation resembles antigenic variation of pathogenic bacteria in many aspects, although these are distinct processes. The best-characterized examples of phase variation involve cellular appendages such as fimbriae and flagellae. Phase variation frequently depends on genetic changes, such as genomic inversions or strand-slippage mechanisms. For instance, the phase-variable expression of both type I fimbriae in *E. coli* (Abraham et al. 1985) and flagellae in *Salmonella* (Zieg et al. 1977) involve the inversion of a DNA element (Fig. 12.2a), whereas the expression of certain outer membrane proteins in *Neisseria* (reviewed in Meyer et al. 1990) and virulence factors in *Bortedella* (Stibitz et al. 1989) is determined by frameshift mutations as the result of strand slippage (Fig. 12.2b).

In addition to genetic changes, the origin of phenotypic variation may reside in modifications of the DNA (Fig. 12.2c). These changes are epigenetic in nature, because they do not involve a change in DNA sequence, yet the phenotypes associated with the modification can be inherited by daughter cells after division. For instance, the phase-variable expression of the *pap* operon in *E. coli* involves methylation of two GATC sequences by the Dam methylase (van der Woude et al. 1996). In this case, the methylation pattern of the DNA affects the position of binding for a transcriptional regulator, resulting in repression or activation of transcription.

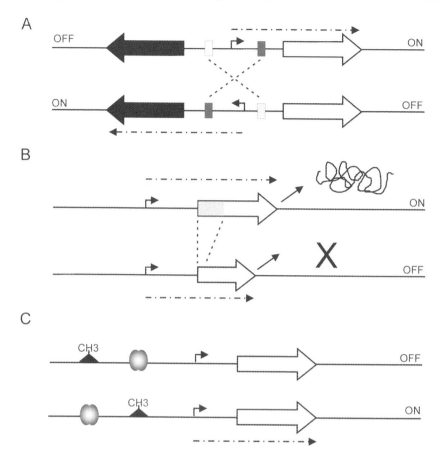

Fig. 12.2 Phase variation. **a** Phase variation as a result of genomic inversion. An element containing a promoter is located in between two divergently oriented genes, driving the expression of one of the genes. After inversion of the element, the other gene is expressed. **b** Phase variation caused by strand slipping. The deletion of a small fragment of the DNA places results in a frame-shift mutation and premature termination of protein synthesis. **c** Phase variation caused by methylation. Methylation of certain sequences of the DNA can prevent the binding of transcriptional regulators, and cause them to bind at different locations. As a result, genes are ON or OFF

However, phenotypic variations need not be accompanied by changes in DNA sequence or modifications of the DNA. Asynchronously growing cells can display different phenotypes depending on the stage of the cell cycle or the growth. Cell cycle-related variation plays an important role in the coordinate expression of genes required for replication and cell division (Holtzendorff et al. 2006; Avery 2006), and has been shown to be responsible for the heterogeneous expression of certain proteins (Sumner and Avery 2002; Rosenfeld et al. 2005). Biological cycles also occur on a longer time scale, as in circadian rhythms (Williams 2007). Because many cellular processes are temporally regulated, cells in different growth phases would automatically result in phenotypic variation at the population level. These forms of phenotypic variation can be distinguished by the synchronization of starter cultures, as well as by correlation with known markers for the cyclic processes.

Cell age can add another level of complexity (Avery 2006). Recently, elegant data were collected suggesting that even symmetrically dividing organisms such as *E. coli* suffer from aging, because cells with older cell poles show lower growth rates compared with cells that have inherited more new poles (Stewart et al. 2005). Although the mechanistic details regarding the role of aging in phenotypic variation are unknown, it implies that cell aging also needs to be taken into account for bacterial research.

It was already pointed out in Sect. 12.1 that gene expression in a homogeneous culture displays a normal distribution in gene expression levels, because of stochastic events (noise). Because, in the absence of a stimulus, inducible transcription factors are, in general, low abundant proteins, it is noteworthy that noise is more significant when fewer molecules are involved (finite number effect). It was shown that noise is most pronounced in the case of low transcription coupled with high translation (Ozbudak et al. 2002), although both transcription and translation levels can have a significant influence. Total noise levels can be broken down into intrinsic noise and extrinsic noise (Swain et al. 2002). Intrinsic noise is inherent to the biochemical process of gene expression, whereas extrinsic noise can be attributed to external factors acting on a single cell (Elowitz et al. 2002). The two types of noise can be discriminated using a system of distinguishable fluorescent reporters that are driven from the same promoter but are located at different positions on the chromosome; intrinsic noise will be reflected by a broad range of ratios in the fluorescence levels of the two reporters in single cells, whereas, for extrinsic noise, the ratios will be similar but the absolute levels of fluorescence may differ per cell (Fig. 12.3). Fluctuations caused by the different types of noise demonstrate different characteristics: intrinsic noise causes rapid fluctuations, whereas extrinsic noise causes fluctuations on a larger time scale (Rosenfeld et al. 2005). Finally, although both types of noise can lead to similar population distributions, it has been reported that, overall, extrinsic noise contributes more to phenotypic variation than intrinsic noise (Elowitz et al. 2002). It is of importance to realize that noise can be propagated in a gene network and that this can significantly affect the final output of a regulatory cascade (Pedraza and van Oudenaarden 2005).

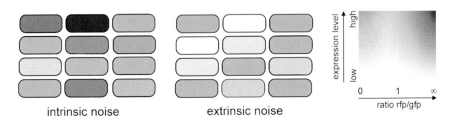

intrinsic noise   extrinsic noise

Fig. 12.3 Noise in gene expression. Intrinsic noise results in a wide range of ratios (from red to green) between the two distinguishable reporter constructs that are driven from the same promoter. Extrinsic noise results in fluctuations in the levels of transcription (from dark to light), whereas the ratio between the two reporters (in this case equal to 1) is the same for all cells. Rfp, red fluorescent protein

In general, high levels of noise are regarded as detrimental for the fitness of cells. In accordance with this, essential genes were reported to exhibit lower levels of noise than non-essential genes (Fraser et al. 2004). One of the most ubiquitous and efficient ways to control the level of noise in gene expression is the introduction of a negative feedback loop (Becskei and Serrano 2000). By limiting the production as the levels of a certain protein increase, negative autoregulation limits the range over which fluctuations in the level of this protein can occur. In addition to this, the intertwinement of regulatory networks can have the ability to filter noise (Hooshangi et al. 2005).

In specific cases, however, noise is exploited to generate phenotypic variability (Rao et al. 2002, Hasty 2000). For example, it was shown that noise in the regulatory cascade governing the chemotactic response of *E. coli* is responsible for behavioral variability of individual cells, as measured by the rotational direction of flagella (Korobkova et al. 2004). Importantly, noise can be amplified through the introduction of a positive feedback loop, resulting in bistability (or more universal: feedback-based multistability [FBM]; see Sect. 12.4 and Smits et al. 2006).

12.4   Bistability in Gene Expression

Here, we refer to bistability as the coexistence of two distinct subpopulations of cells in a culture that does not depend on modifications of the DNA. Each of these subpopulations represents a discrete level of gene expression, referred to as a state. If a gene regulatory network can lead to more than a single stable state, it is said to exhibit multistationarity. Thus, multistationarity at the cellular level can result in a bistable (or multistable) bacterial population (Smits et al. 2006). Switching between these states occurs stochastically, and is usually reversible.

As early as the 1960s, it was postulated that feedback regulation might be responsible for the stable states observed for all-or-none enzyme induction (Novick and Weiner 1957; Monod and Jacob 1961). Pioneering work of Thomas has highlighted the importance of feedback regulation for multistationarity (Thomas 1978, 1998). Subsequently, the prerequisites for a gene network to generate a multistable output were determined by mathematical modeling and the analysis of artificial gene networks (Angeli et al. 2004; Ferrell Jr 2002; Friedman et al. 2006; Gardner et al. 2000; Kobayashi et al. 2004; Hasty et al. 2000; Hasty et al. 2002; Isaacs et al. 2003).

If was found that the system needs to exhibit nonlinear kinetics, implying that the response (e.g., the expression of a reporter gene) is not a linear function of the concentration of a regulator (Ferrell Jr 2002). Frequently, this is apparent from the observation of a threshold concentration of the regulator to elicit a response (Chung et al. 1994). For transcriptional regulators, nonlinearity can for instance be observed as a result of the requirement for multimerization, cooperativity in DNA binding, or phosphorylation of certain amino acid residues.

Without the introduction of complicated mathematical models, the effects of single positive feedback can easily be envisaged (Fig. 12.4). In the absence of

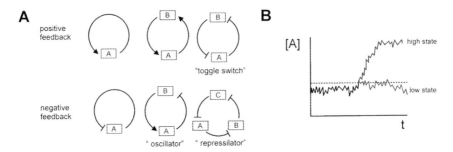

Fig. 12.4 Gene networks. **a** Effective positive feedback can be established by, for instance, a one-component or a two-component positive feedback loop, or a two-component negative feedback loop (toggle switch). *Arrowheads* indicate a positive effect, *perpendicular lines* a negative effect. Negative feedback is displayed by networks harboring a single negative feedback loop. A combination of positive and negative interactions in a two-component loop (oscillator) as well as a three-component negative feedback loop (repressilator) can cause a system to oscillate between states. **b** The effect of positive feedback on noise-induced fluctuations in protein level in time. In the presence of a positive feedback loop, expression levels will increase to the maximum level (high state), whereas, in the absence of positive feedback, fluctuation remain within the low state (**b**, adapted from Smits et al. 2006) with permission of the publisher.

autostimulation, an increase in the level of a regulator will result in a concomitantly increased response (graded response). Autostimulation requires a threshold concentration of a regulator. As a result of noise, certain cells will reach the threshold, initiating a positive feedback loop and causing cells to switch between a low- and high-expressing state (Fig. 12.4b) (Ferrell Jr 2002; Isaacs et al. 2003). Indeed, it was experimentally verified that the introduction of an autostimulatory loop can convert a graded response into a bistable one (Becskei et al. 2001). Increasing the level of the autostimulatory regulator (either by increasing its expression or its activity) will cause more cells to switch from the low to the high state, without affecting the position of the peaks in the bistable distribution pattern (Becskei et al. 2001; Smits et al. 2005; Veening et al. 2005).

Similar behavior is observed with a system of two activators that stimulate the expression of each other. In this double-positive feedback loop, regulator A will stimulate the expression of regulator B once a certain threshold has been reached. In the resulting stable state, the levels of A and B are both high.

A system of two regulators that repress each other, known as a toggle switch, also exhibits two stable states (Ferrell Jr 2002; Tian and Burrage 2004; Lipshtat et al. 2006). As the concentration of regulator A reaches the threshold for repression of gene B, it will indirectly activate its own transcription. The two stable states for a toggle switch, therefore, correspond to reciprocal states of A and B; the presence of high levels of regulator A excludes high levels of B, and vice versa. Experimentally, the functionality of a toggle switch was shown in *E. coli* (Gardner et al. 2000; Kobayashi et al. 2004). The strength of the regulatory interactions is critical, because a toggle switch-like module will not result in multistationarity if one of the components acts much stronger than the other.

Importantly, only gene networks that demonstrate net positive feedback (i.e., including an even number of negative feedback interactions and/or any number of positive feedback loops) seem to be capable of causing multistationarity (Fig. 12.4b) (Angeli et al. 2004). As pointed out before, single-negative feedback reduces noise levels, and leads to homeostatic behavior (Becskei and Serrano 2000). In addition, a combination of a positive and a negative interaction is capable of inducing rhythmic oscillations. This type of regulation is a key component of both cell cycle regulation and circadian rhythms (Holtzendorff et al. 2006; Williams 2007). Finally, it was shown that a three-component negative feedback loop (repressilator) can also cause oscillatory behavior (Elowitz and Leibler 2000).

Although FBM seems to be the predominant form of bistability, one has to realize that the presence of positive feedback is no guarantee for multistationarity (Ferrell Jr 2002; Acar et al. 2005), and it is possible that systems exhibit bistability only within a certain range of parameters, such as inducer concentrations (Santillán et al. 2007, Ozbudak et al. 2004). In addition, mechanisms other than transcriptional autoregulation, such as multisite phosphorylation (Lisman 1985; Ferrell Jr 1996; Ortega et al. 2006), may be capable of inducing a bistable response.

12.5 Examples of "Natural" Multistability

12.5.1 Lactose Use in E. coli

The classic example of bistability or multistability is that of the regulation of the genes responsible for lactose use in E. coli (Novick and Weiner 1957). In fact, regulation of the *lac* operon was the first genetic regulatory mechanism to be elucidated and is often used as the canonical example of prokaryotic gene regulation (Jacob and Monod 1961).

As a result of many years of research, the molecular mechanisms involved in the regulation of lactose use are now well known (for reviews, see Müller-Hill 1996; Laurent and Kellershohn 1999; Smits et al. 2006).

The *lac* operon comprises three genes that are required for the uptake and catabolism of lactose: *lacZ*, encoding β-galactosidase; *lacY*, encoding lactose permease; and *lacA*, encoding a transacetylase.

The expression of the *lac* operon is negatively regulated by the LacI repressor, which is inhibited by allolactose. This molecule is an isomer of lactose, converted from intracellular lactose by the constitutively expressed β-galactosidase (β-galactosidase [LacZ] enzyme in an alternative reaction to the hydrolytic reaction). LacI is also positively regulated by the cyclic-AMP receptor protein (CRP), which is activated under low sugar availability. High levels of the sugar lactose, therefore, inhibit the LacI repressor by a dual mechanism. The system demonstrates a potential positive autostimulatory loop, because the *lac* operon includes the structural gene for the permease for lactose uptake. However, β-galactosidase can metabolize both lactose and allolactose, interrupting the positive feedback loop (Fig. 12.5a),

Fig. 12.5 Natural bistable systems. **a** Simplified scheme of the regulatory network of lactose use in *E. coli*. A gratuitous inducer, IPTG, is taken up by the permease LacY and inhibits the activity of the repressor LacI. As a result, transcription of the entire *lac* operon, including the structural gene encoding the permease, is activated, establishing a positive feedback loop. The transcription of the *lac* operon is additionally modified by the CRP protein, which can act as an activator or repressor. **b** The core of the competence regulatory network in *B. subtilis*. ComK stimulates its own expression, and forms a putative toggle switch with Rok. Experiments have established that only ComK autoregulation is critical for the bistable expression pattern and that the fraction of competent cells in a *rok* mutant is increased. **c** In *C. albicans*, the switch between white and opaque states depends on autostimulation of WOR-1. Switching depends on mating type: a/α cells cannot undergo switching because of repression or WOR-1 transcription. The gene demonstrates a basal level of transcription in a or α cells, which could lead to switching between the white and opaque states

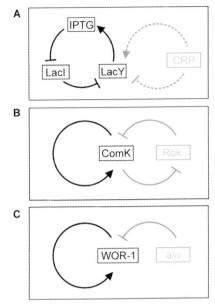

because of which, it is unclear whether the system is capable of demonstrating bistability under natural conditions (van Hoek and Hogeweg 2006). Because of the detail of characterization, however, it makes a good model system to investigate and analyze bistability as it may occur for other processes.

In 1957, Novick and Weiner showed that when a population of *E. coli* is induced at low levels with a gratuitous inducer (i.e., a molecule that cannot be metabolized), reculturing of single cells results in populations that either give high or low *lac* expression. This phenomenon was called all-or-none enzyme induction (Novick and Weiner 1957), and is indicative of the presence of two coexisting subpopulations of cells in a culture (one induced for *lac* expression and one not induced). These subpopulations can occur at concentrations of inducer near the threshold at which stochastic fluctuations can cause part of the cells to initiate an autostimulatory loop. Further characterization of this system by Cohn and Horibata revealed that the fraction of cells that highly expresses the lactose use genes depends on the presence of specific sugars in the growth medium and on the history of the inoculum (Cohn and Horibata 1959a; Cohn and Horibata 1959b). Already in 1961, Monod and Jacob hypothesized that interactions between components of the regulatory network governing lactose use might explain the multistable behavior, by stating "[Moreover,] it is obvious from the analyses of these [regulatory, red.] mechanisms that their known elements could be connected into a wide variety of "circuits", endowed with any desired degree of stability."

Switching from one state to the other (from "ON" to "OFF" or vice versa) requires either an induction or a relief of inducer greater than that required for the reverse

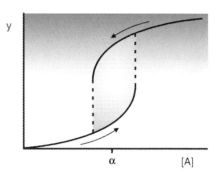

Fig. 12.6 Schematic depiction of hysteresis. Changes in concentration of regulator are indicated by *arrows*. Depending on the starting state of the system, a cell can be in an ON (*grey*) or OFF state (*white*) at intermediate concentrations of the regulator (such as α)

transition. This phenomenon, called hysteresis (Fig. 12.6), is responsible for the observed epigenetic inheritance of the expression state over many generations (Novick and Weiner 1957; Cohn and Horibata 1959a; Cohn and Horibata 1959b). Such an unequal force essentially acts as a buffer, making the switch robust to noise, and minimizing accidental switching of the system. Under certain conditions, however, switching back from an ON state to an OFF state may still be possible. Hysteresis is not only observed for biological, but also for physical systems exhibiting bistability. In Fig. 12.6, at concentration α of inducer, the system can be either in an ON or OFF state, depending on the starting state. The state of a hysteretic system, thus, depends on its recent history. A neat analogy that describes memory by hysteresis was sketched in a review by Casadesus and D'Ari (2002). Microbiological agar is a polymer solution that at 60°C can be either liquid or solid. If the agar is melted by heating it to 100°C and then cooled down to 60°C, it will remain liquid, whereas, if solid agar at room temperature is warmed up to 60°C, it will remain solid. Thus, the state of agar at 60°C is a memory of its history. The origin of hysteresis can, for instance, lie in the stability of component within a gene-regulatory network. The hysteretic behavior of the *lac* operon, for instance, is a consequence of the abundance and stability of the LacY permease (Ozbudak et al. 2004). When little permease is present, the concentration of IPTG required to trigger the autostimulatory loop is high. In contrast, when the level of permease is high (mostly corresponding to an already induced state), cells need little IPTG to maintain high levels of *lac* expression.

Modeling of systems and networks has a long tradition in, for instance, engineering and physics, but did not become common for biological sciences until recently. This can be attributed to the relative complexity of biological systems and the inability to quantitatively analyze the data obtained from biochemical experiments. However, with the development of new experimental methods, such as single-cell analysis using fluorescent reporters, and with the increase in computational power, accurate modeling of biological systems has become possible. As a result, a number of studies combine single-cell analyses with mathematical modeling to describe the behavior of the *lac* operon at both the molecular and population level.

Modeling using (nonlinear) differential equations enabled dynamic simulations of the lactose use network and demonstrated that bistable hysteretic gene expression can be expected on the basis of reasonable biological parameters and indicated the importance of the positive feedback loop in establishing multistability (Laurent and

Kellershohn 1999; Yildirim and Mackey 2003; Yildirim et al. 2004; Ozbudak et al. 2004). Moreover, mathematically, a strong parameter sensitivity was shown. Ozbudak and colleagues demonstrated, using mathematical modeling in combination with single-cell gene expression analyses, that the lactose operon can demonstrate a graded response that can be converted to a binary response (bistability), providing a theoretical scaffold for previous unexplained observations that both states can occur in enzyme induction systems (Ozbudak et al. 2004; Biggar and Crabtree 2001). Santillan and coworkers recently showed that bistability may strongly depend on the concentrations of extracellular sugars (Santillán et al. 2007).

Whether bistability within the *lac* operon is physiologically relevant for *E. coli* remains to be tested because, when the *lac* operon is induced with lactose, bistability is not observed. Recent modeling studies, using *in silico* evolution, indicate that the wild-type *lac* operon indeed is not a bistable switch, but only shows bistability with artificial inducers (van Hoek and Hogeweg 2006).

Taken together, these studies show the quantitative power of using mathematical models to predict and identify key parameters within a (multistable) gene-regulatory network.

12.5.2   Genetic Competence in B. subtilis

Competence for genetic transformation, the active acquisition of extracellular DNA and the heritable incorporation of its genetic information into the host cell, is one of the stationary phase differentiation processes in which the Gram-positive bacterium *B. subtilis* can engage (for reviews, see Dubnau and Lovett 2002; Hamoen et al. 2003). Competence is the transient state governed by an extensive regulatory cascade that, ultimately, results in the activation of ComK, the master competence transcription factor. ComK is responsible for the activation of the late competence genes that constitute the DNA uptake and recombination machinery, and also stimulates its own expression (van Sinderen et al. 1995; van Sinderen and Venema 1994).

Early work has demonstrated that competent cells are physiologically distinct from their noncompetent counterparts. Nester and Stocker (1963) found that there is a lag in the increase of the number of transformants after the addition of transforming DNA to a competent culture of *B. subtilis*. In addition, the expression of a marker on the transforming DNA was not detectable during this period (Nester and Stocker 1963). They elegantly showed that, during the observed lag phase, the number of viable cells could be reduced using penicillin G, whereas the number of transformants stayed constant (Nester and Stocker 1963). These results demonstrate not only a biosynthetic latency of competent cells, but also formed the first indication that a competent culture consists of at least two subpopulations. Several later studies confirmed the presence of these subpopulations using density gradient centrifugation (Cahn and Fox 1968; Hadden and Nester 1968; Singh and Pitale 1968; Singh and Pitale 1967). These experiments demonstrated that a lower buoyant density was an inherent property of the competent cells (Cahn and Fox 1968; Hadden and Nester

1968; Dooley et al. 1971) and did not originate from or require the uptake of DNA. On the basis of density centrifugation, as well as transformation experiments using nonlinked marker genes, it was estimated that competent cells form approximately 10% of the total population (Singh and Pitale 1968; Singh and Pitale 1967; Hadden and Nester 1968; Cahn and Fox 1968). Mutagenesis of *B. subtilis* using a transposon that carries a promoterless copy of the reporter gene, *lacZ*, revealed that many genes whose deletion results in a transformation deficiency when disrupted are preferentially expressed in the band with a lower buoyant density (Albano et al. 1987; Hahn et al. 1987). In addition, the separation of competent and noncompetent cells was found to depend on the first open reading frame of the *comG* operon (Albano et al. 1987; Albano et al. 1989; Hahn et al. 1987; Albano and Dubnau 1989). To date, ComK is the only known regulator of *comG* expression (Hamoen et al. 1998; Susanna et al. 2004), and, indeed, the heterogeneity of competence could be traced back to *comK* transcription. Hahn and coworkers found that *comK-lacZ* activity was still associated with competent cells, whereas the expression of a gene acting upstream of ComK could be detected in both the light and heavy band of cells separated on a density gradient (Hahn et al. 1994). It was found that 5 to 10% of the cells in a wild-type culture express *comK-gfp* (Haijema et al. 2001), which was consistent with previous estimates (Singh and Pitale 1968; Hadden and Nester 1968; Cahn and Fox 1968). Two other important observations were made. First, the presence of the product of a late competence gene, *comEA*, was always found to coincide with *comK* expression. Second, it became clear that there are little or no intermediate levels of ComK–GFP fluorescence. This is consistent with the presence of two distinct bands, rather then a smear, in the density centrifugation experiments. When the fluorescence from a ComK-dependent *gfp* reporter in individual cells is analyzed in a quantitative manner, the resultant curve shows a bimodal distribution indicative of a bistable process (Fig. 12.1).

Recently, the origin of this bistability was experimentally addressed. The regulatory network governing competence development comprises two structural modules that could cause bistability in *comK* expression (Figs. 12.4a and 12.5b). First, ComK stimulates its own expression by directly binding as a tetramer to its own promoter region (van Sinderen and Venema 1994; Hamoen et al. 1998). As such, it forms a single positive feedback loop. Second, ComK seems to be able to form a toggle switch with a repressor of the *comK* gene, *rok* (Hoa et al. 2002). However, on the basis of two independent studies, it was found that ComK autostimulation is the critical determinant for the bistable expression pattern. Smits and coworkers (2005) demonstrated that bistability is still observed in a strain devoid of all other levels of regulation except autostimulation, and competence is still initiated in a bistable manner. In addition, it was reported that replacement of the native copy of *comK* with an inducible copy of the gene results in a graded response that depends on the level of induction (Smits et al. 2005; Maamar and Dubnau 2005). Although these experiments show that ComK autostimulation is indispensable for the bistability of competence, they do not exclude a putative ComK–Rok toggle switch (Fig. 12.5b), because the manipulations described above would also affect such a module. Maamar and Dubnau (2005) addressed this issue by introducing a *rok* mutation in the strain with the inducible ComK. A bistable gene expression pattern was observed both the presence and absence of *rok*, but only in the presence of ComK autostimulation. Together, these data show that competence is initiated in a bistable manner, depending on ComK autostimulation.

Competence, however, is a transient differentiation process (Hadden and Nester 1968; Nester and Stocker 1963; Dubnau 1993), and it is not known how the reduction in cellular ComK levels, necessary for the escape from the competent state, is brought about. At least partially, it was found to rely on an unexplained reduction of *comK* transcription (Leisner et al 2007; Maamar et al 2007). An interesting alternative hypothesis was recently put forward, based on time-lapse microscopy (Suel et al. 2006). The authors showed that the transient differentiation observed in competence development resembles an excitable gene regulatory network and that such characteristics could be explained by a slow-acting negative feedback loop in combination with fast-acting positive autoregulation by ComK. A ComK-dependent negative feedback loop might exist (Hahn et al. 1994; Smits et al 2007), and it provides an attractive hypothesis for the transience of competence. However, investigations that are more detailed are required to address this question.

Modeling the competence regulatory pathway has provided support for the observations that intrinsic noise in *comK* expression destines cells to become competent, and also provides insight into factors that affect the probability of becoming competent and the time spent in the competent state (Maamar et al. 2007; Suel et al. 2007).

Interestingly, *B. subtilis* is not the only organism in which competence is associated with phenotypically distinct subpopulations of an isogenic culture. Competent cultures of *Streptococcus pneumoniae* are comprised of a "donor" and an "acceptor" population (Steinmoen et al. 2002; Steinmoen et al. 2003). The competent *Streptococcus* cells are thought to actively kill noncompetent cells (Guiral et al. 2005; Kreth et al. 2005). This is strikingly similar to the cannibalism described for *B. subtilis* (Gonzalez-Pastor et al. 2003), which depends on the master regulator (Spo0A) of another bistable differentiation process, sporulation (Chung et al. 1995; Veening et al. 2005). It is noteworthy that the bistable response in sporulation can be fine-tuned by regulating the phosphorylation state of the Spo0A protein (Veening et al. 2005). It can be expected that other differentiation processes affected by Spo0A potentially also demonstrate bistable behavior.

12.5.3   *White–Opaque Switching in C. albicans*

In this section, we will summarize recent advancement in the understanding of phenotypic switching in the fungus *Candida albicans*. Although the inclusion of a fungal system in a book on bacterial physiology may seem inappropriate, it is of importance to realize that phenotypic switching occurs in both fungi and bacteria (Burchard et al. 1977; Chantratita et al. 2007; Guerrero et al. 2006). However, because the molecular mechanisms underlying the switching in bacteria are relatively poorly understood, we have chosen to discuss a fungal example.

White–opaque switching, a change between two phenotypically distinct cell types, is one of the best-characterized mechanisms of generating phenotypic diversity of the opportunistic fungal pathogen *Candida albicans* (for reviews, see Bennett and Johnson 2005; Johnson 2003; Lockhart et al. 2003). The white–opaque switch was originally identified in strain WO-1, a clinical isolate of this fungus, in which switching occurs at relatively high frequency (Slutsky et al. 1987). The two phenotypic states can easily be discerned as white cells forming white, dome-shaped colonies on the plate, or opaque cells giving rise to darker and flatter colonies. In addition, white cells appear virtually spherical under the microscope, whereas opaque cells are banana shaped and demonstrate plasma membrane protrusions (Anderson and Soll 1987). It was shown that opaque cells are directly derived from white progenitor cells (Rikkerink et al. 1988). For a long time, it was assumed that *Candida* reproduces asexually, but the identification of a mating type locus similar to that of *S. cerevisiae* (Hull and Johnson 1999) led to the identification of a mating competent state (Hull et al. 2000; Magee and Magee 2000; Tsong et al. 2003). Strikingly, it was found that white–opaque switching is governed by mating-type locus homeodomain proteins (Miller and Johnson 2002). Moreover, it was shown that only opaque cells were able to mate, and that these cells were homozygous for mating type (Lockhart et al. 2002). Profiling of white and opaque cells has revealed numerous differences between the two cell types, most notably the induction of mating-type genes in opaque cells (Lan et al. 2002; Tsong et al. 2003).

Recently, two independent studies identified a single transcriptional regulator responsible for the white to opaque transition (Zordan et al. 2006; Huang et al. 2006). This regulator, WOR-1, is preferentially expressed in opaque cells and, therefore, repressed in a/α cells (Lan et al. 2002; Tsong et al. 2003). It was reported that, on induction of an ectopic copy of the gene encoding the regulator, white cells are converted to opaque cells, suggesting that the WOR-1 bypasses the repression by the a/α repressor in heterozygous cells (Huang et al. 2006; Zordan et al. 2006). WOR-1 is capable of stimulating its own expression by binding directly to its own promoter (Zordan et al. 2006), which results in a bistable expression pattern (Huang et al. 2006). Two important observations were made by Zordan et al. (2006). First, they found that, in the presence of WOR-1 autostimulation, white cells were stably converted into opaque cells by a pulse in expression from an ectopic copy of the regulator gene. Second, they observed that, in the absence of an autostimulatory loop, continuous expression of WOR-1 was required to maintain the opaque phenotype.

In conclusion, white–opaque switching is repressed by the a/α repressor in heterozygous cells (Lockhart et al. 2002; Soll et al. 2003). In cells homozygous for mating type (which may still be white), a basal level of expression from the *WOR-1* locus will cause some cells to switch to the opaque state when the threshold for WOR-1 autostimulation is reached (Fig. 12.5c). It is of importance to realize that the bistable switching only refers to WOR-1 expression states, and mating type DNA rearrangement and homozygosity/heterozygosity merely set the conditions under which bistability can occur.

Feedback-based bistability may be a common mechanism for switches such as the white–opaque switch. *Myxococcus xanthus* displays a switch between a tan and a yellow phenotype, for instance, and, although this is sometimes referred to as phase variation, no genomic inversion could be demonstrated and the molecular mechanisms remain elusive (Burchard et al. 1977; Laue and Gill 1994). Similarly, the molecular mechanisms behind phenotypic switching of *Cryptococcus neoformans* (Guerrero et al. 2006) and *Burkholderia pseudomallei* (Chantratita et al. 2007) remain to be established. Interestingly, developing cells of this organism display a bimodal distribution in the expression of the *dev* locus (Russo-Marie et al. 1993), which has been suggested to originate from a positive feedback loop (Viswanathan et al 2007).

12.6   Perspectives and Implications of Bistability

Phenotypic variation, in general, and bistability, in particular, are widespread phenomena in the bacterial realm. This has implications for medicine, food industry, biotechnology, and bioinformatic analyses.

In medicine and the food industry, the existence of subpopulations of cells resistant to conventional treatments has a large impact on the control of bacterial infections and contaminations. Latent bacterial infections, the occurrence of persisters, and tolerance of pathogens to a multitude of drugs are increasingly problematic (Lewis 2007).

Persistence is the well-known phenomenon that, after treatment of a bacterial strain with a specific antibiotic, rapid killing of the vast majority of cells is observed, followed by a more complex and slow killing of the remaining cells. Eventually, a small proportion of the cells can survive. This behavior has been described for *Staphylococcus aureus* treated with penicillin in 1944 (Bigger 1944). Persistence is a bet-hedging strategy because, under optimal conditions, a majority of cells proliferate quickly and a small subpopulation suppresses growth. During times of stress, e.g., the presence of antibiotics, these persistent cells can prevent extinction of the entire population (Kussell et al. 2005). Persisters exhibit reduced translation and topoisomerase activity, and/or reduced cell wall biosynthesis, as a result of which, the targets of many antibiotics are blocked. Thus, these cells cannot be easily killed, at the cost of nonproliferation (reviewed in Lewis 2007). The molecular mechanism underlying persistence remained obscure until recently. The use of optical microscopy for single-cell analysis showed that persistence in *E. coli* is a phenotypic switch, and that at least two types of persistence can be identified: stationary phase-induced persistence (type I) and spontaneous persistence (type II) (Balaban et al. 2004). The generation of persisters is most probably caused by stochastic fluctuation and exceeding thresholds, just as described in Sect. 12.4 of this review for the mechanisms underlying multistationarity (Lewis 2007). Possibly, the occurrence of persisters can be reduced by factors that enhance the switching rate from persister to normal growth, providing possible solutions for drug use that is more effective (Balaban et al. 2004).

Phenotypic tolerance (resistance caused by a multifactorial phenotypic adaptation) is closely related to persistence. When bacteria are exposed to bactericidal concentrations of antimicrobial compounds, their sensitivity gradually decreases, and certain subpopulations survive. Usually there is also cross-tolerance to other antimicrobials (Wiuff et al. 2005). Acquired nisin resistance in Gram-positive bacteria was shown to be fully reversible when cells were allowed to grow again in media without the bacteriocin, indicating that the increased resistance was not genetically determined (Kramer et al. 2006). Interestingly, phenotypic heterogeneity was found to occur when cells are subjected to subinhibitory (or sublethal) amounts of bacteriocin. *Listeria monocytogenes* challenged by nisin or leucocin 4010 developed two subpopulations, as observed by fluorescence ratio imaging microscopy (Hornbaek et al. 2006). One of the subpopulations showed cells with a dissipated pH gradient (ΔpH), whereas the other subpopulation maintained the ΔpH. The study shows that it is of great importance to use the appropriate dose of the antimicrobial compound, taking into account its bioavailability, when it is applied for food preservation.

In addition to these phenomena, it has been noted that *Salmonella* growing in host macrophages demonstrates fast- and slow-dividing subpopulations (Abshire and Neidhardt 1993). It remains to be established whether this is related to the production of antimicrobial compounds by the macrophages.

Although the mechanisms responsible for the observed phenotypic heterogeneity are not yet elucidated, it is tempting to speculate that epigenetic regulatory mechanisms such as bistability may play a role. It is interesting to note that, after selection of epigenetically determined phenotypes, the survivor cells retain the ability to generate the same phenotypic variability (Fig. 12.7). Indeed, it was shown through mathematical modeling that cells with a variable phenotype have increased fitness under fluctuating environmental conditions (Thattai and van Oudenaarden 2004).

Besides the occurrence of resistant microbial contaminants in production systems, bistability can also affect processes such as protein production. As extensively discussed in Sect. 12.5.1, the all-or-none enzyme induction of the lactose use operon has become a paradigm for bistability because it can occur for enzyme induction, although similar behavior has been also described for lactose use in *Salmonella enterica* serovar Typhimurium (Tolker-Nielsen et al. 1998). Arabinose-inducible systems are widely used for protein overproduction and purification, where the existence of a nonproducing subpopulation is unwanted. However, arabinose use in *E. coli* also demonstrates a heterogeneous population distribution, most likely as a result of bursts in the synthesis of the arabinose permease, because of stochastic changes in the conformation of a transcriptional activator (Morgan-Kiss et al. 2002). To increase the yields of protein production, derivatives of production strains, in which the production of permease is uncoupled from enzyme induction and the positive feedback loop is, thus, effectively removed, have been constructed (Khlebnikov et al. 2002; Khlebnikov et al. 2001; Khlebnikov et al. 2000). In addition, strains devoid of a dedicated arabinose uptake and degradation system were constructed, in which a mutated LacY permease transports arabinose into the cells (Morgan-Kiss et al. 2002).

From a fundamental point of view, the existence of subpopulations has implications for the interpretation of data from bulk assays, as indicated in Sect. 12.1. It

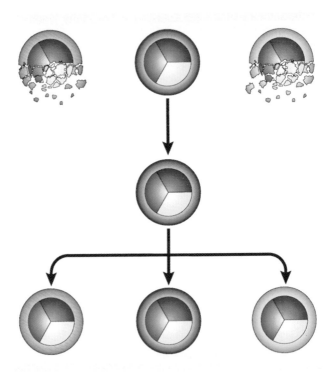

Fig. 12.7 Survival of heterogeneous phenotypes. Three genetically identical cells display pheno-typic variability. The genetic content of the cells is indicated in the inner circle with primary colors. The phenotype is depicted by the colors of the outer rings. In a selective condition, only cells with a particular phenotype will survive. However, because the genetic content has not been altered, the survivor cell has the ability of regenerating the same phenotypic variability (Reproduced from Smits et al. 2006 with permission of the publisher)

would be of great interest to see whether it will become possible to discriminate the behavior of small subpopulations from the large majority of cells with the help of bioinformatics, and, thus, improve the interpretation of high-throughput data, such as DNA microarrays or proteome studies. Additionally, the study of the output from artificial gene networks, modeling studies, and classic molecular and cell biological methods will aid in the reconstruction and connectivity of gene regulatory networks *in silico*.

Highly Recommended Readings

Becskei A, Seraphin B, Serrano L (2001) Positive feedback in eukaryotic gene networks: cell dif-ferentiation by graded to binary response conversion. EMBO J 20:2528–2535

Davey HM, Kell DB (1996) Flow cytometry and cell sorting of heterogeneous microbial popula-tions: the importance of single-cell analyses. Microbiol Rev 60:641–696

Elowitz MB, Levine AJ, Siggia ED, Swain PS (2002) Stochastic gene expression in a single cell. Science 297:1183–1186

Swain PS, Elowitz MB, Siggia ED (2002) Intrinsic and extrinsic contributions to stochasticity in gene expression. Proc Natl Acad Sci USA 99:12795–12800

Ferrell JE Jr (2002) Self-perpetuating states in signal transduction: positive feedback, double-negative feedback and bistability. Curr Opin Cell Biol 14:140–148

Monod J, Jacob F (1961) Teleonomic mechanisms in cellular metabolism, growth, and differentiation. Cold Spring Harb Symp Quant Biol 26:389–401

Smits WK, Kuipers OP, Veening JW (2006) Phenotypic variation in bacteria: the role of feedback regulation. Nat Rev Microbiol 4:259–271

References

Abraham JM, Freitag CS, Clements JR, Eisenstein BI (1985) An invertible element of DNA controls phase variation of type 1 fimbriae of *Escherichia coli*. Proc Natl Acad Sci USA 82:5724–5727

Abshire KZ, Neidhardt FC (1993) Growth rate paradox of *Salmonella typhimurium* within host macrophages. J Bacteriol 175:3744–3748

Acar M, Becskei A, van OA (2005) Enhancement of cellular memory by reducing stochastic transitions. Nature 435:228–232

Albano M, Breitling R, Dubnau DA (1989) Nucleotide sequence and genetic organization of the *Bacillus subtilis comG* operon. J Bacteriol 171:5386–5404

Albano M, Dubnau DA (1989) Cloning and characterization of a cluster of linked *Bacillus subtilis* late competence mutations. J Bacteriol 171:5376–5385

Albano M, Hahn J, Dubnau D (1987) Expression of competence genes in *Bacillus subtilis*. J Bacteriol 169:3110–3117

Amann R, Ludwig W (2000) Ribosomal RNA-targeted nucleic acid probes for studies in microbial ecology. FEMS Microbiol Rev 24:555–565

Amann RI, Ludwig W, Schleifer KH (1995) Phylogenetic identification and in situ detection of individual microbial cells without cultivation. Microbiol Rev 59:143–169

Anderson JM, Soll DR (1987) Unique phenotype of opaque cells in the white-opaque transition of *Candida albicans*. J Bacteriol 169:5579–5588

Angeli D, Ferrell JE Jr, Sontag ED (2004) Detection of multistability, bifurcations, and hysteresis in a large class of biological positive-feedback systems. Proc Natl Acad Sci USA 101:1822–1827

Avery SV (2006) Microbial cell individuality and the underlying sources of heterogeneity. Nat Rev Microbiol 4:577–587

Balaban NQ, Merrin J, Chait R, Kowalik L, Leibler S (2004) Bacterial persistence as a phenotypic switch. Science 305:1622–1625

Barer MR, Harwood CR (1999) Bacterial viability and culturability. Adv Microb Physiol 41:93–137

Becskei A, Seraphin B, Serrano L (2001) Positive feedback in eukaryotic gene networks: cell differentiation by graded to binary response conversion. EMBO J 20:2528–2535

Becskei A, Serrano L (2000) Engineering stability in gene networks by autoregulation. Nature 405:590–593

Bennett RJ, Johnson AD (2005) Mating in *Candida albicans* and the search for a sexual cycle. Annu Rev Microbiol 59:233–255

Biggar SR, Crabtree GR (2001) Cell signaling can direct either binary or graded transcriptional responses. EMBO J 20:3167–3176

Brehm-Stecher BF, Johnson EA (2004) Single-cell microbiology: tools, technologies, and applications. Microbiol Mol Biol Rev 68:538–559

Burchard RP, Burchard AC, Parish JH (1977) Pigmentation phenotype instability in *Myxococcus xanthus*. Can J Microbiol 23:1657–1662

Cahn FH, Fox MS (1968) Fractionation of transformable bacteria from ocompetent cultures of *Bacillus subtilis* on renografin gradients. J Bacteriol 95:867–875

Cai L, Friedman N, Xie XS (2006) Stochastic protein expression in individual cells at the single molecule level. Nature 440:358–362

Carballido-Lopez R (2006) The bacterial actin-like cytoskeleton. Microbiol Mol Biol Rev 70:888–909

Casadesus J, D'Ari R (2002) Memory in bacteria and phage. Bioessays 24:512–518

Chantratita N, Wuthiekanun V, Boonbumrung K, Tiyawisutsri R, Vesaratchavest M, Limmathurosakul D, Chierakul W, Wongratanacheewin S, Pukritiyakamee S, White NJ, Day NP, Peacock SJ (2007) Biological relevance of colony morphology and phenotypic switching by *Burkholderia pseudomallei*. J Bacteriol 189:807–817

Chung JD, Conner S, Stephanopoulos G (1995) Flow cytometric study of differentiating cultures of *Bacillus subtilis*. Cytometry 20:324–333

Chung JD, Stephanopoulos G (1995) Studies of transcriptional state heterogeneity in sporulating cultures of *Bacillus subtilis*. Biotechnol Bioeng 47:234–242

Chung JD, Stephanopoulos G, Ireton K, Grossman AD (1994) Gene expression in single cells of *Bacillus subtilis*: evidence that a threshold mechanism controls the initiation of sporulation. J Bacteriol 176:1977–1984

Cohn M, Horibata K (1959a) Analysis of the differentiation and of the heterogeneity within a population of *Escherichia coli* undergoing induced beta-galactosidase synthesis. J Bacteriol 78:613–623

Cohn M, Horibata K (1959b) Inhibition by glucose of the induced synthesis of the beta-galactoside-enzyme system of *Escherichia coli*. Analysis of maintenance. J Bacteriol 78:601–612

Daniel RA, Errington J (2003) Control of cell morphogenesis in bacteria: two distinct ways to make a rod-shaped cell. Cell 113:767–776

Davey HM, Kell DB (1996) Flow cytometry and cell sorting of heterogeneous microbial populations: the importance of single-cell analyses. Microbiol Rev 60:641–696

Dean DH, Douthit HA (1974) Buoyant density heterogeneity in spores of *Bacillus subtilis*: biochemical and physiological basis. J Bacteriol 117:601–610

Dooley DC, Hadden CT, Nester EW (1971) Macromolecular synthesis in *Bacillus subtilis* during development of the competent state. J Bacteriol 108:668–679

Dubnau D (1993) Genetic exchange and homologous recombination. In: Sonenshein AL, Hoch JA, Losick R (eds) *Bacillus subtilis* and other Gram-positive bacteria. American Society for Microbiology, Washington, D.C., pp 555–584

Dubnau D, Losick R (2006) Bistability in bacteria. Mol Microbiol 61:564–572

Dubnau D, Lovett CM (2002) Transformation and recombination. In: Sonenshein AL, Hoch JA, Losick R (eds) *Bacillus subtilis and its Closest Relatives: from Genes to Cells*. American Society for Microbiology, Washington, pp 453–472

Elowitz MB, Leibler S (2000) A synthetic oscillatory network of transcriptional regulators. Nature 403:335–338

Elowitz MB, Levine AJ, Siggia ED, Swain PS (2002) Stochastic gene expression in a single cell. Science 297:1183–1186

Ferrell JE Jr (1996) Tripping the switch fantastic: how a protein kinase cascade can convert graded inputs into switch-like outputs. Trends Biochem Sci 21:460–466

Ferrell JE Jr (2002) Self-perpetuating states in signal transduction: positive feedback, double-negative feedback and bistability. Curr Opin Cell Biol 14:140–148

Finkel SE, Kolter R (1999) Evolution of microbial diversity during prolonged starvation. Proc Natl Acad Sci USA 96:4023–4027

Fraser HB, Hirsh AE, Giaever G, Kumm J, Eisen MB (2004) Noise minimization in eukaryotic gene expression. PLoS Biol 2:e137

Friedman N, Cai L, Xie XS (2006) Linking stochastic dynamics to population distribution: an analytical framework of gene expression. Phys Rev Lett 97:168302

Gardner TS, Cantor CR, Collins JJ (2000) Construction of a genetic toggle switch in *Escherichia coli*. Nature 403:339–342

Glynn JR Jr, Belongia BM, Arnold RG, Ogden KL, Baygents JC (1998) Capillary electrophoresis measurements of electrophoretic mobility for colloidal particles of biological interest. Appl Environ Microbiol 64:2572–2577

Golding I, Cox EC (2004) RNA dynamics in live *Escherichia coli* cells. Proc Natl Acad Sci USA 101:11310–11315

Golding I, Paulsson J, Zawilski SM, Cox EC (2005) Real-time kinetics of gene activity in individual bacteria. Cell 123:1025–1036

Gonzalez-Pastor JE, Hobbs EC, Losick R (2003) Cannibalism by sporulating bacteria. Science 301:510–513

Greer LF III, Szalay AA (2002) Imaging of light emission from the expression of luciferases in living cells and organisms: a review. Luminescence 17:43–74

Guerrero A, Jain N, Goldman DL, Fries BC (2006) Phenotypic switching in *Cryptococcus neoformans*. Microbiology 152:3–9

Guiral S, Mitchell TJ, Martin B, Claverys JP (2005) Competence-programmed predation of noncompetent cells in the human pathogen *Streptococcus pneumoniae*: genetic requirements. Proc Natl Acad Sci USA 102:8710–8715

Hadden C, Nester EW (1968) Purification of competent cells in the *Bacillus subtilis* transformation system. J Bacteriol 95:876–885

Hahn J, Albano M, Dubnau D (1987) Isolation and characterization of Tn917lac-generated competence mutants of *Bacillus subtilis*. J Bacteriol 169:3104–3109

Hahn J, Kong L, Dubnau D (1994) The regulation of competence transcription factor synthesis constitutes a critical control point in the regulation of competence in *Bacillus subtilis*. J Bacteriol 176:5753–5761

Haijema BJ, Hahn J, Haynes J, Dubnau D (2001) A ComGA-dependent checkpoint limits growth during the escape from competence. Mol Microbiol 40:52–64

Hamoen LW, Van-Werkhoven AF, Bijlsma JJ, Dubnau D, Venema G (1998) The competence transcription factor of *Bacillus subtilis* recognizes short A/T-rich sequences arranged in a unique, flexible pattern along the DNA helix. Genes Dev 12:1539–1550

Hamoen LW, Venema G, Kuipers OP (2003) Controlling competence in *Bacillus subtilis*: shared use of regulators. Microbiology 149:9–17

Hasty J, McMillen D, Collins JJ (2002) Engineered gene circuits. Nature 420:224–230

Hasty J, Pradines J, Dolnik M, Collins JJ (2000) Noise-based switches and amplifiers for gene expression. Proc Natl Acad Sci USA 97:2075–2080

Haugland RP (2002) Handbook of fluorescent probes and research chemicals, 9 edn. Molecular Probes, Inc., Eugene, OR

Hautefort I, Proenca MJ, Hinton JC (2003) Single-copy green fluorescent protein gene fusions allow accurate measurement of *Salmonella* gene expression in vitro and during infection of mammalian cells. Appl Environ Microbiol 69:7480–7491

Henderson IR, Owen P, Nataro JP (1999) Molecular switches—the ON and OFF of bacterial phase variation. Mol Microbiol 33:919–932

Hoa TT, Tortosa P, Albano M, Dubnau D (2002) Rok (YkuW) regulates genetic competence in *Bacillus subtilis* by directly repressing *comK*. Mol Microbiol 43:15–26

Holtzendorff J, Reinhardt J, Viollier PH (2006) Cell cycle control by oscillating regulatory proteins in Caulobacter crescentus. Bioessays 28:355–361

Hooshangi S, Thiberge S, Weiss R (2005) Ultrasensitivity and noise propagation in a synthetic transcriptional cascade. Proc Natl Acad Sci USA 102:3581–3586

Hornbaek T, Brockhoff PB, Siegumfeldt H, Budde BB (2006) Two subpopulations of *Listeria monocytogenes* occur at subinhibitory concentrations of leucocin 4010 and nisin. Appl Environ Microbiol 72:1631–1638

Huang G, Wang H, Chou S, Nie X, Chen J, Liu H (2006) Bistable expression of WOR1, a master regulator of white-opaque switching in *Candida albicans*. Proc Natl Acad Sci USA 103:12813–12818

Hull CM, Johnson AD (1999) Identification of a mating type-like locus in the asexual pathogenic yeast *Candida albicans*. Science 285:1271–1275

Hull CM, Raisner RM, Johnson AD (2000) Evidence for mating of the "asexual" yeast *Candida albicans* in a mammalian host. Science 289:307–310

Isaacs FJ, Hasty J, Cantor CR, Collins JJ (2003) Prediction and measurement of an autoregulatory genetic module. Proc Natl Acad Sci USA 100:7714–7719

Jacob F, Monod J (1961) Genetic regulatory mechanisms in the synthesis of proteins. J Mol Biol 3:318–356

Johnson A (2003) The biology of mating in *Candida albicans*. Nat Rev Microbiol 1:106–116

Joux F, Lebaron P (2000) Use of fluorescent probes to assess physiological functions of bacteria at single-cell level. Microbes Infect 2:1523–1535

Kasten FH (1993) Introduction to fluorescent probes: properties, history and applications. In: Mason WT (ed) Fluorescent and luminescent probes for biological activity: a practical guide to technology for quantitative real-time analysis. Academic Press, Inc., New York, NY, pp 12–33

Kearns DB, Losick R (2005) Cell population heterogeneity during growth of *Bacillus subtilis*. Genes Dev 19:3083–3094

Kidane D, Graumann PL (2005) Intracellular protein and DNA dynamics in competent *Bacillus subtilis* cells. Cell 122:73–84

Khlebnikov A, Datsenko KA, Skaug T, Wanner BL, Keasling JD (2001) Homogeneous expression of the P(BAD) promoter in *Escherichia coli* by constitutive expression of the low-affinity high-capacity AraE transporter. Microbiology 147:3241–3247

Khlebnikov A, Risa O, Skaug T, Carrier TA, Keasling JD (2000) Regulatable arabinose-inducible gene expression system with consistent control in all cells of a culture. J Bacteriol 182:7029–7034

Khlebnikov A, Skaug T, Keasling JD (2002) Modulation of gene expression from the arabinose-inducible *araBAD* promoter. J Ind Microbiol Biotechnol 29:34–37

Kobayashi H, KAErn M, Araki M, Chung K, Gardner TS, Cantor CR, Collins JJ (2004) Programmable cells: interfacing natural and engineered gene networks. Proc Natl Acad Sci USA 101:8414–8419

Korobkova E, Emonet T, Vilar JM, Shimizu TS, Cluzel P (2004) From molecular noise to behavioural variability in a single bacterium. Nature 428:574–578

Kramer NE, van Hijum SA, Knol J, Kok J, Kuipers OP (2006) Transcriptome analysis reveals mechanisms by which *Lactococcus lactis* acquires nisin resistance. Antimicrob Agents Chemother 50:1753–1761

Kreth J, Merritt J, Shi W, Qi F (2005) Co-ordinated bacteriocin production and competence development: a possible mechanism for taking up DNA from neighbouring species. Mol Microbiol 57:392–404

Kussell E, Kishony R, Balaban NQ, Leibler S (2005) Bacterial persistence: a model of survival in changing environments. Genetics 169:1807–1814

Lan CY, Newport G, Murillo LA, Jones T, Scherer S, Davis RW, Agabian N (2002) Metabolic specialization associated with phenotypic switching in *Candida albicans*. Proc Natl Acad Sci USA 99:14907–14912

Lange M, Tolker-Nielsen T, Molin S, Ahring BK (2000) In situ reverse transcription-PCR for monitoring gene expression in individual *Methanosarcina mazei* S-6 cells. Appl Environ Microbiol 66:1796–1800

Laue BE, Gill RE (1994) Use of a phase variation-specific promoter of *Myxococcus xanthus* in a strategy for isolating a phase-locked mutant. J Bacteriol 176:5341–5349

Laurent M, Kellershohn N (1999) Multistability: a major means of differentiation and evolution in biological systems. Trends Biochem Sci 24:418–422

Leisner M, Stingl K, Radler JO, Maier B (2007) Basal expression rate of *comK* sets a 'switching window' into the K-state of *Bacillus subtilis*. Mol Microbiol 63: 1806–1816

Lemon KP, Grossman AD (1998) Localization of bacterial DNA polymerase: evidence for a factory model of replication. Science 282:1516–1519

Lewis K (2007) Persister cells, dormancy and infectious disease. Nat Rev Microbiol 5:48–56

Lipshtat A, Loinger A, Balaban NQ, Biham O (2006) Genetic toggle switch without cooperative binding. Phys Rev Lett 96:188101

Lisman JE (1985) A mechanism for memory storage insensitive to molecular turnover: a bistable autophosphorylating kinase. Proc Natl Acad Sci USA 82:3055–3057

Lockhart SR, Daniels KJ, Zhao R, Wessels D, Soll DR (2003) Cell biology of mating in *Candida albicans*. Eukaryot Cell 2:49–61

Lockhart SR, Pujol C, Daniels KJ, Miller MG, Johnson AD, Pfaller MA, Soll DR (2002) In *Candida albicans*, white-opaque switchers are homozygous for mating type. Genetics 162:737–745

Maamar H, Dubnau D (2005) Bistability in *the Bacillus subtilis* K-state (competence) system requires a positive feedback loop. Mol Microbiol 56:615–624

Maamar H, Raj A, Dubnau D (2007) Noise in gene expression determines cell fate in *Bacillus subtilis*. Science Jun 14 (Epub ahead of print).

Magee BB, Magee PT (2000) Induction of mating in *Candida albicans* by construction of MTLa and MTLalpha strains. Science 289:310–313

Makinoshima H, Nishimura A, Ishihama A (2002) Fractionation of *Escherichia coli* cell populations at different stages during growth transition to stationary phase. Mol Microbiol 43:269–279

Meyer TF, Gibbs CP, Haas R (1990) Variation and control of protein expression in *Neisseria*. Annu Rev Microbiol 44:451–477

Miller MG, Johnson AD (2002) White-opaque switching in *Candida albicans* is controlled by mating-type locus homeodomain proteins and allows efficient mating. Cell 110:293–302

Monod J, Jacob F (1961) Teleonomic mechanisms in cellular metabolism, growth, and differentiation. Cold Spring Harb Symp Quant Biol 26:389–401

Morgan-Kiss RM, Wadler C, Cronan JE Jr (2002) Long-term and homogeneous regulation of the *Escherichia coli araBAD* promoter by use of a lactose transporter of relaxed specificity. Proc Natl Acad Sci USA 99:7373–7377

Moter A, Gobel UB (2000) Fluorescence in situ hybridization (FISH) for direct visualization of microorganisms. J Microbiol Methods 41:85–112

Muller-Hill B (1996) The Lac Operon: A Short History of a Genetic Paradigm. Walter de Gruyter, Berlin

Nebe-von-Caron G, Stephens PJ, Hewitt CJ, Powell JR, Badley RA (2000) Analysis of bacterial function by multi-colour fluorescence flow cytometry and single cell sorting. J Microbiol Methods 42:97–114

Nester EW, Stocker BA (1963) Biosynthetic latency in early stages of deoxyribonucleic acid transformation in *Bacillus subtilis*. J Bacteriol 86:785–796

Nishino T, Nayak BB, Kogure K (2003) Density-dependent sorting of physiologically different cells of *Vibrio parahaemolyticus*. Appl Environ Microbiol 69:3569–3572

Novick A, Weiner M (1957) Enzyme induction as an all-or-none phenomenon. Proc Natl Acad Sci USA 43:553–566

Ortega F, Garces JL, Mas F, Kholodenko BN, Cascante M (2006) Bistability from double phosphorylation in signal transduction. FEBS J 273:3915–3926

Ozbudak EM, Thattai M, Kurtser I, Grossman AD, van Oudenaarden A (2002) Regulation of noise in the expression of a single gene. Nat Genet 31:69–73

Ozbudak EM, Thattai M, Lim HN, Shraiman BI, van Oudenaarden A (2004) Multistability in the lactose utilization network of *Escherichia coli*. Nature 427:737–740

Pedraza JM, van Oudenaarden A (2005) Noise propagation in gene networks. Science 307:1965–1969

Rikkerink EH, Magee BB, Magee PT (1988) Opaque-white phenotype transition: a programmed morphological transition in *Candida albicans*. J Bacteriol 170:895–899

Rosenfeld N, Young JW, Alon U, Swain PS, Elowitz MB (2005) Gene regulation at the single-cell level. Science 307:1962–1965

Russo-Marie F, Roederer M, Sager B, Herzenberg LA, Kaiser D (1993) Beta-galactosidase activity in single differentiating bacterial cells. Proc Natl Acad Sci USA 90:8194–8198

Santillán M, Mackey MC, Zeron ES (2007) Origin of bistability in the *lac* operon. Biophys J 92: 3830–3842

Sentchilo V, Ravatn R, Werlen C, Zehnder AJ, van der Meer Jr (2003a) Unusual integrase gene expression on the *clc* genomic island in *Pseudomonas* sp. strain B13. J Bacteriol 185:4530–4538

Sentchilo V, Zehnder AJ, van der Meer Jr (2003b) Characterization of two alternative promoters for integrase expression in the *clc* genomic island of *Pseudomonas* sp. strain B13. Mol Microbiol 49:93–104

Shaner NC, Campbell RE, Steinbach PA, Giepmans BN, Palmer AE, Tsien RY (2004) Improved monomeric red, orange and yellow fluorescent proteins derived from *Discosoma* sp. red fluorescent protein. Nat Biotechnol 22:1567–1572

Shaner NC, Steinbach PA, Tsien RY (2005) A guide to choosing fluorescent proteins. Nat Methods 2:905–909

Singh RN, Pitale MP (1967) Enrichment of bacillus subtilis transformants by zonal centrifugation. Nature 213:1262

Singh RN, Pitale MP (1968) Competence and deoxyribonucleic acid uptake in *Bacillus subtilis*. J Bacteriol 95:864–866

Slutsky B, Staebell M, Anderson J, Risen L, Pfaller M, Soll DR (1987) "White-opaque transition": a second high-frequency switching system in *Candida albicans*. J Bacteriol 169:189–197

Smits WK, Eschevins CC, Susanna KA, Bron S, Kuipers OP, Hamoen LW (2005) Stripping *Bacillus*: ComK auto-stimulation is responsible for the bistable response in competence development. Mol Microbiol 56:604–614

Smits WK, Kuipers OP, Veening JW (2006) Phenotypic variation in bacteria: the role of feedback regulation. Nat Rev Microbiol 4:259–271

Smits WK, Bongiorni C, Veening JW, Hamoen LW, Kuipers OP, Perego M (2007) Temporal separation of distinct differentiation pathways by a dual specificity Rap-phr in *Bacillus subtilis*. Mol Microbiol 65:103–120

Soll DR, Lockhart SR, Zhao R (2003) Relationship between switching and mating in *Candida albicans*. Eukaryot Cell 2:390–397

Southward CM, Surette MG (2002) The dynamic microbe: green fluorescent protein brings bacteria to light. Mol Microbiol 45:1191–1196

Steinmoen H, Knutsen E, Havarstein LS (2002) Induction of natural competence in *Streptococcus pneumoniae* triggers lysis and DNA release from a subfraction of the cell population. Proc Natl Acad Sci USA 99:7681–7686

Steinmoen H, Teigen A, Havarstein LS (2003) Competence-induced cells of *Streptococcus pneumoniae* lyse competence-deficient cells of the same strain during cocultivation. J Bacteriol 185:7176–7183

Stewart EJ, Madden R, Paul G, Taddei F (2005) Aging and death in an organism that reproduces by morphologically symmetric division. PLoS Biol 3:e45

Stibitz S, Aaronson W, Monack D, Falkow S (1989) Phase variation in Bordetella *pertussis* by frameshift mutation in a gene for a novel two-component system. Nature 338:266–269

Suel GM, Garcia-Ojalvo J, Liberman LM, Elowitz MB (2006) An excitable gene regulatory circuit induces transient cellular differentiation. Nature 440:545–550

Suel GM, Kulkarni RP, Dworkin J, Garcia-Ojalvo J, Elowitz MB (2007) Tunability and noise dependency in differentiation dynamics. Science 315:1716–1719

Sumner ER, Avery SV (2002) Phenotypic heterogeneity: differential stress resistance among individual cells of the yeast Saccharomyces cerevisiae. Microbiology 148:345–351

Susanna KA, van der Werff AF, den Hengst CD, Calles B, Salas M, Venema G, Hamoen LW, Kuipers OP (2004) Mechanism of transcription activation at the *comG* promoter by the competence transcription factor ComK of *Bacillus subtilis*. J Bacteriol 186:1120–1128

Swain PS, Elowitz MB, Siggia ED (2002) Intrinsic and extrinsic contributions to stochasticity in gene expression. Proc Natl Acad Sci USA 99:12795–12800

Thattai M, van Oudenaarden A (2004) Stochastic gene expression in fluctuating environments. Genetics 167:523–530

Thomas R (1998) Laws for the dynamics of regulatory networks. Int J Dev Biol 42:479–485

Thomas R (1978) Logical analysis of systems comprising feedback loops. J Theor Biol 73:631–656

Tian T, Burrage K (2004) Bistability and switching in the lysis/lysogeny genetic regulatory network of bacteriophage lambda. J Theor Biol 227:229–237

Tolker-Nielsen T, Holmstrom K, Boe L, Molin S (1998) Non-genetic population heterogeneity studied by in situ polymerase chain reaction. Mol Microbiol 27:1099–1105

Tsien RY (1998) The green fluorescent protein. Annu Rev Biochem 67:509–544

Tsong AE, Miller MG, Raisner RM, Johnson AD (2003) Evolution of a combinatorial transcriptional circuit: a case study in yeasts. Cell 115:389–399

van der Woude M, Braaten B, Low D (1996) Epigenetic phase variation of the *pap* operon in Escherichia coli. Trends Microbiol 4:5–9

van Hoek MJ, Hogeweg P (2006) In silico evolved *lac* operons exhibit bistability for artificial inducers, but not for lactose. Biophys J 91:2833–2843

van Merode AE, van der Mei HC, Busscher HJ, Waar K, Krom BP (2006a) *Enterococcus faecalis* strains show culture heterogeneity in cell surface charge. Microbiology 152:807–814

van Merode AE, van der Mei HC, Busscher HJ, Krom BP (2006b) Influence of culture heterogeneity in cell surface charge on adhesion and biofilms formation by *Enterococcus faecalis*. J Bacteriol 188:2421–2426

van Sinderen D, Luttinger A, Kong L, Dubnau D, Venema G, Hamoen L (1995) *comK* encodes the competence transcription factor, the key regulatory protein for competence development in *Bacillus subtilis*. Mol Microbiol 15:455–462

van Sinderen D, Venema G (1994) comK acts as an autoregulatory control switch in the signal transduction route to competence in *Bacillus subtilis*. J Bacteriol 176:5762–5770

Veening JW, Hamoen LW, Kuipers OP (2005) Phosphatases modulate the bistable sporulation gene expression pattern in Bacillus subtilis. Mol Microbiol 56:1481–1494

Vidal-Mas J, Resina P, Haba E, Comas J, Manresa A, Vives-Rego J (2001) Rapid flow cytometry—Nile red assessment of PHA cellular content and heterogeneity in cultures of *Pseudomonas aeruginosa* 47T2 (NCIB 40044) grown in waste frying oil. Antonie Van Leeuwenhoek 80:57–63

Viswanathan P, Ueki T, Inouye S, Kroos L (2007) Combinatorial regulation of genes essential for *Myxococcus xanthus* development requires a response regulator and a LysR-type regulator. Proc Natl Acad Sci USA 104(19):7969–7974

Vives-Rego J, Lebaron P, Nebe-von CG (2000) Current and future applications of flow cytometry in aquatic microbiology. FEMS Microbiol Rev 24:429–448

Williams SB (2007) A circadian timing mechanism in the cyanobacteria. Adv Microb Physiol 52:229–296

Wiuff C, Zappala RM, Regoes RR, Garner KN, Baquero F, Levin BR (2005) Phenotypic tolerance: antibiotic enrichment of noninherited resistance in bacterial populations. Antimicrob Agents Chemother 49:1483–1494

Yildirim N, Mackey MC (2003) Feedback regulation in the lactose operon: a mathematical modeling study and comparison with experimental data. Biophys J 84:2841–2851

Yildirim N, Santillan M, Horike D, Mackey MC (2004) Dynamics and bistability in a reduced model of the *lac* operon. Chaos 14:279–292

Zieg J, Silverman M, Hilmen M, Simon M (1977) Recombinational switch for gene expression. Science 196:170–172

Zordan RE, Galgoczy DJ, Johnson AD (2006) Epigenetic properties of white-opaque switching in *Candida albicans* are based on a self-sustaining transcriptional feedback loop. Proc Natl Acad Sci USA 103:12807–12812

Index